Allele Mining for Genomic Designing of Fruit Crops

This book deliberates on the concept, strategies, tools, and techniques of allele mining in fruit crops and its application potential in genome elucidation and improvement including studying allele evolution, discovery of superior alleles, discerning new haplotypes, assessment of intra- and interspecific similarity, and also studies of gene expression and gene prediction. Available gene pools in global germplasm collections specifically consisting of wild allied species and local landraces for almost all major crops have facilitated allele mining. Advanced genomic techniques have been developed including PCR-based allele priming and Eco-TILLING-based allele mining that are being widely used now for mining superior alleles. Allele discovery has become more relevant now for employing molecular breeding to develop designed crop varieties matching with consumer needs and also with genome plasticity to adapt the climate change scenarios. All these concepts and strategies along with precise success stories are presented over the chapters dedicated to the major fruit crops.

The features of this book are as follows:

1. The first book on the novel strategy of allele mining in fruit crops for precise breeding
2. Presents genomic strategies of mining superior alleles underlying agronomic traits from genomic resources
3. Depicts case studies of PCR-based allele priming and Eco-TILLING-based allele mining
4. Elaborates on gene discovery and gene prediction in major fruit crops

This book will be useful to students and faculties in various plant science disciplines including genetics, genomics, molecular breeding, agronomy, and bioinformatics; scientists in seed industries; and also policy makers and funding agencies interested in crop improvement.

Nextgen Agriculture: Novel Concepts and Innovative Strategies

Series Editor: Chittaranjan Kole

Allele Mining for Genomic Designing of Fruit Crops
Chittaranjan Kole, Kenta Shirasawa, and Anil Kumar Singh

Allele Mining for Genomic Designing of Cereal Crops
Chittaranjan Kole, Sharat Kumar Pradhan, and Vijay Kumar Tiwari

For more information, please visit our website: https://www.routledge.com/Nextgen-Agriculture/book-series/CRCNEXAGRINOV

Allele Mining for Genomic Designing of Fruit Crops

Edited by
Chittaranjan Kole, Kenta Shirasawa,
and Anil Kumar Singh

CRC Press
Taylor & Francis Group
Boca Raton London New York

CRC Press is an imprint of the
Taylor & Francis Group, an **informa** business

First edition published 2024
by CRC Press
6000 Broken Sound Parkway NW, Suite 300, Boca Raton, FL 33487-2742

and by CRC Press
4 Park Square, Milton Park, Abingdon, Oxon, OX14 4RN

CRC Press is an imprint of Taylor & Francis Group, LLC

ISBN: 978-1-032-47909-5 (hbk)
ISBN: 978-1-032-47910-1 (pbk)
ISBN: 978-1-003-38649-0 (ebk)

DOI: 10.1201/9781003386490

Typeset in Times
by codeMantra

Dedication

Dr. Norman Ernest Borlaug

The Father of Green Revolution, Nobel Peace Prize winner and the Founder of the World Food Prize

Who inspired my research and academic works with his generous appreciations and advice.

Contents

About the Editors

Prof. Chittaranjan Kole is an internationally reputed scientist with an illustrious professional career spanning over forty years and globally recognized for his original contributions in the fields of plant genomics, biotechnology, and molecular breeding, leading to the publication of more than 150 quality research articles. He has edited over 180 books for the leading publishers of the world including Springer, Wiley-Blackwell, and Taylor and Francis Group. His scientific contributions and editing acumen have been appreciated by seven Nobel Laureates including Profs. Norman Borlaug, Arthur Kornberg, Werner Arber, Phillip Sharp, Günter Blobel, Lee Hartwell, and Roger Kornberg. He has been honored with a number of fellowships, honorary fellowships, and national and international awards including the "Outstanding Crop Scientist Award" conferred by the International Crop Science Society. Prof. Kole has served in all prestigious positions in academia including as Vice-Chancellor of the BC Agricultural University and Project Coordinator of Indo-Russian Center of Biotechnology in India, and Director of Research of the Institute of Nutraceutical Research of Clemson University in the USA. He also worked in the Pennsylvania State University and Clemson University as a Visiting Professor in the USA. He was awarded with the Raja Ramanna Fellow by the Department of Energy, Government of India. He is also heading the International Climate-Resilient Crop Genomics Consortium, International Phytomedomics and Nutriomics Consortium, and Genome India International as their Founder President. Prof. Kole has recently established the Prof. Chittaranjan Kole Foundation for Science and Society, and working as its Founder Chairman.

Dr. Kenta Shirasawa is the head of the Laboratory of Plant Genetics and Genomics at the Kazusa DNA Research Institute in Japan. He holds a Ph.D. in Agricultural Science from Tohoku University, which focused on unraveling the genetic mechanisms underlying important traits in rice, specifically stress tolerance. Dr. Shirasawa specializes in utilizing advanced genomic technologies and bioinformatics tools to analyze and understand the complex genetic composition of more than 50 plant species such as vegetables, fruits, and ornamentals including tomato, pepper, peanut, pear, rhododendron, and cherry. His research is centered on revealing the molecular mechanisms behind plant traits such as disease resistance, stress tolerance, and yield potential. Through the study of diverse plant species' genomes, he aims to identify key genes and genetic variations responsible for these traits, with the ultimate aim of enhancing crop productivity and sustainability. Dr. Shirasawa's work has been extensively published in scientific journals such as *Nature*, *Nature Genetics*, *Nature Plants*, and *DNA Research*, and he actively collaborates with fellow scientists and research institutions to advance the field of plant genomics. His contributions to the comprehension of plant genomes and their implications for agriculture hold immense potential for significant advancements in crop improvement and the global pursuit of food security. In addition, Dr. Shirasawa has organized citizen science projects aimed at developing a DNA testing kit for cherry blossom forecasting and uncovering the molecular mechanisms behind leaf color evolution in oxalis during urbanization.

Dr. Anil Kumar Singh is currently Principal Scientist at ICAR-National Institute for Plant Biotechnology, New Delhi, India. He has been working in the field of plant molecular biology and biotechnology for more than two decades. His group has characterized genomes and transcriptomes of several crop plants and commercially important microbes, and developed gene resource for crop improvement. Many of these genes have been validated in model plants as well as economically important crops like potato and rice using gene pyramiding approach. He has also done pioneering work to understand the molecular regulatory mechanism of decline in yield and quality of apple fruit due to climate change. For the first time, his group showed a significant association between chilling and DNA methylation, and suggested that chilling acquisition during dormancy affects the epigenetic regulation through DNA methylation. Dr. Singh has published more than 85 articles in peer-reviewed international journals with cumulative Impact Factor > 280, >3000 citations, and h-index 32. He has also edited one book, authored 15 book chapters, and delivered invited/keynote talks in >35 national and international conferences in India and abroad. He is serving as editor of various reputed journals such as, *Frontiers in Plant Science*, *PLoS ONE*, *BMC Research Notes*, and *Phyton-International Journal of Experimental Botany* and a guest edited special issues in *Antioxidants*, *Genes*, *Tree Physiology*, and *Physiologia Plantarum*. For his excellent publication record and contribution to plant molecular biology research, he has been conferred membership of the National Academy of Sciences, India (NASI) and Plant Tissue Culture Association-India.

Contributors

Fawad Ali
Plant Breeding and Genetics
Centre for Planetary Health and Food Security, Griffith University
Brisbane, Australia

Raminta Antanynienė
Department of Orchard Plant Genetics and Biotechnology
Lithuanian Research Centre for Agriculture and Forestry
Institute of Horticulture
Akademija, Kėdainiai distr., Lithuania

C. Anuradha
Division of Crop Improvement
ICAR-National Research Centre for Banana
Tiruchirappalli, India

S. Backiyarani
Division of Crop Improvement
ICAR-National Research Centre for Banana
Tiruchirappalli, India

Seunghyun Ban
Department of Horticulture
College of Agriculture and Life Science
Kyungpook National University
Daegu, Republic of Korea

A. Chandrasekar
Division of Crop Improvement
ICAR-National Research Centre for Banana
Tiruchirappalli, India

Cheol Choi
Department of Horticulture
College of Agriculture and Life Science
Kyungpook National University
Daegu, Republic of Korea

Laura Dougherty
USDA-ARS, Genetic Improvement for Fruits & Vegetables Laboratory
Beltsville, Maryland, USA

Rebecca Ford
Crop Breeding, Genetics and Pathogen Genomics
Centre for Planetary Health and Food Security, Griffith University
Brisbane, Australia

Birutė Frercks
Department of Orchard Plant Genetics and Biotechnology
Lithuanian Research Centre for Agriculture and Forestry
Institute of Horticulture
Akademija, Kėdainiai distr., Lithuania

Gilad Gabay
Department of Plant Sciences
University of California
Davis, California
and
French Associates Institute for Agriculture and Biotechnology of Drylands
The Jacob Blaustein Institutes for Desert Research
Ben-Gurion University of the Negev
Sede-Boker, Israel

Filippo Geuna
Department of Agricultural and Environmental Sciences (DISAA)
University of Milan
Milan, Italy

De Mori Gloria
Department of Agricultural, Food, Environmental and Animal Sciences (DI4A)
University of Udine
Udine, Italy

Nandni Goyal
Department of Biotechnology
Panjab University
Chandigarh, India

Cipriani Guido
Department of Agricultural, Food, Environmental and Animal Sciences (DI4A)
University of Udine
Udine, Italy

Hiroyoshi Iwata
Department of Agricultural and Environmental Biology
Graduate School of Agricultural and Life Sciences
The University of Tokyo
Bunkyo, Tokyo, Japan

Augustina Kolytaitė
Department of Orchard Plant Genetics and Biotechnology
Lithuanian Research Centre for Agriculture and Forestry
Institute of Horticulture
Akademija, Kėdainiai distr., Lithuania

Rakesh Kumar
Division of Vegetable Seeds
Syngenta Vegetables
Woodland, California, USA

Gunars Lacis
Latvian Institute of Horticulture
Department of Genetics & Breeding
Dobele, Latvia

Pedro Martinez-Gomez
Department of Plant Breeding
Centro de Edafología y Biología Aplicada del Segura (CEBAS-CSIC)
Espinardo, Spain

Ingrida Mažeikienė
Department of Orchard Plant Genetics and Biotechnology
Lithuanian Research Centre for Agriculture and Forestry
Institute of Horticulture
Akademija, Kėdainiai distr., Lithuania

Edvinas Misiukevičius
Department of Orchard Plant Genetics and Biotechnology
Lithuanian Research Centre for Agriculture and Forestry
Institute of Horticulture
Akademija, Kėdainiai distr., Lithuania

Shigeki Moriya
Division of Fruit Tree Breeding Research
Institute of Fruit Tree and Tea Science
National Agriculture and Food Research Organization
Morioka, Japan

Rytis Rugienius
Department of Orchard Plant Genetics and Biotechnology
Lithuanian Research Centre for Agriculture and Forestry
Institute of Horticulture
Akademija, Kėdainiai distr., Lithuania

Juan Alfonso Salazar
Department of Plant Breeding
Centro de Edafología y Biología Aplicada del Segura (CEBAS-CSIC)
Espinardo, Spain

Roshni R. Samarth
Biotechnology Lab
ICAR-National Research Centre for Grapes
Pune, India

Tokurou Shimizu
Division of Citrus Research
NARO Institute of Fruit Tree Science
Okitsu, Shizuoka, Japan

Kashmir Singh
Department of Biotechnology
Panjab University
Chandigarh, India

Anuradha Upadhyay
Biotechnology Lab
ICAR-National Research Centre for Grapes
Pune, India

De Rosa Valeria
Department of Agricultural, Food, Environmental and Animal Sciences (DI4A)
University of Udine
Udine, Italy

Jurgita Vinskienė
Department of Orchard Plant Genetics and Biotechnology
Lithuanian Research Centre for Agriculture and Forestry
Institute of Horticulture
Akademija, Kėdainiai distr., Lithuania

Abbreviations

1-MCP	1-Methylcyclopropene
2D PAGE	Two-dimensional polyacrylamide gel electrophoresis
2*n*	Diploid number of chromosomes
A	Actinidia
AAT	Aspartate aminotransferase
ABA	Abscisic acid
ABC	Approximate Bayesian computation
ABS	Alternaria brown spot
ABSr	ABS resistance
ACC	1-Aminocyclopropane-1-carboxylate
ACLSV	Apple chlorotic leaf spot virus
ACO	1-Aminocyclopropane-1-carboxylic acid oxidase
ACS	1-Aminocyclopropane-1-carboxylic acid synthase
AFDDD	Allele frequency directional difference, and density
AFLP	Amplified fragment length polymorphism
AGPase	ADP-glucose pyrophosphorylase
ALMT	Aluminum-activated malate transporter
Ank	Ankirin gene
ANR	Anthocyanidin reductase gene
ANS	Anthocyanidin synthase
ARF106	Auxin response factor 106
ARS	Agricultural Research Service
AS	Alternatively spliced
AT	Annealing temperature
ATAF	Arabidopsis transcription activation factor
AVDRC	Asian Vegetable Research and Development Center
BAC	Bacterial artificial chromosome
BBTV	Banana Bunchy Top Virus
bHLH	Basic helix-loop-helix
bHLH74	Basic helix-loop-helix transcription factor 74
BLAST	Basic local alignment search tool
BLUP	Best linear unbiased estimates
BRV	Blackcurrant reversion virus
BSA	Bulked segregant analysis
BSV	Banana Streak Virus
BUSCO	Benchmarking universal single-copy orthologs
BZR1	BRASSINAZOLE RESISTANT1 (gene)
CA	Controlled atmosphere
CAPS	Cleaved amplified polymorphic sequence
Cas9	CRISPR-associated protein 9
CC	Coiled-coil
CCD1/4	Carotenoid cleavage dioxygenase 1 and 4
CCoAOMT	Caffeoyl-CoA O-methyltransferase
cDNA	Complementary-DNA
cDNA-AFLP	cDNA-AFLP
CDS	Coding sequence
CEN-like	CENTRORADIALIS-like (gene)

Chr	Chromosome
CHS	Chalcone synthase
cM	CentiMorgan
COI1	Coronatine-insensitive 1 coding gene
COMTs	Caffeic acid O-methyltransferases
cpDNA	Chloroplast DNA
CPVO	Community Plant Variety Office
CR	Chilling requirement
CRISPR	Clustered regularly interspaced short palindromic repeats
CUC	Cupshaped cotyledon
Cul1a	Cullin 1a protein coding gene
cv.	Cultivar
CVYV	Cucumber Vein Yellowing Virus
CYP	Cytochrome P450
CYSDV	Cucurbit Yellow Stunting Disorder Virus
DAM	Dormancy-associated MADS-box
DArT	Diversity arrays technology
DArT_seq	DArT sequencing
dCAPS	Derived cleaved amplified polymorphic sequence
ddRADseq	Double digest restriction-associated DNA sequencing
DEG	Differentially expressed gene
DFR	Dihydroflavonol 4-reductase
DMAPP	Dimethylallyl diphosphate
Eco-TILLING	Ecotype TI LLING
ECPGR	European Collaborative Programme for Genetic Resources
eIF4E	Eukaryotic initiating factor 4E
EMS	Ethylmethane sulfonate
EPDB	European Prunus Database
EPIC	Exon primed intron-crossing
EPRVs	Endogenous pararetrovirus sequences
eQTL	Expression QTL
ERF	Ethylene response factor
ERF	Ethylene-responsive transcription factor
ERF003/5/12	Ethylene-responsive transcription factor 003/5/12
ESD	Early sex discrimination
EST	Expressed sequence tag
ET	Ethylene
EU.CHERRY	European Union project aimed to align datasets using standard marker sets for cherry germplasm
F	Female
F1	First filial generation
F3'H	Flavonoid 3' hydrolase
FAD2	Delta(12)-fatty-acid desaturase
FAO	Food and Agriculture Organization
FAOSTAT	FAO Corporate Statistical Database
FaRCa1	Fragaria resistance to *C. acutatum* locus 1
FATS	*F. x ananassa* transcriptome-derived SSR markers
F-box	F-box family
FLC	Flowering locus C
FM	Formosana
Foc	*Fusariun oxysporum* fsp. *cubense*

FrBy	Friendly Boy gene
FST	Fixation index
FT	Flowering locus T
FW	Fusarium wilt
GA	Gibberellic acid
GA20ox	Gibberellin 20-oxidase
GA2ox	Gibberellin 2-oxidase
GA3ox	Gibberellin 3-oxidase
GBS	Genotyping-by-sequencing
GCA	General combining ability
GCE	Grapevine color enhancer
GC-MS	Gas chromatography-mass spectrometry
GDDH	'Golden Delicious' doubled haploid
GDDH13	GDDH genome
GDR	Genome Database for Rosaceae
GISH	Genomic *in situ* hybridization
GLK1/2	GOLDEN 2-LIKE transcription factor
GRAS-Di	Genotyping by random amplicon sequencing-direct
GS	Genomic selection
GSA	Genome scanning approach
GSI	Gametophytic self-incompatibility
GST	Glutathione S-transferase
GWA	Genome-wide association
GWAS	Genome-wide association study
GxE	Genotype x environment interaction
H2	Broad-sense heritability
HB	Homeobox
HIDRAS	High-quality disease resistant apples for sustainable agriculture
HIFs	Heterozygous inbred families
HR	Heat requirement
HRM	High-resolution melting
HV	Hypervariable
HVa	Hypervariable region a
HY5	LONG HYPOCOTYL5
IBMP	3-Isobutyl-2-methoxypyrazine
IEF	Isoelectric focusing
IG	Incompatibility group
IGS	Intergenic spacers
incRNA	Intronic noncoding RNA
Indel	Insertion/deletion
INRA	National Institute of Agricultural Research (Institut National de la Recherche Agronomique)
INV1	Invertase 1
IPP	Isopentenyl diphosphate
IRSC	International RosBreed SNP Consortium
ISSR	Inter simple sequence repeat
ITS	Internal transcribed spacer
JA	Jasmonic acid
KASP	Kompetitive allele-specific PCR
KEGG	Kyoto Encyclopedia of Genes and Genomes
KVDS	Kiwifruit Vine Decline Syndrome

LBD	Lateral organ boundaries domain
LD	Leafing date
LD	Linkage disequilibrium
LG	Linkage group
LOD	Logarithm of odds
LRR	Leucine-rich repeat
LRS	Long-read sequencing
LSU	Large sub-unit
LTR	Long terminal repeat
LTR-RT	LTR retrotransposon
LYB	Lycopene beta-cyclase
M	Male
Ma	Malic acid locus
Ma1	Aluminum-activated malate transporter 1
Ma2	Aluminum-activated malate transporter 2
MAB	Marker-assisted breeding
MADS14	MADS-box transcription factor 14
MAFD	Mutant allele frequency and density
MAI	Marker-assisted introgression
MAPS	Marker-assisted parent selection
MAS	Marker-assisted selection
MASS	Marker-assisted seedling selection
MD	Maturity date
MdACO1	*Malus domestica* 1-amino-cyclopropane-1-carboxylate oxidase
MdACS1	*Malus domestica* ACC synthetase
MdAF3	*Malus domestica* α-l-arabinofuranosidase
MdAGO4	*Malus domestica* AGO4
MdCDKA1	*Malus domestica* cyclin-dependent kinase A-1
MdCYCA2	*Malus domestica* cyclin-A2
MdERF109	*Malus domestica* ERF109 transcription factor
MdFT1	*Malus domestica* Flowering locus T like 1 gene
MdFT2	*Malus domestica* Flowering locus T like 2 gene
MdHSP70	*Malus domestica* Heat shock protein 70
MdLNC499	*Malus domestica* long noncoding RNA 499
MdME	*Malus domestica* malate enzyme
MdMYB10	*Malus domestica* MYB10 transcription factor
MdMYB4	*Malus domestica* R2R3-MYB transcription factor
MdMYB44	*Malus domestica* MYB44 transcription factor
MdNuc54	*Malus domestica* nucleoporin gene 54
MdPG1	*Malus domestica* endo-polygalacturonase 1
MdPI	*Malus domestica* PI homolog
MdPP2CH	*Malus domestica* protein phosphatase2C
MdSAUR37	*Malus domestica* small auxin-upregulated RNA gene
MdTAR1	*Malus domestica* tryptophan aminotransferase-related protein 1
MdTFL1	*Malus domestica* Terminal Flower 1
MdWRKY1	*Malus domestica* WRKY1 transcription factor
MdYUCCA6	*Malus domestica* flavin monooxygenase-like enzyme 6 gene
MEP	2-C-Methyl-D-erythritol-4-phosphate
MFT	Mother of *FT*
Mfu10	Fire blight resistance QTL
miRNA	MicroRNA

MNSV	Melon Necrotic Spot Virus
mQTLs	Metabolic QTL
mRNA	Messenger RNA
MusaRgeneDB	*Musa* resistance genes database
MusatransSSRDB	*Musa* Transcriptome SSR database
MVA	Mevalonate
MYB1	Myeloblastosis 1 transcription factor
MYB10	Myeloblastosis 10 transcription factor
MYB1R1/44	Transcription factor MYB1R1/44
NAC2/25	Transcription factor NAC2/25
NADP-ME	NADP-dependent malic enzyme
NAM	No apical meristem
NARO	National Agriculture and Food Research Organization
NB-LRR	Nucleotide-binding LRR
NBS	Nucleotide-binding site
NCBI	National Center for Biotechnology Information
NCED1	9-Cis-epoxycarotenoid dioxygenase
NEPHGE	Non-equilibrium pH gel electrophoresis
NGS	Next-generation sequencing
NIH	National Institute of Health
NLM	National Library of Medicine
N-terminal	Amino terminal
OMTs	O-Methyltransferases
ORF	Open reading frame
PaCYP78A9	*Prunus avium* CYP78A gene
PAL	Phenylalanine ammonia-lyase
PavMADS1	*Prnus avium* dormancy-associated MADS-box 1 gene
PBA	Pedigree-based analysis
PC	Principal component
PCR	Polymerase chain reaction
PD	Pyruvate decarboxylase
PDR	Pierce's disease resistance
PER	Period gene
PEV	Phenotypic explained variation
PGI	Phosphoglucose isomerase
PGL	Pale green lethal
PGM	Phosphoglucomutase
PI	Probability of identity
pi	Pistillata mutant in Arabidopsis
PIC	Polymorphism information content
PIF3/4	Phytochrome interacting factor 3 and 4
PIsibs	Probability of identity among siblings
PKW	Pisang Klutuk Wulung
PmDAM	Dormancy-associated MADS-box gene
PP2A-2	Protein phosphatase 2A-2
PPM	Pollen-part mutant
PPO	Polyphenol oxidase
PPV	Plum Pox Virus
PR	Pathogenesis-related
PRR	Pattern recognition receptor
Psa	*Pseudomonas syringae* pv. *actinidiae*

PSC	Partial self-compatibility
PSY	Phytoene synthase
PTR-MS	Proton transfer reaction-mass spectrometry
QRLs	Quantitative resistance loci
qRT-PCR	Quantitative reverse transcription PCR
QTL	Quantitative trait locus
QTLs	Quantitative trait loci
R gene	Resistance gene
RADseq	Restriction site-associated DNA sequencing
RAP2–12	Ethylene-responsive transcription factor RAP2–12
RAPD	Random amplified polymorphic DNA
RC4	Rosaceae conserved region
Rca2	Anthracnose resistance gene
REML	Restricted maximum likelihood ratio
RFLP	Restriction fragment length polymorphism
RG	Reference genotype
RGA	Resistance gene analog
RGA	Rapid generation advancement
RGTH1	Royal Gala Tetra-haploid genome
RHV	Hypervariable region
RIL	Recombinant inbred line
RLK	Receptor-like kinase
RNA	Ribonucleic acid
RNase	Ribonuclease
RNA-seq	RNA sequencing
Rpf1	Red stele root rot gene
RPW8	Resistance to powdery mildew 8
RRM	Reiterated reproductive meristem
rRNA	Ribosomal RNA
S-allele	*S*-locus allele
SB	Speed breeding
SC	Self-compatibility
SCAR	Sequence characterized amplified region
SCF	Skp1-Cul1-F-box protein
SD	Short day
SDI	Seed development inhibitor
SDR	Sex-determining region
Seq	Sequencing
SFB	S haplotype-specific F-box gene
SFB1/2/C	*S*-locus F-box protein 1/2/C
SFBB	*S*-locus F-box brothers genes
SFBB	*S*-locus F-box brothers
SI	Self-incompatibility
SiRNA	Small interfering RNA
SLF	Seasonal flowering locus gene
SLF	*S*-locus F-box
SLFL	*S*-locus F-box-like genes
***S*-locus**	Self-(in)compatibility locus
SNP	Single nucleotide polymorphism
SPET	Single-primer enrichment technology
SPM	Stylar-part mutation

Spp.	Species
sppIDer	Species identification tool
SPS2	Styrylpyrone synthase 2
***S*-RNase**	*S*-ribonuclease
SSC	Soluble solid content
SSH	Suppression subtractive hybridization
SSK1	Senescence-specific killing compound
SSR	Simple sequence repeat
STS	Sequence-tagged site
SUS1	Sucrose synthase 1
SuSy	Sucrose synthase
SyGI	Shy Girl gene
TA	Titratable acidity
TASSEL	Trait analysis by association, evolution and linkage
TC7	Tai-Chiao No. 7
TD	Threonine deaminase
td	Touch down
TF	Transcription factor
TFL	Terminal flower
TFL1	Terminal flower 1
TILLING	Targeting-induced local lesions in genomes
TIR	Toll interleukin 1 receptor
TPS	Terpene synthase
TR4	*Foc* tropical race4
TSS	Total soluble solids
UBC	Ubiquitin-conjugating enzyme
UFGT	UDP-glucose flavonoid glucosyl transferase
USBP	UNICENTRO's Strawberry Breeding Program
USDA	United States Department of Agriculture
VOS	Volatile organic compound
W	Weeping allele
WGS	Whole-genome sequencing
WRKY6/31/69	Transcription factor WRKY6/31/69
ZEP	Zeaxanthin epoxidase
ZYMV	Zucchini Yellow Mosaic Virus

Series Preface

Agriculture is going through a phase of transformation now. It has both challenges and opportunities. Of course, one of the major challenges it is facing is food security. A population of about 9.3 billion projected necessitates an increase in crop production by 60–70%. Another 10–15% increase would be needed to counter the reduction in crop production due to global warming and climate change. Nutrition security is another challenge with more than 800 million people suffering from malnutrition. Depletion of fossil fuel and increasing dependence on biofuel from crop residues is another challenge. Loss of biodiversity and land degradation also add to the list of challenges. The cultivable area is also decreasing fast due to housing, transport, and industrialization. Above all, climate change and global warming have a huge impact on agriculture, health, and environment.

Several novel concepts and strategies are emerging to address the above challenges. For example, the practice of precision farming will lead to precise and timely use of inputs resulting in higher yield and better quality of produce on the one hand and reduction in loss of inputs on the other hand. Similarly, urban agriculture and vertical agriculture will add a lot of cultivable areas in the future. Protected farming will facilitate growing crops under controlled conditions leading to reduction in biotic and abiotic stresses and precise utilization of inputs.

Agriculture requires maintenance of the philosophy that farmers have been following since inception over thousands of years. These ideas are being sold today in new packages with sophisticatedly coined labels such as traditional agriculture, natural agriculture, sustainable agriculture, conservation agriculture, integrated agriculture, regenerative agriculture, organic agriculture, etc.

In the changing scenarios, next generation agriculture will need designed crop varieties and breeds produced through designed practices. This book series aims at deliberations on several novel concepts and innovative strategies developed for next generation agriculture and will hopefully benefit agricultural education, research, and outreach.

Kolkata
January 10, 2024

Chittaranjan Kole

Preface

The affordability, precision, and high-throughput nature of sequencing technologies in the last two decades have provided an unprecedented understanding of the genomes and target plant traits of economic importance. Further, the advances in the third-generation sequencing technologies including Hi-C especially in the last decade have made it possible for developing high-quality reference genomes, pangenomes, and gene annotations including genome-wide structural variations. These developments have also increased the precision in gene and superior haplotype discovery followed by mining allelic diversity among breeding lines and genebank collections. The high-throughput and cost-effective genotyping platforms have further boosted the adoption rate of genomic tools in the ongoing breeding programs across the world. Deployment of these advances in crops and making them well equipped to scale up the integration of genomic interventions in the breeding programs will facilitate development of future-ready crops.

The above advances have made gene/allele mining more precise and cost-effective for the target traits preferred by farmers, consumers, and industry as well as efforts toward mitigation of climate change impacts and achieving nutritional security prospective. It is important to accelerate the process of genomic designing to develop future-ready crops by bringing new allelic combination(s) at one or more loci underlying desired genes using modern breeding approaches such as marker-assisted breeding, haplotype-based breeding, gene/allele pyramiding, and also now through gene editing.

These new alleles originate through natural or induced mutations at one or more genomic regions at a time. The traditional plant breeding methods practiced in the twentieth century depended solely on phenotypes for the development of new genotypes primarily through selection and hybridization. The advent of DNA-based molecular markers in the 1980s facilitated indirect identification of the alleles by using linked molecular markers using genetic linkage mapping. Molecular breeding as a supplement to traditional breeding has successfully been employed in almost all major crops for their genetic improvement in a short time. It is worthwhile to mention that transgenic breeding has also contributed significantly to crop improvement at least for some major input target traits such as herbicide tolerance and insect resistance; however, this technology is still facing challenges in societal adoption in different parts of the world, especially for food crops. Nevertheless, gene editing has been receiving positive vibes from the policymakers of different countries making a conducive environment for accelerating the use of this technology on large scale.

Despite the huge loss of diversity during domestication, several national and international gene banks have millions of diverse germplasms to explore variability for the traits that are needed to breed climate-resilient and future-ready crops. Genetic diversity played a pivotal role during the entire period of conventional plant breeding, traditional and molecular included. However, the plant breeders have been using mainly the primary gene pool constituting the indigenous germplasm lines and local landraces as a genetic resource of new alleles. Meantime, a drastic dearth of new alleles in the primary gene pool happened due to genetic erosion caused naturally or due to domestication over time but aggravated because of the injudicious and irrational implementation of the so-called green revolution. Precisely, this highlighted the importance of allied wild crop relatives (WCRs) and thereby a mission of collection, conservation, characterization, and utilization of biodiversity. The emergence of advanced bioinformatics and the availability of cheaper and innovative genome sequencing technologies have empowered us to explore germplasm and identify a large number of donor alleles, and allele mining has become a reality!

The strategy of association mapping, the most popular trait mapping approach in animal and human sciences, has received huge adoption in plant genomics studies. The availability of large sequenced diverse panels with multilocation phenotyping data has provided much-needed insights into trait genomics through genome-wide association studies (GWAS) leading to the discovery of genes and superior haplotypes in many crops. This has led to the possibility of mining a large number of genes in the available accessions in crops. Mutations occur in the genome as single nucleotide polymorphisms (SNPs) or insertion/deletions (In/Dels) resulting in the generation of new alleles or changes in existing alleles and allele combinations. Using exactly this principle, the approach of Eco-TILLING has facilitated large-scale allele mining in several major crops. On the other hand, sequence-based allele mining has led to the discovery of allelic diversity both at the target gene and at the whole genome. Recent advances in modern genomic tools and technologies such as genome engineering and speed breeding facilitate the deployment of these new alleles in breeding programs in a relatively short time to accelerate the rate of genetic gains. Truly speaking, allele mining has revolutionized the strategy of utilizing genetic as well as genomic resources and become an important component of next-generation precision breeding and biotechnology!

Allele discovery has become more relevant now for employing molecular breeding to develop designed crop varieties matching consumer needs and with genome plasticity to adapt the climate change scenarios using targeted and precision breeding tools validated by both genomics and phenomics studies. All these concepts and strategies along with precise success stories are presented in over 50 chapters allocated under 5 volumes dedicated to the 5 crop groups including cereals, oilseeds, grain legumes, fruits, and vegetables under this book series.

This volume, entitled *Allele Mining for Genomic Designing of Fruit Crops*, includes 12 chapters focused one Banana, Apple, Citrus, Grapevine, Berries, Cherries, Papaya, Pear, Kiwi Fruit, Apricot, and Melon contributed by 31 scientists from 9 countries including Australia, India, Italy, Japan, Korea, Latvia, Lithuania, Spain, and USA. We express our thanks for their excellent contributions and timely cooperation.

This book will be hopefully useful to the students, research scholars, and teaching faculties in the academia to learn about fundamentals and applications of allele mining; the scientists practicing plant breeding in public and private institutes in improvising their programs; and also, the policymakers and funding agencies in prioritizing research fields.

Chittaranjan Kole
Kolkata, India

Kenta Shirasawa
Chiba, Japan

Anil Kumar Singh
New Delhi, India

1 Allele Mining in Banana

C. Anuradha, S. Backiyarani, and A. Chandrasekar

1.1 INTRODUCTION

Banana and plantain belonging to the Musaceae family is the fourth most important crop worldwide and produces 125 million tons of fruit annually. Nearly one thousand cultivars of banana are being grown globally for consumption purpose. These parthenocarpic cultivars are the derivatives of inter- and intraspecific natural hybrids of *Musa acuminata* and *M. balbisiana* of diploid groups (AA and BB) and various other ploidy groups such as auto- (AAA, AAAA) and allopolyploids (AB, AAB, ABB, AAAB, AABB, ABBB) (Simmonds and Shepherd, 1955). These present-day commercial cultivars are evolved as the results of the recombination and exchange of chromosomal segments between the A and B genomes as well as between the subspecies of *M. acuminata* (Perrier et al., 2009; Jeridi et al., 2011).

Despite the fact that the A and B genomes share a high degree of homology in terms of the number of genes and conserved regions (Davey et al., 2013), differences in codon usage and the dosage effect may contribute to phenotypic variation among cultivars belonging to the same genomic group. According to Symonds et al. (2010), high levels of variation in polyploid crops can either be attributed to the respective diploid parents' mutational divergence or to the multiple origins produced by various polyploidization events between various genotypes of the same diploid species. According to a report by D'Hont et al. (2012), the presence of a large number of multigene families in banana is due to the occurrence of three whole-genome duplications (WGDs) in the ancestral banana genome (Cenci et al., 2014). Other than these, banana has inherent issues like parthenocarpy, low male and female fertility, and perennial nature, making it impossible to identify candidate genes through classical Mendelian inheritance. Allele mining is an alternative method for revealing the genetic diversity in crops (Kaur et al., 2008). By comparing the nucleotide variation of candidate genes and associating with phenotypic variations, allele mining may be used to identify the candidate allele. High-throughput and accuracy detection of nucleotide variation is possible with the advent of next-generation sequencing platforms. Further, it is challenging to detect genetic variation at the allelic level in the complex genome of the banana because of interspecific hybrids, heterozygosity, and different dosage effects (polyploid). High sequence coverage/in-depth sequencing and longer reads using high-throughput sequencing platforms can be used to generate allelic data to address this. Despite this, having more genes makes it harder to distinguish between homologs and homeologs of the genes when the genomic data is assembled. Identifying the homeologs of the gene of interest is the important step in the banana improvement program through genetic modification. This can be achieved by reducing the genomic complexity by sequencing the haploid genome of the ancestor species of the present-day edible banana. Additionally, the nucleotide variation present in the non-coding region could also cause changes in the phenotype through altering the expression (Fiume et al., 2004; Fu et al., 2005). According to evidence provided by Sureshkumar et al. (2009), functional variations in genes are brought on by the insertion of transposons and/or changes to the microsatellite repeat sequences in the intronic region. This emphasizes that examining the sequence variation in candidate genes' regulatory and non-coding regions, in particular, could be regarded as true allele mining. Thus, this chapter covers the various aspects and approaches of allele mining in banana (Figure 1.1).

DOI: 10.1201/9781003386490-1

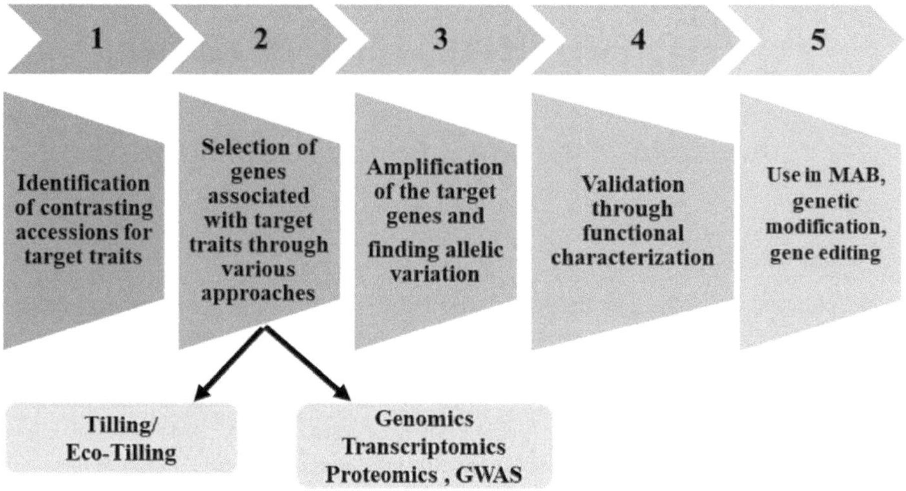

FIGURE 1.1 Step and approaches involved in allele mining from banana.

1.2 EVOLUTION OF NEW ALLELES IN BANANA

Allelic diversity in any crop species exists because of mutation, which is considered to serve as an evolutionary driving force. Single nucleotide polymorphisms (SNPs) or insertion and deletion (InDel) mutations in the genic regions of the genome are crucial for producing new alleles or changing the existing alleles and allelic combinations. While mutations in non-coding regions of a gene are frequently silent and have no impact on the phenotype, those in coding regions and/or regulatory regions of a gene may have a significant impact on the phenotype by changing the structure and/or function of the encoded protein. Despite the fact that most mutations are harmful, 0.1% of mutations are extremely important leading to alterations in gene function which may be highly necessary for the survival of the plant (Singh et al., 2005). Many of these occurrences are typically seen in various crop species, and in banana, Tai-Chiao No. 7′ (TC7) and Formosana (FM) were found to share a deletion region on chromosome 5; however, different TC7 tissues showed different levels of allele ratios in this region, indicating that TC7 and FM may have chimerism (Hou et al., 2022).

In banana, a polyploid genome has a higher allelic diversity than a diploid genome, leading to a high level of natural variation (Udall et al., 2005). Despite having a single genome, autoploids like AA, AAA, AAAA, and BB exhibit multisomic inheritance because they contain pyramided genes with a variety of alleles. Similar to this, the allopolyploidy genomes of AB, AAB, ABB, ABBB, AABB, and AAAB exhibit increased allelic diversity as a result of intergenomic heterozygosity and intra-genomic gene pyramiding (Cenci et al., 2019). Thus, understanding the gene expression of polyploid genome is of paramount importance. The occurrence of huge variation in the polyploid species is not only due to the allelic diversity of the coding regions but also results from the diversity in the regulatory sequences determining the cis- and trans-effects, such as methylation, long non-coding RNAs, miRNAs, and transposon activation.

1.3 BANANA GENETIC RESOURCES FOR ALLELE MINING

The seeded diploid *Musa* species are the ancestors of the modern commercial seedless cultivars. The *M. acuminata* chromosomes underwent spontaneous structural changes and recombination, which led to the emergence of natural reproductive barriers and increased subspecies diversity (Jones, 2000). Hardy *M. balbisiana*, unlike *M. acuminata*, has no such classification at the subspecies level and only exhibits less variability (Uma et al., 2005; Wang et al., 2007, 2008). However, it has a high potential value in breeding programs because it has been reported to be tolerant to biotic stresses like Banana bunchy top virus (BBTV) (Hapsari and Masrum, 2012), Fusarium wilt (FW), and leaf spot diseases (Singh et al., 2001) as well as abiotic stresses like drought (Daniells et al., 2001), osmotic, and cold stresses (Thomas et al., 1998). Various studies have shown that *M. itinerans* is resistant to waterlogging and FW; *M. sikkimensis*, *M. basjoo*, and *M. thomsonii* are resistant to cold; and *M. balbisiana* and *M. nagensium* are the sources of resistance to drought (MusaNet, 2016; García-Bastidas et al., 2019). According to Uma et al. (2005), *M. balbisiana* is resilient to both biotic and abiotic stresses, while *M. ac.* ssp. *zebrina* is resistant to *Fusarium oxysporum* fsp. *cubense* (Foc) race 1 (Arinaitwe et al., 2019). According to Vakili (1965), the resistant sources for Foc race 1 included *M. ac.* ssp. *malaccensis*, *M. ac.* ssp. *burmannica*, *M. ac.* ssp. *microcarpa*, and *M. ac.* ssp. *siamea*. According to Shepherd (1968), the Rhodochlamys section member *M. laterita*, which is resistant to Foc tropical race4 (TR4), is more compatible with the Eumusa section and also used as one of the male parents for passing down the resistance genes. *M. basjoo* and *M. itinerans* were found to be highly resistant based on the *in vitro* and field screening of wild banana species against Foc-TR4, whereas *M. nagensium*, *M. ruiliensis*, *M. velutina*, and *M. yunnanensis* showed some degree of infection against Foc TR4 in the greenhouse but were found to be resistant in the field (Li et al., 2014). Some of the wild *Musa* species, such as *M. ac.* ssp. *burmanicca*, *malaccensis*, and *siamea*, are found to be resistant to black sigatoka (Jones, 2000). Two wild accessions Tuu Gia and Birmanie were found to be resistant to *Pseudocercospora fijiensis* (Ortiz-Vázquez et al., 2005; Rebouças et al., 2018; Nascimento et al., 2020), whereas *M. ac.* ssp. *malaccensis* and *M. ac.* ssp. *microcarpa* are considered moderately resistant category (Nascimento et al., 2020).

One of the best sources of multiple resistance to biotic stresses, including Foc, corm weevil, and nematodes, is *M. ac.* ssp. *burmaniccoides* (Calcutta 4), which is used in resistant breeding programs (Thangavelu et al., 2020). During the evolution and domestication of the present-day cultivated crop species, trait-enhancing alleles were left behind that were better expressed in wild species than in cultivated bananas. The trait performance of elite banana varieties will be improved by transferring these advantageous alleles from wild species to elite genetic backgrounds. It can be argued that such altered/novel alleles of important genes are preserved in numerous related wild species, land races, and cultivated varieties of a crop species, despite the fact that the propagation of these altered/novel alleles (either through natural or artificial selection) depends on their role in the plant's adaptation. As a result, studying the natural variation in key genes is absolutely necessary for current crop improvement programs.

1.4 TILLING/Eco-TILLING IN BANANA

SNPs and small insertions/deletions can be found and characterized using the high-throughput technique known as Eco-TILLING (ecotype-targeting induced local lesions in genomes). Nucleotide polymorphisms were found and described using the Eco-TILLING method in diploid and polyploid accessions of *Musa* spp. The screening of six triploid Grand Nain mutants that had undergone gamma radiation as well as a non-irradiated Grand Nain control can be viewed as a proof-of-concept for banana TILLING. The mutants were previously chosen for their higher tolerance to the toxin produced by *P. fijiensis* (Roux, 2001). In *Musa*, Eco-TILLING method was used for the discovery and characterization of nucleotide polymorphisms of diploid and polyploid accessions. Over 800 novel alleles from 80 accessions were identified as polymorphic using 14 gene-specific primers by Till et al. (2010). In this study, more than 6,000 polymorphisms were detected, and further, sequencing-based validation was performed for the detection of SNP's variation among the accessions. Consequently, sequencing and banding patterns reveal that Eco-TILLING to be a robust and perfect platform for discovery of polymorphisms in homologous and homeologous gene targets in *Musa* accessions. This study also discovered two SNPs that may have negative effects on the operation of a gene that is thought to be crucial for phototropism.

1.5 MAPPING OF IMPORTANT ALLELES AND QTLs IN BANANA

Understanding the inheritance pattern of the genes is the foremost important criteria for the successful durable breeding in banana. Screening of a hybrid population and determination of the segregation ratios can provide information on the inheritance and number of alleles controlling a trait. Banana is the most difficult crop to map a gene(s) for a specific trait than other seeded crops because of its female and male sterility, structural heterozygosity, ploidy, and various genomic combinations (González de León and Fauré, 1993). Additionally, it has been reported that mapping of the *Musa* genome is further hampered by the frequent occurrence of chromosome inversion and translocation (Shepherd, 1999; Vilarinhos et al., 2006). Further, identification and localization of genetic markers linked to gene of interest are required for the molecular mapping of the alleles or the quantitative trait loci (QTLs) in banana. This is usually done using methods like linkage mapping and association mapping, which can show where the alleles controlling a trait are located on the banana genome. The identified QTLs can then be used to develop marker-assisted selection (MAS) methods for breeding programs aimed at improving a trait of interest in banana (Backiyarani et al., 2022). However, because of segregation distortions and the risk of pseudo linkages, it is not possible to create a saturated linkage map in banana. The magnitude of distortions from expected Mendelian segregation varied in populations derived from interspecific and intraspecific combinations because their originators differed, resulting in structural rearrangements (Pardo-Manuel de Villena and Sapienza 2001). According to Rowe and Rosales (1996), one or more dominant alleles regulate the genetic resistance to burrowing nematode. Vakili (1968) came to the conclusion that multiple genes and partially dominant factors conditioned the Yellow sigatoka resistance in *M. acuminata* ssp. *Zebrina*, and Shepherd (1990) reported that the resistance in *M. acuminata* was homozygous recessive. A single major recessive allele and two independent recessive alleles with additive effects have been found to control black sigatoka resistance in plantain hybrids using *M. ac.* ssp. *burmannica* as a resistant parent (Ortiz and Vuylsteke, 1992; Anonymous, 1993). Table 1.1 lists the mapping of alleles and QTLs in banana.

TABLE 1.1
Mapping of Alleles and QTLs in Banana

Linkage Mapping/ QTLs	Parents Used	Population	Gene Action	References
Linkage map	Diploid *M. acuminata*	90 markers		Fauré et al. (1993)
Linkage map				Vilarinhos et al. (2006); Lagoda et al. (1998)
Linkage map	*M. ac.* ssp. *malaccensis*	F1 population		Kayat et al. (2009)
Genetic map	*M. ac.* ssp. *microcarpa* and Pisang Lilin	F1 population		Hippolyte et al. (2010)
Genetic resistance to subtropical race4 (STR4) and TR4	Diploid wild banana *M. ac.* ssp. *malaccensis*	F1 offspring of self-fertilized *M. ac.* ssp. *malaccensis*	Single dominant gene	Fraser-Smith et al. (2016)
Genetic resistance to Black sigatoka	Black sigatoka resistance in plantain hybrids developed using *M. ac.* ssp. *burmannica*	F1 population	One major recessive allele and two independent alleles with additive effects	Ortiz and Vuylsteke (1992); Anonymous (1993)
Inheritance of resistance to Foc Race 1 and TR4	*M. ac.* ssp. *malaccensis*	Selfing of wild banana accession *M. ac.* ssp. *malaccensis*	Single dominant gene	Ahmad et al. (2020)
Inheritance of resistance to Foc Race 1	Monyet (resistant to Foc race 1) and Kokopo (susceptible)	142 F_1 population genotypes	Two dominant genes with epistatic interaction	Arinaitwe et al. (2019)
Inheritance of resistance to Foc Race 1	'SukaliNdizi' and 'TMB2X8075–7'	F_2 populations	Recessive gene	Ssali et al. (2013)
Genetics of resistance to *R. similis*	TMB2x 6142-1 and TMB2x 8075–7	F_1 populations	Two dominant genes	Dochez et al. (2009)

The need for bio-fortification of staple crops has become the concern of modern nutritional genomics in producing staple crops with the ability to accumulate high micronutrients. Beta-carotene (Provitamin A) biosynthesis pathway is a stepwise process that is coded by two main genes: phytoene synthase (*PSY*) and lycopene beta-cyclase (*LYB*). These two important genes *LYB7* and *PSY11* were amplified, sequenced, and analyzed within various diploid and triploid plantain landraces to identify SNPs present and to study their potentials for mapping the QTL in those genes brought about by climatic variation within plantain landraces over time and also reveal their phylogenetic relationship. A total of seven SNPs were found in the two genes from 68 genotypes used for *PSY11* and 35 plantain genotypes for *LYB 7* gene. The high frequency of SNPs detected showed that it is sufficient in mapping the QTLs associated with high β-carotene synthesis and could be incorporated in plantain breeding programs (Mmeka et al., 2013). Mahendhiran et al. (2014) examined 810-base pair amplicons of the ADP-glucose pyrophosphorylase (AGPase) LSU gene in ten Musa accessions. This included representatives of wild ancestors, dessert bananas, plantains, and cooking bananas, revealing a total of 36 SNPs and IndDels. The phylogenetic analysis revealed 15 distinct haplotypes, which were grouped into four variants. He also revealed that this information is used for the development of SNP-based molecular markers for *Musa* germplasm, and alteration of the allosteric properties of AGPase to increase the starch content and manipulate the starch quality of banana fruits.

1.6 ASSOCIATION MAPPING STUDIES FOR ALLELE MINING

Genome-wide association studies (GWAS) are increasingly used to find potential SNPs/genes associated with a variety of biotic and abiotic stresses. However, the application of GWAS in the case of bananas is challenging due to different ploidy levels and the presence of numerous interspecific hybrids, which make SNP mapping in bananas difficult (Sardos et al., 2016). Nyine et al. (2019) identified the clustering of SNPs in the transcription factors and genes involved in cell cycle regulation that primarily regulate the trait fruit filling based on GWAS in the breeding population of bananas. This is consistent with earlier findings from D'Hont et al. (2012) and Liu et al. (2017), who reported that transcription factors MYB, AP2/ERF, and a subfamily of MADs (MCM1, agamous, defensins, SRF) are involved in fruit architecture, development, and ripening.

Simmonds (1953) elucidated that in banana, the trait parthenocarpy is governed by three independent complementary genes, whereas the seeded nature is governed by two homozygous dominant parthenocarpic genes and lacks in one of the three dominant genes. Earlier findings also proved that the trait parthenocarpy is governed by three genes (Simmonds, 1962), whereas the bunch weight is governed by QTLs (Turner and Gibbs, 2018). So far only a limited number of studies have been carried out to explore the genetics of parthenocarpy in banana and plantains. Similarly, based on the segregating pattern, Ortiz and Vuylsteke (1992) also substantiated that parthenocarpy is governed by three genes. Further, it has been postulated that among the ancestor genome (A and B) of the present-day commercial cultivars, A genome contributes to the female sterility which resulted in vegetative parthenocarpy in banana (Dodds and Simmonds, 1948; Simmonds, 1953, 1962). However, the loci or the genetic factors responsible for the trait parthenocarpy are not yet identified in plantains and bananas. This is due to the hindrance in developing mapping population, and association mapping approach is one of the best methods for allele mining in vegetative propagated crop like banana. A set of 5,544 SNP markers were used to study the seedless phenotype through GWAS studies in 33 accessions from Papua. Thirteen candidate genomic regions that may be associated with the seedless phenotype were identified, and a putative orthologous gene to Histidine Kinase *CKI1*, a leading candidate gene for female sterility, was also discovered (Sardos et al., 2018). A genic simple sequence repeat (SSR) marker associated with parthenocarpy was also identified through ddRAD approach and validated across different seeded and parthenocarpic accessions. The identification of markers associated to parthenocarpy would speed up the breeding process by aiding in the selection of parents for crossing or as progeny, thereby reducing the time and resources needed for screening

(Anuradha et al., 2022c). Further, given that complex and low heritable traits in bananas are heavily influenced by environment, it is highly appropriate to use molecular tools in the breeding program to streamline the selection of genotypes for the traits (Tuberosa, 2012). But because the majority of quantitative traits are controlled by numerous minor QTLs, the detection of QTLs through linkage mapping and GWAS is frequently biased (Zhao et al., 2014). A large collection of marker data dispersed throughout the entire genome can be used to get around this restriction.

1.7 GENOMICS, TRANSCRIPTOMICS, AND PROTEOMICS APPROACH FOR ALLELE MINING

Whole-genomic sequence (WGS) data, gene annotations, transcriptome, proteome, and metabolome data are increasingly becoming more widely available. These data can be used to support the identification of potential candidate gene(s) underlying the QTL of interest. WGS data provide a more detailed profile of a QTL region for any trait of interest. Similarly, GWAS QTL and transcriptomic data or fine mapping with transcriptomics, proteomics, and metabolomics can be used to pinpoint the most promising candidate gene underlying the trait of interest (Thomson et al., 2022).

In order to explore the diversity in banana, whole-genome sequence projects were initiated. The genome size of *M. acuminata* subspecies ranged from 591 to 615 Mbp, while *M. balbisiana* is 537 Mbp which are nearly 30% and 15% larger, respectively, than *Oryza sativa* L. ssp. *indica* (Yu et al., 2002). Before the next-generation sequencing era, to understand the genetic architecture of the banana genome, several bacterial artificial chromosome (BAC) libraries have been constructed such as *M. acuminata* cv. Calcutta 4 [MA4 BAC library] (Vilarinhos et al., 2003), *M. acuminata* cv. Tuu Gia [TGBIBAC library] (Ortiz-Vázquez et al., 2005), and *M. balbisiana* cv. PKW [MBP BAC library] (Šafář et al., 2004). BAC library of *M. balbisiana* cv. Pisang Klutuk Wulung (PKW) resulted in developing chromosome-specific BACs which have served as landmarks to identify chromosomes and chromosome arms of *M. balbisiana* as well as to clone the Banana streak virus (BSV) expressed locus, which controls the activation of BSV endogenous pararetrovirus sequences (EPRVs) during hybridization between *M. acuminata* and *M. balbisiana*. Similarly, BACs library of cv. Tuu Gia is known to possess one of the most effective and durable black leaf streak resistance mechanisms in the *M. acuminata* species (Fullerton and Olsen, 1995; Craenen and Ortiz, 2003) and provide the information on regulated defense genes, inducible promotors, phytoalexin biosynthesis genes, or any kind of signal perception and transduction kinases which will facilitate to develop molecular markers for *P. fijiensis* resistance.

With the ever-decreasing costs of genome sequencing, banana genomic resources have greatly expanded from the BAC libraries to the whole-genome sequencing of the progenitors of the cultivated banana genome of *M. ac.* ssp. *malaccensis* (Pahang – A genome) (D'Hont et al., 2012), *M. balbisiana* (PKW – B genome) (Davey et al., 2013), and banana wild relative *M. itinerans* (Wu et al., 2016). The whole-genome sequencing of *M. itinerans*, one of the most Foc-TR4 resistant and cold tolerant species of the genus *Musa* (Li et al., 2015), provides valuable resource for mining of genes specific to disease resistance and cold tolerance in banana. Further WGS of various species of *Musa* genus will facilitate the understanding of the lineage-specific genome evolution in the *Musa* genus. Till date, WGS of ten *Musa* species is available at banana genome hub. These high-quality reference sub-genomes could be used as multiple reference genomes to identify the allelic variation of the polyploid accessions of inter-/intraspecific commercial cultivars for the better understanding of the important traits.

The chloroplast genomes of of *M. ac.* ssp. *malaccensis* (Martin et al., 2013), *M. balbisiana* variety PKW (Ying-Feng Niu et al., 2018), and *M. ornata* (Liu et al., 2018) were annotated and analyzed, refining *Musa*'s phylogeny. The study highlighted the close relationships between *M. balbsiana* and *M. textilis* and *M. ornate*, as well as *M. ornata* with *M. acuminata*, shedding light on the origin of cultivated triploid bananas. Ever since the publication of WGS data for DH Pahang, there has been dramatic increase in QTL analysis, association, and transcriptomic studies, and with the availability of the high-quality reference genome and annotation available, banana has become a model crop for annotating crops belonging to Zingiberales. The new reference genomes of banana will allow the identification of candidate alleles controlling important agronomic traits that are not present in other genomes. Further, the expansion of genomic data has made it possible for researchers to start analyzing potential genes for particular desired traits. Resistance gene analogs are one group of potential genes for host-plant resistance. According to recent studies (Miller et al., 2008; Pei et al., 2007; Peraza-Echeverria et al., 2008; Azhar and Heslop-Harrison, 2008), such banana resistance gene analogs are being investigated. By aligning other genomes to DH Pahang reference genome, we can identify SNPs which will help to explore the genetic diversity of a large collection of germplasm accessions and for search of new alleles associated to biotic and abiotic stresses.

The exploration of *Musa*'s transcriptome and gene expression offers a potent model to unravel fundamental aspects such as parthenocarpy, sterility, and biparental cytoplasmic inheritance. This model stands out, especially when compared to the two conventional species, Arabidopsis and rice, which are not suitable for studying these specific traits. In banana, two approaches, namely, suppression subtractive hybridization (SSH) and RNA-Seq techniques, have been followed for profiling the transcripts expressed under certain physiological condition. The SSH technique is mainly applicable for identification of differentially expressed genes/alleles under various stress conditions, developmental stages, and various tissues (Diatchenko et al., 1999). This technique has the advantage of sequencing differentially expressed genes, which have high, middle, low, or rare abundance transcripts. Initially, this technique has been used in combination with cDNA microarray technique to identify the genes which are differentially expressed. Swarupa et al. (2013), Backiyarani et al. (2014), and Uma et al. (2016) demonstrated the resistant mechanism underlying in banana for various stresses such as Foc, *P. coffeae*, and *P. eumusae*, respectively. Apart from these studies, SSH combined with cDNA microarray technique has been used to identify the genes corresponding to the trait of interest (Table 1.2).

RNA-sequencing (RNA-Seq) is widely used for transcriptome profiling due to its high sensitivity, wider dynamic range, low cost of sequencing, and lack of need for prior sequence knowledge. This method is regarded as a high-throughput technique as it provides the entire transcriptome sequence along with differences in gene expression. Numerous studies using RNA-Seq have been carried out in banana to identify the alleles and molecular mechanisms associated to the trait of interest (Table 1.3).

TABLE 1.2
SSH and cDNA Microarray Technique for the Identification of Pathways/Genes Involved in Various Traits in Banana

Cultivar/Trait	Genes/Pathways Involved	References
Foc-tolerant Cavendish selection and a Foc-susceptible Williams cultivar		Van den Berg et al. (2004)
Foc-tolerant GCTCV-218 and susceptible cv. Williams	Cell wall-associated phenolic compounds	Van den Berg et al. (2007)
Post-harvest ripening process	ATP-dependent degradation of various cellular proteins tagged with a multi-ubiquitin chain as a degradation signal occur during the onset of ethylene biosynthesis in banana fruit	Mbéguié-A-Mbéguiéa et al. (2007); Xu et al. (2007); Jin et al. (2009)
Ethylene-induced ripening of banana	Stress/defense-related genes along with the genes involved in ethylene biosynthesis, cell wall loosening, etc.	Kesari et al. (2007)
Genes responsible for crown rot disease resistance	Cellulose synthase, a CAF1 gene, 2 glycolipid transfer protein and a dopamine-β-monooxygenase	Lassois et al. (2011)
Calcutta 4, black sigatoka (*P. fijiensis*) resistant accession	Putative disease resistance protein RGA1	Eduardo Sánchez Timm et al. (2016)

TABLE 1.3
Transcriptome Analysis for the Identification of Pathways/Genes Responsible for Various Traits in Banana

Trait	Genes/Pathways Involved	References
Foc R1 and TR4	Ethylene biosynthetic enzyme ACC oxidase and several ethylene-responsive transcription factors (ERF)	Li et al. (2013)
Foc TR4	Jasmonic acid (JA) and Ethylene (ET) hormone signaling pathways	Li et al. (2012)
Foc TR4	Enrichment of alpha-linolenic acid metabolism	Wang et al. (2012)
Foc TR4	Flavonoid biosynthesis, flavone and flavonol biosynthesis, alpha-linolenic acid metabolism, starch and sucrose metabolism, and phenylpropanoid biosynthesis	Wang et al. (2017)
Foc TR4	Pattern recognition receptors (PRRs) and disease resistance (R genes), cell wall modification	Li et al. (2020)
Foc TR4	PR proteins, cell wall modification, phenylpropanoid biosynthesis, and plant hormone signal transduction	Bai et al. (2013)
Foc TR4	Plant hormone signal transduction, phenylpropanoid biosynthesis, and flavonoid biosynthesis	Sun et al. (2019)
Foc TR4	Auxin-activated signaling pathway, cellular response, phenylpropanoid pathway, lignin biosynthesis	Cheng et al. (2019)
Foc TR4	Cell wall biosynthesis and degradation, chitinase activity, pectinesterase activity and xyloglucan activity, fructose and mannose metabolism, sphingolipid metabolism, butanoate metabolism, porphyrin and chlorophyll metabolism, carotenoid and ribosome biosynthesis	Niu et al. (2018)
Foc TR4	PR proteins, transcription factors, cell wall modification, phenylpropanoid biosynthesis, plant hormone signal transduction effectors of the immune response	Zhang et al. (2019) Costa (2013)
Foc TR4	Specific receptor-like kinases (RLKs), transcription factors (TFs), secondary metabolites, and plant hormone-related genes	Dong et al. (2019)
Foc R1	Plant–pathogen response, plant hormone signal transduction, starch and sucrose metabolism, cysteine and methionine metabolism, flavonoid biosynthesis, selenocompound metabolism, phenylpropanoid biosynthesis, mRNA surveillance pathway, mannose type *O*-glycan biosynthesis, amino acid and nucleotide sugar metabolism, cyanoamino acid metabolism, and hormone signal transduction	Kaushal et al. (2021)
P. eumusae	ROS, lignin biosynthesis, hormonal pathways, detoxification, and production of secondary metabolites	Uma et al. (2016)
P. fijiensis	Signal transduction like receptor protein kinase containing LRR repeats, serine/threonine protein kinase, mitogen-activated protein kinase, WRKY superfamily transcription factors	Passos et al. (2012)
Procyanidin-treated fruit	Biosynthetic enzymes and regulatory factors controlling PA biosynthesis	Zhou et al. (2021)
High level potassium	Chitinases, thaumatin-like proteins, peroxidases, ICS1, and 2 PR1 genes	He et al. (2019)
Low-potassium stress	ABC transporters, protein kinases, transcription factors, and ion transporters	Xu et al. (2019)
Water deficit	Protein modifications, lipid metabolism, alkaloid biosynthesis, carbohydrate degradation, glycan metabolism, and biosynthesis of amino acid, cofactor, nucleotide sugar, hormone, terpenoids, and other secondary metabolites	Muthusamy et al. (2016)

Although transcriptomics provides a useful overview of global gene expression, proteomics is often used as a complementary technique that provides a comprehensive insight into the protein profile. Proteomics is one of the cutting-edge approaches for understanding genes' expression, their function, and post-translational modifications (Wilkins et al., 1996). Protein profiling provides base information to short list the genes involved in expression of specific trait. In banana, efforts were taken to identify pathways and the proteins expressed under various stresses/physiological stages which might play a major role in identification of alleles associated with traits of interest (Table 1.4).

TABLE 1.4
Proteomic Analysis for the Identification of Genes/Pathways Responsible for Various Traits in Banana

Trait	Proteins/Pathways	References
Gigantism	Glutamine synthetase	Livramento et al. (2018)
Somatic embryo (early developmental stage)	Ankyrin protein kinase, tubulin alpha chain	Kumaravel et al. (2019)
Late developmental stage	Small HSP, LEA, and APOX	
Fruit quality	Pectate lyase, malate dehydrogenase, and starch phosphorylase	Toledo et al. (2012)
Cold tolerance of fruits during storage	24-Epibrassinolide	Li et al. (2018)
Foc TR4	Beta-1,3-glucanase and chitinase	Lu et al. (2013)
	Chalcone isomerase, glycine-rich RNA-binding protein, carboxylesterase, and GDSL lipase	Zhang et al. (2020)
P. musicola	Chitinases, thaumatin, glutathione transferases, putative avrRpt2-induced AIG2 protein, Germin/oxalate oxidase, regulator of pathogen resistance responses of RPS2 and RPM1 genes, and IAA-amino acid hydrolase	Miller et al. (2016)
BBrMV	Eukaryotic initiating factor 4E (eIF4E)	Anuradha et al. (2021)
Drought	Phosphoglycerate kinase, phosphoglucomutase, UDP-glucose pyrophosphorylase) and proteins that are associated with stress adaptation (e.g., OSR40-like protein, abscisic stress ripening protein-like protein)	Carpentier et al. (2007)
	Heat shock protein	Mattos-Moreira et al. (2018)
Foc TR4	Proteins related to PR response, cell wall strengthening, and antifungal compound synthesis	Li et al. (2013)
Fruit ripening	Energy metabolism, oxidation-reduction, cell wall metabolism, and starch degradation pathways besides ethylene signaling, auxin signaling	Li et al. (2019)
Post-climacteric (P2) banana fruit	Cell wall metabolism, stress and defense response, signaling, and protein metabolism and modification	Xiao et al. (2020)
Drought tolerant	Respiration, metabolism of ROS and several dehydrogenases involved in NAD/NADH homeostasis play an important role	Vanhove et al. (2012)
Cold stress tolerance	increased antioxidation capability	Feng et al. (2015)

1.8 BIOINFORMATICS TOOLS FOR ALLELE MINING IN BANANA

The South Green Bioinformatics Platform has supported the development of banana genome hub, an online genome database construction toolkit that contains several significant banana data sets for genome analysis (https://banana-genome-hub.southgreen.fr). It includes genomic profiles of different *Musa* species (including *M. acuminata* DH Pahang, *M. balbisiana* PKW, *M. itinerans*, *M. schizocarpa*, *Musa troglodytarum*, *Musa beccarii*, *Ensete glaucum*, and *M. acuminate* subspecies such as *banksii*, *zebrina*, and *burmannica*) and provides access to genomic data for download. It has the capacity to store enormous volumes of new information, such as re-sequencing of related samples, transcriptomics (RNA-Seq), transcriptional regulation profiling (e.g., Chip-Seq), epigenetic studies, high-throughput genotyping, and other related whole-genome functional studies. It contains transcriptome data for various tissues and stresses, making it easier to use the search tool to retrieve molecular data for a given query genome ID. For a query sequence, the search tools in this database, including BLAST, locus, keywords or coordinates, gene families (GreenPhyl), metabolic pathway (MusaCyc), SNP markers, and transcriptomic search, assist in retrieving data on alleles, annotation, molecular function, SNPs and SSR markers, and the expression pattern in particular tissues.

MusatransSSRDB (http://bioinfnrcb.byethost7.com/nrcbbio/) was created using the transcriptome data of various cultivars under various physiological conditions. The information on polymorphic SSRs (for repeat length and motif region) between the contrasting genotypes for stress resistance is one of the distinctive features of this database. By avoiding monomorphic SSRs, the *in silico* polymorphic SSRs described in this database could increase the effectiveness of markers and allele development while saving money on laboratory testing. Additionally, since this database has a connection to the SSR-containing genes involved in metabolic pathways, one can retrieve information on SSRs for the required metabolic pathway, speeding up the development of markers/alleles using the candidate gene approach. This database has also been updated with the genomic SSRs of various *Musa* species which can be used for genetic diversity studies and to develop fingerprints (Backiyarani et al., 2019). MusaRgeneDB is complemented with complete details of 3,598 resistant (R) genes identified from eight *Musa* spp. This will aid in the identification of alleles and markers associated, cloning of full-length R genes, and genetic transformation or gene editing of the R genes in susceptible cultivars. This database would facilitate the understanding of defense mechanisms in bananas against biotic and abiotic stresses leading to the development of promising disease-resistant varieties (Anuradha et al., 2022a, b). Apart from these banana-specific databases, other bioinformatics tools such as PLACE, plantCARE, TRANSFAC, JASPAR, MEME, Plantprom DB, DCPD, SCPD, BioEdit, and ClustalW can be exploited for allele mining (Table 1.5).

1.9 APPLICATIONS OF ALLELE MINING IN BANANA

Banana allele mining will help identify alleles linked to resistance to a variety of diseases, such as Foc and leaf spot, and these genes can be pyramided in commercial cultivars to produce horizontal resistance. Additionally, it decreases the use of pesticides and prolongs the emergence of new races and strains. The development of climate-resilient varieties, which are urgently needed in the current climate change scenario, will be facilitated by the identification of alleles that are stress-tolerant to drought, cold, salinity, and high temperature conditions. The molecular breeder will be able to identify the key QTLs that will be used to create varieties with improved yield by mining alleles specific to traits in bananas that are related to yield. Reduced post-harvest loss and improved nutritional security will result from the identification of the alleles causing delayed ripening and improved nutritional quality of bananas. Finding alleles that alter plant architecture, such as dwarfness and duration, will increase productivity and lower cultivation costs. In the long run, allele mining has the potential to significantly enhance the quality, productivity, and sustainability of bananas, thereby assisting in ensuring a reliable food supply for a population that is expanding worldwide (Figure 1.2).

TABLE 1.5
Bioinformatics Tools for Allele Mining

Description	Name of the Tools and Database	URL	References
Reads sequence alignment NGS tool	Hisat 2	http://daehwankimlab.github.io/hisat2/download/	Kim et al. (2019)
Genome Analysis Tool kit	GATK	https://gatk.broadinstitute.org/hc/en-us	McKenna et al. (2010)
SNP identification and annotation	SNPEff	http://pcingola.github.io/SnpEff/	Cingolani et al. (2012)
Annotation software	BLAST2GO	https://www.blast2go.com/	Conesa and Götz (2008)
Protein Databank	PDB	https://www.rcsb.org/	Berman et al. (2000)
Identification, characterization, and quantitation of proteins	Mascot Server	https://www.matrixscience.com/	Sancho et al. (2005)
Trait Analysis by aSSociation, Evolution, and Linkage	TASSEL	https://www.maizegenetics.net/tassel	Bradbury et al. (2007)
Genome association and prediction integrated tool	GAPIT	http://www.maizegenetics.net/GAPIT	Lipka et al. (2012)
Multiple Em for motif Elicitation	MEME	https://meme-suite.org/meme/meme_5.5.0/tools/meme	Bailey et al. (2006)
The high-quality transcription factor binding profile database	JASPAR	https://jaspar.genereg.net/	Bryne et al. (2008)
Database of Plant Cis-acting Regulatory DNA Elements	New PLACE	https://www.dna.affrc.go.jp/PLACE/?action=newplace	Higo et al. (1999)
Database of Plant Cis-acting Regulatory DNA Elements	plantCARE	https://bioinformatics.psb.ugent.be/webtools/plantcare/html/	Lescot et al. (2002)
Database of eukaryotic transcription factors and binding sites	TRANSFAC	https://genexplain.com/transfac/	Wingender (2008)
Plant promoter database	Plantprom DB	http://linux1.softberry.com/berry.phtml?topic=plantprom&group=data&subgroup=plantprom	Shahmuradov et al. (2003)
Drosophila Core Promoter Database	DCPD	http://www.biology.ucsd.edu/labs/Kadonaga/DCPD.html	Kutach and Kadonaga (2000)
Saccharomyces cerevisiae promoter database	SCPD	http://rulai.cshl.edu/SCPD/ http://cgsigma.cshl.org/jian	Zhu and Zhang (1999)
Sequence alignment and homology search	BioEdit	https://thalljiscience.github.io/	Ahmed (2011)
Multiple sequence alignment	Clustal Omega	https://www.ebi.ac.uk/Tools/msa/clustalo/	Sievers and Higgins (2018)
Molecular Evolutionary genetic analysis	MEGA	https://www.megasoftware.net/	Kumar et al. (2018)
Primer Designing software	Primer3 plus	https://www.bioinformatics.nl/cgi-bin/primer3plus/primer3plus.cgi	Rozen and Skaletsky (1999)
In silico PCR tool	FASTPCR	https://primerdigital.com/fastpcr.html	Kalendar et al. (2017)

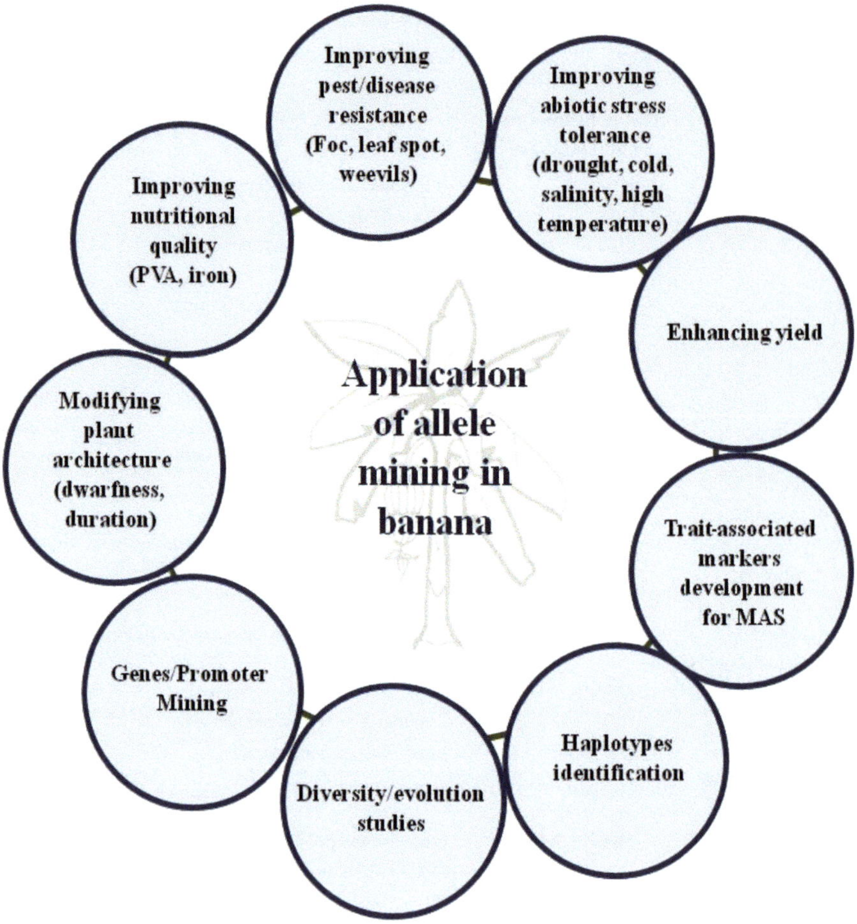

FIGURE 1.2 Applications of allele mining in banana.

1.10 FUTURE PERSPECTIVES

The success of the allele mining depends on the accurate phenotyping for a specific trait. As a result, greater emphasis should be placed on phenotypic characterization of accessions using robust methodologies for screening a specific trait. Information on SNPs in the exonic region of the gene/s governing the specific trait will not suffice to detect the allele that has a significant impact on the expression of the phenotype. Understanding nucleotide variation at the regulatory and promoter regions is also essential for developing strategies for improving specific traits in banana. Overall, allele mining has a great deal of potential to enhance banana traits like increased yield, disease resistance, and stress tolerance. Future developments in genome editing methods, like CRISPR-Cas9, will probably have a significant impact on allele mining by making it possible to validate the alleles and precisely introduce desired alleles into bananas. Furthermore, improvements in high-throughput sequencing and data analysis technologies will make it possible to identify and characterize valuable alleles within plant populations more quickly and effectively. In general, allele mining has the potential to significantly impact agriculture and food security by enhancing the productivity and sustainability of bananas.

REFERENCES

Ahmad F, Martawi NM, Poerba YS, de Jong H, Kema GHJ (2020) Genetic mapping of Fusarium wilt resistance in a wild banana *Musa acuminata* ssp. *malaccensis* accession. *Theor Appl Genet* 133:3409–3418.

Ahmed MA (2011) BioEdit: an important software for molecular biology. *GERF Bull Biosci* 2(1):60–61.

Anonymous (1993) ICAR-NRCB Annual Report 1993.

Anuradha C, Chandrasekar A, Backiyarani S, Thangavelu R, Uma S (2022a) Genome-wide identification, characterization, and evolutionary analysis of NBS genes and their association with disease resistance in *Musa* spp. *Funct Integr Genom* 23(1): 1–18.

Anuradha C, Chandrasekar A, Backiyarani S, Uma S (2022b) MusaRgeneDB: an online comprehensive database for disease resistance genes in *Musa* spp. *3 Biotech* 12:222.https://doi.org/10.1007/s13205-022-03285-1.

Anuradha C, Chandrasekar A, Prashina M, Backiyarani S, Uma S (2022c) A whole genome based reduced representation approach for identification of seedless phenotype in banana (*Musa* spp.). In: Rajendiran S, Muralidhara BM, Rani AT, Madhu G (eds) *Souvenir and Abstracts, National Symposium on Horticultural Crops of Humid Tropics for Nutritional and Livelihood security*, ICAR-IIHR, Bengaluru, 02-03 Dec 2022, Kodagu, Karnataka. pp 106.

Anuradha C, Selvarajan R, Jebasingh T, SankaraNaynar P (2021) Evidence of viral genome linked protein of banana bract mosaic virus interaction with translational eukaryotic initiation factor 4E of plantain cv. Nendran based on yeast two hybrid system study. *Virus Dis* 32(1):123–130. https://doi.org/10.1007/s13337-021-00672-9.

Arinaitwe IK, Teo CH, Kayat F, Tumuhimbise R, Uwimana B et al. (2019) Evaluation of banana germplasm and genetic analysis of an F1 population for resistance to *Fusarium oxysporum* f. sp. *cubense* race 1. *Euphytica* 215:1–11.

Azhar M, Heslop-Harrison JS (2008) Genomes, diversity and resistance gene analogues in *Musa* species. *Cytogenet Genome Res* 121:59–66.

Backiyarani S, Anuradha C, Uma S (2022) Genomic designing for biotic stress resistant banana. In: Kole C (eds) *Genomic Designing for Biotic Stress Resistant Fruit Crops*. Springer Nature, Cham, pp. 25–74.

Backiyarani S, Chandrasekar A, Uma S, Saraswathi MS (2019) MusatransSSRDB (a transcriptome derived SSR database) - an advanced tool for banana improvement. *J Biosci* 44(1):4.

Backiyarani S, Uma S, Arunkumar G, Saraswathi MS, Sundararaju P (2014) Differentially expressed genes in incompatible interactions of *Pratylenchus coffeae* with *Musa* using suppression subtractive hybridization. *Physiol Mol Plant Pathol* 86:11–18.

Bai TT, Xie WB, Zhou PP, Wu ZL, Xiao WC et al. (2013) Transcriptome and expression profile analysis of highly resistant and susceptible banana roots challenged with *Fusarium oxysporum* f. sp. *cubense* tropical race 4. *PLoS One* 8(9):e73945.

Bailey TL, Williams N, Misleh C, Li WW (2006) MEME: discovering and analyzing DNA and protein sequence motifs. *Nucleic Acids Res* 34:W369–W373.

Berman HM, Westbrook J, Feng Z, Gilliland G, Bhat TN et al. (2000) The protein data bank. *Nucleic Acids Res* 28:235–242.

Bradbury PJ, Zhang Z, Kroon DE, Casstevens TM, Ramdoss Y, Buckler ES (2007) TASSEL: software for association mapping of complex traits in diverse samples. *Bioinformatics* 23:2633–2635.

Bryne JC, Valen E, Tang MHE, Marstrand T, Winther O et al. (2008) JASPAR, the open access database of transcription factor-binding profiles: new content and tools in the 2008 update. *Nucleic Acids Res* 36:D102–D106.

Carpentier SC, Witters E, Laukens K, Van Onckelen H, Swennen R, Panis B (2007) Banana (*Musa* spp.) as a model to study the meristem proteome: acclimation to osmotic stress. *Proteomics* 7(1):92–105. https://doi.org/10.1002/pmic.200600533.

Cenci A, Guignon V, Roux N, Rouard M (2014) Genomic analysis of NAC transcription factors in banana (*Musa acuminata*) and definition of NAC orthologous groups for monocots and dicots. *Plant Mol Biol* 85:63–80. https://doi.org/10.1007/s11103-013-0169-2.

Cenci A, Hueber Y, Zorrilla-Fontanesi Y, van Wesemael J, Kissel E et al. (2019) Effect of paleopolyploidy and allopolyploidy on gene expression in banana. *BMC Genomics* 20(1):244. https://doi.org/10.1186/s12864-019-5618-0.

Cheng C, Liu F, Sun X, Tian N, Mensah RA, Li D, Lai Z (2019) Identification of *Fusarium oxysporum* f. sp. *cubense* tropical race 4 (Foc TR4) responsive miRNAs in banana root. *Sci Rep* 9:13682.

Cingolani P, Platts A, Wang LL, Coon M, Nguyen T et al. (2012) A program for annotating and predicting the effects of single nucleotide polymorphisms, SnpEff: SNPs in the genome of *Drosophila melanogaster* strain w1118; iso-2; iso-3. *Fly (Austin)* 6(2):80–92.

Conesa A, Götz S (2008) Blast2GO: a comprehensive suite for functional analysis in plant genomics. *Int J Plant Genom* 2008:619832. https://doi.org/10.1155/2008/619832.

Costa JL (2013) Estudos histológicos e moleculares da interação *Musa* spp. x *Fusarium oxysporum* f. sp. *cubense*. Piracicaba, São Paulo: Universidade de São Paulo, Dissertação (Mestrado em ciências).

Craenen K, Ortiz R (2003) Genetic improvement for a sustainable management of resistance. In: Jacome L, Lepoivre P, Marin D, Ortiz R, Romero R, Escalant JV (eds) *Mycosphaerella Diseases of Bananas: Present Status and Outlook*. INIBAP, Montpellier. pp. 181–198.

D'Hont A, Denoeud F, Aury JM, Baurens FC, Carreel F et al. (2012) The banana (*Musa acuminata*) genome and the evolution of monocotyledonous plants. *Nature* 488 (7410):213–217. https://doi.org/10.1038/nature11241.

Daniells J, Jenny C, Karamura D, Tomekpe K (2001) *Musalogue: A Catalogue of Musa Germplasm. Diversity in the Genus Musa*. International Network for the Improvement of Banana and Plantain, Montpellier, France.

Davey MW, Gudimella R, Harikrishna JA, Sin LW, Khalid N, Keulemans J (2013) A draft *Musa balbisiana* genome sequence for molecular genetics in polyploid, inter- and intra-specific Musa hybrids. *BMC Genomics* 14:683. https://doi.org/10.1186/1471-2164-14-683.

Diatchenko L, Lukyanov S, Lau YF, Siebert PD (1999) Suppression subtractive hybridization: a versatile method for identifying differentially expressed genes. *Methods Enzymol* 303:349–380. https://doi.org/10.1016/s0076-6879(99)03022-0.

Dochez C, Tenkouano A, Ortiz R, Whyte J, De Waele D (2009) Host plant resistance to *Radopholus similis* in a diploid banana hybrid population. *Nematology* 11:329–335.

Dodds KS, Simmonds NW (1948) Sterility and parthenocarpy in diploid hybrids of Musa. *Heredity* 2:101–117.

Dong H, Li Y, Fan H, Zhou D, Li H (2019) Quantitative proteomics analysis reveals resistance differences of banana cultivar 'Brazilian' to *Fusarium oxysporum* f. sp. *cubense* races 1 and 4. *J Proteomics* 203:103376. https://doi.org/10.1016/j.jprot.2019.05.004.

Fauré S, Noyer JL, Horry JP, Bakry F, Lanaud C, Goñzalez de León D (1993) A molecular marker-based linkage map of diploid bananas (*Musa acuminata*). *Theor Appl Genet* 87:517–526. https://doi.org/10.1007/BF00215098.

Feng R, Zhang L, Wang J, Luo J, Peng M et al. (2015) Proteomic analysis of cold stress responses in banana leaves. *J Amer Soc Hortic Sci* 140(3):214–222. https://doi.org/10.21273/JASHS.140.3.214.

Fiume E, Christou P, Gianı S, Breviario D (2004) Introns are key regulatory elements of rice tubulin expression. *Planta* 218:693–703.

Fraser-Smith S, Czislowski E, Daly A, Meldrum R, Hamill S et al. (2016) Single gene resistance to *Fusarium oxysporum* f. sp. *cubense* Race 4 in the wild banana *Musa acuminata* subsp. *malaccensis*. *Acta Hort* 1114:95–100.

Fu D, Szucs P, Yan L, Helguera M, Skinner JS et al. (2005) Large deletions within the first intron in VRN-1 are associated with spring growth habit in barley and wheat. *Mol Genet Genom* 273:54–65.

Fullerton RA, Olsen TL (1995) Pathogenic variability in *Mycosphaerella fijiensis* Morelet cause of black Sigatoka in banana and plantain. *New Zealand J. Crop Hortic Sci* 23:39–48.

García-Bastidas FA, Van der Veen AJT, Nakasato-Tagami G, Meijer HJG, Arango-Isaza RE, Kema GHJ (2019) An improved phenotyping protocol for Panama disease in banana. *Front Plant Sci* 10:1–12.

González de León D, Fauré S (1993) Genetic mapping of the banana diploid genome-toward an integrated approach to the study of the *Musa* genome and the use of molecular marker technologies in *Musa* breeding. In: *Proceedings of the INIBAP Worshop on Biotechnologies for Banana and Plantain*. San José, Costa-Rica. (in press).

Hapsari L, Masrum A (2012) Preliminary screening resistance of *Musa* germplasms for banana bunchy top disease in Purwodadi Botanic garden, Pasuruan, East Java. *Buletin Kebun Raya* 15(2):57–70.

He Y, Li R, Lin F, Xiong Y, Wang L et al. (2019) Transcriptome changes induced by different potassium levels in banana roots. *Plants (Basel, Switzerland)* 9(1):11.

Higo K, Ugawa Y, Iwamoto M, Korenaga T (1999) Plant cis-acting regulatory DNA elements (PLACE) database: 1999. *Nucleic Acids Res* 27:297–300.

Hippolyte I, Bakry F, Séguin M, Gardès L, Rivallan R et al. (2010) A saturated SSR/DArT linkage map of *Musa acuminata* addressing genome rearrangements among bananas. *BMC Plant Biol* 10:65. https://doi.org/10.1186/1471-2229-10-65.

Hou BH, Tsai YH, Chiang MH, Tsao SM, Huang SH, Chao CP, Chen HM (2022) Cultivar-specific markers, mutations, and chimerisim of Cavendish banana somaclonal variants resistant to *Fusarium oxysporum* f. sp. *cubense* tropical race 4. *BMC Genomics* 23(1):470. https://doi.org/10.1186/s12864-022-08692-5.

Jeridi M, Bakry F, Escoute J, Fondi E, Carreel F et al. (2011) Homoeologous chromosome pairing between the A and B genomes of Musa spp. revealed by genomic in situ hybridization. *Ann Bot* 108(5):975–981.

Jin ZQ, Xu BY, Liu JH, Zhang JB, Jia CH (2009) Combination of suppression-subtractive hybridisation with CDNA microarray, a novel way to identify genes from banana involved in fruit ripening. *Acta Hortic* 897:195–206.

Jones DR (2000) *Diseases of Banana, Abaca and Enset*. CABI, Oxfordshire.

Kalendar R, Khassenov B, Ramankulov Y, Samuilova O, Ivanov KI (2017) FastPCR: an in silico tool for fast primer and probe design and advanced sequence analysis. *Genomics* 109:312–319. https://doi.org/10.1016/j.ygeno.2017.05.005.

Kaur N, Street K, Mackay M, Yahiaoui N, Keller B (2008) Molecular approaches for characterization and use of natural disease in wheat. *Eur J Plant Pathol* 121:387–397.

Kaushal M, Mahuku G, Swennen R (2021) Comparative transcriptome and expression profiling of resistant and susceptible banana cultivars during infection by *Fusarium oxysporum*. *Int JMol Sci* 22(6):3002.

Kayat F, Bonar N, Waugh R, Rajinder S, Rahimah AR et al. (2009) Development of a genetic linkage map for genes associated with resistance and susceptibility to *Fusarium oxysporum* f. sp. *cubense* from an F1 hybrid population of *Musa acuminata* ssp. *malaccensis*. *Acta Hort* 828:333–340.

Kesari R, Trivedi PK, Nath P (2007) Ethylene-induced ripening in banana evokes expression of defense and stress related genes in fruit tissue. *Post harvest Biol Technol* 6:136–143.

Kim D, Paggi JM, Park C, Bennett C, Salzberg SL (2019) Graph-based genome alignment and genotyping with HISAT2 and HISAT-genotype. *Nat Biotechnol* 37:907–915.

Kumar S, Stecher G, Li M, Knyaz C, Tamura K (2018) MEGA X: molecular evolutionary genetics analysis across computing platforms. *Mol Biol E* 35(6):1547–1549.

Kumaravel M, Uma S, Backiyarani S, Saraswathi MS (2019) Molecular analysis of somatic embryogenesis through proteomic approach and optimization of protocol in recalcitrant *Musa* spp. *Physiol Planta* 167(3):282–301. https://doi.org/10.1111/ppl.12966.

Kutach AK, Kadonaga JT (2000) The downstream promoter element DPE appears to be as widely used as the TATA box in Drosophila core promoters. *Mol Cell Biol* 20:4754–4764.

Lagoda PJL, Noyer JL, Baurens FC (1998) Results of a *Musa* mapping project. No. IAEA-TECDOC--1047.

Lassois L, Frettinger P, de Lapeyre de Bellaire L, Lepoivre P, Jijakli H (2011) Identification of genes involved in the response of banana to crown rot disease. *Mol Plant-Microbe Interact* 24(1):143–153. https://doi.org/10.1094/MPMI-01-10-0020.

Lescot M, Déhais P, Thijs G, Marchal K, Moreau Y et al. (2002) PlantCARE, a database of plant cis-acting regulatory elements and a portal to tools for in silico analysis of promoter sequences. *Nucleic Acids Res* 30:325–327.

Li C, Shao J, Wang Y, Li W, Guo D et al. (2013) Analysis of banana transcriptome and global gene expression profiles in banana roots in response to infection by race 1 and tropical race 4 of *Fusarium oxysporum* f. sp. *cubense*. *BMC Genomics* 14:851.

Li CY, Deng GM, Yang J, Viljoen A, Jin Y et al. (2012) Transcriptome profiling of resistant and susceptible Cavendish banana roots following inoculation with *Fusarium oxysporum* f. sp. *cubense* tropical race 4. *BMC Genomics* 13:374. https://doi.org/10.1186/1471-2164-13-374.

Li MH, Xie XL, Lin XF, Shi JX, Ding ZJ et al. (2014) Functional characterization of the gene FoOCH1 encoding a putative α-1,6-mannosyltransferase in *Fusarium oxysporum* f. sp. *cubense*. *Fungal Genet Biol* 65:1–13.

Li T, Yun Z, Wu Q, Qu H, Duan X, Jiang Y (2019) Combination of transcriptomic, proteomic, and metabolomic analysis reveals the ripening mechanism of banana pulp. *Biomol Ther* 9(10):523. https://doi.org/10.3390/biom9100523.

Li T, Yun Z, Wu Q, Zhang Z, Liu S et al. (2018) Proteomic profiling of 24-epibrassinolide-induced chilling tolerance in harvested banana fruit. *J Proteom* 187:1–12. https://doi.org/10.1016/j.jprot.2018.05.011.

Li WM, Dita M, Wu W, Hu GB, Xie JH, Ge XJ (2015) Resistance sources to *Fusarium oxysporum* f. sp. *cubense* tropical race 4 in banana wild relatives. *Plant Pathol* 64(5):1061–1067.

Li X, Bai T, Li Y, Ruan X, Li H (2013) Proteomic analysis of *Fusarium oxysporum* f. sp. *cubense* tropical race 4-inoculated response to Fusarium wilts in the banana root cells. *Proteome Sci* 11(1):41. https://doi.org/10.1186/1477-5956-11-41.

Li Z, Wang T, He CL, Cheng K, Zeng R, Song Y (2020) Control of Panama disease of banana by intercropping with Chinese chive (*Allium tuberosum* Rottler): cultivar differences. *BMC Plant Biol* 20:432. https://doi.org/10.1186/s12870-020-02640-9.

Lipka AE, Tian F, Wang Q, Peiffer J, Li M et al. (2012) GAPIT: genome association and prediction integrated tool. *Bioinformatics (Oxford, England)* 28(18):2397–2399. https://doi.org/10.1093/bioinformatics/bts444.

Liu J, Gao CW, Niu Y F (2018) The complete chloroplast genome sequence of flowering banana, *Musa ornata*. *Mitochondrial DNA B Resour* 3(2):960–961.

Liu J, Zhang J, Zhang J, Miao H, Wang J, Gao P, Hu W, Jia C, Wang Z, Xu B, Jin Z (2017) Allele mining in tomato for increased fruit size. *Plant Sci* 260:12–18.

Livramento KG, Fialho L, Santos A, do Livramento D, Cardoso T, Paiva L (2018) Proteomic analysis reveals differentially accumulated proteins in banana somaclonal variants. *J Exp Agric Int* 21(6):1–13. https://doi.org/10.9734/JEAI/2018/39857.

Lu Y, Liao D, Jinji Pu J, Qi Y, Xie Y (2013) Proteome analysis of resistant and susceptible Cavendish banana roots following inoculation with *Fusarium oxysporum* f. sp. cubense. *Physiol Mol Plant Pathol* 84:163–171.

Mahendhiran M, Ramírez-Prado JH, Medrano RE, Canto-Canché, BB, Tzec-Simá M et al (2014) Single nucleotide polymorphisms in partial sequences of the gene encoding the large sub-units of ADP-glucose pyrophosphorylase within a representative collection of 10 Musa genotypes. *Elect J Biotechnol* 17:137–147.

Martin G, Baurens F-C, Cardi C, Aury J-M, D'Hont A (2013) The complete chloroplast genome of banana (*Musa acuminata*, Zingiberales): insight into plastid monocotyledon evolution. *PLoS One* 8(6):e67350. https://doi.org/10.1371/journal.pone.0067350.

Mattos-Moreira LA, Ferreira CF, Amorim EP, Pirovani CP, Andrade EM et al. (2018) Differentially expressed proteins associated with drought tolerance in bananas (*Musa* spp.). *Acta Physiol Plant* 40:60. https://doi.org/10.1007/s11738-018-2638-3.

Mbéguié-A-Mbéguié D, Hubert O, Sabau X, Chillet M, Fils-Lycaon B, Baurens FC (2007) Use of suppression subtractive hybridization approach to identify genes differentially expressed during early banana fruit development undergoing changes in ethylene responsiveness. *Plant Sci* 172(5):1025–1036.

McKenna A, Hanna M, Banks E, Sivachenko A, Cibulskis K et al. (2010) The Genome Analysis Toolkit: a MapReduce framework for analyzing next-generation DNA sequencing data. *Genome Res* 20(9):1297–1303. https://doi.org/10.1101/gr.107524.110.

Miller RNG, Bertioli DJ, Baurens FC, Santos CMR, Alves PC et al. (2008) Analysis of non-TIR NBS-LRR resistance gene analogs in *Musa acuminata* colla: isolation, RFLP marker development, and physical mapping. *BMC Plant Biol* 8:1–15.

Miller RNG, Passos MAN, Emediato FL, Cruz VO, Silva Jr O, Franco OL, Oliveira-Junior NG (2016) Gene expression and proteome analysis during the *Musa acuminata-Mycosphaerella musicola* interaction. *Acta Hortic* 1114:149–152. https://doi.org/10.17660/ActaHortic.2016.1114.21.

Mmeka EC, Adesoye AI, Vroh BI, Ubaoji KI (2013) Single nucleotide polymorphism (SNP) markers discovery within *Musa* spp (Plantain landraces, AAB Genome) for use in beta carotene (Provitamin A) trait mapping. *Amer J Biol Life Sci* 1(1):11–19.

MusaNet (2016) *Global Strategy for the Conservation and Use of Musa Genetic Resources, B. Laliberté, Compiler*. Bioversity International, Montpellier, France.

Muthusamy M, Uma S, Backiyarani S, Saraswathi MS, Chandrasekar A (2016) Transcriptomic changes of drought-tolerant and sensitive banana cultivars exposed to drought stress. *Front Plant Sci* 7:1609.

Nascimento FS, Sousa, YM, Rocha AJ, Ferreira CF, Haddad F, Amorim EP (2020) Sources of black Sigatoka resistance in wild banana diploids. *Rev Bras Fruticul* 42(4):e–038.

Niu Y, Hu B, Li X, Chen H, Takáč T, Šamaj J, Xu C (2018) Comparative digital gene expression analysis of tissue-cultured plantlets of highly resistant and susceptible banana cultivars in response to *Fusarium oxysporum*. *Int J Mol Sci* 19(2):350.

Niu Y-F, Gao C-W, Liu J (2018) The complete chloroplast genome sequence of wild banana, *Musa balbisiana* variety 'Pisang Klutuk Wulung' (Musaceae). *Mitochondrial DNA Part B Res* 3(1):460–461. https://doi.org/10.1080/23802359.2018.1462123.

Nyine M, Uwimana B, Akech V, Brown A, Ortiz R et al. (2019) Association genetics of bunch weight and its component traits in East African highland banana (*Musa* spp AAA group). *Theor Appl Genet* 132:3295–3308. https://doi.org/10.1007/s00122-019-03425-x.

Ortiz R, Vuylsteke D (1992) Inheritance of black sigatoka resistance and fruit parthenocarpy in triploid AAB plantain. In: *Agronomy Abstracts*. American Society of Agronomy, Madison, WI, p. 109.

Ortiz-Vázquez E, Kaemmer D, Zhang HB, Muth J, Rodríguez-Mendiola M et al. (2005) Construction and characterization of a plant transformation-competent BIBAC library of the black Sigatoka resistant banana *Musa acuminata* cv. Tuu Gia (AA). *Theor Appl Genet* 110(4):706–713. https://doi.org/10.1007/s00122-004-1896-1.

Pardo-Manuel De Villena F, Sapienza C (2001) Nonrandom segregation during meiosis: the unfairness of females. *Mamm Genome* 12:331–339. https://doi.org/10.1007/s003350040003.
Passos MAN, Cruz VO, Emediato FL, Teixeira CC, Souza MT et al. (2012) Development of expressed sequence tag and expressed sequence tag-simple sequence repeat marker resources for *Musa acuminata*. *AoB Plants* 2012:pls030.
Pei X, Li S, Jiang Y, Wang Z, Jia S (2007) Isolation, characterisation and phylogenetic analysis of the resistance gene analogues (RGAs) in banana (*Musa* spp). *Plant Sci* 172:1166–1174.
Peraza-Echeverria S, Dale J, Harding R, Smith M, Collet C (2008) Characterization of resistance gene candidates of the nucleotide binding site (NBS) type from banana and correlation of a transcriptional polymorphism with resistance to FOC race 4. *Mol Breed* 22:565–579.
Perrier X, Bakry F, Carreel F, Jenny C, Horry JP et al. (2009) Combining biological approaches to shed light on the evolution of edible bananas. *Ethnobot Res Appl* 7:199–216.
Rebouças TA, Haddad F, Ferreira CF, Oliveira SAS, Ledo CAS et al. (2018) Identification of banana genotypes resistant to Fusarium wilt race 1 under field and greenhouse conditions. *Sci Hort* 239:308–313.
Roux NS (2001) Mutation induction in Musa review. In: Jain SM, Swennen R (eds) *Banana Improvement: Cellular, Molecular Biology, Induced Mutations*. Science Publishers, Enfield, pp. 23–32.
Rowe P, Rosales F (1996) Bananas and plantains. In: Janick J, Moore JN (eds) *Fruit Breeding, Tree and Tropical Fruits*. Wiley, New York, Vol I, pp. 167–211.
Rozen S, Skaletsky H (1999) Primer3 on the WWW for general users and for biologist programmers. In: Clifton NJ (ed) *Bioinformatics Methods and Protocols*. Springer, pp. 365–386.
Šafář J, Bartoš J, Janda J, Bellec A, Kubaláková M et al. (2004) Dissecting large and complex genomes: flow sorting and BAC cloning of individual chromosomes from bread wheat. *Plant J Cell Mol Biol* 39(6):960–968.
Sánchez Timm E, Hidalgo Pardo L, Pacheco Coello R, Chávez Navarrete T, Navarrete Villegas O, Santos Ordóñez E (2016) Identification of differentially-expressed genes in response to *Mycosphaerella fijiensis* in the resistant musa accession 'Calcutta-4' using suppression subtractive hybridization. *PLoS One* 11(8):e0160083. https://doi.org/10.1371/journal.pone.0160083.
Sancho PP, Juiz C, Puigjaner R (2005) Evaluating the performance of architectures in MASCOT. *Sci Comput Program* 57:45–60. https://doi.org/10.1016/j.scico.2004.10.004.
Sardos J, Christelová P, Cizková J, Paofa J, Sachter-Smith G et al. (2018) Collection of new diversity of wild and cultivated bananas (*Musa* spp.) in the autonomous region of Bougainville, Papua New Guinea. *Genet Resour Crop E* 65:2267–2286.
Sardos J, Rouard M, Hueber Y, Cenci A, Hyma KE et al. (2016) A genome-wide association study on the seedless phenotype in banana (*Musa* spp.) reveals the potential of a selected panel to detect candidate genes in a vegetatively propagated crop. *PLoS One* 11:e0154448. https://doi.org/10.1371/journal.pone.0154448.
Shahmuradov IA, Gammerman A, Hancock JM, Bramley PM, Solovyev VV (2003) PlantProm: a database of plant promoter sequences. *Nucleic Acids Res* 31: 114–117.
Shepherd K (1968) Banana breeding in the West Indies. *Pest Artic News Summ* 14:370–379.
Shepherd K (1990) Genetic improvement of bananas in Brazil: aspects related to resistance to the genus *Mycosphaerella*. In: Fullerton RA, Stover RH (eds) *Sigatoka Leaf Spot Diseases of Bananas. Proceedings of an International Workshop Held at San José, Costa Rica*, March 28- April 1, 1989. INIBAP, Montpellier, France, pp. 237–242.
Shepherd K (1999) *Cytogenetics of the Genus Musa*. IPGRI Montpellier, France. ISBN: 2-910810-30-5.
Sievers F, Higgins DG (2018) Clustal Omega for making accurate alignments of many protein sequences. *Protein Sci* 27(1):135–145. https://doi.org/10.1002/pro.3290.
Simmonds NW (1953) The development of the banana fruit. *J Expt Bot* 4(10):87–105.
Simmonds NW (1962) *The Evolution of the Bananas*. Tropical Science Series. Longmans, London (GBR), 170 p.
Simmonds NW, Shepherd K (1955) The taxonomy and origins of the cultivated bananas. *J Linn Soc (Bot)* 55:302–312.
Singh HP, Uma S, Sathiamoorthy S (2001) *A Tentative Key for Identification and Classification of Indian Bananas*. National Research Centre for Banana (ICAR), Tiruchirapalli.
Singh S, Singh AK, Singh HP, Singh VN, Singh RS (2005) Studies on combining ability and heterobeltiosis of organogenesis for salt tolerance in rice under in vitroconditions. *Oryza* 42(4):260–267.
Ssali RT, Kiggundu A, Lorenzen J, Karamura E, Tushemereirwe W et al. (2013) Inheritance of resistance to *Fusarium oxysporum* f. sp.*cubense* race 1 in bananas. *Euphytica* 194:425–430. https://doi.org/10.1007/s10681-013-0971-6.
Sun J, Zhang J, Fang H, Peng L, Wei S et al. (2019) Comparative transcriptome analysis reveals resistance-related genes and pathways in *Musa acuminata* banana 'Guijiao 9' in response to Fusarium wilt. *Plant Physiol Biochem* 141:83–94.

Sureshkumar S, Todesco M, Schneeberger K, Harilal R, Balasubramanian S, Weigel D (2009) A genetic defect caused by a triplet repeat expansion in *Arabidopsis thaliana*. *Science* 323:1060–1063.

Swarupa V, Ravishankar K, Rekha A (2013) Characterization of tolerance to *Fusarium oxysporum* f.sp. *cubense* infection in banana using suppression subtractive hybridization and gene expression analysis. *Physiol Mol Plant Pathol* 83:1–7.

Symonds VV, Soltis PS, Soltis DE (2010) Dynamics of polyploid formation in Tragopogon (Asteraceae): recurrent formation, gene flow, and population structure. *Evolution* 64(7):1984–2003. https://doi.org/10.1111/j.1558-5646.2010.00978.x.

Thangavelu R, Muthukathan G, Pushpakanth P, Murugan L, Edwin Raj E et al. (2020) First report of *Fusarium oxysporum* f.sp. cubense VCG 0125 and VCG 01220 of Race 1 infecting Cavendish bananas (*Musa* sp. AAA) in India. *Plant Dis* 2. https://doi.org/10.1094/PDIS-09-20-2052-PDN.

Thomas DS, Turner DW, Eamus D (1998) Independent effects of the environment on the leaf gas exchange of three banana 8 *Musa* sp cultivars of different genomic constitution. *Sci Hort* 75:41–57.

Thomson MJ, Biswas S, Tsakirpaloglou N, Septiningsih EM (2022) Functional allele validation by gene editing to leverage the wealth of genetic resources for crop improvement. *Int J Mol Sci* 23(12):6565.

Till BJ, Jankowicz-Cieślak J, Sági L, Huynh OA, Utsushi H et al. (2010) Discovery of nucleotide polymorphisms in the *Musa* gene pool by ecotilling. *Theor Appl Genet* 121(7):1381–1389.

Toledo TT, Nogueira SB, Cordenunsi BR, Gozzo FC, Pilau EJ, Lajolo FM, Nascimento JRO (2012) Proteomic analysis of banana fruit reveals proteins that are differentially accumulated during ripening. *Postharvest Biol Technol* 70:51–58.

Tuberosa R (2012) Phenotyping for drought tolerance of crops in the genomics era. *Front Physiol* 3:347.

Turner DW, Gibbs DJ (2018) A functional approach to bunch formation in banana. In: Kema GHJ, Drenth A (eds) *Achieving Sustainable Cultivation of Bananas: Volume 1-Cultivation Techniques*. Burleigh Dodds Science Publishing, Cambridge. ISBN 978 1 78676 156 9.

Udall JA, Quijada PA, Osborn TC (2005) Detection of chromosomal rearrangements derived from homologous recombination in four mapping populations of *Brassica napus* L. *Genetics* 169(2):967–979. https://doi.org/10.1534/genetics.104.033209.

Uma S, Backiyarani S, Saravanakumar AS, Chandrasekar A, Thangavelu R, Saraswathi MS (2016) Identification of *Mycosphaerella eumusae* responsive unique genes/transcripts from a resistant banana cultivar. *Acta Hort* 1114:111–118.

Uma S, Siva SA, Saraswathi MS, Durai P, Sharma TVRS et al. (2005) Studies on the origin and diversification of Indian wild banana (*Musa balbisiana*) using arbitrarily amplified DNA markers. *J Hortic Sci Biotechnol* 80:575–580.

Vakili NG (1965) Fusarium wilt resistance in seedlings and mature plants of *Musa* species. *Phytopathology* 55:135–140.

Vakili NG (1968) Responses of *Musa acuminata* species and edible cultivars to infection by *Mycosphaerella musicola*. *Trop Agric (Trinidad)* 45:13–22.

Van Den Berg N, Berger DK, Hein I, Birch PR, Wingfield MJ, Viljoen A (2007) Tolerance in banana to Fusarium wilt is associated with early upregulation of cell wall-strengthening genes in the roots. *Mol Plant Pathol* 8(3):333–341. https://doi.org/10.1111/j.1364-3703.2007.00389.x.

Van den Berg N, Crampton BG, Hein I, Birch PR, Berger DK (2004) High-throughput screening of suppression subtractive hybridization cDNA libraries using DNA microarray analysis. *Biotechniques* 37(5):818–824. https://doi.org/10.2144/04375RR02.

Vanhove AC, Vermaelen W, Panis B, Swennen R, Carpentier SC (2012) Screening the banana biodiversity for drought tolerance: can an in vitro growth model and proteomics be used as a tool to discover tolerant varieties and understand homeostasis. *Front Plant Sci* 3:176. https://doi.org/10.3389/fpls.2012.00176.

Vilarinhos A, Carreel F, Rodier M, Hippolyte I, Benabdelmouna A et al. (2006) Characterization of translocations in banana by FISH of BAC clones anchored to a genetic map. In: *Plant and Animal Genome XIV Conf*, San Diego, CA, Abst W4, p. 8.

Vilarinhos AD, Piffanelli P, Lagoda PJ, Thibivilliers S, Sabau X, Carreel F, D'Hont A (2003) Construction and characterization of a bacterial artificial chromosome library of banana (*Musa acuminata* Colla). *Theor Appl Genet* 106:1102–1106.

Wang M, Allefs S, Berg RGVD, Vleeshouwers VGAA, Vossen EAGVD, Vosman B (2008) Allele mining in *Solanum*: conserved homologues of Rpi-blb1 are identified in *Solanum stoloniferum*. *Theor Appl Genet* 116:933–943.

Wang XL, Chiang TY, Roux N, Hao G, Ge XJ (2007) Genetic diversity of wild banana (*Musa balbisiana* Colla) in China as revealed by AFLP markers. *Genet Resour Crop E* 54:1125–1132.

Wang Y, Xia Q, Wang G, Zhang H, Lu X, Sun J, Zhang X (2017) Differential gene expression in banana roots in response to Fusarium wilt. *Can J Plant Pathol* 39(2):163–175. https://doi.org/10.1080/07060661.2017.1342693.

Wang Z, Zhang J, Jia C, Liu J, Li Y et al. (2012) De novo characterization of the banana root transcriptome and analysis of gene expression under *Fusarium oxysporum* f. sp. *cubense* tropical race 4 infection. *BMC Genomics* 13:650. https://doi.org/10.1186/1471-2164-13-650.

Wilkins M, Pasquali C, Appel RD, Ou K, Golaz O et al. (1996) From proteins to proteomes: large scale protein identification by two-dimensional electrophoresis and arnino acid analysis. *Nat Biotechnol* 14:61–65. https://doi.org/10.1038/nbt0196-61.

Wingender E (2008) The TRANSFAC project as an example of framework technology that supports the analysis of genomic regulation. *Brief Bioinformat* 9(4):326–332. https://doi.org/10.1093/bib/bbn016.

Wu W, Yang YL, He WM, Rouard M, Li W-M et al. (2016) Whole genome sequencing of a banana wild relative *Musa itinerans* provides insights into lineage-specific diversification of the Musa genus. *Sci Rep* 6:31586. https://doi.org/10.1038/srep31586.

Xiao H, Jedrychowski MP, Schweppe DK, Huttlin EL, Yu Q et al. (2020) A quantitative tissue-specific landscape of protein redox regulation during aging. *Cell* 180(5):968–983.e24. https://doi.org/10.1016/j.cell.2020.02.012.

Xu BY, Su W, Liu JH, Wang JB, Jin ZQ (2007) Differentially expressed cDNAs at the early stage of banana ripening identified by suppression subtractive hybridization and cDNA microarray. *Planta* 226:529–539.

Xu M, Zeng C-B, He R, Yan Z, Qi Z et al. (2019) Transcriptome analysis of banana (*Musa acuminat*a L.) in response to low-potassium stress. *Agronomy* 9(4):169.

Yu J, Hu S, Wang J, Wong GK, Li S et al. (2002) A draft sequence of the rice genome (*Oryza sativa* L. ssp. *indica*). *Science (New York, NY)* 296(5565):79–92. https://doi.org/10.1126/science.1068037.

Zhang L, Liu L, Li S, Cenci A, Rouard M et al. (2020) Proteomic analysis of banana vascular sap provides insight into resistance mechanisms to *Fusarium oxysporum* F. sp. *cubense* Tropical Race 4. 2020:1–18. https://doi.org/10.21203/rs.3.rs-307891/v1.

Zhang L, Yuan L, Staehelin C, Li Y, Ruan J et al. (2019) The lysin motif-containing receptor-like kinase 1 protein of banana is required for perception of pathogenic and symbiotic signals. *New Phytol* 223:1530–1546. https://doi.org/10.1111/nph.15888.

Zhao Y, Mette MF, Gowda M, Longin CFH, Reif JC (2014) Bridging the gap between marker assisted and genomic selection of heading time and plant height in hybrid wheat. *Heredity* 112:638–645. https://doi.org/10.1038/hdy.2014.1.

Zhou C, Zhang H, Li Y, Li F, Chen J, Yuan D, Hong K (2021) Transcriptome analysis reveals the procyanidin treatment-responsive genes involved in regulating procyanidin accumulation during banana ripening and senescence. *J Amer Soc Hortic Sci* 146(2):109–117.

Zhu J, Zhang, MQ (1999) SCPD: a promoter database of the yeast *Saccharomyces cerevisiae*. *Bioinformatics* 15:607–611.

2 Practices of Allele Mining in Apple

Shigeki Moriya

2.1 INTRODUCTION

Apples have been a part of human life since the prehistoric age, and the industry is continuously expanding today. The global production of apples was 86.4 M metric tons in 2020 and remain the second most cultivated fruit tree crop after banana (Food and Agriculture Organization Corporate Statistical Database 2022). China is the largest producer among the major temperate-zone producing nations, followed by the United States of America (USA), Turkey, Poland, India, Italy, Iran, and Russia. Notably, Asia dominated production with a share of 64.4%, followed by Europe (20.3%) and the Americas (10.6%). Over the past 30 years, the quantity of apples produced has doubled, from 41.0 to 86.4 M metric tons. This increase is majorly attributed to China (mainland), where 10 times increase in production occurred from 4.3 to 40.0 M metric tons. Since 'Fuji' accounts for 65% of apple output in China (Matsumoto et al. 2018), it may be the most popular cultivar globally.

From the ancient to the modern age, humans have unintentionally or, to some extent intentionally, improved the attributes of fruit crops including cultivated apples (*Malus × domestica* Borkh.) by selecting trees that could fulfill their demands and then planting the seeds obtained from their fruits. They may have exchanged their "excellent" budwoods and fruits between communities, subsequently grafted them, or planted the seeds resulting in inducing new landraces. This marks the beginning of domestication and the initiation of breeding since these behaviors made apples unintentionally introduce genes and/or alleles beneficial to humans from their wild apple relatives and/or already existing landraces. Grafting has a long history dating back more than 2000 years (Webster and Wertheim 2003), implying that landraces could be maintained by the clonal propagation for a long time. This trend might have occurred in each region, which is mainly categorized according to geographical and climatic conditions, as a potential contribution of differential adaptation to a distinct environmental condition based on genetic diversity and structure analysis using the European apple genetic resources (Urrestarazu et al. 2016). Interestingly, the divergence of gene structure between dessert and cider apples has been proposed (Leforestier et al. 2015), which supports the assumption that selecting apples according to their usage has been conducted. However, these behaviors are based on an empirical view rather than genetic knowledge such as what we have today. Furthermore, several cultivars have been recorded so far due to these "unintentional" breeding processes. Some cultivars, which date back to the 16th century, are still alive and well conserved in the genebank repository, such as 'Alexander' and 'Gravenstein' (Morgan and Richards 1993), due to the long-lived perennial nature of apple and conservation efforts (Volk and Bramel 2021).

Modern breeding approaches that target improving specific traits of interest have emerged because of advancements in genetics, such as Mendel's law of heredity, resulting in a recognized phenomenon where a particular feature can be transferred to descendants. A good example is the controlled breeding, which Thomas Andrew Knight pioneered in the late 1700s (Morgan and Richards 1993). The controlled crossing approach has been the driving force of breeding toward clear breeding objectives and has been highly successful to date. Nowadays, most novel cultivars are developed adopting this technique.

 DOI: 10.1201/9781003386490-2

Furthermore, molecular biology techniques have been established and implemented in plant breeding fields for the past three decades. This advancement has provided an opportunity in breeding that targets a specific gene or alleles. A *Malus* gene list published by Alston et al. (2000) has been a compass for apple breeders and researchers seeking to improve horticultural traits and understand their underlying mechanisms.

In the past decades, advances in molecular genetics have rendered molecular-assisted breeding pivotal. Breeders can now validate their materials by employing DNA marker(s), which can either detect causative mutation of the gene of interest or polymorphisms that are in linkage disequilibrium (LD) or at least linked to the target gene/allele. This development suggests that breeders can design the optimum genome to achieve their goal when they understand the effect of the target allele in their breeding pool. The number of traits for which molecular tools are available has been increasing yearly since the first attempt at molecular-assisted breeding in the early 1990s. However, due to apple's lengthy juvenility compared to those of annual crops, breeding new apple cultivars involves an extraordinarily long-term effort, usually more than 20 years. Therefore, although quite a few examples of producing cultivars employing molecular-assisted breeding exist, no doubt that cultivar development using this technique has already started and will obtain significant success in apple breeding in the next decades.

Researchers have been investigating genes that affect the phenotype and evaluating the effect of alleles of interest in order to apply acquired knowledge into breeding through molecular tools, including DNA markers. Consequently, forward genetics techniques have been applied to apples. Until the implement of practical molecular tools in apple breeding, the following steps, depending on the traits to be analyzed, could be considered: (1) discovery and studying of the phenotypic variation, (2) validation and evaluation of the variation's transmissibility, frequently also needed to clarify its inheritance pattern, (3) genetic and/or physical mappings of the phenotypic variation's causal locus, (4) positional cloning of the causal gene/mutation, (5) complementary and/or silencing experiments using model plants/apple, (6) assessment of allele distribution and frequency in the breeding material, (7) determining the appropriate type of markers to be applied in breeding, and (8) preparing seedling progeny when necessary and selecting parent/seedlings using those markers. Although steps 4–6 are optional, other steps are prerequisites. Steps 1 and 2 have often been performed in previous studies since they do not require a molecular biology approach. The first instances of this attempt are two jointly apple breeding programs conducted in collaboration with three universities in the USA working for the resistance to apple scab: the PRI apple breeding program initiated in 1945 (Janick 2006) and for the resistance to woolly apple aphid at the John Innes and East Malling research station in the UK (Webster and Tobutt 2001; Webster and Wertheim, 2003). Since steps 3–7 require molecular biology methodologies, they had to wait for their evolution. However, after its implement, application and advancement have accelerated until now. Developing pseudo-test cross strategy for constructing genetic map in outcrossed species (Grattapaglia and Sederoff 1994) introduced mapping study (step 3) due to its compatibility with apple, after which bi-parental mapping became the basis of genetic mapping study in apple. Subsequently, the release of the first version of the apple genome (Velasco et al. 2010) and sequencing cost reduction by next-generation sequencing (NGS) technology have significantly boosted the recent explosive advances in these procedures. Currently, in addition to the bi-parental approach, pedigree-based analysis (PBA) and genome-wide association study (GWAS) have been developed, which are the practical options thanks to genome-wide single-nucleotide polymorphism (SNP) genotyping arrays (Chagné et al. 2012a; Bianco et al. 2014, 2016) and genotyping-by-sequencing (GBS) technologies. Conversely, positional cloning (step 4) remains challenging to perform due to the difficulties in handling numerous plant materials for fine mapping and developing bacterial artificial chromosome (BAC) libraries. A recently emerged option is the candidate gene approach, which relies on the apple genome and gene annotation information. Therefore, by using public genomic information, identifying candidate genes from candidate gene-located regions based on their annotation is a powerful option. Additionally, resequencing material(s) that contains allele(s) of interest and subsequent assembly to the reference genome will enable researchers to identify sequences of the region of interest from the target material. Silencing, or complementary test, or both (step 5) is optional and the most difficult to perform among the abovementioned steps. First, the genetic transformation of apple via *Agrobacterium* remains challenging due to the varying transformation success rate. Second, employing model plants that are easier to obtain genetically modified plants than apples may not always possess corresponding organs to investigate. For example, the surrogate of a short internode length of shoot conferred by *MdDOX-Co*, the candidate gene of columnar tree shape was the floral internode length of transformed *Arabidopsis thaliana* in the overexpression study (Wolters et al. 2013). Assessing allele distribution and frequency (step 6) is crucial, particularly when an allele of interest has been widely distributed in common cultivars, such as the *MdMYB1-1* allele that controls fruit skin color. Conversely, the importance of this step was reduced when an allele of interest was found in a wild relative that had never been used in previous breeding programs, such as *Fb_E*, a quantitative trait locus (QTL) for fire blight resistance from the ornamental cultivar 'Evereste,' suggesting that the allele is probably unique to the breeding program. Regarding steps 7 and 8, there are two approaches, which include parent and seedling selections. For the marker-assisted selection (MAS), sequence characterized amplified region (SCAR), and simple sequence repeat (SSR) markers are the two major options. However, SNP markers and related tools are getting general due to their availability on high-throughput marker detection platforms. The details on how to develop a diagnostic DNA test are described by Vanderzande et al. (2018).

Although the abovementioned research is pivotal in performing allele mining and designing the genome in apple breeding, research techniques performed toward mining depend on the studied trait. This chapter provides the efforts and advances of allele mining histories of agriculturally and/or horticulturally important traits, how they were dissected, and pursued to allele(s) that are applicable in breeding with genome designing.

2.2 RESISTANCE TO DISEASES AND PESTS

Apples are infected with at least 70 diseases and 100 pests (Grove et al. 2003; Beers et al. 2003). Therefore, fungicides and pesticides are sprayed during the growing season, especially between spring and early summer, to control these diseases and pests to an economically acceptable level. However, such chemical inputs in agriculture should be reduced to mitigate their environmental impact. One solution to address this issue involves using disease and pest-resistant cultivars. Since these resistances are usually found in wild species, they should be introgressed through multiple backcrossing to the cultivated species, which takes a long period in the conventional breeding of fruit species. Recent reviews have discussed the subject of breeding disease-resistant cultivars with high-quality fruits (Luo et al. 2020a; Khan and Korban 2022).

Resistance genes are typically involved in resistance, whose effect can be either qualitative or quantitative. These genes, which demonstrate significant effects, are generally dominant and can be mapped as a single gene and treated as monogenic resistance, enabling the potential of positional cloning of the causal gene. Conversely, although QTL mapping can be used to track those with minor effects, precise location delimitation may be challenging.

Since pathogens, such as fungi, bacteria, and insects, can evolve more rapidly than perennial crops including apples, single cultivars (genotypes) have been planted over decades. Therefore, to achieve resistance not easily overcome by pathogens (i.e., durable resistance), characterizing these resistance genes and introducing multiple resistance genes in one genotype (resistance gene pyramiding) is necessary. Specifically, the resistance gene effectiveness (major or minor) and spectrum (broad or narrow) are two major aspects of interest (Bus et al. 2011). Generally, the following ideas are well recognized: "vertical resistance," which indicates a drastic response to specific strains, such as a hypersensitive reaction susceptible to being overcome, and "horizontal resistance," which shows partial resistance to several strains not vulnerable to infestation (Vanderplank 1968). Therefore, to understand the gene-for-gene interaction between the resistance gene and pathogen, several pathogens with varying virulence (races if the pathogen is well characterized) should be used for evaluating effect and spectrum of the resistant gene of interest. This step enables us to determine the type and number of genes to be pyramided to achieve durable resistance. Furthermore, molecular markers are robust tools for resistance gene pyramiding because they can sense any allele of interest even when their phenotypes (the resistance reaction) are masked by the effect of other resistance genes. This section concentrates on resistance genes and quantitative loci mining for major diseases and some minor ones which are well-studied.

2.2.1 Apple Scab

Apple scab, which is one of the most prominent fungal diseases in the region where apples are produced on a global scale, is caused by *Venturia inaequalis* (Cooke) G. Winter. Lesions are observed on leaves, petiole, and fruit, among others (Biggs and Stensvand 2014). Young lesions on leaves are velvety brown to olive green and feathery, in distinct margins, and later the margins become distinct, but they may be obscured if several lesions coalesce. When there are numerous lesion, eventually, leaves become curled, swarmed, and distorted and can drop early, resulting in direct loss of marketable fruits and causing problems in reproduction (Biggs and Stensvand 2014).

Breeding activity toward scab resistance in apples was initiated in the USA as a collaborative program between Purdue University and the University of Illinois in 1945 (Janick 2006). Subsequently, Rutgers University joined, currently known as the PRI breeding program, which has already retired. However, this group had been at the forefront of global scab-resistance breeding projects for a long time. So far, twenty scab-resistance genes and several quantitative resistance loci (QRLs) have been discovered (Khajuria et al. 2018).

2.2.1.1 Resistance Genes

The systematic nomenclature system of apple scab-resistance genes has been developed based on the interaction between apples and *V. inaequalis*, providing the gene name as prefix *Rvi* (resistance to *Venturia inaequalis*) suffixed with the number of specific interactions (Bus et al. 2011). A large-scale survey that monitors the breakdown of each resistance gene has been conducted with the collaboration of >24 partners from >14 countries for more than 10 years, which have identified the five most promising resistance genes to develop cultivars with durable resistance to date, including *Rvi5* (previously called *Vm*), *Rvi11* (*Vbj*), *Rvi12* (*Vb*), *Rvi14* (*Vdr1*), and *Rvi15* (*Vr2*) that are either not or rarely overcome (Patocchi et al. 2020). These five genes in addition to *Rvi6* (*Vf*) are primarily described.

2.2.1.1.1 Rvi6

Rvi6 is currently the most researched and frequently used gene in breeding. This gene is the first to be successfully applied in positional cloning in apple. This gene is derived from a crabapple accession, *Malus floribunda* 821, and causes chlorosis in most cases, while variations are detected based on likely genetic background and environmental conditions (Chevalier et al. 1991). Several cultivars carrying *Rvi6* have been released (Crosby et al. 1992), whereas the breakdown of this gene already reported (Parisi et al. 1993). Moreover, sporulation was observed on the *Rvi6* cultivar 'Priscilla' in multiple sites in large-scale monitoring (Patocchi et al. 2020). Nevertheless, cloning and the complementary test are complete, and its application in genetic engineering has been conducted. Interestingly, a recent study showed that *V. inaequalis* isolates, which are *Rvi6*-compatible, are genetically distinct from the *Rvi6*-incompatible isolates and have been reproductively confined for several thousand years (Lemaire et al. 2016).

According to history, the epidemic of apple scab in 1943 motivated the research on *Rvi6* resistance. Hough (1944) observed a clear segregation of immune and susceptible plants in an F2 population from crossing *M. floribunda* 821 and 'Rome Beauty.' Later, in 1970, the first scab-resistant apple cultivar 'Prima' was developed from the PRI breeding program through successive back-crossing with cultivated apples (Janick 2006). It is four generations apart from the *Rvi6* source *M. floribunda* 821. The first marker associated with *Rvi6* is the isozyme marker *Pgm-1*, which is

8 cM apart from the gene (Manganaris et al. 1994). Subsequently, random amplified polymorphic DNA (RAPD) markers created using bulked segregant analysis (BSA) will shortly be published (Durham and Korban 1994; Koller et al. 1994). The genetic location of *Rvi6* has been verified in the middle of the linkage group (LG) 1 in the first apple consensus map (Maliepaard et al. 1998). Additionally, additional studies focused on fine mapping and positional cloning of *Rvi6* have been comprehensively reviewed in Gardiner et al. (2007) and Han and Korban (2010). Briefly, fine mapping and BAC clone sequencing was conducted by two independent research groups (Swiss+Italian researchers and the University of Illinois) and led to the discovery of a cluster of genes encoding leucine-rich repeat (LRR) receptor-like family protein (Vinatzer et al. 2001; Xu and Korban 2002). The transformation of *HcrVf2*, which is a gene from this cluster, into susceptible 'Gala' resulted in a resistance reaction, confirming that *HcrVf2* is *Rvi6* (Belfanti et al. 2004). Moreover, *Rvi6* has been employed in genetic modification studies for additional characterization (Malnoy et al. 2008; Szankowski et al. 2009) and cisgenesis applying different cultivars (Vanblaere et al. 2011; Würdig et al. 2015). Recently, it was confirmed that *Rvi6* transformed pear cultivar 'Conference' induced resistance to *Venturia pyrina*, the causal agent of pear scab (Perchepied et al. 2021).

DNA markers including the SCAR markers (ACS-3, ACS-7, and ACS-9) created through positional cloning (Xu et al. 2001) and an SSR (CH-Vf1), which originated from the sequence of the BAC clone (Vinatzer et al. 2004) can be applied in breeding. Later, Jänsch et al. (2015) recognized SNPs from amplicons of the markers associated with *Rvi6*; therefore, promoting the development of high-throughput detection systems, including KASP™ (Kompetitive allele-specific PCR; Baumgartner et al. 2016) and International RosBREED SNP consortium OpenArray v1.0 (IRSCOA v1.0) platform (Chagné et al. 2019).

2.2.1.1.2 Rvi5

Rvi5 originates from *Malus micromalus* 245-38 and *Malus atrosanguinea* 804 (Dayton and Williams 1968, 1970). This gene conditions a hypersensitive reaction. Some cultivars, including 'Murray,' 'Rouville,' and 'Orlovim,' have been released to date (Crosby et al. 1992). Although several *V. inaequalis* race five isolates that are compatible with the *Rvi5* carrier have been identified (Bus et al. 2011), the large-scale international scab monitoring has reported that this gene is rarely overcome so far (Patocchi et al. 2020). Therefore, *Rvi5* is expected to be highly valuable for use in breeding.

The first molecular marker created for *Rvi5* was from BSA using the F1 population of 'Empire'×NY74828-12 (Cheng et al. 1998). The RAPD marker OPB12 was then converted into the $OPB12_{687}$ SCAR. This marker is linked to *Rvi5* in ~8 cM. *Rvi5* has been assigned to the bottom of LG 17, resulting from the genome scanning approach (GSA), using an F1 population derived from 'Murray' (Patocchi et al. 2005). After the 'Golden Delicious' genome assembly (Velasco et al. 2010), the candidate region of *Rvi5* was delimited to 228 kb in the 'Golden Delicious' genome assembly using markers obtained from its sequence (Cova et al. 2015b). Notably, this region consists of genes that code one toll interleukin1 receptor-nucleotide binding site-LRR (TIR-NBS-LRR) class and three LRR-like plant resistance proteins. Additionally, SSR and sequence-tagged site (STS) markers, which were developed by Cova et al. (2015b), can be used in breeding. Therefore, additional study to develop a BAC library from a material carrying *Rvi5* is warranted since the above-listed candidate genes are examined from the susceptible cultivar 'Golden Delicious.' Moreover, the physical location of *Rvi5* is close to the self-incompatibility locus *S*. The S_{41} haplotype was observed to be coupled with *Rvi5* in 'Murray' (De Franceschi et al. 2016). They proposed an effective cross combination design that can result in almost all descendants carrying resistance genes.

2.2.1.1.3 Rvi11

The *Rvi11* is derived from *Malus baccata* Jackii (Dayton and Williams 1968), and it conditions stellate necrosis and chlorosis. Although limited sporulation with resistance reaction has been observed, no race overcomes this gene (Bus et al. 2011), and resistance breakdown has not been observed (Patocchi et al. 2020). Patocchi et al. (2020) also mentioned the likelihood of underestimating the resistance of *M. baccata* Jackii because it carries more than one scab-resistant gene. Additionally, introducing this gene into a cultivated apple would take a long time since the resistance source of *Rvi11* is a wild accession.

Gygax et al. (2004) conducted a mapping study of *Rvi11* using an F1 population generated from the cross A722–7×'Golden Delicious.' They developed three SCAR markers (K08-SCAR, T06-SCAR, Z13-SCAR) by BSA and confirmed their linkage with SSR markers on LG 2 (Liebhard et al. 2002) through a comparative mapping approach. Ultimately, the *Rvi11* was mapped 0.6 cM away from the CH05e03 SSR marker. Gygax et al. (2004) identified plants with genotype-phenotype-incongruence in their material. Additionally, Jänsch et al. (2015) have revisited this mapping population and developed SSR and SNP markers that are more closely linked to *Rvi11.* Of these markers, SNP markers, FBsn*Rvi11*-1 and FBsn*Rvi11*-2, will be an excellent starting point for positional cloning of *Rvi11.* Moreover, these markers can be a source of high-throughput detection system development.

2.2.1.1.4 *Rvi12*

According to Dayton and Williams (1968), Hansen's baccata #2 is a source of *Rvi12*, which promotes a chlorosis reaction with slight sporulation. Since this source is also a wild relative bearing small fruits with poor fruit quality, its improvement will take a long time by conventional breeding. Notably, no race has overcome this gene (Bus et al. 2011), and the breakdown has rarely been observed in field monitoring (Patocchi et al. 2020). However, the resistance may also be overestimated (Patocchi et al. 2020).

The first mapping study by BSA identified an RAPD marker linked to *Rvi12*, which was assigned to LG 1 (Hemmat et al. 2003); however, later studies did not align with this result. Erdin et al. (2006) mapped *Rvi12* on LG 12 through GSA using an F1 population from the cross between 'Golden Delicious' and Hansen's baccata #2. Padmarasu et al. (2014) performed fine mapping using 1,285 F1 individuals from the cross of 'Gala'×Hansen's baccata #2. They developed markers using 'Golden Delicious' genome assembly and ultimately recognized a candidate region spanning 958 kb between SNP_23.523 and SNP_24.482. These SNPs can be applied in breeding since the *Rvi12* coupled alleles from Hansen's baccata #2 are unique among breeding materials. Moreover, the candidate region was defined further, and a BAC clone of Hansen's baccata #2 covering the entire candidate region was discovered (Padmarasu et al. 2018). Finally, a 62 kb region was identified as the candidate region of *Rvi12* using PacBio RSII sequencing technology. The gene prediction and domain search identified a potential candidate gene named *Rvi12_Cd5* that encodes LRR receptor-like serine/threonine-protein kinase.

2.2.1.1.5 *Rvi14*

Rvi14 originated from a German heirloom cultivar, 'Dülmener Rosenapfel' (Laurens et al. 2004), which conditions chlorosis with slight sporulation. A few breakdown events have been observed during the international monitoring (Patocchi et al. 2020). Since 'Dülmener Rosenapfel' can carry more than one resistance gene (Soufflet-Freslon et al. 2008), the monitoring result may overestimate its breeding value (Patocchi et al. 2020). Nevertheless, *Rvi14* has been found in cultivated apples, which signifies that fruit quality improvement is easier than using resistance genes from wild relatives with the conventional breeding approach.

Rvi14 has been mapped on the top of LG 6 using an F1 population generated from the cross of 'Gala'×'Dülmener Rosenapfel' (Soufflet-Freslon et al. 2008). An SSR marker HB09-SSR was linked to *Rvi14* in 4 cM. Therefore, a detailed characterization of *Rvi14* is necessary since no further study has not been conducted to date.

2.2.1.1.6 *Rvi15*

Rvi15 was derived from a crabapple accession GMAL2473, which was thought to be a clone of a Russian seedling (Patocchi et al. 2004). This gene conditions slow pinpoint pit for approximately 6 days after inoculation, while the *Rvi5* exhibits it approximately 3 days after inoculation (Patocchi et al. 2004; Galli et al. 2010a). However, the breakdown of *Rvi15* has not been observed using large-scale field monitoring (Patocchi et al. 2020).

The BSA was applied to conduct a mapping study of *Rvi15* with the amplified fragment length polymorphism (AFLP) technique (Patocchi et al. 2004). *Rvi15* was then assigned to LG 2 using the F1 population made from the cross between 'Idared' and GMAL2473. Two AFLP markers (EA35MA41, EA37MA39) and the CH02c02a SSR marker (Liebhard et al. 2002) cosegregated with *Rvi15*. Galli et al. (2010b) conducted the fine mapping of *Rvi15* using 989 individuals produced from the cross between 'Golden Delicious' and GMAL2473. They employed markers established from resistance gene analogous sequences (Baldi et al. 2004) and a BAC library of 'Florina' that was created for *Rvi6* cloning (Vinatzer et al. 1998), resulting in the delimitation of a genetic distance of 0.5 cM for the *Rvi15* candidate region. Subsequently, Galli et al. (2010c) developed a BAC library from GMAL2473 and performed chromosome walking. They identified and sequenced a single clone that encompassed the entire candidate region. Gene prediction identified three TIR-NBS-LRR class resistance genes named *Vr2-A*, *Vr2-B*, and *Vr2-C*, respectively. Furthermore, Schouten et al. (2014) introduced each of these genes, including native promoters, introns, and terminators, to the susceptible 'Gala.' Notably, a scab inoculation test of these transformed lines induced resistance when *Vr2-C* was introduced. Contrarily, the other two did not display resistance reactions. Therefore, these results show that *Vr2-C* is *Rvi15* and is the second example of resistance gene cloning for apple scab.

The markers developed during positional cloning can be an excellent source of markers used in breeding. Jänsch et al. (2015) found GMAL2473 unique SNPs from the amplicons of 9C10T7 and FBsnRvi15-1 markers, signifying the possibility of developing a high-throughput detection system for *Rvi15*.

2.2.1.1.7 Other Resistance Genes

Some apple scab-resistance genes have been targeted for positional cloning in addition to those described above. *Rvi1* (*Vg*) is a narrow spectrum gene from 'Golden Delicious' (Benaouf and Parisi 2000) that display incompatibility with the *V. inaequalis* race 7. It has been fine mapped at the bottom of LG 12, with six predicted genes that encode TIR-NBS-LRR class of plant resistance protein identified as candidate genes (Cova et al. 2015a). *Rvi18* (*V25*), which shows a broad-spectrum effect, was recently identified from the breeding accession 1980-015-025 in LG 11 (Soriano et al. 2014). Additionally, it has been fine mapped in a region that spans 34 kb in the susceptible 'Golden Delicious' genome, not the *Rvi18* carrier. From the region, an L-type lectin-domain containing receptor kinase was identified as a candidate gene. Until recently, the number of scab-resistance genes has increased as research develops. The current research status of other resistant genes has been described in a review (Khajuria et al. 2018).

2.2.1.2 QRLs

Resistance genes generally show high levels of resistance, but its strong selection pressure to pathogen can lead to the emergence of new races, resulting in the breakdown of resistance genes. The adoption of QRLs that confer a partial resistance to plants (Bus et al. 2011) is essential to achieve a durable resistance because the utilization of QRLs that have a broad-spectrum and complemental mode of action contributes to achieving durability (Pilet-Nayal et al. 2017). Additionally, combining these QRLs with resistance genes that have major effects is a valuable approach to enhance the effectiveness of major genes (Pilet-Nayal et al. 2017).

To date, QRLs for apple scab have been identified on 11 of the 17 LGs by bi-parental QTL analysis (Khajuria et al. 2018). The spectrum, which can be referred to as race specificity, is a major interest of these QRLs. Thus, research has often been conducted using isolates of *V. inaequalis*, whose pathogenicity had been previously characterized (Durel et al. 2003; Calenge et al. 2004). So far, the QRLs identified on LGs 11 and 17 have been generally detected in same position and have been regarded as broad-spectrum QRLs as well as being expressed in various genetic backgrounds and environmental conditions (Table 2.1). These QRLs showed synergistic effects when combined into one genotype (Durel et al. 2003; Liebhard et al. 2003c; Soufflet-Freslon et al. 2008; Bénéjam et al. 2021). QRLs from

TABLE 2.1
Partial Scab-Resistance QRLs Detected on LGs 11 and 17

QRL	Population	Inoculum	Race(s)	Source(s)	Contribution%	References
LG 11	'Prima' × 'Fiesta'	302	6	Prima, Fiesta	18.7–22.6	Durel et al. (2003)
		EU-D42	6	Prima, Fiesta	16.5–22.8	
	'Fiesta' × 'Discovery'	Not specified[a]	–	Fiesta, Discovery	11	Liebhard et al. (2003c)
	'Gala' × 'Dülmener Rosenapfel'	Mixture of isolates[b]	1 + 6 + 7	Gala	18.3–22.4	Soufflet-Freslon et al. (2008)
	TN10–8 × 'Fiesta'	EU-B04	1	Fiesta	18.1	Bénéjam et al. (2021)
		09BCZ014	–	Fiesta	7.9	
LG 17	'Prima' × 'Fiesta'	302	6	Fiesta	11.0–13.4	Durel et al. (2003)
		EU-D-42	6	Fiesta	10.3	
	'Fiesta' × 'Discovery'	Not specified[a]	–	Fiesta, Discovery	8–23	Liebhard et al. (2003c)
	'Discovery' × TN10–8	EU-B04	1	Discovery	6.3	Calenge et al. (2004)
		104	1	Discovery	19.4	
		302	6	Discovery	13.1	
		EU-D42	6	Discovery, TN-10–8	9.0	
		EU-NL24	–	Discovery, TN-10–8	10.0	
		MLI[a]	–	Discovery	4.7	
		Gd-CH	–	Discovery	19.0	
	'Gala' × 'Dülmener Rosenapfel'(DR)	Mixture of isolates[b]	1 + 6 + 7	Gala, DR	9.4–14.4 (Gala), 4.4–5.7 (DR)	Soufflet-Freslon et al. (2008)
	'Orin' × 'Akane'	Natural infection	–	Orin, Akane	5.2 (Orin), 20.0 (Akane)	Moriya et al. (2019b)
	TN10–8 × 'Fiesta'	EU-B04	1	Fiesta	29.0	Bénéjam et al. (2021)
		09BCZ014	–	Fiesta	10.7	

[a] *Venturia inaequalis* taken from scabbed leaves in the field.
[b] Mixture of EU-B04, EU-NL19, EU-D42, and EU-NL24.

'Fiesta' have been suggested to be involved in the initiation of basal resistance (Laloi et al. 2017). It should be noted that these QRLs may have originated from multiple sources that do not share a common founder cultivar, suggesting that multiple alleles from these loci can contribute to scab resistance. Additionally, the QRL on LG 1 precisely located on *Rvi6* position showed a significant effect in terms of scab resistance; nevertheless, the spectrum is narrower than QRLs on LGs 11 and 17 (Durel et al. 2003; Calenge et al. 2004; Bénéjam et al. 2021). Bénéjam et al. (2021) suggested that this QRL is an allele of *Rvi6*.

A definitive DNA test that can select or discriminate these QRLs has not been developed yet. However, as these QRLs were detected in cultivated apples, their utilization and pyramiding are likely to be easier than those of resistance genes, which are often found in wild relatives. The durability of these QRLs should be carefully monitored in long-term research due to the risk of erosion has been reported (Caffier et al. 2014).

2.2.2 Fire Blight

Fire blight is a destructive disease caused by the bacterium *Erwinia amylovora* (Burrill) Winslow et al. It was first observed in New York State, USA, in the late 1700s (Sundin 2014) and it has now spread to at least 46 countries, including Europe, the Middle East, and New Zealand (Sundin 2014). In countries where it has not been widespread (e.g., East Asia), fire blight is a quarantine disease. Fire blight symptoms can be observed on flowers, leaves, petioles, fruit, young shoots, scaffold limbs, and rootstocks (Sundin 2014). The symptoms begin with blossom blight, progress into shoot blight and finally into rootstock blight, which can kill the rootstock (Sundin 2014). Thus, breeding fire blight-resistant cultivars is crucial for the management of this disease.

The great majority of cultivated apples (*M.×domestica*) are highly susceptible to fire blight, although to different degrees (Aldwinckle et al. 1976; Schöneberg et al. 2017). Because the degree of resistance/susceptibility is evaluated quantitatively based on the length of lesions on young shoots, the bi-parental QTL analysis is generally conducted. In cultivated apple, a moderate QRL explaining 35%–40% of the phenotypic variance has been identified on LG 7 in 'Fiesta,' called *FB_F7* (Calenge et al. 2005; Khan et al. 2006). In a recent study, PBA using 314 individuals from 32 full-sib families identified reduced-susceptibility alleles that collectively explained ~28% of the phenotypic variance on LGs 6, 7, and 15 (Kostick et al. 2021). GWAS using historical data obtained from 566 apples has identified the association with blossom blight and shoot blight on all LGs (Thapa et al. 2021). Thus, in cultivate apples, no decisive resistance factor has been discovered, despite detection of multiple minor QRLs. On the other hand, highly resistant wild apple relatives have been observed (Gardner et al. 1980a). Efforts including the positional cloning approach for major effect QRLs and genetic engineering have been undertaken to elucidate genetic mechanisms of resistances observed in these plants. These major QRLs are described in the following sections.

2.2.2.1 *FB_MR5*

FB_MR5 which was originally identified in *Malus robusta* 5, is currently the most studied QRL against fire blight (Gardner et al. 1980a). It was suggested that a few dominant, additive genes are involved in the inheritance of fire blight resistance in *M. robusta* 5 (Gardiner et al. 1980b). The inoculation of *E. amylovola* to *M. robusta* 5 via hypodermic needle was shown to produce a retro tic reaction, which is a hypersensitive reaction. However, the resistance of *FB_MR5* has already been overcome by aggressive strains (Vogt et al. 2013a; Wöhner et al. 2014).

In a previous study, QRLs have been identified by bi-parental QTL analysis using an F1 population obtained from the cross 'Idared'×*M. robusta* 5 (Peil et al. 2007). As a result, a QRL was identified at the bottom of LG 3 and it explained 80% of the phenotypic variance. In a recent study, bi-parental QTL analysis demonstrated the effectiveness of *FB_MR5* against blossom blight (Peil et al. 2019). The position of *FB_MR5* was refined by Gardiner et al. (2012), although multiple genes near the region were suggested to be involved in resistance. Fahrentrapp et al. (2013) conducted a fine mapping study using 2,133 individuals from six crosses and BAC clone identification, which resulted in the positioning of *FB_MR5* between markers FEM47 and rp16k15, spanning a 0.24-cM interval. A BAC clone obtained from *M. robusta* 5 encompassed the entire candidate region and was therefore sequenced. *In silico* gene prediction identified a coiled-coil (CC)-NBS-LRR class resistance gene as a promising candidate. This gene was transformed into 'Gala' with the CaMV 35S promoter and the ocs terminator or native promoter and terminator construct, inducing resistance in the transformed plantlet; therefore, *FB_MR5* was cloned (Broggini et al. 2014).

The SSR markers including FEM47 developed by Fahrentrapp et al. (2013) can be utilized for breeding purposes. Additionally, the SNPs identified from the nucleotide sequences of markers including rp16k15 (Jänsch et al. 2015) have been employed in the OpenArray platform, a high-throughput SNP detection system (Chagné et al. 2019).

FB_MR5 has been utilized in rootstock breeding in Germany and the USA as well as in apple-scion breeding in New Zealand and Switzerland (Gardiner et al. 2012; Bassett et al. 2015; Kellerhals et al. 2017). In particular, the fast-track breeding approach, which promotes early flowering of seedling plants within three years without genetic modification, is currently practiced in Switzerland (Kellerhals et al. 2017). In terms of genetic engineering approaches, the cisgenic 'Gala Galaxy' transformed *FB_MR5* has been developed (Kost et al. 2015).

2.2.2.2 *FB_E*

The ornamental cultivar 'Evereste' represents a source of *FB_E*, a QRL with major effects against fire blight (Durel et al. 2009). Although the 'Evereste' breeding lineage suggests that the source of *FB_E* may be the pollen parent, which can be a wild relative, the potential pollen parent has not been identified yet (Durel et al. 2009).

A Bi-parental QTL analysis using 194 individuals obtained from the 'MM106' × 'Evereste' cross, resulted in the identification of a major effect QRL at the bottom of LG 12, accounting for 57% of the phenotypic variance (Durel et al. 2009). Then, fine mapping and map-based cloning were conducted using 2,703 individuals from the reciprocal crosses between 'MM106' and 'Evereste' (Parravicini et al. 2011). BAC clones harboring the temporary candidate region were subjected to marker development and they were then used to delimitate the candidate region. The final candidate region spanned 78 kb and eight predicted genes potentially involved in disease resistance, including seven genes coding a protein kinase (*MdE-EaK1–7*) and one gene coding a CC-NBS-LRR class disease resistance protein (*MdE-EaN*). A comparison of nucleotide sequences between resistant and susceptible allele-located chromosomes and an expression analysis concluded that *MdE-EaK7* and *MdE-EaN* are the best candidate genes underlying *FB_E*. Moreover, *M. floribunda* 821 also shows resistance to fire blight and the QRL responsible for it has been identified at the bottom of LG 12, close to *FB_E* (Durel et al. 2009).

DNA markers developed during fine mapping studies are suitable for breeding purposes (Parravicini et al. 2011). Jänsch et al. (2015) have identified SNPs uniquely associated with *FB_E* resistance from the sequences of FBsnFBE markers tightly linked to *FB_E* and those SNPs were implemented in the OpenArray platform (Chagné et al. 2019).

In contrast to *FB_MR5* which employs the fast-track breeding approach for introgression, *FB_E* has been employed the early flowering approach, which uses transgenic plants showing shortened juvenile period. The early flowering transgenic cultivar 'Pinova' (the T1190 line), which overexpresses *BpMADS4* from silver birch, has been crossed with 'Evereste' and the subsequent backcrossing resulted in the establishment of advanced selections of the fifth generation from 'Evereste' within 7 years (Le Roux et al. 2012; Schlathölter et al. 2018). In the half number of the fifth-generation plants, only 4% of the 'Evereste' genome was left (Schlathölter et al. 2018). Because the QRL against fire blight derived from 'Fiesta' (*FB_F7*) has been pyramided in a single plant in this project, if plants that does not carry the early flowering cassette (null segregants) are obtained, they will be a very promising parent for breeding (Schlathölter et al. 2018).

2.2.2.3 *FB_Mar12*

Recently, *FB_Mar12* has been identified as a major QRL against fire blight obtained from *Malus×arnoldiana* (accession number MAL0004) via bi-parental QTL analysis (Emeriewen et al. 2017). This gene is located at the bottom of LG 12 and it is valuable because it displays a significant resistance against *E. amylovora* strain Ea3049, which can overcome *FB_MR5* (Emeriewen et al. 2017); therefore, it is a good candidate for gene pyramiding. Fine mapping studies have been conducted using 892 individuals from three F1 populations and the *FB_Mar12* locus was identified in a 0.57-cM interval (Emeriewen et al. 2021a). Because *M.×arnoldiana* is a hybrid between *M. baccata* and *M. floribunda*, the draft genome of the former species (Chen et al. 2019) was used for subsequent gene prediction, which identified the serine/threonine kinase and LRR domains (Emeriewen et al. 2021a). To accurately analyze the candidate genes, gene prediction using the original *M.×arnoldiana* MAL0004 is necessary. Additionally, as the bottom of LG 12 is a hot spot of fire blight resistance genes (*FB_E*, *M. floribunda*, *FB_Mar12*), it is of interest to establish whether these genes are allelic.

2.2.2.4 *FB_Mfu10*

Gardner et al. (1980a) have identified a highly resistant accession in *Malus fusca*; however, no further research has been conducted until recently. Emeriewen et al. (2014) conducted a bi-parental QTL analysis using an F1 population obtained from the cross between *M. fusca* (accession number MAL0045) and 'Idared', which identified a major effect QRL (*FB_Mfu10*) on LG 10. *FB_Mfu10* accounted for 66% of the phenotypic variance against *E. amylovora* strain Ea222 (Emeriewen et al. 2014) and showed a significant effect on strains that can overcome *FB_MR5*; specifically, it accounted for 41% and 89% of the phenotypic variance against the Ea3049 and ZYRKD3-1 strains (Emeriewen et al. 2020).

The fine mapping of *FB_Mfu10* was conducted using 1,888 individuals derived from reciprocal crosses between *M. fusca* MAL0045 and 'Idared' (Emeriewen et al. 2018). DNA markers developed from sequences of the 'Golden Delicious' double haploid 13 (GDDH13) genome (Daccord et al. 2017), short read sequences of *M. fusca*, or identified BAC clones of *M. fusca*, delimited the candidate *FB_Mfu10* location in a 0.33-cM interval corresponding to 187 kb in the GDDH13 genome (Emeriewen et al. 2018). Gene prediction from a BAC clone spanning the *FB_Mfu10* region identified a candidate gene encoding the B-lectin and serine/threonine kinase domains. By comparing the sequences of resistant and susceptible alleles, an 8-bp insertion unique to the first exon of the susceptible allele was identified (Emeriewen et al. 2018). Recently, this gene was transformed into the 'Pinova' cultivar with the native promoter and terminator (Emeriewen et al. 2021b).

Because the QRLs for fire blight described above have been identified in wild apple strains, they are comparatively easier to utilize for rootstock breeding that does not aim to obtain an excellent fruit quality. For scion cultivar breeding, the application of conventional breeding approaches is apparently an obstacle because recurrent backcrossing is necessary to improve fruit quality and that process is time-consuming. Fast-track breeding and early flowering approaches have already been applied (Kellerhals et al. 2017) to overcome this obstacle, with promising results.

2.2.3 Powdery Mildew

Powdery mildew is caused by *Podosphaera leucotricha* (Ellis & Everh.) E. S. Salmon and is present in most areas where apples are grown (Hickey and Yoder 2014). The initial symptoms include whitish, feltlike patches of fungal mycelium and spores, which then cover the entire leaf (Hickey and Yoder 2014). Fruit infections may develop on severely diseased trees, resulting in small, russeted fruit (Hickey and Yoder 2014).

Because *P. leucotricha* is an obligate biotrophic fungus, using the controlled inoculum is always challenging for researchers who study about interactions between apples and *P. leucotricha*. Although race development has been suggested (Urbanietz and Dunemann 2005;

Bus et al. 2010), the evaluation of resistance level, indeed, owes to natural infection or spraying conidia taken from the infected leaves. Therefore, multiple-year assessments are required for accurate phenotyping. Although research has been conducted since the 1970s', only a few significant reports have been published after the release of the apple genome (Velasco et al. 2010); therefore, fine mapping and positional cloning studies on powdery mildew resistance have not been conducted so far.

No resistance gene with major effects has been identified in cultivated apples (*M.×domestica*). Minor effect QRLs, collectively explaining 29%–37% of the variation, have been detected by biparental QTL analysis using an F1 population obtained from the 'Discovery'×'TN10–8' cross (Calenge and Durel 2006). To date, six major effect genes have been identified in wild relatives and have been all analyzed as single dominant genes. Because these major genes originated from apple accessions bearing small- and poor-quality fruits, introducing them into the genetic background of cultivated apples via conventional cross breeding will take a long time. Therefore, the positional cloning of causal genes is desired to allow their use in genetic engineering approaches. These six major effect genes are described below.

2.2.3.1 *Pl-1* and *Pl-d* on LG 12

Pl-1 is derived from *M. robusta* 5 (Knight and Alston 1968). The first DNA marker for *Pl-1*, AT20-SCAR, was developed as a RAPD marker by BSA using the F1 population of 85/23-2 (*Pl-1* carrier: resistant)×81/19–35 (susceptible) and was converted into SCAR (Markussen et al. 1995). AT20-SCAR was located 4.5 cM away from *Pl-1*. Subsequently, Dunemann et al. (2007) assigned *Pl-1* at the bottom of LG 12, although fluctuations in the map position had been observed depending on the location (glasshouse or field) where the resistance was evaluated. AT20-SCAR has been used as an indicator in the resistance gene pyramiding project (Baumgartner et al. 2015).

Pl-d is derived from cultivar 'D12,' one of the 'D' series of seedlings (Visser and Verhaegh 1980). James et al. (2004) have mapped this gene at the bottom of LG 12 by BSA using the F1 population of 'Fiesta'×A871-14 ('Worcester Pearmain'×'D12'). *Pl-1* and *Pl-d* are located almost at the same position based on the comparison of flanking markers. It should be elucidated whether these genes are allelic.

2.2.3.2 *Pl-2* and *Pl-m* on LG 11

Pl-2 has been identified in *Malus zumi* (Knight and Alston 1968). Several SCAR markers and expressed-sequence tag-derived markers have been mapped near this gene, and have been located at the top of LG 11 using an F1 population generated from the cross of 'Royal Gala'×A689-24 (Gardiner et al. 2003, 2007). Subsequently, Fernández-Fernández et al. 2008) also mapped *Pl-2* at the top of LG 11 using the F1 population of 'Fiesta'×'Totem.' Jänsch et al. (2015) mentioned that *Pl-2* had already been cloned in a patent specification by Rikkerink et al. (2007). In breeding programs, CH04h02 (Liebhard et al. 2002) is used for the seeding selection (Baumgartner et al. 2015). In addition, Jänsch et al. (2015) identified many SNPs uniquely associated with *Pl-2* from the promoter sequence of this gene. Using this information, SNP-based high-throughput detection platforms have been developed (Baumgartner et al. 2016; Chagné et al. 2019).

Pl-m is derived from MIS developed by Dayton (1977). Markers used to map *Pl-2* have also been linked to *Pl-m*, which was thus located on LG 11 using the F1 population of 'Fuji'×MIS OP 93.051 G02–054 (Gardiner et al. 2003, 2007; Bus et al. 2010). As *Pl-2* and *Pl-m* share common flanking markers, their physical position is likely to be the same or very close. Future research will clarify whether their relationship is allelic.

2.2.3.3 *Pl-w* on LG 8 and *Pl-bj* on LG 10

Pl-w has been identified in the ornamental crabapple 'White Angel' (Gallot et al. 1985). Evans and James (2003) have mapped this gene on the middle of LG 8 by BSA and comparative mapping with SSR markers using an F1 population derived from the cross 'Fiesta'×E295-4

('Gloster 69' × 'White Angel'). Luo et al. (2019) have constructed a novel genetic map around *Pl-w* and identified CH01e12 (Liebhard et al. 2002) as the nearest marker. In addition, the study confirmed the involvement of the unmapped second locus that complements the expression of resistance of *Pl-w* (Batle and Alston 1996).

Pl-bj has been identified in *M. baccata* jackii (Dunemann and Schuster 2009), and through AFLP-based BSA and SSR screening using the F1 population of *M. baccata* jackii × 'Pinova,' its map position was determined on LG 10 (Dunemann and Schuster 2009). The PLBJ-SCAR marker showed perfect cosegregation with *Pl-bj* in the initial population.

2.2.4 Other Diseases and Pests

In addition to the three the most important diseases described above, many other minor or local diseases and pests affect apple production. Some of them are well characterized to date.

Three resistance genes have been reported for the rosy leaf-curling aphid *Dysaphis devecta* Wlk., namely *Sd-1*, *Sd-2*, and *Sd-3*. Of these, *Sd-1* and *Sd-2* have been studied using molecular techniques. *Sd-1* was identified on the cultivar 'Cox's Orange Pippin' and was mapped on LG 7 using an F1 population from the cross between 'Prima' and 'Fiesta' (Maliepaard et al. 1998). The effect of this gene was also confirmed by bi-parental QTL analysis (Stoeckli et al. 2008). Fine mapping studies and subsequent BAC contig construction has been carried out (Cevik and King 2002a, 2002b). However, the BAC library used in Cevik and King (2002b) was developed from 'Florina,' which does not carry *Sd-1*. Therefore, the BAC library of a cultivar carrying *Sd-1* should be developed. *Sd-2* was identified in the 'Northern Spy' cultivar and was mapped on LG 7 in the same position as *Sd-1*, suggesting that these two genes are closely linked or allelic (Cevik and King 2002a).

Dp-fl is known as a resistance gene against the rosy apple aphid *Dysaphis plantaginea* Passerini, and it was identified in 'Florina' (Pagliarani et al. 2016). This gene was mapped on LG 8 using the F1 population of 'Perico' × 'Florina' (Pagliarani et al. 2016). A candidate region spanning 95 kb was delimited via fine mapping using seven populations and chromosome walking using the BAC library of 'Florina' (Dall'Agata et al. 2018). It was suggested that four of the 11 genes predicted in the candidate region are involved in resistance mechanisms (Dall'Agata et al. 2018).

To date, four resistance genes against the wooly apple aphid (*Eriosoma lanigerum* Hausm.) have been identified, namely *Er-1* (derived from 'Northern Spy'), *Er-2* (*M. robusta* 5), *Er-3* ('Aotea 1'), and *Er-4* (Mildew Immune Selection OP 93.051 G02–54) (Bus et al. 2008, 2010). *Er-1* and *Er-3* were mapped almost at the same position on LG 8, therefore, they are likely to be closely linked genes or alleles (Bus et al. 2008). These genes are located on the upstream side from the CH01h01 marker and near *Pl-w*, while *Dp-fl* is located on the downstream side from the CH01h01. A fine mapping approach will clarify the relationship between them. *Er-4* was mapped on LG 7, 17.9 cM away from the CH-Sd marker that closely links to *Sd-1* (Bus et al. 2010). Although neither fine mapping had been conducted nor candidate gene had been identified, markers linked to *Er-2* and *Er-3* were applied to rootstock breeding program for pyramiding resistance genes (Basset et al. 2015).

Alternaria blotch is a leaf disease caused by the *Alternaria alternata* apple pathotype, which produces a host-specific toxin known as AM-toxin. The *Alt* gene confers susceptibility to the AM-toxin and consequently controls the susceptibility to Alternaria blotch; therefore, the susceptibility to the disease is dominantly inherited (Saito and Takeda 1984; Abe et al. 2010). *Alt* was mapped at the top of LG 11 using F1 populations generated from the reciprocal crosses between 'Starking Delicious' and 'Jonathan' (Moriya et al. 2019a). Fine mapping using 797 individuals from 15 crosses was conducted and gene prediction from a BAC clone encompassing the candidate region of *Alt* identified a CC-NBS-LRR class plant disease resistance gene as a promising candidate (Moriya et al. 2019a). A DNA marker applicable to breeding was developed from the insertion/deletion sequence identified from the promoter sequence of candidate gene.

The postharvest disease known as blue mold is caused by *Penicillium expansum* Link. A QTL (pM-*Pe3.1*) accounting for 27.5% of the contribution was identified in *Malus sieversii* (accession number PI613981) and was mapped on LG 3 (Norelli et al. 2017). A locus-specific DNA test to detect pM-*Pe3.1* has been established (Luo et al. 2020b). Because this QTL was identified in a wild progenitor, long-term effort has to be undertaken for introgression of this gene into cultivated apple via the conventional breeding approach. Therefore, rapid cycle breeding employing the early flowering transgenic line T1190 was attempted to shorten the juvenile period (Luo et al. 2020b).

Wood rot diseases are a serious problem for apple production because they severely damage trees or even kill them occasionally. The *Rnd1* QRL acting against European canker caused by *Neonectria ditissima* was identified in *M. robusta* 5 using the F1 population of 'M. 9'×*M. robusta* 5 (Bus et al. 2019). Bus et al. (2019) also suggested the existence of several undetected resistance genes in *M. robusta* 5. Previous studies have evaluated the resistance level of *Malus* species against Valsa canker caused by (*Valsa ceratosperma*) and have suggested possible breeding sources (Abe et al. 2007, 2011). Two QRLs on LGs 9 and 16 were identified from the *Malus asiatica* accession 'Zisai Pearl' (Tan et al. 2017). More research is necessary to elucidate genetic architectures of wood rot diseases.

2.3 TREE SHAPE

Apple tree shape, or branching habit, is a key determinant of planting density and training technique in commercial orchards. Therefore, studying tree shapes has been of interest since long ago. Tree shape and fruiting characteristics have been categorized into several types (Lespinasse and Delort 1986). Although tree shapes exhibiting clear transmissibility to offspring are rare, both polygenic and environmental effects should be involved in tree shape forming. Two rare visibly transmissive tree shape examples, such as columnar and weeping (pendulous) habits are provided in this section.

2.3.1 COLUMNAR HABIT

It could be argued that studies on columnar growth habits are unique in the history of allele mining in apples, implying that other alleles studied and implemented in breeding are variants that are already in existence in nature (in cultivated apples or wild relatives) and that the occurrence of columnar habit as a spontaneous mutant is relatively recent and has recorded its emergence.

Short internode length, lower number of lateral brunches, and increased spurs that display "column"-like tree shape characterize a columnar-type tree (Figure 2.1a). This shape could benefit intensive orchards requiring less training and pruning, facilitating less laborious and high-yield orchards. Furthermore, recent advances in agricultural robotics (Zhang et al. 2019b) would enable a columnar tree shape to be more valuable.

The spontaneous mutation of columnar habit was discovered in British Columbia, Canada, in the 1960s. An apple grower, Mr. Wijcik, noticed a sport showing stunted growth shoot on his 50-year-old 'McIntosh' tree in 1961 (Fisher 1995). He contacted Dr. Fisher in 1963, who eventually visited his orchard in 1965 and recognized a four-foot-long single shoot with a tight fruit cluster. This discovery instigated the studies on the 'McIntosh Wijcik' columnar apple and its descendants.

The first attempt was to clarify whether this distinct tree shape was transmissible. Since it is agronomically attractive, breeding programs have been initiated in several countries (Lapins 1976; Tobutt 1985; Ognjanov et al. 1999; Brown et al. 2004; Moriya et al. 2009; Savelyeva et al. 2018), and new cultivars with columnar shape have been released since the 1980s' (Tobutt 1988a, 1988b). The tree shape segregation ratio in progenies resulting from the crosses between standard×columnar, columnar×columnar, and standard×standard was labeled as the single dominant gene *Columnar* (*Co*) (Lapins 1969, 1976). These experiments also suggested modifier genes since the number of columnar trees is always less than that predicted.

Co is one of the target genes in a very early stage of apple genetic linkage map construction studies. Research using molecular biology techniques for columnar-type apples began in the 1990s. The first genetic map, including the *Co* locus, was constructed mainly by RAPD markers and contained 19 LGs, identifying LG 10, where *Co* is located (Conner et al. 1997). LG 10 was coincidentally the

FIGURE 2.1 The tree shape of (a) the columnar cultivar 'McIntosh Wijcik', (b) the weeping cultivar 'Red Jade'.

same number of LG subsequently used in the apple consensus map (Maliepaard et al. 1998). Hemmat et al. (1997) have identified a DNA sequence of a *Co*-linked RAPD marker named $OA11_{1000}$, indicating that this marker contained an SSR sequence comprising 17 repeats of "GA." They converted this SSR into an SSR marker called SSR_{co} located 6–15 cM away from *Co*. This marker was the second trait-linked SSR marker in the plant after HSP176L, a soybean mosaic virus resistance (*Rsv1*) linked marker in soybean (Yu et al. 1996). Unfortunately, SSR_{co} generated a null allele from the *Co*-coupling chromosome; therefore, Gianfranceschi et al. (1998) redesigned primer sequence from SSR_{co} and developed the COL marker, whose discriminability was improved. This marker exhibited polymorphism in other cultivars (Gianfranceschi et al. 1998) due to the transferability of the SSR marker, resulting in identifying LG of *Co* in the apple consensus map that, nevertheless, neither cultivars ('Prima' and 'Fiesta') carried *Co* (Maliepaard et al. 1998).

Subsequently, several studies have been published concerning developing DNA markers using the BSA strategy. Kim et al. (2003) identified the $WB82_{670}$ RAPD marker 1.8 cM away from *Co;* they converted it into a SCAR marker that was designated as $SCB82_{670,}$ using 110 F1 individuals which were clearly scored and produced from the 'Fuji' (*co/co*)×'Tuscan' (*Co/co*) cross, which initially contained 227 individuals. Tian et al. (2005) also conducted BSA using a generated population from the cross of 'Telamon' (*Co/co*)×'Spur Fuji' (*co/co*). They constructed a genetic map of partial LG 10, which contains SSR markers from a saturated reference map (Liebhard et al. 2003b), $SCAR_{682}$ (2.9 cM), and the $SCAR_{216}$ (12.3 cM) they developed. Although several additional mapping studies have been conducted using SSR, AFLP, and inter-simple sequence repeat (ISSR) markers (Kenis and Keulemans 2007; Zhu et al. 2007; Fernández-Fernández et al. 2008), *Co* locations consistently had some contradictions based on the order among markers and *Co*, which could have resulted from the unreliable phenotyping and marker reproducibility. QTL analysis for growth characteristics in a population that segregates columnar and standard trees showed that the QTL of several growth indicators collocates with *Co* (Kenis and Keulemans 2007). They also highlighted that the root system influences QTL stability from the results of the self-rooted and grafted trees. Moriya et al. (2009) also revealed the more precise location of *Co* using majorly grafted trees that enable accurate phenotyping. SSR markers CH03d11 (Liebhard et al. 2002) and Hi01a03 (Sifverberg-Dilworth et al. 2006) were identified as the most closely located markers on either side of *Co* (1.4 and 7.6 cM, respectively). The association of $SCAR_{682}$ and $SCAR_{216}$ with *Co* was confirmed, although $SCB82_{670}$ was unverified.

The introduction of the first version of the apple genome (Velasco et al. 2010) has induced an increasing trend for fine mapping and positional cloning of *Co.* Interestingly, three fine mapping studies were published successively (Bai et al. 2012a; Moriya et al. 2012a; Baldi et al. 2013). All three articles used genome sequences as the origin for developing markers in targeted regions. Moriya et al. (2012a) applied 1,000 F1 plants from 31 populations and observed a candidate region of *Co* spanning 196 kb. Bai et al. (2012a) also applied markers 528 F1 plants and 290 columnar-type breeding selections and demarcated the *Co* region in 193 kb. Lateral Organ Boundaries Domain (LBD) containing transcription factor genes were listed as candidate genes from this region. Furthermore, Baldi et al. (2013) identified a 393-kb region that originated from the fine mapping using 1,551 F1 individuals, which comprised five populations. BAC clones sequencing developed from columnar cultivars has been conducted after these fine mapping studies, which identified an insertion sequence on 'McIntosh Wijcik' genome annotated as a Ty3/Gypsy or a long terminal repeat (LTR) retrotransposon, although inserted nucleotide lengths vary depending on researchers (Wolters et al. 2013; Otto et al. 2014; Okada et al. 2016). The insertion length is probably ~8.2 Kb according to the long PCR using a primer pair adjacent to the insertion sequence, generating 16 and 8 kb product from *Co/Co* and *co/co* materials, respectively (Okada et al. 2016). Although RNA-seq analyses have demonstrated that numerous gene expression levels significantly vary between 'McIntosh' and 'McIntosh Wijcik' (Krost et al. 2012, 2013; Otto et al. 2014; Okada et al. 2016), *MdCo31* (*2OG-Fe(II) oxygenase*)/ *Downy Mildew Resistant6/91071*-gene, which is a common gene, has been identified among studies as a gene expressed in 'McIntosh Wijcik' and unexpressed in 'McIntosh' shoots (Wolters et al. 2013; Petersen et al. 2015; Okada et al. 2016). Henceforth, this gene is referred to as *MdDOX-Co* based on the functional analysis described below (Watanabe et al. 2021).

Overexpression of *MdDOX-Co* in *Arabidopsis thaliana* causes a short floral internode length (Wolters et al. 2013). Similarly, *MdDOX-Co* overexpression in apples caused short internodes similar to columnar apples (Okada et al. 2016). These results contrast the insertion mutation, which upregulated the expression of the adjacent *MdDOX-Co*, resulting in a columnar phenotype (Okada et al. 2016). In further studies, ectopic expression of *MdDOX-Co* in shoot apices and leaves in columnar apples was observed rather than in non-columnar apples, where it was detected only in roots (Wada et al. 2018). Additionally, the functional analysis using *Arabidopsis thaliana*, which overexpressed *MdDOX-Co*, revealed that *MdDOX-Co* metabolizes active gibberellin to its inactive variant through OH-bound addition, inducing dwarfing phenotype due to gibberellin deficiency (Watanabe et al. 2021). Low endogenous gibberellin levels in columnar apples compared to non-columnar apples have been reported (Lee and Looney 1977; Okada et al. 2020). However, an active gibberellin compound GA_3 application partially restored the dwarf phenotype in transgenic tobacco (Okada et al. 2020; Wang et al. 2021) and columnar apples (Lee and Looney 1977; Okada et al. 2020).

Co is a unique allele in breeding and has recently emerged through spontaneous mutation. Therefore, a DNA marker should be developed from a mutation point for DNA testing in breeding to discriminate the *Co* chromosome from the original one of 'McIntosh.' Therefore, for this purpose, a DNA marker that detects causal insertion has been designed (Okada et al. 2016) and applied in the breeding program at the National Agriculture and Food Research Organization (NARO) in Japan.

Modifier genes described in previous studies until the 2000s' (Lapins 1976; Kim et al. 2003) were not probably mentioned since the study developments in columnar apples enabled researchers to phenotype and differentiate columnar and non-columnar apples; however, Dougherty et al. (2020) recently reemphasized the influence of modifier genes. They observed standard and standard-like tree forms from apple plants, both carrying *Co.* Deep sequencing of two pooled DNAs of columnar and standard apples both carrying *Co*, and subsequent comparing the allele frequency directional difference, and density (AFDDD) mapping approach identified two loci of age-dependent recessive suppressors on chromosomes 10 and 9. *MdDOX-Co* suppression was also confirmed on the standard apple carrying *Co*. However, the lineage of plant materials in this study has not been described in detail, and it is unclear whether these suppressors are identical to modifiers mentioned in earlier studies. Therefore, the origin of these suppressor alleles and their distribution in the breeding programs should be clarified. Moreover, identifying these suppressor genes will help to elucidate the mechanisms that regulate tree shape.

2.3.2 Weeping Habit

The weeping habit in apples, sometimes called the pendulous habit, is characterized by branches that exhibit downward growth (Figure 2.1b). These characteristics frequently exist in crabapples and crabapple-derived source due to their ornamental value, while the columnar tree shape is expected to be agronomically valuable. Popular weeping apple cultivars include the 'Red Jade' (offspring of 'Exzellenz Thiel' [*Malus prunifolia* 'Pendula'×*M. floribunda*]) (Fiala 1994). The inheritance manner of the weeping habit was firstly investigated by Sampson and Cameron (1965). They produced two crosses of *M. baccata* 'Gracilis' (weeping)× 'Geneva' and *M. baccata* 'Gracilis'×*M. niedzwetzkyana*; they observed 1:1 segregation between weeping and standard, although the population sizes were small (only 28 seedlings). Additionally, they concluded that a single dominant gene confers weeping habit and that *M. baccata* 'Gracilis' is heterozygous for this gene (they used the gene name *Ph*, subsequently renamed *W* by Alston et al. (2000)). Dominant weeping gene is very rare in tree crop; however, only a Japanese chestnut example (Terakami et al. 2021) in addition to *W*. Weeping, has been typically controlled using a single recessive gene in other crops, including peach (*Prunus persica*), mulberry (*Morus alba*), and Japanese apricot (*Prunus mume*) (Monet et al. 1988; Scorza et al. 2002; Werner and Chaparro 2005; Yamanouchi et al. 2009; Zhang et al. 2015). Therefore, *W* is used in horticulture and is also valuable for the physiological study of tree architecture.

Research on *W* has not been active probably because its priority is not high compared to *Co* due to its low agronomic importance. There have been some descriptions of segregating weeping habits and intermediate phenotypes occurring in plants with both *Co* and *W* (Brown et al. 2004; Moriya et al. 2012a). Recent developments in massive sequencing techniques have brought this level to the molecular level in a moment. Dougherty et al. (2018) constructed two DNA pools of weeping and standard trees selected from the F1 population obtained from the 'Cheal's Weeping' (weeping)× 'Evereste' (standard) cross. The two pools were sequenced using the NGS platform, which was later mapped on the reference genome of Velasco et al. (2010). Furthermore, the variation among the weeping pool, the standard pool, and the reference genome were evaluated using the mutant allele frequency and density (MAFD) and AFDDD mapping strategies. These strategies resulted in identifying four chromosomes exhibiting a difference in allele frequency. Among them, chromosome 13, which has the largest difference, was detected and considered the major locus: *W*. Moreover, *W* locus was effective in two other populations (NY-051 [standard]× 'Louisa' [weeping] and NY-011 [standard]×NY-100 [weeping]) that do not share the precise origin of the weeping habit. The second locus (*W2*) on chromosome 2 was ineffective in the NY-051× 'Louisa' population, implying a different control mechanism of weeping. They also mentioned candidate genes for *W* resulting from RNA-seq analysis to include two GXMT-like genes: a LAZY1-like gene and a transcription factor TCP20-like gene. To our knowledge, studies on the weeping habit have evolved rapidly by applying proper molecular techniques; however, further genetic and physiological characterization is needed for a better insight into the weeping trait.

2.4 FRUIT QUALITY

The determining factors of consumers' preference for apple fruits are as follows: fruit size, sweetness, sourness, juiciness, flesh texture, aroma, and appearance. Most apple-scion cultivar breeding programs target several or all of those traits, and they are dominated by multiple genes often influenced by the environment. Fruit skin color and sourness – measured by the amount of malic acid content in fruit juice, are traits typically governed by a few key genes according to early genetic studies (Wilcox and Angelo 1936; Nybom 1959). Additionally, recent studies have elucidated that several major QTLs control those traits instead of the polygene model. Therefore, this implies that it is possible to seek a breeder's ideal genotype that meets their objective using this genetic information. This section primarily focuses on the quality and physiological disorder traits of the fruit.

2.4.1 Sweetness

The sweetness of apple fruit is an essential attribute, although genetic analysis of sweetness has been challenging. For example, depending on the materials employed, QTL analyses for sweetness sensation and total soluble solid contents (SSC), also referred to as Brix, cannot identify stable QTLs. Regarding QTLs for SSC, Liebhard et al. (2003a) employed the F1 population derived from the 'Fiesta' × 'Discovery' cross and identified QTLs on LGs 3, 6, 8, and 9. Kenis et al. (2008) identified those on LGs 2 and 10 using the 'Telamon' × 'Braeburn' population. Kunihisa et al. (2014) have detected those on LGs 15 and 16, resulting from the QTL analysis using the 'Orin' × 'Akane' population. Conversely, Potts et al. (2014) could not detect any QTL using the Co-op17×Co-op16 population, whereas Guan et al. (2015) identified QTLs in LGs 2, 3, and 15 through PBA. Additionally, QTLs in LGs 4 and 16 were identified by Amyotte et al. (2017) and Minamikawa et al. (2021) by GWAS, respectively. These results indicate that genetic control of SSC is quite complicated, indicating the existence of an additive effect and epistatic interaction. Narrow-sense heritability of SSC was relatively low compared to firmness and acidity (Kouassi et al. 2009). Furthermore, the correlation between sensory sweetness and SSC is weak, which suggests that these descriptors have different meanings (Kouassi et al. 2009; Oraguzie et al. 2009; Amyotte et al. 2017). Other tastes can also affect sensory sweetness; for example, negative correlations were observed between sensory sweetness and malic acid content (Kouassi et al. 2009; Rymenants et al. 2020); therefore, this might hamper QTL analysis to dissect the actual genetic contribution. Indeed, Rymenants et al. (2020) conducted PBA and identified sensory QTLs for sensorial sweetness perception on LGs 8 and 16, whose position coincide with sensorial acidity. Moreover, no QTL was detected by haplotype-based GWAS (Minamikawa et al. 2021).

Apple fruits accumulate individual sugars comprising fructose, glucose, sucrose, and sorbitol. Since SSC approximates the sum of these sugars, several independent/interactive genetic mechanisms may affect SSC control. Therefore, it could be better to analyze individual sugar content instead of SSC and sensory sweetness to elucidate the mechanism underlying sugar accumulation in fruit. Table 2.2 shows the QTLs for individual sugars previously identified by bi-parental QTL analysis, PBA, or GWAS. Colocalization of QTLs is suggested on LGs 1, 3, 11, 15, 16, and 17, with their contribution ranging from 5% to 49.9%. Interestingly, a single locus is always associated with multiple individual sugars rather than just one. Complex networks that constitute sugar metabolism and the enzymes that convert one sugar into another may have caused these colocations. For instance, a gene coding vacuolar invertase (MDP0000149570), which metabolizes sucrose to fructose and glucose, has been suggested as a strong candidate gene of QTL on LG 1 (Larsen et al. 2019) and has been characterized (Hyun et al. 2011). A QTL for sucrose content in Japanese pear has been identified on the same locus, and its causal gene *PPAIV3*, an invertase, has been suggested (Nishio et al. 2018). This feature may be a perfect example of conserved gene function between apple and pear, which belong to the Pyrinae.

To design the flavor come from sugar, it should be noted that each individual sugar has a distinct sensory perception: fructose and sucrose are sweeter than glucose or sorbitol, and fructose perception becomes sweeter at low temperatures due to the increase of the sweetest isomer β-D-fructopyranose (Belitz et al. 2009). Therefore, a future goal of breeding for sugar content will be that breeders can design desirable sweetness by regulating the content of individual sugar. Additional research should be conducted to elucidate the genetic architecture of sugar accumulation and practical allele introduction in future breeding.

TABLE 2.2
QTLs Reported for Individual Sugars

	Individual Sugar	Chrom./LG	Position		LOD/-log_{10}P	Contribution %	Material	Method	References
			cM	Mb					
*	Sorbitol	1	16.3		4.1	19.8	'Prima' × '95P6'	Bi-parental QTL mapping	Kunihisa et al. (2019)
*	Sucrose	1	17.3		4.1	26.1	'Prima' × '95P6'	Bi-parental QTL mapping	Kunihisa et al. (2019)
*	Fructose	1	39.4		9.3	49.9	'Prima' × '95P6'	Bi-parental QTL mapping	Kunihisa et al. (2019)
*	Fructose	1	59.8–85.7		–	34–45	233 sdl + 32 cult	PBA	Guan et al. (2015)
*	Fructose%	1		28–29 M	8.94	47	108 Danish cult	GWAS	Larsen et al. (2019)
*	Glucose	1	60.9–85.7		–	19–22	233 sdl + 32 cult	PBA	Guan et al. (2015)
*	Sorbitol	1	62.9–85.7		–	21	233 sdl + 32 cult	PBA	Guan et al. (2015)
*	Sucrose	1	54.6–76.3		–	17–36	233 sdl + 32 cult	PBA	Guan et al. (2015)
*	Sucrose	1		28–29 M	5.58	27	108 Danish cult	GWAS	Larsen et al. (2019)
*	Sucrose%	1		28–29 M	6.82	34	108 Danish cult	GWAS	Larsen et al. (2019)
	Glucose	2	11.4		6.2	13.4	'Prima' × '95P6'	Bi-parental QTL mapping	Kunihisa et al. (2019)
	Fructose	3	5–28.9		–	10–15	233 sdl + 32 cult	PBA	Guan et al. (2015)
*	Fructose	3	50.6		5.75	20.6	'Jiguan' × 'Wangshanhong'	Bi-parental QTL mapping	Ma et al. (2016)
*	Glucose	3	56.1		14.3	42.6	'Prima' × '95P6'	Bi-parental QTL mapping	Kunihisa et al. (2019)
*	Glucose	3	49.6		4.47	17.1	'Jiguan' × 'Wangshanhong'	Bi-parental QTL mapping	Ma et al. (2016)
*	Sorbitol	3	49.6		7.73	28	'Jiguan' × 'Wangshanhong'	Bi-parental QTL mapping	Ma et al. (2016)
*	Sucrose	3	48.6		3.41	11.7	'Jiguan' × 'Wangshanhong'	Bi-parental QTL mapping	Ma et al. (2016)
	Glucose	4	44.4		4.49	16.7	'Jiguan' × 'Wangshanhong'	Bi-parental QTL mapping	Ma et al. (2016)
	Sucrose	4	7.9–16.1		–	8	233 sdl + 32 cult	PBA	Guan et al. (2015)
	Glucose	5	0		3.34	12.4	'Orin' × 'Akane'	Bi-parental QTL mapping	Kunihisa et al. (2014)
	Sorbitol	5	3.2–10.8		–	2	233 sdl + 32 cult	PBA	Guan et al. (2015)
	Fructose	6	8.6		3.27	10.9	'Orin' × 'Akane'	Bi-parental QTL mapping	Kunihisa et al. (2014)
	Sorbitol	9	40.2–49.1		–	1	233 sdl + 32 cult	PBA	Guan et al. (2015)
	Sucrose	9	0.7–19.6		–	24	233 sdl + 32 cult	PBA	Guan et al. (2015)

(Continued)

TABLE 2.2 (*Continued*)
QTLs Reported for Individual Sugars

	Individual Sugar	Chrom./LG	Position		LOD/-log_{10}P	Contribution %	Material	Method	References
			cM	Mb					
	Sucrose	10	26.3		4.21	14.3	'Orin' × 'Akane'	Bi-parental QTL mapping	Kunihisa et al. (2014)
*	Glucose	11	62.5		3.3	6.3	'Prima' × '95P6'	Bi-parental QTL mapping	Kunihisa et al. (2019)
*	Sorbitol	11	39.2–57.5		–	7	233 sdl + 32 cult	PBA	Guan et al. (2015)
	Sorbitol	12	23.6		2.79	10.5	'Orin' × 'Akane'	Bi-parental QTL mapping	Kunihisa et al. (2014)
	Glucose	15	22.9		4.2	8.3	'Prima' × '95P6'	Bi-parental QTL mapping	Kunihisa et al. (2019)
*	Fructose	15	94.7–104.3		–	5	233 sdl + 32 cult	PBA	Guan et al. (2015)
*	Sucrose	15	92.5		3.06	10.1	'Orin' × 'Akane'	Bi-parental QTL mapping	Kunihisa et al. (2014)
*	Fructose	16	18.5		3	10	'Orin' × 'Akane'	Bi-parental QTL mapping	Kunihisa et al. (2014)
*	Glucose	16	2.4–8.8		–	6	233 sdl + 32 cult	PBA	Guan et al. (2015)
*	Sorbitol	16	15.5		3.74	13.8	'Orin' × 'Akane'	Bi-parental QTL mapping	Kunihisa et al. (2014)
*	Sorbitol	16	6.6		5.3	26.6	'Prima' × '95P6'	Bi-parental QTL mapping	Kunihisa et al. (2019)
*	Sorbitol	16		3.9M	–	–	185 cult + 'Orin' × 'Akane'	GWAS (BayesB)	Minamikawa et al. (2021)
*	Fluctose	17		32.5 M	–	–	185 cult + 'Orin' × 'Akane'	GWAS (BayesB)	Minamikawa et al. (2021)
*	Glucose	17		32.6 M	–	–	185 cult + 'Orin' × 'Akane'	GWAS (BayesB)	Minamikawa et al. (2021)
*	Sucrose	17		32.7 M	–	–	185 cult + 'Orin' × 'Akane'	GWAS (BayesB)	Minamikawa et al. (2021)

*Colocation of QTLs.
GWAS, genome-wide association study; PBA, pedigree-based analysis.

2.4.2 Sourness

Sourness, as the acid concentration of fruit juice, affects palatability alone or when combined with sugar. Consumer preference depends on the right balance of sweetness and acidity, with too much or too little acid being disliked (Iwanami 2011). In contrast to the weak correlation between the sensorial perception of sweetness and SSC, sensorial sourness strongly correlates with the malic acid content (Kouassi et al. 2009). Malic acid constitutes most of the accumulating acids in apple fruit, followed by citric acid (Petkovsek et al. 2007; Zhang et al. 2010). At harvest, the level of malic acid gradually drops to a baseline value, which is common to all cultivars (Verma et al. 2019). Therefore, QTLs for malic acid content in harvested fruit is described.

Genetic analysis for malic acid content at harvest has been conducted since the 1950s. Nybom (1959) measured the pH of fruit juice using multiple populations from the controlled crosses and found that specific cross combinations possess individuals with high pH (i.e., low malic acid content). The proportion of high pH to low pH individuals fit the 1:3 ratio, enabling the designation of a single dominant gene *A* that causes low pH in fruits. This gene is also called *Malic acid* (*Ma*).

Ma is a beneficial gene due to its enormous contribution to malic acid content and relatively easy-to-map process, as well as its breeding importance; therefore, it has been a target of early mapping studies as a Mendelian trait locus (Conner et al. 1997). Additionally, *Ma* has been mapped on LG 16 as a Mendelian trait locus on the first consensus map of apple constructed from the 'Prima' × 'Fiesta' population (Maliepaard et al. 1998). Afterward, QTL analyses were conducted considering a quantitative genetic control model for malic acid content; therefore, identifying QTLs on the equivalent position of two linkage maps constructed from the populations derived from the 'Fiesta' × 'Discovery' and 'Telamon' × 'Braebauen' crosses (Liebhard et al. 2003a; Kenis et al. 2008). In addition to the *Ma* locus, a novel locus with almost identical contribution as the *Ma* locus has been identified on LG 8 (LG 8 (46%) vs *Ma* (42%)) (Liebhard et al. 2003a). Subsequently, several studies have attempted to identify the QTL for malic acid content or sensory sourness, and successfully identified both or either a QTL through bi-parental QTL analysis (Zhang et al. 2012; Kunihisa et al. 2014, 2019; Ma et al. 2016), GWAS (Kumar et al. 2013; Kunihisa et al. 2016; Minamikawa et al. 2021) and PBA (Verma et al. 2019; Rymenants et al. 2020); however, some minor QTLs have also been reported (Kenis et al. 2008; Xu et al. 2012; Rymenants et al. 2020).

Moreover, fine mapping was attempted after the release of the apple genome (Velasco et al. 2010). Xu et al. (2012) used two F1 populations (438 individuals) and conducted QTL analyses for both malic acid content and pH. They established numerous markers around the *Ma* region and identified an SSR marker 18695.28-2 that cosegregates with either malic acid content, pH, or both. Ultimately, a candidate region spanning 150 kb was determined based on the genotypes of six critical recombinants. Furthermore, Bai et al. (2012b) included plant materials in this study and delimited the candidate region to 65 kb. Consequently, this region was subjected to BAC clone identification and has resulted in the discovery of two clones from the BAC library of 'G.41,' which were later sequenced. Notably, two aluminum-activated maleate transporter-like genes (*MdALMTII*) from this region were identified and designated as *Ma1* and *Ma2*. The expression level was higher for *Ma1* than the *Ma2*, and that of *Ma1* positively correlates with malic acid content. Using other materials (the 'Prima' × 'Fiesta' population), the identical cluster of genes was identified, and a positive correlation between the expression level of *Ma1* and malic acid content was confirmed (Khan et al. 2013). Bai et al. (2012b) further analyzed the amino acid sequences of both *Ma1* and *ma1* alleles and found a probable causal mutation at the 1,455th nucleotide on *ma1*, which induces a stop codon. Therefore, *ma1* is a truncated protein that is probably non-functional. Furthermore, cleaved amplified polymorphic sequence (CAPS) markers that employ this polymorphism have been established (Bai et al. 2012b; Ma et al. 2015).

QTL on LG 8, designated as *Ma3*, has been genetically characterized by the PBA using 16 populations of 200 individuals (Verma et al. 2019). They concluded that an additive dosage model of *Ma1* and *Ma3* is appropriate, and one high acidity-associated allele contributed 1.8 mg/L malic acid increase. Additionally, they found a rare haplotype that induces a high malic acid content for both *Ma1* and *Ma3* loci. Conversely, Jia et al. (2018) proposed a higher hierarchical genetic interaction model for determining acidity. They applied both classical bi-parental QTL analysis and NGS-based BSA, which delimited the candidate region of *Ma3* in three adjacent chromosomal segments. Candidate genes, such as *MdPP2CH*, *MdSAUR37*, and *MdMYB44*, were identified from each segment. Additionally, the polymorphic associations of these genes and malic acid content were confirmed using 92 cultivars and two full-sib populations. They also showed interactions among these proteins: MdPP2CH dephosphorylates MdALMTII (*Ma1* candidate), which inactivated the MdALMTII; MdSAUR37 suppressed the dephosphorylation activity of MdPP2CH. They also proposed that the malic acid content of specific material is expected from the genotypes of *Ma1* and *Ma3* loci. Furthermore, Jia et al. (2021) showed that MdMYB44 inhibited the transcription of MdALMTII by influencing the upstream region of that gene, suggesting the control mechanism for the malic acid content of this gene. However, Ban and Xu (2020) discovered that *MdSAUR37* polymorphism had no significant effect on their plant materials. Therefore, since current findings regarding the genetic make-up of malic acid content vary, further efforts should be undertaken in the future. Particularly, there are multiple candidate genes mentioned for *Ma3*. It should be clarified which gene(s) is responsible for the trait setup.

In the breeding context, the trait-predictive DNA test has somewhat become feasible for *Ma1* and *Ma3*. Additionally, markers and polymorphism data described in articles could be applied in actual breeding.

Variations among haplotypes of major loci influencing acidity can be a next subject. Minamikawa et al. (2021) performed a haplotype-based GWAS and predicted the effects of haplotypes from founders. For the *Ma1* locus, a haplotype from 'Ralls Janet' exhibited the strongest influence on the high acidity, whereas, for the *Ma3* locus, one from 'Jonathan' displayed the highest effect. Furthermore, recent studies have reported the minor QTLs as *Ma4*, *Ma5*, and *Ma6* on LGs 6, 1, and 4 using PBA (Rymenants et al. 2020) and GBS-based BSA, also called AFDDD (Ban and Xu 2020). Although *Ma1* alleles have been binarily scored, it would be advantageous to evaluate the exemplary contribution of each allele to malic acid content and include the effect of minor QTLs in the model. Consequently, this will enable breeders to advance phenotypic prediction and genotypic selection of malic acid content.

2.4.3 Flesh Texture and Associated Traits

Apple fruit flesh texture greatly influences sensory taste perception, of which crispness, firmness, and juiciness are essential. A good texture is required at harvest and after a particular storage period. Therefore, flesh texture is an important breeding objective in breeding programs globally (Iwanami et al. 2005; Laurens et al. 2018; Teh et al. 2021). Storability is also a critical feature in keeping the fruit fresh during distribution. Since the texture is a complex quantitative trait, many genes may contribute to its composition. For the genetic analysis, two types of measurement has been used to evaluate texture. The first one is the mechanical measurements predominantly applicable to firmness. The other is the sensory evaluation applied to crispness and juiciness, among others. Trained experts are indispensable for conducting sensory evaluation; however, the evaluation scale may sometimes vary among research groups, which is challenging. Mechanical measurements are simpler than sensory ones, although they remain challenging due to variations in techniques except for a classical firmness evaluation using a penetrometer. Due to the difficulty in comparing results obtained from different measurements, we will focus on well-established QTL loci that are remarkably important and/or described thus far. Additionally, fruit texture QTLs have frequently exhibited possible pleiotropy on other traits, including storability, harvest date, and preharvest fruit drop. We also discuss these aspects in this section.

2.4.3.1 *MdPG1* on LG 10

MdPG1 encodes polygalacturonase protein and modifies pectin structure in the cell walls. This gene is genetically mapped at the middle of the LG 10 (Costa et al. 2010) and could be among the most beneficial and applied genes in recent molecular-assisted breeding for flesh texture.

MdPG1 was initially studied using a reverse genetics approach (Wakasa et al. 2006). A gene expression analysis using 14 cultivars indicated that cultivars with less firmness loss during storage showed relatively low expression levels of *MdPG1*. Moreover, *MdPG1* expression was suppressed by applying an ethylene inhibitor, 1-methylcyclopropene (1-MCP) (Wakasa et al. 2006). Furthermore, *MdPG1*-suppressed 'Royal Gala' showed firmer fruits than the control after ripening and higher intercellular adhesion (Atkinson et al. 2012).

In the forward genetics approach, many QTLs for texture traits have been mapped near *MdPG1*. The QTLs include firmness at harvest (Kenis et al. 2008; Chagné et al. 2014; Kunihisa et al. 2014; Ben Sadok et al. 2015), and after storage (Costa et al. 2010; Chagné et al. 2014), sensory texture attributes (Ben Sadok et al. 2015), mechanical and acoustic indices analyzed using a texture analyzer, which reflects the crispness (Longhi et al. 2012, 2013b), and loss of firmness after storage or mealiness (Chagné et al. 2014; Moriya et al. 2017a). GWAS and PBA have also been employed in QTLs for firmness at harvest (Kumar et al. 2013), firmness after 2 months of cold storage (Bink et al. 2014), mealiness (Moriya et al. 2017a; Minamikawa et al. 2021), and mechanical and acoustic indices (Di Guardo et al. 2017). Although McClure et al. (2018) found QTL for loss of firmness near this locus, they suggested that the ethylene response factor (ERF) gene was involved because of the no correlation observed with *MdPG1* polymorphism.

QTLs for harvest time have also been mapped near *MdPG1* in addition to texture traits (Chagné et al. 2014; Kunihisa et al. 2014). Migicovsky et al. (2021) assessed the contribution of *MdPG1* to harvest date, firmness at harvest and after 3 months of cold storage, and softening using ~850 cultivars for harvest date or ~500 cultivars for softening and firmness. *MdPG1* contributed 8.2% and 7.3% of the phenotypic variation for softening and firmness after storage, respectively, while 1.2% and 1.3% for the firmness at harvest and harvest date, respectively.

The Md-PG1$_{SSR}$10kd SSR marker (Longhi et al. 2012) is a robust tool for the detailed analysis between alleles and phenotypes. This marker was designed from the 10 kb downstream of *MdPG1*, identifying three alleles coded 1, 2, and 3. Longhi et al. (2012, 2013a, 2013b) examined the correlation between mechanical and acoustic indices and genotypes of Md-PG1$_{SSR}$10kd. The "3" allele-homozygous cultivars exhibited an overall low texture behavior, those that lacked the "3" allele displayed a favorable behavior, and the heterozygous ones showed intermediate texture behavior. Moriya et al. (2017a) discovered a relationship between mealiness after 4 weeks of storage at ambient temperature and Md-PG1$_{SSR}$10kd allelotype. The "2" allele showed the most anti-mealiness effect, followed by the "1" allele, which is the anti-mealy one, and the "3" was the mealiest one (Figure 2.2).

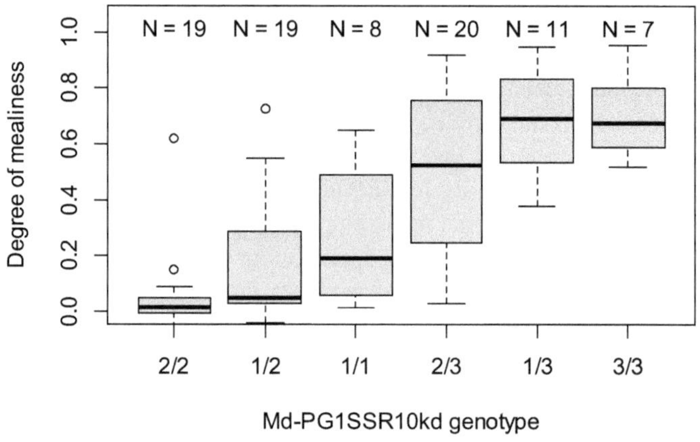

FIGURE 2.2 Distribution of degree of mealiness among apple cultivars based on Md-PG1$_{SSR}$10kd genotype. Degree of mealiness of 0.3 is the threshold of mealiness perception. 0: not mealy, 1: mealy (Moriya et al. 2017a).

Several techniques for detection can be selected for the breeding purpose, in addition to Md-PG1$_{SSR}$10kd, based on the user's preference, which includes the "3" allele-specific primer set (Costa et al. 2010) and SNP-based high-throughput genotyping platforms called KASP™ and the OpenArray assay (Baumgartner et al. 2016; Chagné et al. 2019) that differentiate the "3" allele from the others.

Although implementing the MAS of *MdPG1* into the breeding program has been performed with the achievement described above, phenotypic differences regulated by each allele have not been released onto causal polymorphism of the gene. Therefore, additional studies are warranted to clarify the mechanism that controls texture attributes by *MdPG1*.

2.4.3.2 *MdACS1* on LG 15

Ethylene is a gaseous phytohormone that controls several physiological phenomena, such as senescence and ripening. Mostly, it is synthesized from *S*-adenosyl-L-methionine via the 1-aminocyclopropane-1-carboxylic acid (ACC) pathway catalyzed by the ACC synthase (ACS). *MdACS1* was found to encode ACS and is located on LG 15 (Costa et al. 2005).

Previous studies on this gene via the reverse genetics approach resulted in the significant achievement of unifying molecular studies and breeding strategies. Sunako et al. (1999) cloned *MdACS1* from 'Golden Delicious' and found *MdACS1-1* and *MdACS1-2* alleles. Notably, the latter has a 162-bp insertion in the 5′-flanking region of the coding sequence. At the climacteric stage, *MdACS1-1* is dominantly expressed in the 'Golden Delicious' fruit, whereas its expression level was deficient in the low-ethylene producing cultivar 'Fuji,' which is homozygous of *MdACS1-2*. This association has been further verified in several cultivars, with the amount of ethylene production ranked as *MdACS1-1/1*>*MdACS1-1/2*>*MdACS1-2/2* (Harada et al. 2000). Additionally, breeding interest has developed through the studies that reported the association between this polymorphism and fruit softening after storage (Oraguzie et al. 2004, 2007; Costa et al. 2005) or firmness at harvest (Zhu and Barrit 2008). Migicovsky et al. (2021) clarified using more than 800 cultivars that *MdACS1* was responsible for 1.8% phenotypic variation for softening; however, a significant effect was not observed for harvest date, firmness at harvest, or after storage.

The *MdACS1-2* allele is connected with fruits retained on the tree. Although the significance of this gene on the abovementioned traits is not as high as expected in the previous studies, this locus was reportedly associated with preharvest drop through simple association analysis, bi-parental QTL analysis, and GWAS (Sato et al. 2004; Kunihisa et al. 2014; Minamikawa et al. 2021). For the physiological support of ethylene involvement in fruit abscission, exposure to the ethylene inhibitor 1-MCP has decreased the preharvest drop rate of 'Tsugaru' (*MdACS1-1/2*), an ethylene producer (Sato et al. 2004). Therefore, it should be noted that the selection of *MdACS1* could affect fruit storability and other horticultural characteristics.

The STS marker initially developed by Sunako et al. (1999) remains a breeder-friendly tool for application, although numerous SNP-based platforms have been designed (Baumgartner et al. 2016; Chagné et al. 2019). To our knowledge, 'Shinano Lip,' which was developed by the local government of Nagano Prefecture, Japan, in 2018, is probably the first registered cultivar selected by the *MdACS1* marker in the seedling stage (Kanamaru and Yamaguchi, unpublished). The cross raised 'Shinano Lip' was made in 2000, immediately after the publication of Sunako et al. (1999) and Harada et al. (2000).

2.4.3.3 *MdACO1* on LG 10

ACC oxidase (ACO) is involved in the last step of the ethylene synthesis pathway. *MdACO1* has been associated with the internal ethylene concentration and mapped on LG 10, which is the bottom side of *MdPG1* (Costa et al. 2005). Two alleles of *MdACO1* (1 and 2) have been described and characterized with or without the 62 bp sequence in the third intron. *MdACO1-1*, which is the allele with no 62 bp sequence, is linked with low internal ethylene concentration and less softening compared to *MdACO1-2* (Costa et al. 2005, 2008). Interestingly, among the cultivars that harbor *MdACS1-2/2*, cultivars harboring *MdACO1-1/1* showed firmer fruits at harvest and after cold storage than the other genotypes (Zhu and Barritt 2008). Bi-parental QTL analysis studies identified QTLs near this locus for firmness after cold storage (Costa et al. 2010; Chagné et al. 2014). Conversely, no relationship was found between the *MdACO1* genotypes and firmness at harvest or loss of firmness after storage using 127 cultivars (Nybom et al. 2013). Furthermore, Migicovsky et al. (2021) revealed that *MdACO1* exhibited no significant effect on harvest date, firmness, or softening. Overall, *MdACO1* contributed less to firmness at harvest and loss of firmness than *MdPG1* and *MdACS1*. However, STS markers for detecting 62 bp insertion (Costa et al. 2005) and high-throughput SNP-based platforms are available regarding the selection tool (Baumgartner et al. 2016; Chagné et al. 2019).

2.4.3.4 *NAC18.1* on LG 3

NAC (No apical meristem [NAM], Arabidopsis transcription activation factor [ATAF], cup-shaped cotyledon [CUC]) domain transcription factors are involved in regulating development, defense, and senescence (Mathew and Agarwal, 2018). These are involved in the ripening process in many species, such as tomatoes (Kumar et al. 2018), bananas (Shan et al. 2012), and peach (Pirona et al. 2013). In apples, *NAC18.1,* which is situated on chromosome 3, was discovered from a GWAS for harvest date and firmness using historical data of 689 cultivars (Migicovsky et al. 2016). The principal SNP associated with harvest date and firmness has been detected in the coding sequence of *NAC18.1*. This SNP causes an aspartate (D) to tyrosine (Y) substitution at the fifth amino acid (also called the D5Y mutation) in a highly conserved region of this gene, therefore, inferring functional modulation. Additionally, QTLs for the harvest date located near this locus were detected through the bi-parental QTL analysis (Kenis et al. 2008; Kunihisa et al. 2014) and GWAS (Urrestarazu et al. 2017; McClure et al. 2018; Larsen et al. 2019; Jung et al. 2020; Minamikawa et al. 2021), suggesting the implication of this locus in harvest date rather than firmness. Although this gene has been indicated to be a stronger predictor of harvest date and firmness at harvest than the *MdPG1*, *MdACS1*, and *MdACO1*, it is a less strong or even weak predictor of firmness after 3 months of cold storage and softening rate compared to *MdPG1* (Migicovsky et al. 2021). Additionally, Migicovsky et al. (2021) investigated the *NAC18.1* genotype among modern breeding sources, including 'Granny Smith,' 'McIntosh,' 'Red Delicious,' 'Golden Delicious,' 'Fuji,' 'Empire,' 'Gala,' 'Pink Lady,' and 'Honey Crisp.' All these cultivars possess a favorable (i.e., firm) homozygous state of the D5Y mutation in *NAC18.1*, signifying that this favorable allele may already be under selection by breeders (Migicovsky et al. 2021). Urrestarazu et al. (2017) conducted a GWAS using a multi-locus mixed model for harvest time applying 1,168 European cultivars and detected multiple SNPs on chromosome 3. Furthermore, they suggested the presence of heterogeneity, which is the presence of more than two functional alleles of a given gene that influences a phenotypic trait (Wood et al. 2011). Moreover, they found two tandemly located NAC genes on the GDDH13 genome (Daccord et al. 2017). Urrestarazu et al. (2017) also proposed the effect of geographical adaptation on this locus where the genetically early-ripe cultivars were common in the Northern and Eastern origin cultivars.

According to previous studies, *NAC18.1* is essentially a strong predictor of harvest date and firmness at harvest. However, the D5Y mutation has been selected as the favorable allele in major breeding resources, indicating the less significant importance of this locus in the recent breeding program. Consequently, additional studies are necessary whether this locus should be implemented in the current breeding program.

2.4.3.5 Implication of the Current Genetic Status in These Loci

Traits such as storability and less fruit drop characteristics are expected to have gained their importance during the era of the modern apple industry in which people transported apples from farms to distant consuming regions, sometimes across continents, was established. Moreover, considering the medieval apple production, apples were possibly consumed in a comparatively narrow region in a short time, leading to the less importance of storability. Consequently, this circumstance could have exempted *MdPG1* from certain selection pressure or even that toward the Md-PG1$_{SSR}$10kd-3 allele due to its association with rich aroma (Farneti et al. 2017), presumably developing increased flavor. Similarly, lower fruit drop characteristics, which resulted from the *MdACS1-2* allele, were probably not preferred in medieval apple production. Until one hundred years ago, apple trees were mainly grafted on seedling rootstock, which makes them tall and big. People may have had an obstacle in picking fruits from the top of the tree canopy, leading to a preference for the ease of fruit dropping.

Specifically, cider apple fruits, even now, are frequently gathered from the ground after fruit drop. Since the *MdACS1-2* allele was observed in the wild germplasms (Sunako et al. 1999; Sun et al. 2009), its appearance may be quite ancient and weakly selected in the cultivated apple gene pool. Therefore, until recently, positive selection pressure toward the *MdACS1-1* allele was probable. However, everything is opposite in the current apple production and industry, comprising growers, distributors, and consumers. Fruits should be appropriately harvested when they remain on the tree; storability is required for long-term storage to transport to distant regions and facilitate year-round distribution. Research and development of technologies associated with these subjects are currently among the major focuses of horticultural science. Currently, various breeding programs are focusing on firm and crisp texture; therefore, a drastic change in selection pressure regarding these loci may have been occurring. Nybom et al. (2008) studied the allele frequency of *MdACS1* using old to modern apple cultivars. They resolved that *MdACS1-2* frequency in the apple cultivars is historically increasing, indicating a significant impact of a few *MdACS1-1/2* heterozygous cultivars on modern apple breeding. Lastly, it should be noted that fruit firmness at harvest and after the storage is influenced by harvest time (Nybom et al. 2013; Migicovsky et al. 2021), which shows that environmental factors contribute to firmness. Therefore, environmental factors should be considered, as well as genetic factors, to understand the genetic architecture of texture, storability, and harvest time.

2.4.4 Aroma

Apple aromas are an essential component of apple flavor. When we eat apples, the perception of the aroma comes through retronasal olfaction, building the flavor combined with sweetness and sourness, among others. Although the importance of the aroma is unquestionable, there is currently no concrete strategy for obtaining a "good" aroma due to insufficient information and evidence on how to increase consumer preference by modifying apple aroma and which genetic factor requires greater focus. Interestingly, the importance of aroma for sensory perception has been highlighted in strawberries. Adding strawberry aromas to sugar solution has been proven to enhance the perception of sweetness (Frank and Byram 1988; Schifferstein and Verlegh 1996). Additionally, in strawberries, the influence of volatile compounds on sweetness perception has been investigated using large sensory datasets and chemical analyses, which identified 20 volatile compounds that positively contributed to increasing sweetness perception (Fan et al. 2021).

Similarly, Iwanami (2011) described that the combined effects of different volatiles in apples, rather than the total amount of volatiles, may be crucial for the fruit quality. Although more than 300 volatile compounds have been detected in apple fruits, 20 are characterized as "character impact" aromas, mainly consisting of esters followed by alcohols and aldehydes (Dixon and Hewett 2000). Since each compound has a different threshold for sensory perception (Dixon and Hewett 2000), the absolute content of the compound does not probably correspond to its significance. The role of these "character impact" compounds includes the characteristics of the typical aroma of apples and the intensity and quality of aromas. Additionally, of the biosynthetic pathways of volatiles, the fatty acid and isoleucine pathways are associated with the synthesis of straight chain and the branched-chain esters, respectively (Schaffer et al. 2007).

Many factors influence volatile biosynthesis and release. The apple volatiles increase after ripening, and the maximum endogenous volatile concentration occurs at the climacteric peak (Dixon and Hewett 2000). External ethylene application has increased volatile production through a climacteric rise in *ACO*-suppressed 'Royal Gala' that produces no detectable levels of ethylene (Schaffer et al. 2007). Moreover, tissue damage such as mastication changes the release of volatiles through cell disruption (Cappellin et al. 2015). Anaerobic biosynthesis in the watercored cortex also contributes to volatile production since watercored 'Fuji' has more methyl and ethyl esters than the non-watercored ones (Tanaka et al. 2016). Therefore, each study's sample preparation procedure and cultivation practice are critical factors.

Currently, studies to elucidate the genetic architecture of the contents of volatile compounds, particularly esters, are actively conducted. However, variations in sample preparation and measurement equipment of volatiles depend on the research groups and their objectives. For instance, ripening duration ranges from 2 weeks to 2 months. Fruits can also be intact, peeled, crushed, or processed into juice. The following two methods are mainly used for measuring volatiles: gas chromatograph-mass spectrometry (GC-MS), which quantifies the amount of each volatile compound, and the proton transfer reaction-mass spectrometry (PTR-MS), which can effectively analyze samples although constrained by analytical power, i.e., each peak does not correspond to a particular compound. Due to the limitations above, the results obtained by bi-parental QTL analysis or GWAS cannot be directly compared and used to consider the benefit of LOD value and the contribution of QTLs. However, a QTL locus in LG 2, in which *MpAAT1* (*MdAAT1*) is located, has been repeatedly reported in numerous publications and recognized as pivotal genetic factors for volatile composition (Table 2.3). *AAT* encodes alcohol acyl CoA transferase and catalyzes the final step of ester synthesis pathways. It can use a range of alcohol substrates from short to medium straight chain (C3–C10),

TABLE 2.3
QTLs Reported for Volatile Organic Compounds Detected on the *MpAAT1* Region

Compound(s)	Population	Sample Preparation	VOC Detection Method	References
Esters alcohols	'Fiesta' × 'Discovery'	1 months at 4°C then 1 h at room temperature	PTR-MS	Zini et al. (2005)
2-Methylbutyl acetate	'Royal Gala' × 'Granny Smith'	15 days at 0.5°C then 5 days at 20°C	GC-MS	Rowan et al. (2009)
Esters (acetate, butanoate, hexanoate) Alcohols (hexanol, iso-pentanol)	'Discovery' × 'Prima'	Immediately after harvest Slices without core homogenized with an immersion blender	GC-MS	Dunemann et al. (2009)
Ethanol Esters	'Fiesta' × 'Discovery'	1 months at 4°C then 1 h at room temperature	PTR-MS	Costa et al. (2013)
Esters (acetate, propanoate, butanoate, pentanoate, hexanoate, heptanoate, octanoate, 2-methylbutanoate) Alcohols (butanol, pentanol, hexanol)	'Royal Gala' × 'Granny Smith'	15 days at 0.5°C then 5 days at 20°C	GC-MS	Souleyre et al. (2014)
Acetate esters	230 accessions	42 days at 0.5°C then 1 day at 20°C	GC-MS	Kumar et al. (2015)
Esters	'Fuji' × 'Delearly'	2 months cold storage then room temperature overnight	PTR-ToF-MS	Cappellin et al. (2015)
Esters The second principal component of volatile profile	162 accessions	2 months cold storage then room temperature overnight Crushed by an artificial chewing device	PTR-ToF-MS	Farneti et al. (2017)
Acetate esters Butyl acetate/Butanol Hexyl acetate/Hexanol	145 Danish cultivars	Fruit juice at 37°C taken immediately after harvest	GC-MS	Larsen et al. (2019)

GC-MS, gas chromatography-mass spectrometry; PTR-ToF-MS, proton-transfer-reaction time-of-flight mass spectrometry; VOC, volatile organic compound.

branched chain, and aromatic and terpene alcohols (Souleyre et al. 2005). Therefore, we will examine the volatile studies of apples regarding *MpAAT1*.

The first attempt at the bi-parental QTL analysis of apple volatiles was conducted by Zini et al. (2005). They used 57 individuals from the F1 population derived from 'Fiesta' × 'Discovery,' and volatiles was analyzed using the PTR-MS on fruits after 1 month of cold storage. The QTL analysis results identified a QTL cluster for esters and alcohols at the top of LG 2, where *MpAAT1* is located. Table 2.3 shows the results and methods applied to the bi-parental QTL analysis, PBA, and GWAS that identified the QTLs for volatile compounds in this region. As shown in the table, QTLs at the top of LG 2 were identified regardless of the detection method (GC-MS or PTR-MS related methods) and sample preparation procedure, indicating the clear association between volatile synthesize and *MpAAT1* in various genetic backgrounds.

Meanwhile, *MpAAT1* alleles have been subjected to be characterized. Dunemann et al. (2012) inferred the *MpAAT1* haplotypes of 102 cultivars and attempted to identify the association between volatile content and *MpAAT1* polymorphisms. They designed primers that can amplify a partial *MpAAT1* sequence, and direct sequencing of amplicons was conducted. Based on four SNPs at locations 62, 110, 425, and 459 of the nucleotide amplicon, haplotypes of cultivars were deduced. Among the eight haplotypes, the H1 haplotype displayed the lowest content of butyl, hexyl, 2-methylbutyl, and pentyl acetates, while the H8 haplotype showed the highest. However, it is plausible that their method for determining haplotypes could not account for cultivars polymorphism. For example, the findings concerning 'Golden Delicious' showed inconsistencies. Although 'Golden Delicious' was heterozygous for all four SNPs, 'Jonagold' and 'Mutsu,' triploid cultivars whose seed parent is 'Golden Delicious,' demonstrated homozygous for all four SNPs. Since the triploid cultivar inherits $2n$-gamete from the seed parent (Evans et al. 2011; Moriya et al. 2011), 'Jonagold' and 'Mutsu' should be heterozygous for all four SNPs. Additionally, the haplotype composition of 'Golden Delicious' using the parentage information from 'Horei,' 'Rubinett,' and 'Elster' inferred all different haplotype sets for 'Golden Delicious.' Therefore, although the characterization method by Dunemann et al. (2012) could estimate the volatile content to a certain extent, it could not be applied to estimate the haplotype of the cultivar of interest accurately due to its hereditary contradiction of polymorphism. Cappellin et al. (2015) developed the $MdAAT1_{SSR}$ SSR marker from the 36 kb downstream of *MpAAT1* and investigated 124 cultivars for association analysis. Of the eight alleles observed, allele 201, whose frequency was 0.05, was connected to volatile content. Moreover, $MdAAT1_{SSR}$-201 presence was associated with an increase of volatile production other than acetate esters, which displayed a reduction.

Nonetheless, specific alleles of *MpAAT1* have been functionally characterized. Souleyre et al. (2014) identified many variants for *MpAAT1* from 'Royal Gala' and 'Granny Smith,' but only one allele from each cultivar was functional. Although these functional alleles are highly expressed in the cortex and skin of ripe fruit, *MpAAT1* protein was observed primarily on the skin. They also developed *MpAAT1* suppressed transgenic lines of 'Royal Gala' and investigated the influence on volatile content and related gene expression. The total ester production was significantly lower in transgenic lines while total alcohols and aldehydes, precursors of esters, increased. Additionally, threonine deaminase (*TD1*) and pyruvate decarboxylase (*PD1*) that are involved in the two ester biosynthesis pathways were upregulated in transgenic lines, whereas lipoxygenase (*LOX3*) and *AAT*s, in addition to *MpAAT1* such as *MpAT16* and *MdAP6* were down-regulated, probably resulting from the increased substrate pool.

The volatile content and flesh texture interactions have been illustrated through GWAS. Farneti et al. (2017) used 162 cultivars to conduct a multiple factor analysis between volatile profile and texture indices, including mechanical and acoustic measures. They found that cultivars that produce high aromatic volatile have low texture performance, while a contrary trend was observed in the low aromatic volatile production cultivars. Subsequently, volatile profiles were analyzed using the multivariate functional principal component analysis, indicating two main principal components (PC1 and PC2), which account for 50.1% and 12.3% of total phenotypic variance, respectively. GWAS for

PC1 and PC2 revealed that QTLs were associated with *MdPG1* and *MpAAT1*, respectively, signifying that these two genes independently contribute to the genetic regulation of volatile content and profile. Consequently, for this connection, Farneti et al. (2017) deduced that firm apples have intact and solid cell walls, which prevent the release of volatiles due to the titer cell compartment.

The QTL mapping approach, complementary study, and association analysis have gradually clarified the relationship between *MpAAT1* and the volatile compound. However, studies to develop a selection technique for apple aroma are still scarce. Therefore, we may have a long way to go in establishing molecular-assisted breeding of apple aroma. A significant drawback to realizing an ideal apple that harbors a good aroma is the negative correlations of volatile quantity and ethylene production/texture properties. Furthermore, a well-trained sensory evaluation panel will be required to harmonize sensorial perception and chemical volatile compound property. A dynamic measurement system with artificial chewing can be a practical approach to simulate the actual aroma pattern perceived in our mouth (Farneti et al. 2017). Although genes other than *AAT*, such as *LOX* and carboxylesterase (*CXE*), have been highlighted to be involved in volatile components (Dunemann et al. 2009; Vogt et al. 2013b; Kumar et al. 2015), including these genes in the genetic model of volatile metabolism is necessary.

2.4.5 Appearance

Consumers' first impression of the appearance of apples is influenced by their visual perception of the fruit's skin and flesh color, which helps their decision on whether or not the apple is delicious. From this background, apple fruits with proper and uniformly colored skin are traded at a high price in Japan, motivating growers to lay reflective sheets on the surface of their orchards and remove the leaves that cover fruit a few weeks before harvesting to stimulate coloring. The red flesh color is attractive to consumers due to its novelty and expected human health benefits due to the antioxidant activity. Although flesh browning is believed to be a stalking characteristic of apples, browning-resistant cultivars can expand the usage of processed apples. Moreover, physiological disorders such as russet and bitter pit can depreciate market value and consumer preference. Therefore, improving the appearance is a crucial breeding target and efforts for understanding the genetic architecture of these characteristics have been undertaken.

2.4.5.1 Fruit Skin Color

In red apple fruit skin, cyanidin 3-galactoside is the primary pigment accounting for more than 85% of the anthocyanin present and whose synthesis pathway has been reviewed in Honda and Moriya (2018). Before the introduction of molecular genetics in apple, multiple genetic models were proposed, which were built from one to a few genes. BSA brought significant progress in elucidating the genetic architecture of apple skin color determination. Cheng et al. (1996) conducted a BSA using the 'Rome Beauty' × 'White Angel' population and developed the BC226 SCAR marker, whose amplicon of 1,160 bp (A_1) was linked with red skin color in four segregating populations. This result supports the hypothesis that one dominant gene, *Rf*, determines the skin color. However, since the green-skinned fruit cultivar 'Granny Smith' showed an A_1 allele, the association with the BC226 SCAR marker is incomplete. Subsequently, BC226 was mapped onto the bottom of LG 9 of the first apple consensus genetic map (Maliepaard et al. 1998), revealing the LG of *Rf*. Moreover, GWAS repeatedly identified associations between fruit skin color and this locus (Kumar et al. 2013; Migicovsky et al. 2016; Amyotte et al. 2017; Moriya et al. 2017b; McClure et al. 2018; Minamikawa et al. 2021), demonstrating the strong driving force of *Rf* on fruit skin color.

The reverse genetics approach is employed to identify the causal gene of *Rf*. R2R3-MYB transcription factors are known as a regulator of anthocyanin biosynthesis in many plants (Naing and Kim 2018). Takos et al. (2006) and Ban et al. (2007) found an R2R3-type MYB transcription factor *MdMYB1* (*MdMYBA*) from the fruit skin of 'Cripp's Pink' and 'Tsugaru,' respectively. *MdMYB1* was mapped at the bottom of LG 9, close to the BC226 SCAR marker (Ban et al. 2007). The allele that contributes to the red coloration is known as *MdMYB1-1*. Complementary experiments on Arabidopsis and tobacco also verified their function in regulating anthocyanin biosynthesis. Moreover, it was elucidated that *MdMYB1* expression in apple upregulates the structural genes of the anthocyanin biosynthesis pathway, which results in red coloration on the fruit skin (Takos et al. 2006). The translated protein of *MdMYB1-3* allele that neither expresses in the fruit skin nor contributes to red skin coloration are also functionally indistinguishable from *MdMYB1-1*; therefore, the difference in the expression level of *MdMYB1* alleles principally influences skin color (Takos et al. 2006), which suggests that the regulatory mechanism of *MdMYB1* expression is pivotal for the anthocyanin synthesis. After translation, MYB forms a complex with a basic helix-loop-helix (bHLH) protein and a WD-repeat protein (Gonzalez et al. 2008).

Recently, a high-quality genome assembly of HFTH1, which is a trihaploid red-skinned cultivar 'Hanfu,' has been released (Zhang et al. 2019a). Comparing the genomic sequence of 'Hanfu' and GDDH13, a yellow fruit skin material (Daccord et al. 2017), was identified as a gypsy-like LTR retrotransposon insertion (4,097 bp) from the 3.3 kb upstream of the translation start site of the *MdMYB1-1* allele. The LTR retrotransposon has been regarded as an enhancer that regulates the development of red coloration by lowering the threshold of the light response (Zhang et al. 2019a). This insertion is also identified from *M. sieversii* accession, which is a wild ancestor of the cultivated apple from Xinjiang, China, indicating the ancient origin of this mutation. Interestingly, the enhancer activity of LTR retrotransposon to anthocyanin biosynthesis through the upregulation of R2R3 MYBs has also been observed in blood orange (Butelli et al. 2012) and grapevine (Azuma and Kobayashi 2022).

The analysis of *MdMYB1* expression revealed that this gene could explain various phenomena in fruit skin coloration. A chromosomal deletion, encompassing the *MdMYB1-1*, has been identified from 'Kogane Fuji,' a yellow-skinned spontaneous mutant of 'Fuji' (Igarashi et al. 2014). *MdMYB1-1* homozygotes develop redder fruit skin than heterozygotes (Chagné et al. 2016; Moriya et al. 2017b). Light and ambient temperature influences the red coloration of apple fruit skin (Ubi 2004) due to the induction of *MdMYB1* expression by UV-B and low temperature (Ban et al. 2007). *MdMYB1* expression level is also influenced by the degree of methylation, causing color variation among spontaneous mutations (El-Sharkawy et al. 2015), reddening of bag removed fruits after bagging treatment (Bai et al. 2016; Ma et al. 2019; Cho et al. 2022), and red stripe/solid surface formation due to anthocyanin concentration (Telias et al. 2011; Zhang et al. 2019a). Although the mechanism is still unknown, the differences in coloration ability have been suggested within the *MdMYB1-1* allele. A QTL for skin color on the *MdMYB1* locus has been identified from 'Akane,' which is an *MdMYB1-1* homozygous cultivar (Kunihisa et al. 2014). Furthermore, haplotype-based GWAS has placed a high genomic breeding value for coloration on the *MdMYB1-1* haplotype from 'Worcester Pearmain' (Minamikawa et al. 2021).

Several markers applicable to breeding have been developed. Although after the BC226 SCAR marker (Cheng et al. 1996), derived CAPS (dCAPS), and dominant STS markers were developed (Takos et al. 2006; Yuan et al. 2014; Zhang et al. 2014); however, these markers could not discriminate fruit colors from diverse collections. Subsequently, SNP-based markers and an SSR marker have also been developed (Chagné et al. 2016, 2019; Moriya et al. 2017b). So far, the most accurate marker may be a marker targeting the *MdMYB1-1* specific retrotransposon insertion, which behaves as a dominant marker (Zhang et al. 2019a). Therefore, a new primer set that behaves as a codominant marker should be developed. This conversion will enable breeders to select *MdMYB1-1* homozygous individuals with high adaptability to the warm climate region (Chagné et al. 2016).

2.4.5.2 Flesh Color

Generally, the flesh color of cultivated apples is white to yellowish white; however, there are numerous red flesh apples in genetic resources. Notably, two red flesh origins have been mentioned as follows: *Malus niedzwetzkyana* (type 1) and the *M.×domestica* cultivar 'Surprise' (type 2). Possible reasons for maintaining this natural mutation were their visual impact on humans and their rich antioxidant molecules, rewarding seed dispersers with improved fitness and increased longevity due to a healthier diet, enabling them more likely to visit these trees and disperse more seeds (Chagné et al. 2013).

The type 1 red flesh exhibits red pigment in many plant organs, including leaf, stem, petal, anther, fruit skin, the cortex, and core. Lewis and Crane (1938) described the inheritance of this unique phenotype as a single dominant gene responsible for this trait. Additionally, a causal gene has been identified through the reverse genetics approach. *MdMYB10* was identified from an expressed R2R3 MYB gene on the cortex of 'Red Field' (Espley et al. 2007). *MdMYB10* was later proven to be located on the bottom of LG 9 and was allelic to *MdMYB1* (Chagné et al. 2007; Lin-Wang et al. 2010). A specific polymorphism observed in *MdMYB10* is a tandemly repeated 23-bp mini-satellite motif, i.e., *MdMYB1* has one repeat, but *MdMYB10* has another five times repeats (Espley et al. 2009). Consequently, this 23-bp motif interacts with MdMYB10 and enhances its transcription, which upregulates the genes of the anthocyanin biosynthesis pathway, causing ectopic anthocyanin accumulation.

A DNA test is unnecessary in the breeding context when *MdMYB10* is targeted alone since this phenotype can be easily identified through the cotyledon color. Nevertheless, a marker using three primers has been developed (Chagné et al. 2007). In the actual breeding, consideration should be given to breeding type 1 red flesh since a correlation between type 1 red flesh and browning symptoms after cold storage has been reported from GWAS (Kumar et al. 2013). Moreover, a transgenic study on *MdMYB10*-expressed 'Royal Gala' has deduced a pleiotropic effect between flesh color and browning (Volz et al. 2013). Therefore, browning-resistant type 1 red flesh cultivars is likely a challenge for breeders.

In contrast to the type 1 red flesh, the type 2 red flesh accumulates anthocyanin only in the fruit cortex. The origin of type 2 red flesh was reviewed in Wang et al. (2018). The first genetic information about this phenotype is its linkage to the self-incompatibility locus *S* on LG 17 and the association between red flesh and the S_3 allele of self-incompatibility (Sekido et al. 2010). Two research groups pursued to identify causal gene. Chagné et al. (2013) have conducted a bi-parental QTL analysis using a progeny developed from crossing a type 2 red flesh cultivar 'Sandrago.' They listed 74 non-overlapping genes in the candidate region and focused on two R2R3 MYB genes (*MdMYB110a* and *MdMYB110b*) using genome assembly data (Velasco et al. 2010). Expression analysis that showed only *MdMYB110a* to be associated with the red flesh trait and functional characterization in tobacco indicated that *MdMYB110a* upregulates the anthocyanin biosynthesis pathway. Interestingly, when *MdMYB10*, a type 1 red flesh causal gene, was analyzed through BLAST to the 'Golden Delicious' genome assembly (Velasco et al. 2010), these two *MYB*s were the best matches. LGs 9 and 17 are derived from an ancient chromosomal duplication (Velasco et al. 2010), which suggests that these genes are duplicated from a single origin. Umemura et al. (2013) identified *MdMYB110a*_JP, which showed 99.3% homology with *MdMYB110a* from expressed genes of the cortex of 'Pink Pearl.' However, the molecular basis that constitutes this gene cortex-specific expression is still unknown.

Since type 2 red flesh apples possess better-eating quality than type 1 red flesh apples, and red-fleshed apples with yellow skin are only possible in type 2 red flesh, breeding efforts have been conducted in Germany, New Zealand, and Japan (Wang et al. 2018). A high-resolution melting (HRM) and dCAPS markers have been developed to be applied in breeding (Chagné et al. 2013; Sato et al. 2017). Moreover, the genetic linkage between *MdMYB110a* and the self-incompatibility locus *S* can contribute to the efficient cross combinations that reveal the vast majority of progeny, which display the red flesh trait (Umemura et al. 2013). Furthermore, the cross of type 1 × type 2 red flesh cultivars did not display individuals carrying both *MdMYB10* and *MdMYB110a*, suggesting the lethality of this allele combination (Hamada et al. 2015).

2.4.5.3 Enzymatic Browning

Enzymatic browning can occur on damaged apple fruits, such as cut surface, grating, and bruising. This phenomenon is triggered by the conversion of polyphenols to polymerized quinones and browning pigments due to the oxidation activity of polyphenol oxidase (PPO) (Murata et al. 1995). Browned organs can reduce consumer appeal and causes loss to the market value. Therefore, controlling browning is a critical subject in the fruit processing industry.

The development of browning-resistant apple cultivars has been conducted through genetic engineering and conventional breeding. Although bi-parental QTL analysis has been performed to identify QTLs controlling enzymatic browning on flesh or juice (Kenis et al. 2008; Mellidou et al. 2012; Di Guardo et al. 2013; Kunihisa et al. 2014, 2019; Sun et al. 2014), their validation is inadequate. Regarding genetic engineering, browning-resistant 'Golden Delicious,' 'Granny Smith,' and 'Fuji' have been developed by suppressing the PPO gene family using the RNAi technique (Xu 2015), and they are currently used commercially in the USA and Canada (Arctic™ apples 2022). In contrast, there is no example of a browning-resistant cultivar through project-based breeding. Here, 'Aori 27' shows exceptional browning-resistant characteristics, implying that browning does not occur for more than several days after being sliced or grated (Igarashi et al. 2016). It was developed by a crossing between 'Kinsei' ('Golden Delicious' × 'Delicious') × MaHe 7 (['Golden Delicious' × 'Indo'] × 'RedGold') at the Aomori Apple Experiment Station, Japan. Therefore, it is a useful material to elucidate the physiological and genetic mechanisms of browning because of its unique characteristic. Therefore, we will specifically focus on allele mining research on the non-browning trait of 'Aori 27.'

Polyphenol content and PPO activity have been investigated to identify the biochemical mechanism for the non-browning of 'Aori 27' (Holderbaum et al. 2010). The total polyphenol content of 'Aori 27' was the lowest among 'Fuji,' 'Elster,' and 'Mellow,' and PPO activity was also the weakest in the harvest period. Notably, no detectable level of epicatechin and the lowest level (1/130th of 'Fuji,' the second lowest among the investigated materials) of procyanidin B2 were found in 'Aori 27.'

The genetic analysis of the non-browning trait on 'Aori 27' has been conducted using bi-parental QTL analysis and GWAS. Specifically, the first attempt was the bi-parental QTL analysis using two relatively small (~50 individuals) populations (Tazawa et al. 2019). Assessing browning using grated flesh showed that one-third of individuals from the first population (F1–165 × 'Aori 27') displayed non-browning character; however, only two individuals from the second population ('Kotaro' × 'Aori 27') showed non-browning. QTL analyses for the degree of browning (15 min and 24 h after grating), total polyphenol content, and PPO activity resulted in identifying the colocation of all QTLs on the top of LG 16 of F1–165 and 'Kotaro,' where the *leucoanthocyanidin reductase 1* (*LAR1*) that controls polyphenol content is situated (Khan et al. 2012). Notably, contribution of the QTL for the degree of browning 24 h after grating was 73% in the first population. These results suggested that (1) the non-browning trait of 'Aori 27' is recessively controlled, (2) the *LAR1* locus significantly contributes to non-browning, and (3) genetic factors other than *LAR1* can be involved.

Next, GWAS was employed to examine the non-browning trait in populations with larger genetic backgrounds and cultivars (Kunihisa et al. 2021). Consequently, associations were identified on chromosomes 5, 16, and 17 using the degree of browning data (15 min and 24 h after grating) from 468 individuals, primarily from 24 full-sib populations. Additionally, PPO activity and total polyphenol content were also examined using 86 cultivars, leading to the identification of QTLs for the PPO activity (chromosomes 5 and 16), and total polyphenol content (chromosome 16). The major QTL region for the PPO activity on chromosome 5 contains seven to ten members of the PPO gene family. An early study found a QTL for chlorogenic acid content, which is a significant substrate of PPO, near the QTL on chromosome 17 (Chagné et al. 2012b). The causal gene(s) of this QTL was presumed to be *hydroxycinnamoyl CoA shikimate/quinate hydroxycinnamoyl transferase* (*HCT/HQT*), which is involved in the chlorogenic acid synthesis. Homozygous recessive alleles at these three QTLs were linked with minimal browning, and 'Aori 27' was the only cultivar with recessive homozygous alleles on all three QTLs. The origin of these recessive alleles is only the 'Golden Delicious' for QTL on chromosome 5, although there are multiple sources for those on chromosomes 16 and 17. This study indicates that browning-resistant apple is no more miraculous, namely, well-designed cross combinations can obtain non-browning individuals from materials with extensively available genetic resources such as 'Golden Delicious' and 'Delicious.' However, breeders could not achieve browning-resistant cultivar so far due to the tight linkage with other relevant traits (Kunihisa et al. 2021). For example, the QTL on chromosome 16 (*LAR1*) and *Ma*, a causative locus for fruit juice acidity, are closely linked (Khan et al. 2012). Currently, SNPs that associated with QTLs for browning have not been converted into breeder-friendly markers or formats for the high-throughput detection system; therefore, the development of DNA test method is required.

2.4.5.4 Other Physiological Disorders

Physiological disorders on apple fruits are poorly studied compared to disease resistance and fruit quality. Therefore, MAS for them is probably not widely used. A reason could be the interactive determination of disorder expression between genetic factors and environmental conditions such as precipitation and storage operation. Especially, the contribution of the storage operation makes mitigation treatment possible and loosening the genetic marker selection relevance. Moreover, it could be challenging to design proper experiments to obtain high-quality data for genetic analysis of physiological disorders since multiple factors are involved in disorder expression as described above and various fruits, trees, and location replicates are necessary, which are labor-intensive and time-consuming. Nevertheless, reducing the genetic risk of disorders using MAS generates vast benefits and is highly cost-effective for future breeding.

2.4.5.4.1 Bitter Pit

A bitter pit is defined as soft discrete pitting of the cortical flesh, which occurs during the pre-maturing stage and throughout the storage duration. Its occurrence causes a loss of market value. The susceptibility of the bitter pit is cultivar specific; however, the environment and cultural practices also influence the expression of this disorder (Ferguson and Watkins 1989). Applying calcium to fruits can partly control symptom expression but does not entirely prevent the symptoms (Lötze et al. 2008).

Previous studies have proposed two loci model where the dominant allele of both loci must be present to confer at least partial resistance (Korban and Swiader 1984). Currently, two examples of molecular genetic analysis of bitter pit inheritance exist. Kumar et al. (2013) conducted GWAS using 1,200 seedlings and identified a QTL on LG 16 that accounted for 11% of the phenotypic variance. Buti et al. (2015) conducted a bi-parental QTL analysis using the 'Redchief' (susceptible)×'X6688' (resistant) family. They detected a QTL at the exact position of LG 16, accounting for 50.9% of the phenotypic variance. The most significant SNP identified by Kumar et al. (2013) is located in the *LAR1* locus, which is a strong candidate gene for the polyphenol content of fruit (Khan et al. 2012). Additionally, transcriptome analyses have been performed to identify causal genes from this region. *MDP0000263725* and *MDP0000300083* (Buti et al. 2018) and *MdMADs5* and *MdIDL1* (Orcheski et al. 2021) have been found due to the different expression levels between intact and damaged fruits. Collectively, because only one locus has been detected for bitter pit, further studies are needed to elucidate whether the second locus is existed. The causal genetic factor located on LG 16 is of interest to understand the mechanism conferring bitter pit. More studies are necessary to develop the MAS method for this trait.

2.4.5.4.2 Fruit Cracking

Fruit cracking that occurs before or after harvesting reduces the marketability of fruits. However, no concrete cultural practices are available, and the molecular mechanism is poorly understood (Khadivi-Khub 2015). QTLs have been identified using bi-parental QTL analysis (Kunihisa et al. 2019) and GWAS (Kumar et al. 2013; Minamikawa et al. 2021). The contribution of QTLs is generally high. Specifically, a QTL identified by Kunihisa et al. (2019), who employed an F1 population obtained from crossing a fertile double haploid line, accounted for 83% of the phenotypic variation. These studies have identified QTLs at the *LAR1* location on LG 16, indicating that this gene or at least this region is involved in cracking.

2.4.5.4.3 Watercore

Watercore is a disorder with a dual nature. From a disadvantaged viewpoint, it induces a browning cortex during cold storage. Contrarily, the watercore-induced anaerobic conditions facilitate the production of ethyl esters, leading to an increased consumer preference compared to the no watercored fruits (Tanaka et al. 2016). Therefore, depending on industrial requirements and consumer demand and acceptance, two breeding directions (exclude or select watercored apples) are possible. Currently, two molecular genetic studies have been conducted, and both studies focused on the watercore origin 'Fuji.' Kunihisa et al. (2016) have tracked the inheritance of chromosomes among relatives of 'Fuji' using SNP markers heterozygous in 'Fuji.' They found an association with watercore on LG 14. Minamikawa et al. (2021) identified a major QTL on LG 14 and a minor QTL on LG 2 through haplotype-based GWAS. The QTL on LG 14 is colocated and originated from 'Delicious,' a watercore susceptible cultivar. Therefore, this QTL can be applied to at least the 'Delicious' relatives; however, neither the developing breeder-friendly tools nor identifying causal genes have been conducted so far.

2.4.5.4.4 Russet

Russet on fruit skin damages its appearance, resulting in the loss of market value. Bagging treatments have been conducted in some specific cultivars such as 'Kinsei' to prevent skin russeting. Genetic factors, climate conditions, and spraying of agrochemicals influence the occurrence of russet, although its genetic control is poorly elucidated. Falginella et al. (2015) have identified a causal locus of skin russeting of 'Renetta Grigia di Torriana,' which is an entirely russeting cultivar. The skin-russeting trait from 'Renetta Grigia di Torriana' was stable in its offspring, implying that Pearson's correlation coefficient value among 3 years of phenotyping ranged from 0.96 to 0.99. QTL analysis identified a major QTL (*Ru_RGT*) on LG 12, which explains the maximum 90.1% phenotypic variation. This high contribution enabled its fine mapping, delimiting the candidate region of the causal gene to 400 kb. The ABCG11 transporter gene involved in cuticle development was identified as a candidate gene from the gene annotation of the 'Golden Delicious' genome (Velasco et al. 2010). Interestingly, 17 cultivars with skin russeting were examined by SSR markers located in the candidate region, resulting in the identification of cultivars both sharing the russeting haplotype and those not sharing it. This result indicates multiple determinants in the full-skin-russeting trait in the genetic resources. On the other hand, Kunihisa et al. (2014), in a study using normal cultivars that cause skin russeting in a particular condition, identified QTLs for russeting on fruit calyx and pedicel on LGs 2 and 15, respectively, which resulted from the bi-parental QTL analysis of F1 population derived from the cross of 'Orin' × 'Akane.' Lashbrooke et al. (2015) also identified QTLs for russeting on LGs 2 and 15 using the F1 population of 'Golden Delicious' × 'Braeburn.' Moreover, Minamikawa et al. (2021) conducted a haplotype-based GWAS and identified associations with russeting on the body, pedicel, and calyx on chromosomes 1, 2, and 4, respectively. Therefore, more studies are necessary since validation and characterization of these QTLs are incomplete, and developing DNA test tools would take a long time.

2.4.5.4.5 Soft Scald and Soggy Breakdown

Soft scald on the skin and soggy breakdown on the cortex occur during cold storage. 'Honeycrisp' bred by the Minnesota University apple breeding program and recently planted actively in the USA and Canada, is vulnerable to these two disorders. Studies have been conducted on cultivation and storage practices to prevent these disorders. For example, late harvest date and storage temperature of 0°C–0.5°C are connected to the development of these disorders (Watkins et al. 2005). However, research interest in their genetic control has been growing due to the high narrow-sense heritability shown by soft scald (Alspach et al. 2000), and 'Honeycrisp' is currently a vital breeding material because of its crispy texture (Howard et al. 2018). Recent QTL identification studies have identified QTLs for soft scalds located at the top of LG 2 through bi-parental QTL analysis (McClure et al. 2016) and PBA (Howard et al. 2018). Howard et al. (2018) also identified QTL for the soggy breakdown in the same position on LG 2. These results show that soft scald and soggy breakdown may be under the same genetic control. Furthermore, Howard et al. (2018) identified a haplotype, which was consistently associated with the occurrence of these disorders, and 'Honeycrisp' was homozygous for this haplotype. This haplotype is also found in other breeding founders/materials such as 'Golden Delicious' and 'Cripps Pink.' Leisso et al. (2016) have observed an increased risk of soft scald occurrence associated with higher levels of hexanol in 'Honeycrisp.' These results suggest that *MpAAT1*, which regulates volatile metabolism, may be a promising candidate gene for QTL on LG 2 (McClure et al. 2016; Howard et al. 2018). Miller et al. (2021) developed a DNA test method based on the KASP™ assay using SNPs from a 20 K Infinium SNP array (Bianco et al. 2014). Since 'Honeycrisp' is a vital breeding parent, this marker will be relevant in a breeding program that adopts 'Honeycrisp' as a parent.

2.5 OTHER TARGETS

2.5.1 Self-Incompatibility

Apples show a self-incompatibility nature regulated by the gametophytic self-incompatibility mechanism (McClure 2004; Sassa 2016). One factor is located in the pistil (*S*-RNase), inhibiting pollen tube growth, and the other in the pollen (S-locus F-box brothers; SFBBs), which degrades *S*-RNase. A compatible reaction occurs when SFBBs successfully degrade all *S*-RNases in the pistil, while an incompatible reaction occurs when SFBBs' degradation is incomplete. The SFBB set of particular *S*-haplotype can degrade all *S*-RNases other than the self *S*-RNase. Therefore, they are called *S*-haplotypes since *S-RNase* and *SFBB*s are closely located in a chromosome and are empirically inherited concurrently. The self-incompatibility locus *S* was assigned to the bottom of LG 17 (Maliepaard et al. 1998).

The influence of self-incompatibility in apple cultivation is that growers must plant cultivars harboring different sets of *S*-haplotype in their orchards to guarantee fruit sets. In breeding, breeders should be concerned with the *S*-haplotype of the parents to prevent incompatible crossing. Therefore, keeping the diversity of *S*-haplotype in the breeding program is valuable to release new cultivars that harbor various *S*-haplotypes, produce fruit sets easily, and have the flexibility of crossing design. Moreover, self-incompatibility influences genetic research because researchers cannot design a self-fertilized segregating population, hampering to genetically analyze the recessive traits.

Since the molecular biology technique was introduced, *S*-haplotype determination has become drastically easier and faster. Traditionally, *S*-haplotype is determined using a pollination test that requires conducting several crosses. Nowadays, the polymorphism of nucleotide sequences in *S-RNase* usually determines the *S*-haplotypes. Generally, allele-specific PCR or CAPS systems described in various studies and summarized in the *S-RNase* database of apple have been conducted (http://www.agr.nagoya-u.ac.jp/~0hort/apple/; Matsumoto et al. 2011). *S*-haplotypes of more than 1,100 cultivars/selections are currently registered on the database. The number of haplotypes registered in the database is more than 40. Furthermore, improving the *S*-haplotype determination method has been further attempted by employing a non-allele-specific CAPS system (Kim et al. 2009), multiplex PCR followed by CAPS analysis combined with a fragment analysis detection method (Larsen et al. 2016), direct sequencing of PCR amplicons (Kasajima et al. 2017), closely located SSRs (López-Girona et al. 2021), and the HRM technique (López-Girona et al. 2021).

2.5.2 Pale Green Lethal (PGL) Disorder

PGL is characterized as a severe deficiency in chlorophyll (Klein et al. 1961). PGL-expressed plants stop growing at the stage of two to three true leaves and die less than 1 month after germination. PGL is exhibited when the plant harbors the causal gene *l* at the homozygous state (Klein et al. 1961); this implies that 25% of the seedling death depends on the cross combination. Essential breeding founders, including 'Golden Delicious,' 'Jonathan,' and 'Fuji,' carry *l* in the heterozygous state, making *l* prevalent in the breeding program. However, information about the cultivars/breeding materials that have never been used as breeding parents is scarce.

Genetic and physiological characterization of PGL has been conducted using a segregating population and genetic resources. Orcheski et al. (2015) conducted genetic mapping of *l* using the 'Cameo' × 'Jonathan' population, resulting in the mapping of *l* in a 1.7 cM (= ~700 kb) interval at the top of LG 16. They also screened the gene annotations in this region and recognized *MdPHYLLO1* as a candidate gene. The amino acid substitution of leucine (wild type) to proline (PGL) at position 1,502 caused by T to C transition was associated with the 70 PGL carrier apple cultivars. The PGL allele of *MdPHYLLO1* cannot produce phylloquinone (vitamin K1), essential for proper functioning of plants. Therefore, phylloquinone application can rescue PGL seedlings from chlorophyll deficiency (Orcheski et al. 2015). They developed a CAPS marker to detect the causal mutation, enabling breeders to assess whether the material of interest harbor *l*.

2.6 ROOTSTOCK TRAITS

Using rootstock in a commercial apple orchard is a prerequisite because the cultivated apple cultivars (*M.* × *domestica* Borkh.) cannot or rarely develop adventitious roots via conventional propagation techniques such as cutting and layering. Therefore, cultivated apples are propagated by grafting budwood onto the rootstock. Until the mid-1800s, cultivated apple seedlings were used as roostock; however, there are drawbacks such as very vigorous tree and poor uniformity and yield. Therefore, clonal rootstocks have been bred and selected in a modern orchard (Webster and Wertheim 2003; Wertheim and Webster 2003). The essential characteristics of apple rootstocks are the dwarfing and precocity-inducing effect of the scion cultivar, abiotic stress tolerance, and biotic stress resistance. In the 19th century, several clonal rootstocks, also known as 'Paradise,' were available in Europe; however, they were often a mixture of different genotypes and misclassified. Therefore, clarifying this confusion and breeding effort have been initiated at East Malling Research station by R. G. Hatton, which resulted in the identification of nine types of clonal rootstocks through phenotypic characterization (Hatton 1917, 1919). Of the identified clones, Type IX (Jaune de Metz, 'Malling 9,' or 'M. 9') remains the leading rootstock cultivar worldwide.

Apple rootstock breeding timescale is longer than that of the scion apple, which is usually more than 25 years from crossing to cultivar release. Because evaluating rootstock performance should be conducted with training and management the same as the commercial orchard, a long-term investigation using a broad area is required (Fazio 2021). Regarding the generation of rootstock, because of the abovementioned lengthy generation time, the latest cultivar, 'G.213' released from an active breeding program conducted in United States Department of Agriculture-Agricultural Research Service (USDA-ARS) in Geneva in the USA, is only two generations apart from 'M. 9' (Fazio 2021). Furthermore, the intensive labor required during rootstock performance tests is also a challenge for effective evaluation. Therefore, implementing a genome-designed breeding strategy including MAS is more beneficial than scion cultivar breeding based on skipping the initial phenotyping stage and reducing the number of trees investigated. This section describes several rootstock-related traits of which well-evaluated allele(s) have been identified, including the effects for dwarfing and precocious-inducing, easy propagation, soil-borne disease resistance, and harmful leafy decoloration.

2.6.1 Dwarfing/Precocity/Root Bark Ratio

Apple rootstock affects various attributes of the scion cultivar. An important factor is tree vigor that affects the labor and cropping efficiency through planting density. For example, an intensive planting of more than 3,000 trees per hectare combined with the tall spindle pruning and training system using dwarfing rootstock ('M. 9') enables 15–20 metric tons per hectare yield in the second year after planting feathered trees (Robinson et al. 2011). However, the inducing mechanism of the dwarfing on the grafted scions has been considered under polygenic control (Tydeman 1933). Different factors can be involved in dwarfing, such as endogenous factors, including phytohormone, photosynthesis, mineral transportation, water relation, and anatomical differences reviewed by Hayat et al. (2021) and the environment (Foster et al. 2016).

In the early stage of the mapping study, the dwarfing effect of the dwarfing rootstocks had been recognized as comparatively simply inheritable than assumed (Fazio and Mazzola 2004). The first mapping study was conducted with BSA and SSR markers using an F1 population (146 individuals) derived from the 'M. 9' (dwarfing) × 'Robusta 5' (vigorous) cross (Rusholme-Pilcher et al. 2008). This study postulated a single dominant gene, *Dwarfing 1* (*Dw1*), and prepared two replicates of DNA bulks from extremely dwarf and vigorous individuals according to their growth data until the fourth year. RAPD primers were screened, and 10 markers possibly linked to *Dw1* were obtained. Of these 10 markers, two were converted into SCAR markers. Additionally, 74 SSR markers (Liebhard et al. 2002) were selected to cascade almost the entire genome and subsequently screened their association with *Dw1*. Notably, possible linkage to *Dw1* was detected for the four markers on LG 5. However, the accurate mapping of *Dw1* was difficult with the extreme 100 individuals because the 12 individuals scored as vigorous by expert observation were carriers of a possible dwarfing allele from 'M. 9.' After separating those individuals, the remaining 88 individuals enabled a reliable positioning of *Dw1* on the middle of LG 5 near the CH03a09 SSR marker. Although this study was a significant achievement in studying the dwarfing effect of 'M. 9,' the mapping process implied that multiple genes are involved in the dwarfing effect.

Two research groups have identified the second locus through QTL mapping using genetic maps covering the entire LGs (Fazio et al. 2014; Foster et al. 2015). Fazio et al. (2014) applied an F1 population derived from the 'Ottawa 3' (dwarfing) × 'Robusta 5' cross containing 186 individuals. A QTL analysis confirmed *Dw1* and identified a new locus *Dw2* at the top of LG 11. In this population, *Dw1* and *Dw2* independently decreased the tree vigor. The second population was designed as both parents carrying *Dw1* and *Dw2* heterozygously (*Dw1/dw1* and *Dw2/dw2*), namely 'Geneva 935' × 'Budagovsky 9.' The analysis of growth characteristic on the second population detected an interaction of the two loci. The mean tree diameter of individuals homozygous for non-dwarfing alleles at either *Dw1* or *Dw2* locus did not differ from those of the non-dwarfing homozygous at either locus. Foster et al. (2015) employed the same population used in Rusholme-Pilcher et al. (2008), whose linkage map was improved by Celton et al. (2009). The QTL analysis performed for growth data obtained up to the seventh year from grafting resulted in the identification of *Dw1* and *Dw2* loci which was almost colocated with that of Fazio et al. (2014) and minor QTLs on LG 6, 9, 10 and 12. *Dw1* and *Dw2* accounted for 50.2% and 11.4% of phenotypic variation in this population. Moreover, the phenotypic evaluation using an enlarged population found individuals which showed dwarfing trees carrying *Dw1* alone; however, no individual showed dwarfing trees carrying *Dw2* alone. Therefore, they concluded that *Dw1* has a stronger effect than *Dw2*, and *Dw2* could be an enhancer of *Dw1*. In both studies, individuals carrying *Dw1* and *Dw2* showed a more potent dwarfing ability than other genotypes (Fazio et al. 2014; Foster et al. 2015).

Furthermore, a three-locus model has been proposed via the analysis of the root bark percentage (Harrison et al. 2016), which is a classical indicator of the dwarfing effect of apple rootstock (Beakbane and Thompson 1947). Root bark percentage data were obtained from a backcross-derived population (140 individuals of the 'M. 27' × 'M. 116' ['M. M. 106' × 'M. 27'] cross). Genetic maps of this population were constructed (Antanaviciute et al. 2012), and subsequently root bark percentage data was analyzed using non-parametric QTL mapping procedures. It identified three QTLs on *Rb1*, *Rb2*, and *Rb3* on LGs 5, 11, and 13, respectively. *Rb1* and *Rb2* were colocated with *Dw1* and *Dw2* previously identified, and *Rb3* was newly identified. Three newly developed SSR markers (EM_Rb1, EM_Rb2, and EM_Rb3) linked with these loci indicated allelic interactions, such as those with dominance and epistasis. Although the experiment was performed with a small subset of the same population, a negative correlation between root bark percentage and dwarfing intensity was detected. The dwarfing model, which includes the *Rb3* effect, also fits better than the excluded one. *Rb3* is suggested to be analogous to *Dw3*, according to these results. However, more examples are required for these three loci models to accurately understand the association between root bark percentage and dwarfing effect.

Regarding breeding use, the EM_Rb series markers provide suitable tools for marker-assisted breeding. Investigating the allelic status of *Rb1* to *Rb3* in other common rootstocks using EM_Rb1 to EM_Rb3 identified only 'M. 9', 'M. 26', and 'M. 27' showed dwarfing alleles on all three loci, proposing the practicality of these cultivars in breeding parents for dwarfing rootstocks (Harrison et al. 2016). Although 'Mac 9' and 'Ottawa 3' were not detected in any of the three loci, 'Ottawa 3' was applied as a dwarfing parent by Fazio et al. (2014). Therefore, additional studies should be conducted to clarify whether recombination occurred near the QTL loci in these materials and whether another allele involved in dwarfing existed. Recently, a study on dwarfing loci identification using the largest population has been posted in a preprint repository (Yang et al. 2021).

A factor that decides the precocious-inducing effect of rootstock is its tendency to photoassimilate and lead to reproductive or vegetative growth. Among the abovementioned dwarfing loci, *Dw1* and *Dw2* have been linked with precocious characteristics, facilitating the designation of QTLs as *Early bearing 1* (*Eb1*) and *Eb2* (Fazio et al. 2014; Foster et al. 2015). According to both studies, the parallel relationship implies that the more the dwarfing, the more precocious they become. It also indicates that plants harboring both *Dw1* (*Eb1*) and *Dw2* (*Eb2*) tend to photoassimilate reproductive growth rather than vegetative.

Although three loci already identified so far may have complicated interaction mechanisms, causal gene identification and functional annotation will shed light on detailed mechanisms for dwarfing. In contrast, tools for MAS have been well developed for dwarfing since the seedling dwarfing level can be estimated to some extent. *Dw1* was implemented in a MAS scheme of a rootstock breeding program in New Zealand (Bassett et al. 2015; Bus et al. 2017) and Japan. It is noteworthy that Fazio et al. (2014) provided molecular evidences of the availability of classical phenotypic indices, such as root bark percentage and mother plant vigor, that allow to estimate dwarfing effect. Although dwarfing and precocity-inducing effects are the most crucial breeding targets for apple rootstock breeding, they are the most time-consuming traits to evaluate. Therefore, current molecular markers are powerful enough for genome-design breeding, and an accurate understanding of the dwarfing mechanism will provide more power to them.

2.6.2 Propagation

The ease of propagation is listed as one of the most vital features of a nursery (Webster and Wertheim 2003). However, the mechanism underlying adventitious rooting in apple rootstock is commonly unknown. Despite the *Adventitious rooting related oxygenase-1* (*ARRO-1*) gene that is upregulated in rooting stem discs of the 'Jork 9' rootstock caused by auxin-treatment (Butler and Gallagher 1999), later studies have implied that this gene is involved in auxin homeostasis than rooting (Smolka et al. 2009; Li et al. 2012). This section will explain an adventitious rooting from hardwood cutting, a distinctive characteristic of *M. prunifolia*.

M. prunifolia Marubakaido is a clonal rootstock used in Japan for more than a century. It still has a 67% share by area of the Japanese apple orchard (Ministry of Agriculture, Forest and Fisheries 2022). Although Marubakaido is a vigorous rootstock resulting in decreasing cropping efficiency, growers in regions with heavy snowfall in winter prefer its use because it can prevent damage from snow in winter, i.e., invigorating trees are not damaged by snow. Additionally to the merit above, it can be easily propagated by hardwood cutting, which is valuable for fruit tree nurseries and growers who prepare saplings by themselves. NARO has concentrated on this attribute and has bred cutting-capable dwarfing rootstocks named 'JM1,' 'JM2,' 'JM5', 'JM7,' and 'JM8' selected from the Kiminoinuringo (synonym Morioka Seishi [JP No.169673])×'M. 9' cross (Soejima et al. 2010, 2013). Since cutting capability has been recognized as an inheritable acknowledgment of this breeding program, identification of QTL controlling hardwood cutting capability has been conducted (Moriya et al. 2015). They used the 120 individuals derived from the 'JM7'×*Malus sieboldii* Sanashi 63 (cutting incapable) cross and evaluated the rooting rate and root quantity from hardwood cutting. Rooting rate and root quantity demonstrated a significant positive correlation. QTL analysis was applied to a genetic linkage map constructed by SSR and AFLP markers (Moriya et al. 2012b), which led to the identification of a major QTL on LG 17, accounting for 40%–57% of the phenotypic variation. Further graphical genotype analysis, such as JM rootstocks and relatives, delimited the candidate region by approximately 535 kb on the 'Golden Delicious' genome (Velasco et al. 2010). In this region, two auxin response-related genes were recognized as candidate genes.

The cutting-capable allele of this locus obtained from Kinimoinuringo is very distinctive in the apple rootstock breeding program. Therefore, SSR markers used in this study can easily be applied to breeding. In the NARO apple rootstock breeding program, MAS has been conducted on this trait, and the stability of cutting capability has been confirmed on three generations apart from Kiminoinuringo (unpublished). The next step is the identification of the causal gene/mutation of cutting capability. Conversely, the quantitative trait nature of cutting ability and overlapping phenotypic value between plants carrying cutting-capable and -incapable alleles noticed in phenotypic assessment might make accurate mapping challenging. Complement studies regarding candidate genes may assist in explaining the mechanisms underlying hardwood cutting capability.

2.6.3 Crown Gall Resistance

Crown gall disease in apples is caused by soil-borne bacteria *Rhizobium radiobacter* (Ti) and *Rhizobium rhizogenes* (Ti). These bacteria are conventionally, or even currently in bioengineering area, known as *Agrobacterium tumefaciens* based on their tumor-inducing characteristics on a plant, which is the artificial classification. However, recent studies regarding their natural classification have revealed that tumorigenicity depends on the tumor-inducing (Ti) plasmid that can be transmitted among *Rhizobium* clade species, deducing the abovementioned naming system (Young et al. 2001). Generally, crown gall infection causes poor root development and blackening of stem tissue (Moore and Burr 2014). In particular, this disease significantly affects commercial nurseries since the infected saplings lose their value. Therefore, the breeding of crown gall-resistant rootstock is crucial for controlling this disease.

The plant genetic source of resistance has been screened from several rootstocks and wild relatives through an inoculation test (Moriya et al. 2008). Since a positive correlation between the rate of gall formation on the shoot and the root was demonstrated, evaluation using the shoot as opposed to the labor-intensive root can be adopted. Sanashi 63 and Mo-15, accessions from *M. sieboldii* were relatively resistant to all the three tested tumor-inducing *Rhizobium* isolates compared to other materials. An inoculation test on individuals obtained from the 'JM7'×Sanashi 63 cross was conducted, and monogenic or polygenic inheritance was suggested to be dependent on the isolate.

GSA (Patocchi et al. 2005) has been applied to genetically map monogenic factors inducing resistance to the Peach CG8331 isolate by simplifying gall-formation rate as binary data (resistant or susceptible) (Moriya et al. 2010). Consequently, crown gall resistance gene *Rrr1* (resistance to *Rhizobium rhizogenes* [Ti]) was mapped on LG 2 of Sanashi 63. Subsequently, QTL analysis was performed using the genetic map reconstructed using marker enrichment through a GBS technique called Genotyping by Random Amplicon Sequencing-Direct (GRAS-Di) (Moriya et al. 2021). However, for the Peach CG8331 isolate, no significant locus was detected other than the already detected *Rrr1* locus, which explains 57%–83% of phenotypic variation. For the other two isolates (Nagano 1 and Nagano 2), the *Rrr1* locus was detected again, accounting for ~45% of phenotypic variation. Additive QTLs were identified in the other three isolates on both LG 11 and 15 (for the ARAT-001 isolate) or only LG 11 (for ARAT-002 and Kazuno 2 isolates), which explains 15%–25% of the phenotypic variation. Furthermore, all QTLs were colocated and stable for the tested years. Since QTLs on LG11 and 15 were detected from the susceptible parent 'JM7,' recessive control of this resistance has been suggested.

The *Rrr1* locus has been fine mapped and sequenced (Moriya et al. 2021). Fine mapping was conducted using 34 plants that showed recombination near the locus selected from 3,306 plants derived from eight crosses. The estimated physical distance of the candidate region of *Rrr1* according to the GDDH13 genome (Daccord et al. 2017) was 196 kb. Additionally, two clones were identified from a BAC library constructed from an *Rrr1* carrying individual and sequenced using NGS technology, which resulted in identifying the complete nucleotide sequence of the candidate region spanning 217 kb. In this region, four TIR-NBS-LRR class disease resistance genes were predicted. Notably, this is the first report on the candidate resistance genes against *R. rhizogenes* (Ti) in a plant. Based on these results, MAS for *Rrr1* has been conducted in the NARO apple rootstock breeding program.

2.6.4 Virescent Locus

Chlorotic leaves at germination or bud break are the characteristics of virescent traits in apples, and subsequently, those restored green in the growing season. This feature is harmful because these plants will likely die or show weak growth due to poor photosynthesis. The trait was observed in plants obtained from the cross between the Malling series rootstocks and reported as recessive (Tydeman 1933; Crane and Lawrence 1934; Way et al. 1976; Haniuda et al. 1985).

An attempt to genetically map a *virescent* (*vir*) locus has been performed by Fernández-Fernández et al. (2014). The population applied was obtained from the 'M. 27' × 'M. 116' cross. Marker segregation data were also scored with the CP method (for outbreeder full-sib family) to map recessive traits. Only markers segregated in both parents were used for the analysis. Additionally, the phenotype was scored as h- and kk for the wild and virescent plants, respectively. Finally, the *vir* locus was mapped on LG 12 at the interval of the CH01g12 SSR marker and SNP 475880474 from the 8K IRSC apple SNP typing array (Chagné et al. 2012a). Alleles associated with *vir* were not only detected from 13 of the 27 rootstock accessions but all heterozygous, proposing the close genetic relationship among classical Malling series rootstocks. These two markers are very useful for breeding, and positional cloning is almost feasible when the plant materials are ready. However, it should be noted that the hidden advantage, which is linked with *vir*, might exist and may have resulted in the *vir* being retained in the current gene pool (Fernández-Fernández et al. 2014).

2.6.5 Future Targets

Many attributes of rootstock have not been studied or identified as useful alleles for breeding. Several fundamental studies have been reviewed (Fazio 2021). For example, resistance to apple replant disease, abiotic stress tolerance such as drought, and cold hardiness are vital in the specific apple-producing regions. Therefore, future studies will lead to developments in these areas.

Candidate genes controlling essential traits of apple rootstock are currently being identified. The next step will be to conduct complementary and/or knock-out studies on these genes and to understand the mechanisms underlying the trait of interest. These studies will promote molecular-based genomic designing and help reduce the number of plants evaluated and shorten the breeding period in apple rootstocks.

2.7 CLUSTERS OF ALLELES OF INTEREST

To date, numerous genes of interest have been mapped, identified, or cloned, including genes involved in disease and pest resistances and fruit quality traits. Several of these genes form clusters in a small region spanning a few megabases (Table 2.4). These colocations could result from pleiotropy, different alleles of the same gene, or closely linked genes. The trade-off and limitation induced by these gene clusters are pivotal for future molecular-assisted breeding aiming to achieve high productivity, consumer acceptance, and sustainable production.

In the case of pleiotropy, the allele of interest affects multiple traits simultaneously. For example, a situation in which harvest date and fruit firmness at harvest are both controlled by the *NAC18.1* gene is the case (Migicovsky et al. 2021). The trade-off derived from *NAC18.1* is that the earlier the harvest is, the softer the fruits. Therefore, to break the relationship between harvest date and fruit firmness conferred by *NAC18.1*, feasibility should be assessed for the mutation breeding of *NAC18.1* so that it alters only one attribute (harvest date or firmness). An alternative strategy to overcome this trade-off would be to employ other genes contributing to the trait of interest via genomic selection.

Alleles of the same gene can be expected when genes conferring similar traits are mapped in the same region, are originated from different sources, and are functionally different. In apple, examples of disease resistance genes are often found, such as *Er-1* and *Er-3* against the wooly apple aphid on LG 8 (Bus et al. 2008) and *Pl-2* and *Pl-m* against powdery mildew on LG 11 (Fernåndez-Fernåndez et al. 2008; Bus et al. 2010). To reveal the correct relationship, fine mapping and positional cloning are necessary. To achieve durable resistance, pyramiding resistance genes are regarded as crucial (Patocchi et al. 2020). In the breeding context, these (putative) allelic resistance genes are difficult to maintain and carry in one genotype through multiple generations without proper crossing design. A possible solution to this problem is to use genetic engineering approaches, such as genetic modification and genome editing; nevertheless, consumer acceptance remains a challenge (Lucht 2015). The mechanisms underlying allelic resistance genes should be checked because allelic resistance genes may have a common mechanism inducing resistance, which is discourage from pyramiding to avoid resistance breakdown.

Different genes that are closely linked are another cause of trait association. A cluster of disease resistance genes has been identified at the top of LG 11 of apple (Table 2.4) and, among them, *Rvi18* and *Alt* were conducted fine mapping (Soriano et al. 2014; Moriya et al. 2019a). The SSR sourced to develop the Rvi-18-SSR marker for *Rvi18* (Soriano et al. 2014) is coincidentally identical to that used to develop Mdo.chr11.3 for *Alt* (Moriya et al. 2019a), which is located at 2.693 Mb on chromosome 11 of the GDDH13 genome (Daccord et al. 2017). *Rvi18* and *Alt* are located on the proximal and distal side of Rvi-18-SSR/Mdo.chr11.3 and are within ~200 kb at a maximum. Resistance gene analogues are clustered in specific regions of the apple genome (Perazzolli et al. 2014), suggesting that the colocation of functional resistance genes is common in genus *Malus*. Another example is the colocation of *MdPG1* (Longhi et al. 2012) and *Co* (Wolters et al. 2013) in approximately 700 kb on chromosome 10, which affect the flesh texture/storability and columnar tree shape, respectively. In these cases, trait association is separable by meiotic recombination. In rice, *pi21*, a rice blast disease resistance gene, and Os04g0401400, a gene conferring poor grain flavor, were closely linked (26.6 kb) (Fukuoka et al. 2009). These genes were separated by meiotic recombination, resulting in the emergence of a new valuable haplotype capable of conferring resistance without decreasing the grain quality (Fukuoka et al. 2009). Once the new haplotype was developed in these closely linked genes, a positive linkage drag easily keeps the ideal haplotype through multiple generations and will facilitate breeding.

Attention should be paid to complexity of clusters on the LGs 16 and 17 in conventional breeding. Previous studies have revealed negative associations among traits for the LG 16 cluster. Kunihisa et al. (2019) reported that the low juice browning allele of *LAR1* is associated with the skin splitting allele and *Ma1*, which confers high acidity of juice. The presence of QTL conferring bitter pit

TABLE 2.4
Genes or QTLs Formed Clusters in the Genome

LG/Chrom. (Relationship)	Gene/QTL[a]	Trait	References
LG2 (ALG)	*Rvi4*	Apple scab resistance	Jänsch et al. (2015)
	Rvi15	Apple scab resistance	Galli et al. (2010c)
	Rrr1	Crown gall resistance	Moriya et al. (2021)
LG2 (PL?)	*MpAAT1*	Fruit aroma	Farneti et al. (2017) and others in Table 3.3
	Not specified	Soft scald	McClure et al. (2016)
	Not specified	Soggy breakdown	Howard et al. (2018)
LG2 (ALG)	*Rvi2*	Apple scab resistance	Jänsch et al. (2015)
	Rvi8	Apple scab resistance	Bus et al. (2005)
	Rvi9	Apple scab resistance	Bus et al. (2011)
	Rvi11	Apple scab resistance	Jänsch et al. (2015)
LG 3 (PL)	*NAC18.1*	Fruit firmness	Migicovsky et al. (2021)
	NAC18.1	Harvest date	Migicovsky et al. (2021)
LG 5 (PL)	*Dw1*	Dwarfing	Rusholme-Pilcher et al. (2008)
	Eb1	Early fruit bearing	Fazio et al. (2014)
	Rb1	Root bark percentage	Harrison et al. (2016)
LG 8 (ALG)	*Pl-w*	Powdery mildew resistance	Luo et al. (2019)
	Er-1	Wooly apple aphid resistance	Bus et al. (2008)
	Er-3	Wooly apple aphid resistance	Bus et al. (2008)
LG 10 (PL, ALG)	*MdPG1*	Texture, storability	Longhi et al. (2013b) and Farneti et al. (2017)
	MdPG1	Fruit aroma	Farneti et al. (2017)
	Co (*MdDOX-Co*)	Columnar tree form	Wolters et al. (2013)
LG 11 (PL, ALG)	*Rvi18*	Apple scab resistance	Sorinano et al. (2014)
	Apple scab QRLs	Apple scab resistance	Bénéjam et al. (2021) and others in Table 3.1
	Pl-2	Powdery mildew resistance	Fernández-Fernández et al. (2008)
	Pl-m	Powdery mildew resistance	Bus et al. (2010)
	Alt	Alternaria blotch resistance	Moriya et al. (2019a)
	qJ11	Crown gall resistance	Moriya et al. (2021)
	Dw2	Dwarf inducing ability	Fazio et al. (2014)
	Eb2	Early fruit bearing	Fazio et al. (2014)
	Rb2	Root bark percentage	Harrison et al. (2016)
LG 12 (ALG)	*Rvi1*	Apple scab resistance	Cova et al. (2015a)
	Pl-1	Powdery mildew resistance	Dunemann et al. (2007)
	Pl-d	Powdery mildew resistance	James et al. (2004)
	FB_E	Fire blight resistance	Parravicini et al. (2011)
	Not specified[b]	Fire blight resistance	Durel et al. (2009)
LG 16 (PL, ALG)	*LAR1*	Polyphenol content	Khan et al. (2012)
	LAR1	Flesh browning	Kunihisa et al. (2021)
	LAR1	Juice browning	Kunihisa et al. (2019)
	Bp-2	Bitter pit	Kumar et al. (2013) and Buti et al. (2015)
	Not specified	Skin splitting	Kumar et al. (2013) and Kunihisa et al. (2019)
	Ma1 (*MdALMTII*)	Malic acid content	Bai et al. (2012b)
	l (*MdPHYLLO1*)	Pale green lethal	Orcheski et al. (2015)

(Continued)

TABLE 2.4 (*Continued*)
Genes or QTLs Formed Clusters in the Genome

LG/Chrom. (Relationship)	Gene/QTL[a]	Trait	References
LG 17 (ALG)	*MdMYB110a*	Red flesh (Type 2)	Chagné et al. (2013)
	Rvi5	Apple scab resistance	Cova et al. (2015b)
	Not specified	Flesh browning	Kunihisa et al. (2021)
	S (*S-RNase*)	Self-incompatibility	Maliepaard et al. (1998)

[a] Including candidate genes.
[b] *Source*: *Malus floribunda* 821.
ALG, likely alleles or closely linked genes; PL, likely pleiotropy.

(Kumar et al. 2013; Buti et al. 2015) and *MdPHYLLO1* (Orcheski et al. 2015), a lethal factor, in this region further complicates the associations among traits and their outcomes. Two promising strategies to address the issue of this cluster are to seek valuable haplotypes resulting from historic recombination events (Tazawa et al. 2019) and the intentional selection of a novel haplotype from seedlings. The self-incompatibility locus (*S*) on LG 17 can either assist or hinder in obtaining the ideal genotype through a pollen-recognition mechanism. Cultivars showing the type 2 red flesh caused by *MdMYB110a* generally have the S_3 haplotype coupled with *MdMYB110a*. When utilizing it as a pollen donor in a crossing, selecting a maternal parent that shares one *S*-haplotype with a pollen donor other than S_3 (i.e., $S_xS_y \times S_xS_3$, where $x \neq 3$, $y \neq 3$) can lead to the majority of offspring carrying both S_3 and *MdMYB110a*, thus showing the red flesh trait. In contrast, the selection of a maternal parent carrying the S_3 haplotype should be avoided because all S_3 pollens are rejected and almost no red flesh individuals can be obtained.

2.8 CONCLUSIONS

The identification of genes controlling or contributing to a certain trait and the development of molecular marker tools from these genes for breeding purposes are of particular importance because apple breeding is a time-consuming and laborious process. Allele mining in apple has been conducted to investigate disease and pest resistances, fruit quality attributes, physiological disorders, tree behavior, and rootstock characteristics mainly through forward genetics approaches. Reverse genetics approaches employed for fruit skin and flesh color and storability have also been very successful in causal gene identification. These genes were confirmed their significance by the use of forward genetics approaches, such as QTL analysis and GWAS. The released genome assemblies and rapid advancement of sequencing technologies have allowed the fine mapping of numerous genes and identification of candidate genes, contributing to elucidating the genetic architectures and the mechanisms underlying the traits of interest. On the other hand, cloning of causal gene and their functionally complement and/or knock-out studies are still in progress. DNA markers developed through mapping studies can be directly used in breeding and be used to develop SNP-based high-throughput detection systems. Consequently, marker-assisted breeding is currently a standard option for apple breeders and will result in significant achievements in the next decades. Moreover, many clusters of genes of interest have been identified to date. A thorough understanding of the relationships among traits and revealing the linkage phase contribute to designing an ideal allele constitution of the cluster and avoid adverse results caused by linkage drags or associations among traits. Genetic engineering techniques, including cisgenesis and genome editing, will provide new insights to achieve new cultivars/breeding materials with the desired characteristics; nevertheless, consumer acceptance remains a challenge.

REFERENCES

Abe K, Iwanami H, Kotoda N, Moriya S, Takahashi S (2010) Evaluation of apple genotypes and *Malus* species for resistance to Alternaria blotch caused by *Alternaria alternata* apple pathotype using detached-leaf method. *Plant Breed* 129:208–218.

Abe K, Kotoda N, Kato H, Soejima J (2007) Resistance sources to Valsa canker (*Valsa ceratosperma*) in germplasm collection of diverse *Malus* species. *Plant Breed* 126:449–453.

Abe K, Kotoda N, Kato H, Soejima J (2011) Genetic studies on resistance to Valsa canker in apple: genetic variance and breeding values estimated from intra- and inter-specific hybrid progeny populations. *Tree Genet Genom* 7:363–372.

Aldwinckle HS, Way RD, Livermore KG (1976) Fire blight in the Geneva apple collection. *Fruit Varieties J* 30:42–55.

Alspach PA, White AG, Ferguson IB, Volz RK (2000) Genetic variability in apple fruit storage disorders. *Acta Hortic* 553:241–244.

Alston FH, Phillips KL, Evans KM (2000) A *Malus* gene list. *Acta Hortic* 538:561–570.

Amyotte B, Bowen AJ, Banks T, Rajcan I, Somers DJ (2017) Mapping the sensory perception of apple using descriptive sensory evaluation in a genome wide association study. *PLoS One* 12:e0171710.

Antanaviciute L, Fernández-Fernández F, Jansen J, Banchi E, Evans KM, Viola R, Velasco R, Dunwell JM, Troggio M, Sargent DJ (2012) Development of a dense SNP-based linkage map of an apple rootstock progeny using the *Malus* infinium whole genome genotyping array. *BMC Genom* 13:203.

Arctic™ Apples (2022) https://arcticapples.com/ Accessed on 5 July 2022.

Atkinson RG, Sutherland PW, Johnston SL, Gunaseelan K, Hallett IC, Mitra D, Brummell DA, Schroder R, Johnston JW, Schaffer RJ (2012) Down-regulation of *POLYGALACTURONASE1* alters firmness, tensile strength and water loss in apple (*Malus* × *domestica*) fruit. *BMC Plant Biol* 12:129.

Azuma A, Kobayashi S (2022) Demethylation of the 3′ LTR region of retrotransposon in *VvMYBA1*^{BEN} allele enhances anthocyanin biosynthesis in berry skin and flesh in 'Brazil' grape. *Plant Sci* 322:111341.

Bai S, Tuan PA, Saito T, Honda C, Hatsuyama Y, Ito A, Moriguchi T (2016) Epigenetic regulation of *MdMYB1* is associated with paper bagging-induced red pigmentation of apples. *Planta* 244:573–586.

Bai TH, Zhu YD, Fernández-Fernández F, Keulemans J, Brown S, Xu KN (2012a) Fine genetic mapping of the *Co* locus controlling columnar growth habit in apple. *Mol Genet Genom* 287:437–450.

Bai Y, Dougherty L, Li MJ, Fazio G, Cheng LL, Xu KN (2012b) A natural mutation-led truncation in one of the two aluminum-activated malate transporter-like genes at the *Ma* locus is associated with low fruit acidity in apple. *Mol Genet Genom* 287:663–678.

Baldi P, Patocchi A, Zini E, Toller C, Velasco R, Komjanc M (2004) Cloning and linkage mapping of resistance gene homologues in apple. *Theor Appl Genet* 109:231–239.

Baldi P, Wolters PJ, Komjanc M, Viola R, Velasco R, Salvi S (2013) Genetic and physical characterisation of the locus controlling columnar habit in apple (*Malus* × *domestica* Borkh.). *Mol Breed* 31:429–440.

Ban S, Xu KN (2020) Identification of two QTLs associated with high fruit acidity in apple using pooled genome sequencing analysis. *Hortic Res* 7:171.

Ban Y, Honda C, Hatsuyama Y, Igarashi M, Bessho H, Moriguchi T (2007) Isolation and functional analysis of a *MYB* transcription factor gene that is a key regulator for the development of red coloration in apple skin. *Plant Cell Physiol* 48:958–970.

Bassett H, Malone M, Ward S, Foster T, Chagné D, Bus V (2015) Marker assisted selection in an apple rootstock breeding family. *Acta Hortic* 1100:25–28.

Batle I, Alston FH (1996) Genes determining luecine aminopeptidase and mildew resistance from the ornamental apple, 'White Angel'. *Theor Appl Genet* 93:179–182.

Baumgartner IO, Kellerhals M, Costa F, Dondini L, Pagliarani G, Gregori R, Tartarini S, Leumann L, Laurens F, Patocchi A (2016) Development of SNP-based assays for disease resistance and fruit quality traits in apple (*Malus* × *domestica* Borkh.) and validation in breeding pilot studies. *Tree Genet Genom* 12:35.

Baumgartner IO, Patocchi A, Frey JE, Peil A, Kellerhals M (2015) Breeding elite lines of apple carrying pyramided homozygous resistance genes against apple scab and resistance against powdery mildew and fire blight. *Plant Mol Biol Rep* 33:1573–1583.

Beakbane AB, Thompson EC (1947) Anatomical studies of stems and roots of hardy fruit trees. IV. The root structures of some new clonal apple rootstocks budded with Cox's Orange Pippin. *J Pomol* 23:206–211.

Beers EH, Suckling DM, Prokopy RJ, Avilla J (2003) Ecology and management of apple arthropod pests. In: Ferree DC, Warrington IJ (eds) *Apples: botany, production and uses*. CABI, Wallingford, pp. 91–124.

Belfanti E, Silfverberg-Dilworth E, Tartarini S, Patocchi A, Barbieri M, Zhu J, Vinatzer BA, Gianfranceschi L, Gessler C, Sansavini S (2004) The *HcrVf2* gene from a wild apple confers scab resistance to a transgenic cultivated variety. *Proc Natl Acad Sci U S A* 101:886–890.

Belitz H-D, Grosch W, Schieberle P (eds) (2009) Carbohydrates. In: *Food chemistry*. Springer, Berlin, Heidelberg, pp. 248–339.

Ben Sadok I, Tiecher A, Galvez-Lopez D, Lahaye M, Lassere-Zuber P, Bruneau M, Hanteville S, Robic R, Cournol R, Laurens F (2015) Apple fruit texture QTLs: year and cold storage effects on sensory and instrumental traits. *Tree Genet Genom* 11:119.

Benaouf G, Parisi L (2000) Genetics of host-pathogen relationships between *Venturia inaequalis* races 6 and 7 and *Malus* species. *Phytopathol* 90:236–242.

Bénéjam J, Ravon E, Gaucher M, Brisset MN, Durel CE, Perchepied L (2021) Acibenzolar-S-methyl and resistance quantitative trait loci complement each other to control apple scab and fire blight. *Plant Dis* 105:1702–1710.

Bianco L, Cestaro A, Linsmith G, Muranty H, Denancé C, Théron A, Poncet C, Micheletti D, Kerschbamer E, Di Pierro EA, Larger S, Pindo M, Van de Weg E, Davassi A, Laurens F, Velasco R, Durel C-E, Troggio M (2016) Development and validation of the Axiom(r)Apple480K SNP genotyping array. *Plant J* 86:62–74.

Bianco L, Cestaro A, Sargent DJ, Banchi E, Derdak S, Di Guardo M, Salvi S, Jansen J, Viola R, Gut I, Laurens F, Chagné D, Velasco R, van de Weg E, Troggio M (2014) Development and validation of a 20K single nucleotide polymorphism (SNP) whole genome genotyping array for apple (*Malus × domestica* Borkh). *PLoS One* 9:e110377.

Biggs AR, Stensvand A (2014) Apple scab. In: Sutton TB, Aldwinckle HS, Agnello AM, Walgenbach JF (eds) *Compendium of apple and pear diseases and pests*, second edition. APS Press, St Paul, MN, pp. 8–11.

Bink MCAM, Jansen J, Madduri M, Voorrips RE, Durel CE, Kouassi AB, Laurens F, Mathis F, Gessler C, Gobbin D, Rezzonico F, Patocchi A, Kellerhals M, Boudichevskaia A, Dunemann F, Peil A, Nowicka A, Lata B, Stankiewicz-Kosyl M, Jeziorek K, Pitera E, Soska A, Tomala K, Evans K, Fernández FF, Guerra W, Korbin M, Keller S, Lewandowski M, Plocharski W, Rutkovski K, Zurawicz E, Costa F, Sansavini S, Tartarini S, Komjanc M, Mott D, Antofie A, Lateur M, Rondia A, Gianfranceschi L, Van de Weg EW (2014) Bayesian QTL analyses using pedigreed families of an outcrossing species, with application to fruit firmness in apple. *Theor Appl Genet* 127:1073–1090.

Broggini GAL, Wöhner T, Fahrentrapp J, Kost TD, Flachowsky H, Peil A, Hanke MV, Richter K, Patocchi A, Gessler C (2014) Engineering fire blight resistance into the apple cultivar 'Gala' using the *FB_MR5* CC-NBS-LRR resistance gene of *Malus × robusta* 5. *Plant Biotechnol J* 12:728–733.

Brown SK, Maloney KE, Hemmat M, Aldwinckle HS (2004) Apple breeding at Cornell: genetic studies of fruit quality, scab resistance and plant architecture. *Acta Hortic* 663:693–697.

Bus V, Singla G, Ward S, Brewer L, Morgan C, Bowatte D, Bassett H, Attfield B, Colhoun K, Bastiaanse H, Walter M, Scheper R, Fisher B, Won K, Montanari S, Volz R, Chagné D, Gardiner S (2017) Progress in pipfruit resistance breeding and research at Plant & Food Research. *Acta Hortic* 1172:7–14.

Bus VGM, Bassett HCM, Bowatte D, Chagné D, Ranatunga CA, Ulluwishewa D, Wiedow C, Gardiner SE (2010) Genome mapping of an apple scab, a powdery mildew and a woolly apple aphid resistance gene from open-pollinated Mildew Immune Selection. *Tree Genet Genom* 6:477–487.

Bus VGM, Chagné D, Bassett HCM, Bowatte D, Calenge F, Celton JM, Durel CE, Malone MT, Patocchi A, Ranatunga AC, Rikkerink EHA, Tustin DS, Zhou J, Gardiner SE (2008) Genome mapping of three major resistance genes to woolly apple aphid (*Eriosoma lanigerum* Hausm.). *Tree Genet Genom* 4:223–236.

Bus VGM, Rikkerink EHA, Caffier V, Durel C-E, Plummers KM (2011) Revision of nomenclature of the differential host-pathogen interactions of *Venturia inaequalis* and *Malus*. *Ann Rev Phytopathol* 49:391–413.

Bus VGM, Rikkerink EHA, van de Weg WE, Rusholme RL, Rikkerink E, Gardiner S (2005) The *Vh8* locus of a new gene for gene interaction between *Venturia inaequalis* and the wild apple *Malus sieversii* is closely linked to the *Vh2* locus in *Malus pumila* R12740–7A. *New Phytol* 166:1035–1049.

Bus VGM, Scheper RWA, Walter M, Campbell RE, Kitson B, Turner L, Fisher BM, Johnston SL, Wu C, Deng CH, Singla G, Bowatte D, Jesson LK, Hedderley DI, Volz RK, Chagné D, Gardiner SE (2019) Genetic mapping of the European canker (*Neonectria ditissima*) resistance locus *Rnd1* from *Malus* 'Robusta 5'. *Tree Genet Genomes* 15:25.

Butelli E, Licciardello C, Zhang Y, Liu J, Mackay S, Bailey P, Reforgiato-Recupero G, Martin C (2012) Retrotransposons control fruit-specific, cold-dependent accumulation of anthocyanins in blood oranges. *Plant Cell* 24:1242–1255.

Buti M, Poles L, Caset D, Magnago P, Fernández FF, Colgan RJ, Velasco R, Sargent DJ (2015) Identification and validation of a QTL influencing bitter pit symptoms in apple (*Malus × domestica*). *Mol Breed* 35:29.

Buti M, Sargent DJ, Bianco L, Magnago P, Velasco R, Colgan RJ (2018) A study of gene expression changes at the *Bp-2* locus associated with bitter pit symptom expression in apple (*Malus pumila*). *Mol Breed* 38:85.

Butler ED, Gallagher TF (1999) Isolation and characterization of a cDNA encoding a novel 2-oxoacid-dependent dioxygenase which is up-regulated during adventitious root formation in apple (*Malus domestica* 'Jork 9') stem discs. *J Exp Bot* 50:551–552.

Caffier V, Lasserre-Zuber P, Giraud M, Lascostes M, Stievenard R, Lemarquand A, van de Weg E, Expert P, Denancé C, Didelot F, Le Cam B, Durel C-E (2014) Erosion of quantitative host resistance in the apple × *Venturia inaequalis* pathosystem. *Infect Genet E* 27:481–489.
Calenge F, Drouet D, Denancé C, Van de Weg WE, Brisset MN, Paulin JP, Durel CE (2005) Identification of a major QTL together with several minor additive or epistatic QTLs for resistance to fire blight in apple in two related progenies. *Theor Appl Genet* 111:128–135.
Calenge F, Durel CE (2006) Both stable and unstable QTLs for resistance to powdery mildew are detected in apple after four years of field assessments. *Mol Breed* 17:329–339.
Calenge F, Faure A, Goerre M, Gebhardt C, Van de Weg WE, Parisi L, Durel CE (2004) Quantitative trait loci (QTL) analysis reveals both broad-spectrum and isolate-specific QTL for scab resistance in an apple progeny challenged with eight isolates of *Venturia inaequalis*. *Phytopathology* 94:370–379.
Cappellin L, Farneti B, Di Guardo M, Busatto N, Khomenko I, Romano A, Velasco R, Costa G, Biasioli F, Costa F (2015) QTL analysis coupled with PTR-ToF-MS and candidate gene-based association mapping validate the role of Md-AAT1 as a major gene in the control of flavor in apple fruit. *Plant Mol Biol Rep* 33:239–252.
Celton JM, Tustin DS, Chagné D, Gardiner SE (2009) Construction of a dense genetic linkage map for apple rootstocks using SSRs developed from *Malus* ESTs and *Pyrus* genomic sequences. *Tree Genet Genomes* 5:93–107.
Cevik V, King GJ (2002a) High-resolution genetic analysis of the *Sd-1* aphid resistance locus in Malus spp. *Theor Appl Genet* 105:346–354.
Cevik V, King GJ (2002b) Resolving the aphid resistance locus *Sd-1* on a BAC contig within a sub-telomeric region of *Malus* linkage group 7. *Genome* 45:939–945.
Chagné D, Carlisle CM, Blond C, Volz RK, Whitworth C, Oraguzie NC, Crowhurst RN, Allan AC, Espley RV, Hellens RP, Gardiner S (2007) Mapping a candidate gene (MdMYB10) for red flesh and foliage colour in apple. *BMC Genom* 8:212–222.
Chagné D, Crowhurst RN, Troggio M, Davey MW, Gilmore B, Lawley C, Vanderzande S, Hellens RP, Kumar S, Cestaro A, Velasco R, Main D, Rees JD, Lezzoni A, Mockler T, Wilhelm L, Van de Weg E, Gardiner SE, Bassil N, Peace C (2012a) Genome-wide SNP detection, validation, and development of an 8K SNP array for apple. *PLoS One* 7:e31745.
Chagné D, Dayatilake D, Diack R, Oliver M, Ireland H, Watson A, Gardiner SE, Johnston JW, Schaffer RJ, Tustin S (2014) Genetic and environmental control of fruit maturation, dry matter and firmness in apple (*Malus × domestica* Borkh.). *Hortic Res* 1:14046.
Chagné D, Kirk C, How N, Whitworth C, Fontic C, Reig G, Sawyer G, Rouse S, Poles L, Gardiner SE (2016) A functional genetic marker for apple red skin coloration across different environments. *Tree Genet Genom* 12:1–9.
Chagné D, Krieger C, Rassam M, Sullivan M, Fraser J, Andre C, Pindo M, Troggio M, Gardiner SE, Henry RA, Allan AC, McGhie TK, Laing WA (2012b) QTL and candidate gene mapping for polyphenolic composition in apple fruit. *BMC Plant Biol* 12:12.
Chagné D, Kui LW, Espley RV, Volz RK, How NM, Rouse S, Brendolise C, Carlisle CM, Kumar S, De Silva N, Micheletti D, McGhie T, Crowhurst RN, Storey RD, Velasco R, Hellens RP, Gardiner SE, Allan AC (2013) An ancient duplication of apple MYB transcription factors is responsible for novel red fruit-flesh phenotypes. *Plant Physiol* 161:225–239.
Chagné D, Vanderzande S, Kirk C, Profitt N, Weskett R, Gardiner SE, Peace CP, Volz RK, Bassil NV (2019) Validation of SNP markers for fruit quality and disease resistance loci in apple (*Malus × domestica* Borkh.) using the OpenArray(r) platform. *Hortic Res* 6:30.
Chen XL, Li SM, Zhang D, Han MY, Jin X, Zhao CP, Wang SB, Xing LB, Ma JJ, Ji JJ, An N (2019) Sequencing of a wild apple (*Malus baccata*) genome unravels the differences between cultivated and wild apple species regarding disease resistance and cold tolerance. *G3-Gene Genom Genet* 9:2051–2060.
Cheng FS, Weeden NF, Brown SK (1996) Identification of co-dominant RAPD markers tightly linked to fruit skin color in apple. *Theor Appl Genet* 93:222–227.
Cheng FS, Weeden NF, Brown SK, Aldwinckle HS, Gardiner SE, Bus VG (1998) Development of a DNA marker for V_m, a gene conferring resistance to apple scab. *Genome* 41:208–214.
Chevalier M, Lespinasse Y, Renaudin S (1991) A microscopic study of the different classes of symptoms coded by the *Vf* gene in apple for resistance to scab (*Venturia inaequalis*). *Plant Pathol* 40:249–256.
Cho HJ, Han AR, Choi C (2022) Effect of paper-bagging on apple skin patterning associated with *MdMYB10* promoter methylation. *Int J Mol Sci* 23:3319.
Conner PJ, Brown SK, Weeden NF (1997) Randomly amplified polymorphic DNA-based genetic linkage maps of three apple cultivars. *J Am Soc Hortic Sci* 122:350–359.

Costa F, Cappellin L, Zini E, Patocchi A, Kellerhals M, Komjanc M, Gessler C, Biasioli F (2013) QTL validation and stability for volatile organic compounds (VOCs) in apple. *Plant Sci* 211:1–7.

Costa F, Peace CP, Stella S, Serra S, Musacchi S, Bazzani M, Sansavini S, Van de Weg WE (2010) QTL dynamics for fruit firmness and softening around an ethylene-dependent polygalacturonase gene in apple (*Malus* × *domestica* Borkh.). *J Exp Bot* 61:3029–3039.

Costa F, Stella S, Van de Weg WE, Guerra W, Cecchinel M, Dallavia J, Koller B, Sansavini S (2005) Role of the genes *Md-ACO1* and *Md-ACS1* in ethylene production and shelf life of apple (*Malus domestica* Borkh). *Euphytica* 141:181–190.

Costa F, Van de Weg WE, Stella S, Dondini L, Pratesi D, Musacchi S, Sansavini S (2008) Map position and functional allelic diversity of *Md-Exp7*, a new putative expansin gene associated with fruit softening in apple (*Malus* × *domestica* Borkh.) and pear (*Pyrus communis*). *Tree Genet Genom* 4:575–586.

Cova V, Bandara NL, Liang W, Tartarini S, Patocchi A, Troggio M, Velasco R, Komjanc M (2015b) Fine mapping of the *Rvi5* (*Vm*) apple scab resistance locus in the 'Murray' apple genotype. *Mol Breed* 35:200.

Cova V, Lasserre-Zuber P, Piazza S, Cestaro A, Velasco R, Durel CE, Malnoy M (2015a) High-resolution genetic and physical map of the *Rvi1* (*Vg*) apple scab resistance locus. *Mol Breed* 35:16.

Crane MB, Lawrence WJC (1934) Genetical studies in cultivated apples. *J Genet* 28:265–296.

Crosby JA, Janick J, Pecknold PC, Korban SS, O'Conner PA, Ries SM, Goffreda J, Voordeckers A (1992) Breeding apples for scab resistance: 1945–1990. *Acta Hortic* 317:43–70.

Daccord N, Celton JM, Linsmith G, Becker C, Choisne N, Schijlen E, van de Geest H, Bianco L, Micheletti D, Velasco R, Di Pierro EA, Gouzy J, Rees DJG, Guerif P, Muranty H, Durel CE, Laurens F, Lespinasse Y, Gaillard S, Aubourg S, Quesneville H, Weigel D, van de Weg E, Troggio M, Bucher E (2017) High-quality *de novo* assembly of the apple genome and methylome dynamics of early fruit development. *Nat Genet* 49:1099–1106.

Dall'Agata M, Pagliarani G, Padmarasu S, Troggio M, Bianco L, Dapena E, Minarro M, Aubourg S, Lespinasse Y, Durel CE, Tartarini S (2018) Identification of candidate genes at the *Dp-fl* locus conferring resistance against the rosy apple aphid *Dysaphis plantaginea*. *Tree Genet Genom* 14:12.

Dayton DF (1977) Genetic immunity to apple mildew incited by *Podosphaera leucotricha*. *HortScience* 12:225–226.

Dayton DF, Williams EB (1968) Independent genes in *Malus* for resistance to *Venturia inaequalis*. *Proc Am Soc Hortic Sci* 92:89–94.

Dayton DF, Williams EB (1970) Additional allelic genes in *Malus* for scab resistance of two reaction types. *J Am Soc Hortic Sci* 95:735–736.

De Franceschi P, Cova V, Tartarini S, Dondini L (2016) Characterization of a new apple *S*-RNase allele and its linkage with the *Rvi5* gene for scab resistance. *Mol Breed* 36:7.

Di Guardo M, Bink M, Guerra W, Lozano TLL, Busatto N, Poles L, Tadiello A, Bianco L, Visser RGF, van de Weg E, Costa F (2017) Deciphering the genetic control of fruit texture in apple by multiple family-based analysis and genome-wide association. *J Exp Bot* 68:1451–1466.

Di Guardo M, Tadiello A, Farneti B, Lorenz G, Masuero D, Vrhovsek U, Costa G, Velasco R, Costa F (2013) A multidisciplinary approach providing new insight into fruit flesh browning physiology in apple (*Malus* × *domestica* Borkh.). *PLoS One* 8:e78004.

Dixon J, Hewett EW (2000) Factors affecting apple aroma/flavour volatile concentration: a review. *N Z J Crop Hortic Sci* 28:155–173.

Dougherty L, Bai T, Brown S, Xu KN (2020) Exploring DNA variant segregation types enables mapping loci for recessive phenotypic suppression of columnar growth in apple. *Front Plant Sci* 11:692.

Dougherty L, Singh R, Brown S, Dardick C, Xu KN (2018) Exploring DNA variant segregation types in pooled genome sequencing enables effective mapping of weeping trait in *Malus*. *J Exp Bot* 69:1499–1516.

Dunemann F, Peil A, Urbanietz A, Garcia-Libreros T (2007) Mapping of the apple powdery mildew resistance gene *Pl1* and its genetic association with an NBS-LRR candidate resistance gene. *Plant Breed* 126:476–481.

Dunemann F, Schuster M (2009) Genetic characterization and mapping of the major powdery mildew resistance gene *Plbj* from *Malus baccata jackii*. *Acta Hortic* 814:791–797.

Dunemann F, Ulrich D, Boudichevskaia A, Grafe C, Weber WE (2009) QTL mapping of aroma compounds analysed by headspace solid-phase microextraction gas chromatography in the apple progeny 'Discovery' × 'Prima'. *Mol Breed* 23:501–521.

Dunemann F, Ulrich D, Malysheva-Otto L, Weber WE, Longhi S, Velasco R, Costa F (2012) Functional allelic diversity of the apple alcohol acyl-transferase gene *MdAAT1* associated with fruit ester volatile contents in apple cultivars. *Mol Breed* 29:609–625.

Durel CE, Denancé C, Brisset MN (2009) Two distinct major QTL for resistance to fire blight co-localize on linkage group 12 in apple genotypes 'Evereste' and *Malus floribunda* clone 821. *Genome* 52:139–147.

Durel CE, Parisi L, Laurens F, Van de Weg WE, Liebhard R, Jourjon MF (2003) Genetic dissection of partial resistance to race 6 of *Venturia inaequalis* in apple. *Genome* 46:224–234.
Durham RE, Korban SS (1994) Evidence of gene introgression in apple using RAPD markers. *Euphytica* 79:109–114.
El-Sharkawy I, Liang D, Xu K (2015) Transcriptome analysis of an apple (*Malus × domestica*) yellow fruit somatic mutation identifies a gene network module highly associated with anthocyanin and epigenetic regulation. *J Exp Bot* 66:7359–7376.
Emeriewen O, Richter K, Kilian A, Zini E, Hanke MV, Malnoy M, Peil A (2014) Identification of a major quantitative trait locus for resistance to fire blight in the wild apple species *Malus fusca*. *Mol Breed* 34:407–419.
Emeriewen OF, Flachowsky H, Peil A (2021b) Characterization of genomic DNA sequence of the candidate gene for *FB_Mfu10* associated with fire blight resistance in *Malus* species. *BMC Res Notes* 14:291.
Emeriewen OF, Peil A, Richter K, Zini E, Hanke MV, Malnoy M (2017) Fire blight resistance of *Malus × arnoldiana* is controlled by a quantitative trait locus located at the distal end of linkage group 12. *Eur J Plant Pathol* 148:1011–1018.
Emeriewen OF, Richter K, Berner T, Keilwagen J, Schnable PS, Malnoy M, Peil A (2020) Construction of a dense genetic map of the *Malus fusca* fire blight resistant accession MAL0045 using tunable genotyping-by-sequencing SNPs and microsatellites. *Sci Rep* 10:16358.
Emeriewen OF, Richter K, Flachowsky H, Malnoy M, Peil A (2021a) Genetic analysis and fine mapping of the fire blight resistance locus of *Malus × arnoldiana* on linkage group 12 reveal first candidate genes. *Front Plant Sci* 12:667133.
Emeriewen OF, Richter K, Piazza S, Micheletti D, Broggini GAL, Berner T, Keilwagen J, Hanke MV, Malnoy M, Peil A (2018) Towards map-based cloning of *FB_Mfu10*: identification of a receptor-like kinase candidate gene underlying the *Malus fusca* fire blight resistance locus on linkage group 10. *Mol Breed* 38:106.
Erdin N, Tartarini S, Broggini GAL, Gennari F, Sansavini S, Gessler C, Patocchi A (2006) Mapping of the apple scab-resistance gene *Vb*. *Genome* 49:1238–1245.
Espley RV, Brendolise C, Chagné D, Kutty-Amma S, Green S, Volz R, Putterill J, Schouten HJ, Gardiner SE, Hellens RP, Allan AC (2009) Multiple repeats of a promoter segment causes transcription factor autoregulation in red apples. *Plant Cell* 21:168–183.
Espley RV, Hellens RP, Putterill J, Stevenson DE, Kutty-Amma S, Allan AC (2007) Red colouration in apple fruit is due to the activity of the MYB transcription factor, MdMYB10. *Plant J* 49:414–427.
Evans KM, James CM (2003) Identification of SCAR markers linked to *Pl-w* mildew resistance in apple. *Theor Appl Genet* 106:1178–1183.
Evans KM, Patocchi A, Rezzonico F, Mathis F, Durel CE, Fernández-Fernández F, Boudichevskaia A, Dunemann F, Stankiewicz-Kosyl M, Gianfranceschi L, Komjanc M, Lateur M, Madduri M, Noordijk Y, van de Weg WE (2011) Genotyping of pedigreed apple breeding material with a genome-covering set of SSRs: trueness-to-type of cultivars and their parentages. *Mol Breed* 28:535–547.
Fahrentrapp J, Broggini GAL, Kellerhals M, Peil A, Richter K, Zini E, Gessler C (2013) A candidate gene for fire blight resistance in *Malus × robusta* 5 is coding for a CC-NBS-LRR. *Tree Genet Genom* 9:237–251.
Falginella L, Cipriani G, Monte C, Gregori R, Testolin R, Velasco R, Troggio M, Tartarini S (2015) A major QTL controlling apple skin russeting maps on the linkage group 12 of 'Renetta Grigia di Torriana'. *BMC Plant Biol* 15:150.
Fan Z, Hasing T, Johnson TS, Garner DM, Barbey CR, Colquhoun TA, Sims CA, Resende MFR, Whitaker VM (2021) Strawberry sweetness and consumer preference are enhanced by specific volatile compounds. *Horti Res* 8:66.
Food and Agriculture Organization Corporate Statistical Database (2022) Food and Agriculture Organization of the United Nations https://www.fao.org/faostat/en/#data Accessed on 25 May 2022.
Farneti B, Di Guardo M, Khomenko I, Cappellin L, Biasioli F, Velasco R, Costa F (2017) Genome-wide association study unravels the genetic control of the apple volatilome and its interplay with fruit texture. *J Exp Bot* 68:1467–1478.
Fazio G (2021) Genetics, breeding, and genomics of apple rootstocks. In: Korban SS (ed) *The apple genome, compendium of plant genomes*. Springer Nature Switzerland, Cham, pp. 105–130.
Fazio G, Mazzola M (2004) Target traits for the development of marker assisted selection of apple rootstocks - prospects and benefits. *Acta Hortic* 663:823–827.
Fazio G, Wan YZ, Kviklys D, Romero L, Adams R, Strickland D, Robinson T (2014) *Dw2*, a new dwarfing locus in apple rootstocks and its relationship to induction of early bearing in apple scions. *J Am Soc Hortic Sci* 139:87–98.

Ferguson IB, Watkins CB (1989) Bitter pit in apple fruit. *Hortic Rev* 11:289–355.
Fernández-Fernández F, Evans KM, Clarke JB, Govan CL, James CM, Maric S, Tobutt KR (2008) Development of an STS map of an interspecific progeny of *Malus*. *Tree Genet Genomes* 4:469–479.
Fernández-Fernández F, Padmarasu S, Surbanovski N, Evans KM, Tobutt KR, Sargent DJ (2014) Characterisation of the virescent locus controlling a recessive phenotype in apple rootstocks (*Malus pumila* Mill.). *Mol Breed* 33:373–383.
Fiala JL (1994) *Flowering crabapples*. Timber Press, Portland, OR.
Fisher DV (1995) The 'Wijick Spur McIntosh'. *Fruit Varieties J* 49:212–213.
Foster TM, Celton JM, Chagné D, Tustin SD, Gardiner SE (2015) Two quantitative trait loci, *Dw1* and *Dw2*, are primarily responsible for rootstock-induced dwarfing in apple. *Hortic Res* 2:15001.
Foster TM, van Hooijdonk BM, Friend AP, Seleznyova1 AN, McLachlan ARG (2016) Apple rootstock-induced dwarfing is strongly influenced by growing environment. *J Hortic* 3:3.
Frank RA, Byram J (1988) Taste-smell interactions are tastant and odorant dependent. *Chem Senses* 13:445–455.
Fukuoka S, Saka N, Koga H, Ono K, Shimizu T, Ebana K, Hayashi N, Takahashi A, Hirochika H, Okuno K, Yano M (2009) Loss of function of a proline-containing protein confers durable disease resistance in rice. *Science* 325:998–1001.
Galli P, Broggini GAL, Gessler C, Patocchi A (2010a) Phenotypic characterization of the *Rvi15* (*Vr2*) apple scab resistance. *J Plant Pathol* 92:219–226.
Galli P, Broggini GAL, Kellerhals M, Gessler C, Patocchi A (2010b) High-resolution genetic map of the *Rvi15* (*Vr2*) apple scab resistance locus. *Mol Breed* 26:561–572.
Galli P, Patocchi A, Broggini GAL, Gessler C (2010c) The *Rvi15* (*Vr2*) apple scab resistance locus contains three TIR-NBS-LRR genes. *Mol Plant Microbe Interact* 23:608–617.
Gallott JC, Lamb RC, Aldwinckle HS (1985) Resistance to powdery mildew from some small-fruited *Malus* cultivars. *HortScience* 20:1085–1087.
Gardiner S, Bus V, Rusholme R, Chagné D, Rikkerink E (2007) Apple. In: Kole C (ed) *Genome mapping and molecular breeding*, vol 4. Springer, Heidelberg, pp. 1–62.
Gardiner SE, Murdoch J, Bus V, Rikkerink E, Gleave A, Warrinton I (2003) Candidate resistace genes from an EST database prove a rich source of markers for major genes conferring resistance to improvement apple pest and diseases. *Acta Hortic* 622:141–151.
Gardiner SE, Norelli JL, de Silva N, Fazio G, Peil A, Malnoy M, Horner M, Bowatte D, Carlisle C, Wiedow C, Wan YZ, Bassett CL, Baldo AM, Celton JM, Richter K, Aldwinckle HS, Bus VGM (2012) Putative resistance gene markers associated with quantitative trait loci for fire blight resistance in *Malus* 'Robusta 5' accessions. *BMC Genet* 13:25.
Gardner RG, Cummins JN, Aldwinckle HS (1980a) Fire blight resistance in the Geneva apple rootstock breeding program. *J Am Soc Hortic Sci* 105:907–912.
Gardner RG, Cummins JN, Aldwinckle HS (1980b) Inheritance of fire blight resistance in *Malus* in relation to rootstock breeding. *J Am Soc Hortic Sci* 105:912–916.
Gianfranceschi L, Seglias N, Tarchini R, Komjanc M, Gessler C (1998) Simple sequence repeats for the genetic analysis of apple. *Theor Appl Genet* 96:1069–1076.
Gonzalez A, Zhao M, Leavitt JM, Lloyd AM (2008) Regulation of the anthocyanin biosynthetic pathway by the TTG1/bHLH/Myb transcriptional complex in Arabidopsis seedlings. *Plant J* 53:814–827.
Grattapaglia D, Sederoff R (1994) Genetic-linkage maps of *Eucalyptus grandis* and *Eucalyptus urophylla* using a pseudo-testcross: mapping strategy and RAPD markers. *Genetics* 137:1121–1137.
Grove GG, Eastwell KC, Jones AL, Sutton TB (2003) Diseases of apple. In: Ferree DC, Warrington IJ (eds) *Apples: botany, production and uses*. CABI, Wallingford, pp. 459–488.
Guan YZ, Peace C, Rudell D, Verma S, Evans K (2015) QTLs detected for individual sugars and soluble solids content in apple. *Mol Breed* 35:135.
Gygax M, Gianfranceschi L, Liebhard R, Kellerhals M, Gessler C, Patocchi A (2004) Molecular markers linked to the apple scab resistance gene *Vbj* derived from *Malus baccata jackii*. *Theor Appl Genet* 109:1702–1709.
Hamada Y, Sato H, Otagaki S, Okada K, Abe K, Matsumoto S (2015) Breeding depression of red flesh apple progeny containing both functional *MdMYB10* and *MYB110a_JP* genes. *Plant Breed* 134:239–246.
Han YP, Korban SS (2010) Strategies for map-based cloning in apple. *Crit Rev Plant Sci* 29:265–284.
Haniuda T, Yoshida TY, Sanada T (1985) Some characteristics and inheritance of pseudoalbino seedlings derived from the cross of dwarf rootstocks, M9 and M27. *Bull Fruit Tree Res Stn Ser C (Morioka)* 12:21–34.
Harada T, Sunako T, Wakasa Y, Soejima J, Satoh T, Niizeki M (2000) An allele of the 1-aminocyclopropane-1-carboxylate synthase gene (*Md-ACS1*) accounts for the low level of ethylene production in climacteric fruits of some apple cultivars. *Theor Appl Genet* 101:742–746.

Harrison N, Harrison RJ, Barber-Perez N, Cascant-Lopez E, Cobo-Medina M, Lipska M, Conde-Ruiz R, Brain P, Gregory PJ, Fernández-Fernández F (2016) A new three-locus model for rootstock-induced dwarfing in apple revealed by genetic mapping of root bark percentage. *J Exp Bot* 67:1871–1881.
Hatton RG (1917) Paradise apple stocks. *J Roy Hortic Soc* 42:361–399.
Hatton RG (1919) Paradise apple stocks: their fruit and blossom described. *J Roy Hortic Soc* 44:89–94.
Hayat F, Iqbal S, Coulibaly D, Razzaq MK, Nawaz MA, Jiang W, Shi T, Gao Z (2021) An insight into dwarfing mechanism: contribution of scion-rootstock interaction toward fruit crop improvement. *Fruit Res* 1:3.
Hemmat M, Brown SK, Aldwinckle HS (2003) Identification and mapping of markers for resistance to apple scab from 'Antonovka' and 'Hansen's baccata #2'. *Acta Hortic* 622:153–161.
Hemmat M, Weeden NF, Conner PJ, Brown SK (1997) A DNA marker for columnar growth habit in apple contains a simple sequence repeat. *J Am Soc Hortic Sci* 122:347–349.
Hickey KD, Yoder KS (2014) Poedery mildew. In: Sutton TB, Aldwinckle, HS, Agnello, AM, Walgenbach, JF (eds) *Compendium of apple and pear diseases and pests*, second edition. APS Publications, St. Paul, MN, pp. 12–15.
Holderbaum DF, Kon T, Kudo T, Guerra MP (2010) Enzymatic browning, polyphenol oxidase activity, and polyphenols in four apple cultivars: dynamics during fruit development. *HortScience* 45:1150–1154.
Honda C, Moriya S (2018) Anthocyanin biosynthesis in apple fruit. *Hortic J* 87:205–314.
Hough LF (1944) A survey of the scab resistance of the foliage on seedlings in selected apple progenies. *Proc Am Soc Hortic Sci* 44:26–72.
Howard NP, van de Weg E, Tillman J, Tong CBS, Silverstein KAT, Luby JJ (2018) Two QTL characterized for soft scald and soggy breakdown in apple (*Malus × domestica*) through pedigree-based analysis of a large population of interconnected families. *Tree Genet Genom* 14:2.
Hyun TK, Eom SH, Kim J-S (2011) Genomic analysis and gene structure of the two invertase families in the domesticated apple (*Malus × domestica* Borkh.). *Plant Omics J* 4:391–399.
Igarashi M, Hatsuyama Y, Fukawsawa-Akada T (2014) Kogane Fuji is a bud sport of Fuji lacking a chromosome region including a gene of skin color regulation. *Hortic Res (Japan)* 13(Suppl 1): 258.
Igarashi M, Hatsuyama Y, Harada T, Fukasawa-Akada T (2016) Biotechnology and apple breeding in Japan. *Breed Sci* 66:18–33.
Iwanami H (2011) Breeding for fruit quality in apple. In: Jenks MA, Bebeli PJ (eds) *Breeding for fruit quality*. John Wiley & Sons, Hoboken, pp. 175–200.
Iwanami H, Moriya S, Kotoda N, Takahashi S, Abe K (2005) Influence of mealiness on the firmness of apples after harvest. *HortScience* 40:2091–2095.
James CM, Clarke JB, Evans KM (2004) Identification of molecular markers linked to the mildew resistance gene *Pl-d* in apple. *Theor Appl Genet* 110:175–181.
Janick J (2006) The PRI apple breeding program. *HortScience* 41:8–10.
Jänsch M, Broggini GAL, Weger J, Bus VGM, Gardiner SE, Bassett H, Patocchi A (2015) Identification of SNPs linked to eight apple disease resistance loci. *Mol Breed* 35:45.
Jia DJ, Shen F, Wang Y, Wu T, Xu XF, Zhang XZ, Han ZH (2018) Apple fruit acidity is genetically diversified by natural variations in three hierarchical epistatic genes: *MdSAUR37*, *MdPP2CH* and *MdALMTII*. *Plant J* 95:427–443.
Jia DJ, Wu P, Shen F, Li W, Zheng XD, Wang YZ, Yuan YB, Zhang XZ, Han ZH (2021) Genetic variation in the promoter of an R2R3-MYB transcription factor determines fruit malate content in apple (*Malus domestica* Borkh.). *Plant Physiol* 186:549–568.
Jung M, Roth M, Aranzana MJ, Auwerkerken A, Bink M, Denancé C, Dujak C, Durel CE, Forcada CFI, Cantin CM, Guerra W, Howard NP, Keller B, Lewandowski M, Ordidge M, Rymenants M, Sanin N, Studer B, Zurawicz E, Laurens F, Patocchi A, Muranty H (2020) The apple REFPOP-a reference population for genomics-assisted breeding in apple. *Hortic Res* 7:189.
Kasajima I, Kikuchi T, Yoshikawa N (2017) Rapid identification of apple (*Malus × domestica* Borkh.) *S* alleles using sequencing-based DNA marker *APPLid*. *Plant Biotechnol* 34:97–106.
Kellerhals M, Schutz S, Patocchi A (2017) Breeding for host resistance to fire blight. *J Plant Pathol* 99:37–43.
Kenis K, Keulemans J (2007) Study of tree architecture of apple (*Malus × domestica* Borkh.) by QTL analysis of growth traits. *Mol Breed* 19:193–208.
Kenis K, Keulemans J, Davey MW (2008) Identification and stability of QTLs for fruit quality traits in apple. *Tree Genet Genom* 4:647–661.
Khadivi-Khub A (2015) Physiological and genetic factors influencing fruit cracking. *Acta Physiol Plant* 37:1718.
Khajuria YP, Kaul S, Wani AA, Dhar MK (2018) Genetics of resistance in apple against *Venturia inaequalis* (Wint.) Cke. *Tree Genet Genom* 14:16.

Khan A, Korban SS (2022) Breeding and genetics of disease resistance in temperate fruit trees: challenges and new opportunities. *Theor Appl Genet* 135:3961–3985.

Khan MA, Duffy B, Gessler C, Patocchi A (2006) QTL mapping of fire blight resistance in apple. *Mol Breed* 17:299–306.

Khan SA, Beekwilder J, Schaart JG, Mumm R, Soriano JM, Jacobsen E, Schouten HJ (2013) Differences in acidity of apples are probably mainly caused by a malic acid transporter gene on LG16. *Tree Genet Genom* 9:475–487.

Khan SA, Chibon PY, de Vos RCH, Schipper BA, Walraven E, Beekwilder J, van Dijk T, Finkers R, Visser RG, van de Weg EW, Bovy A, Cestaro A, Velasco R, Jacobsen E, Schouten HJ (2012) Genetic analysis of metabolites in apple fruits indicates an mQTL hotspot for phenolic compounds on linkage group 16. *J Exp Bot* 63:2895–2908.

Kim H, Kakui H, Kotoda N, Hirata Y, Koba T, Sassa H (2009) Determination of partial genomic sequences and development of a CAPS system of the *S*-RNase gene for the identification of 22 S haplotypes of apple (*Malus × domestica* Borkh.). *Mol Breed* 23:463–472.

Kim MY, Song KJ, Hwang, HJ, Shin Y-U, Lee HJ (2003) Development of RAPD and SCAR markers linked to the *Co* gene conferring columnar growth habit in apple (*Malus pumila* Mill.). *J Hortic Sci Biotechnol* 78:512–517.

Klein LG, Way RD, Lamb RC (1961) The inheritance of a lethal factor in apples. *Proc Am Soc Hort Sci* 77:50–53.

Knight RL, Alston FH (1968) Sources of field immunity to mildew (Podosphaera leucotricha) in apple. *Can J Genet Cytol* 10:294–298.

Koller B, Gianfranceschi L, Seglias N, McDermott J, Gessler C (1994) DNA markers linked to *Malus floribunda* 821 scab resistance. *Plant Mol Biol* 26:597–602.

Korban SS, Swiader JM (1984) Genetic and nutritional status in bitter pit-resistant and pit-susceptible apple seedlings. *J Am Soc Hortic Sci* 109:428–432.

Kost TD, Gessler C, Jänsch M, Flachowsky H, Patocchi A, Broggini GAL (2015) Development of the first cisgenic apple with increased resistance to fire blight. *Plos One* 10:e0143980.

Kostick SA, Teh SL, Norelli JL, Vanderzande S, Peace C, Evans KM (2021) Fire blight QTL analysis in a multi-family apple population identifies a reduced-susceptibility allele in 'Honeycrisp'. *Hortic Res* 8:28.

Kouassi AB, Durel CE, Costa F, Tartarini S, van de Weg E, Evans K, Fernández-Fernández F, Govan C, Boudichevskaja A, Dunemann F, Antofie A, Lateur M, Stankiewicz-Kosyl M, Soska A, Tomala K, Lewandowski M, Rutkovski K, Zurawicz E, Guerra W, Laurens F (2009) Estimation of genetic parameters and prediction of breeding values for apple fruit-quality traits using pedigreed plant material in Europe. *Tree Genet Genom* 5:659–672.

Krost C, Petersen R, Lokan S, Brauksiepe B, Braun P, Schmidt ER (2013) Evaluation of the hormonal state of columnar apple trees (*Malus × domestica*) based on high throughput gene expression studies. *Plant Mol Biol* 81:211–220.

Krost C, Petersen R, Schmidt ER (2012) The transcriptomes of columnar and standard type apple trees (*Malus × domestica*) - a comparative study. *Gene* 498:223–230.

Kumar R, Tamboli V, Sharma R, Sreelakshmi Y (2018) NAC-NOR mutations in tomato Penjar accessions attenuate multiple metabolic processes and prolong the fruit shelf life. *Food Chem* 259:234–244.

Kumar S, Garrik DJ, Bink MCAM, Whitworth C, Chagné D, Volz RK (2013) Novel genomic approaches unravel genetic architecture of complex traits in apple. *BMC Genom* 14:393.

Kumar S, Rowan D, Hunt M, Chagné D, Whitworth C, Souleyre E (2015) Genome-wide scans reveal genetic architecture of apple flavour volatiles. *Mol Breed* 35:118.

Kunihisa M, Hayashi T, Hatsuyama Y, Fukasawa-Akada T, Uenishi H, Matsumoto T, Kon T, Kasai S, Kudo T, Oshino H, Yamamoto T, Tazawa J (2021) Genome-wide association study for apple flesh browning: detection, validation, and physiological roles of QTLs. *Tree Genet Genom* 17:11.

Kunihisa M, Moriya S, Abe K, Okada K, Haji T, Hayashi T, Kawahara Y, Itoh R, Itoh T, Katayose Y, Kanamori H, Toshimi M, Mori S, Sasaki H, Matsumoto T, Nishitani C, Yamamoto T (2016) Genomic dissection of a 'Fuji' apple cultivar: re-sequencing, SNP marker development, definition of haplotypes, and QTL detection. *Breed Sci* 66:499–515.

Kunihisa M, Moriya S, Abe K, Okada K, Haji T, Hayashi T, Kim H, Nishitani C, Terakami S, Yamamoto T (2014) Identification of QTLs for fruit quality traits in Japanese apples: QTLs for early ripening are tightly related to preharvest fruit drop. *Breed Sci* 64:240–251.

Kunihisa M, Takita Y, Yamaguchi N, Okada H, Sato M, Komori S, Nishitani C, Terakami S, Yamamoto T (2019) The use of a fertile doubled haploid apple line for QTL analysis of fruit traits. *Breed Sci* 69:410–419.

Laloi G, Vergne E, Durel CE, Le Cam B, Caffier V (2017) Efficiency of pyramiding of three quantitative resistance loci to apple scab. *Plant Pathol* 66:412–422.

Lapins KO (1969) Segregation of compact growth types in certain apple seedling progenies. *Can J Plant Sci* 49:765–768.

Lapins KO (1976) Inheritance of compact growth type in apple. *J Am Soc Hortic Sci* 101:133–135.

Larsen B, Migicovsky Z, Jeppesen AA, Gardner KM, Toldam-Andersen TB, Myles S, Orgaard M, Petersen MA, Pedersen C (2019) Genome-wide association studies in apple reveal loci for aroma volatiles, sugar composition, and harvest date. *Plant Genome* 12:18010.

Larsen B, Ørgaard M, Toldam-Andersen TB, Pedersen C (2016) A high-throughput method for genotyping S-RNase alleles in apple. *Mol Breed* 36:24.

Lashbrooke J, Aharoni A, Costa F (2015) Genome investigation suggests *MdSHN3*, an APETALA2-domain transcription factor gene, to be a positive regulator of apple fruit cuticle formation and an inhibitor of russet development. *J Exp Bot* 66:6579–6589.

Laurens F, Aranzana MJ, Arus P, Bassi D, Bink M, Bonany J, Caprera A, Corelli-Grappadelli L, Costes E, Durel CE, Mauroux JB, Muranty H, Nazzicari N, Pascal T, Patocchi A, Peil A, Quilot-Turion B, Rossini L, Stella A, Troggio M, Velasco R, van de Weg E (2018) An integrated approach for increasing breeding efficiency in apple and peach in Europe. *Hortic Res* 5:11.

Laurens F, Chevalier M, Dolega E, Gennari F, Goerre M, Fischer C, Kellerhals M, Lateur M, Lefrancq B, Parisi L, Schouten HJ, Tartarini S (2004) Locan European cultivars as sources of durable scab resistance in apple. *Acta Hortic* 663:115–121.

Le Roux PM, Flachowsky H, Hanke MV, Gessler C, Patocchi A (2012) Use of a transgenic early flowering approach in apple (*Malus × domestica* Borkh.) to introgress fire blight resistance from cultivar Evereste. *Mol Breed* 30:857–874.

Lee JM, Looney NE (1977) Branching habit and apical dominance of compact and normal apple seedlings as influenced by TIBA and GA_3. *J Am Soc Hortic Sci* 102:619–622.

Leforestier D, Ravon E, Muranty H, Cornille A, Lemaire C, Giraud T, Durel CE, Branca A (2015) Genomic basis of the differences between cider and dessert apple varieties. *Evol Appl* 8:650–661.

Leisso RS, Gapper NE, Mattheis JP, Sullivan NL, Watkins CB, Giovannoni JJ, Schaffer RJ, Johnston JW, Hanrahan I, Hertog MLATM, Nicolaï BM, Rudell DR (2016) Gene expression and metabolism preceding soft scald, a chilling injury of 'Honeycrisp' apple fruit. *BMC Genom* 17:798.

Lemaire C, De Gracia M, Leroy T, Michalecka M, Lindhard-Pedersen H, Guerin F, Gladieux P, Le Cam B (2016) Emergence of new virulent populations of apple scab from nonagricultural disease reservoirs. *New Phytol* 209:1220–1229.

Lespinasse JM, Delort JF (1986) Apple tree management in vertical axis: appraisal after ten years of experiments. *Acta Hortic* 160:139–155.

Lewis D, Crane MB (1938) Genetical studies in apples. II. *J Genet* 37:119–128.

Li TY, Wang Y, Zhang XZ, Han ZH (2012) Isolation and characterization of *ARRO-1* genes from apple rootstocks in response to auxin treatment. *Plant Mol Biol Rep* 30:1408–1414.

Liebhard R, Gianfranceschi L, Koller B, Ryder CD, Tarchini R, Van de Weg E, Gessler C (2002) Development and characterisation of 140 new microsatellites in apple (*Malus × domestica* Borkh.). *Mol Breed* 10:217–241.

Liebhard R, Kellerhals M, Pfammatter W, Jertmini M, Gessler C (2003a) Mapping quantitative physiological traits in apple (*Malus × domestica* Borkh.). *Plant Mol Biol* 52:511–526.

Liebhard R, Koller B, Gianfranceschi L, Gessler C (2003b) Creating a saturated reference map for the apple (*Malus × domestica* Borkh.) genome. *Theor Appl Genet* 106:1497–1508.

Liebhard R, Koller B, Patocchi A, Kellerhals M, Pfammatter W, Jermini M, Gessler C (2003c) Mapping quantitative field resistance against apple scab in a 'Fiesta' × 'Discovery' progeny. *Phytopathol* 93:493–501.

Lin-Wang K, Bolitho K, Grafton K, Kortstee A, Karunairetnam S, McGhie TK, Espley RV, Hellens RP, Allan AC (2010) An R2R3 MYB transcription factor associated with regulation of the anthocyanin biosynthetic pathway in Rosaceae. *BMC Plant Biol* 10:50.

Longhi S, Cappellin L, Guerra W, Costa F (2013a) Validation of a functional molecular marker suitable for marker-assisted breeding for fruit texture in apple (*Malus × domestica* Borkh.). *Mol Breed* 32:841–852.

Longhi S, Hamblin MT, Trainotti L, Peace CP, Velasco R, Costa F (2013b) A candidate gene based approach validates *Md-PG1* as the main responsible for a QTL impacting fruit texture in apple (*Malus × domestica* Borkh). *BMC Plant Biol* 13:37.

Longhi S, Moretto M, Viola R, Velasco R, Costa F (2012) Comprehensive QTL mapping survey dissects the complex fruit texture physiology in apple (*Malus* x *domestica* Borkh.). *J Exp Bot* 63:1107–1121.

López-Girona E, Bowatte DR, Smart MEM, Alvares S, Brancher TL, Chagné D, Volz RK (2021) A high-throughput *S-RNase* genotyping method for apple. *Fruit Res* 1:10.

Lötze E, Joubert J, Theron KI (2008) Evaluating pre-harvest foliar calcium applications to increase fruit calcium and reduce bitter pit in 'Golden Delicious' apples. *Sci Hortic* 116:299–304.
Lucht JM (2015) Public acceptance of plant biotechnology and GM crops. *Viruses* 7:4254–4281.
Luo F, Sandefur P, Evans K, Peace C (2019) A DNA test for routinely predicting mildew resistance in descendants of crabapple 'White Angel'. *Mol Breed* 39:33.
Luo FX, Evans K, Norelli JL, Zhang ZW, Peace C (2020a) Prospects for achieving durable disease resistance with elite fruit quality in apple breeding. *Tree Genet Genom* 16:21.
Luo FX, Norelli JL, Howard NP, Wisniewski M, Flachowsky H, Hanke MV, Peace C (2020b) Introgressing blue mold resistance into elite apple germplasm by rapid cycle breeding and foreground and background DNA-informed selection. *Tree Genet Genom* 16:28.
Ma BQ, Liao L, Zheng HY, Chen J, Wu BH, Ogutu C, Li SH, Korban SS, Han YP (2015) Genes encoding aluminum-activated malate transporter II and their association with fruit acidity in apple. *Plant Genom* 8:3.
Ma BQ, Zhao S, Wu BH, Wang DM, Peng Q, Owiti A, Fang T, Liao L, Ogutu C, Korban SS, Li SH, Han YP (2016) Construction of a high density linkage map and its application in the identification of QTLs for soluble sugar and organic acid components in apple. *Tree Genet Genom* 12:1.
Ma CQ, Liang BW, Chang B, Yan JY, Liu L, Wang Y, Yang YZ, Zhao ZY (2019) Transcriptome profiling of anthocyanin biosynthesis in the peel of 'Granny Smith' apples (*Malus domestica*) after bag removal. *BMC Genom* 20:353.
Ministry of Agriculture, Forest and Fisheries (2022) Ministry of Agriculture, Forest and Fisheries https://www.maff.go.jp/j/tokei/index.html Accessed on 6 June 2022.
Maliepaard C, Alston FH, Arkel GV, Brown LM, Chevreau E, Dunemann F, Evans KM, Gardiner S, Guilford P, Heusden AWV, Janse J, Laurens F, Lynn JR, Manganaris AG, Nijs APMD, Periam N, Rikkerink E, Roche P, Ryder C, Sansavini S, Schmidt H, Tartarini S, Verhaegh JJ, Ginkel MV-V, King GJ (1998) Aligning male and female linkage maps of apple (*Malus pumila* Mill.) using multi-allelic markers. *Theor Appl Genet* 97:60–73.
Malnoy M, Xu M, Borejsza-Wysocka E, Korban SS, Aldwinckle HS (2008) Two receptor-like genes, *Vfa1* and *Vfa2*, confer resistance to the fungal pathogen *Venturia inaequalis* inciting apple scab disease. *Mol Plant Microbe Interact* 21:448–458.
Manganaris AG, Alston FH, Weeden NF, Aldwinckle HS, Gustafson HL, Brown SK (1994) Isozyme locus *Pgm-1* is tightly linked to a gene (*Vf*) for scab resistance in apple. *J Am Soc Hortic Sci* 119:1286–1288.
Markussen T, Kruger J, Schmidt H, Dunemann F (1995) Identification of PCR-based markers linked to the powdery-mildew-resistance gene *Pl1* from *Malus robusta* in cultivated apple. *Plant Breed* 114:530–534.
Mathew IE, Agarwal P (2018) May the fittest protein evolve: favoring the plant-specific origin and expansion of NAC transcription factors. *BioEssays* 40:1800018.
Matsumoto S, Li TZ, Otagaki S, Li Y, Bai SL (2018) Efficient breeding and cultivation of type 2 red-fleshed apple cultivars using a search system for suitable apple cultivar combination. *Hortic Plant J* 4:219–225.
Matsumoto S, Okada K, Kojima A, Shiratake K, Abe K (2011) *S-RNase* genotypes of apple (*Malus domestica* Borkh.) including new cultivars, lineages, and triploid progenies. *J Hortic Sci Biotechnol* 86: 654–660.
McClure B (2004) *S-RNase* and *SLF* determine *S*-haplotype-specific pollen recognition and rejection. *Plant Cell* 16:2840–2847.
McClure KA, Gardner KM, Douglas GM, Song J, Forney CF, DeLong J, Fan LH, Du LN, Toivonen PMA, Somers DJ, Rajcan I, Myles S (2018) A genome-wide association study of apple quality and scab resistance. *Plant Genome* 11:1.
McClure KA, Gardner KM, Toivonen PMA, Hampson CR, Song J, Forney CF, DeLong J, Rajcan I, Myles S (2016) QTL analysis of soft scald in two apple populations. *Hortic Res* 3:16043.
Mellidou I, Chagné D, Laing WA, Keulemans J, Davey MW (2012) Allelic variation in paralogs of GDP-L-galactose phosphorylase is a major determinant of vitamin C concentrations in apple fruit. *Plant Physiol* 160:1613–1629.
Migicovsky Z, Gardner KM, Money D, Sawler J, Bloom JS, Moffett P, Chao CT, Schwaninger H, Fazio G, Zhong GY, Myles S (2016) Genome to phenome mapping in apple using historical data. *Plant Genome* 9:2.
Migicovsky Z, Yeats TH, Watts S, Song J, Forney CF, Burgher-MacLellan K, Somers DJ, Gong YH, Zhang ZQ, Vrebalov J, van Velzen R, Giovannoni JG, Rose JKC, Myles S (2021) Apple ripening is controlled by a NAC transcription factor. *Front Genet* 12:671300.
Miller BA, Tillman JR, Howard NP, Kostick SA, Evans KM, Luby JJ (2021) A DNA test for high incidence of soft scald and soggy breakdown postharvest disorders in *Malus domestica* Borkh. *Mol Breed* 41:57.
Minamikawa MF, Kunihisa M, Noshita K, Moriya S, Abe K, Hayashi T, Katayose Y, Matsumoto T, Nishitani C, Terakami S, Yamamoto T, Iwata H (2021) Tracing founder haplotypes of Japanese apple varieties: application in genomic prediction and genome-wide association study. *Hortic Res* 8:49.

Monet R, Bastard Y, Gibault B (1988) Genetic-study of the weeping habit in peach. *Agronomie* 8:127–132.

Moore LW, Burr TJ (2014) Crown gall. In: Sutton TB, Aldwinckle, HS, Agnello, AM, Walgenbach, JF (eds) *Compendium of apple and pear diseases and pests*, second edition. APS Publications, St. Paul, MN, pp. 93–94.

Morgan J, Richards A (1993) *The book of apples*. Ebury Press, London.

Moriya S, Goto S, Kubo T, Kunihisa M, Tazawa J, Kudo T, Kasai S, Kudo H, Okada K, Yamamoto T, Fukazawa-Akada T, Hatsuyama Y, Abe K (2019b) Identification of *Venturia inaequalis* races in Morioka, Japan and identification of a quantitative trait locus associated with resistance to apple scab in 'Akane' apples. *Hort Res (Japan)* 18:349–361.

Moriya S, Iwanami H, Haji T, Okada K, Shimizu T, Suzaki K, Kitamoto N, Katayose Y, Wu JZ, Yamamoto T, Abe K (2021) QTL analysis of crown gall disease resistance in apple: first plant *R* gene candidates effective against *Rhizobium rhizogenes* (Ti). *Tree Genet Genom* 17:25.

Moriya S, Iwanami H, Haji T, Okada K, Yamada M, Yamamoto T, Abe K (2015) Identification and genetic characterization of a quantitative trait locus for adventitious rooting from apple hardwood cuttings. *Tree Genet Genom* 11:59.

Moriya S, Iwanami H, Kotoda N, Haji T, Okada K, Terakami S, Mimida N, Yamamoto T, Abe K (2012b) Aligned genetic linkage maps of apple rootstock cultivar 'JM7' and *Malus sieboldii* 'Sanashi 63' constructed with novel EST-SSRs. *Tree Genet Genom* 8:709–723.

Moriya S, Iwanami H, Kotoda N, Takahashi S, Yamamoto T, Abe K (2009) Development of a marker-assisted selection system for columnar growth habit in apple breeding. *J Jpn Soc Hortic Sci* 78:279–287.

Moriya S, Iwanami H, Okada K, Yamamoto T, Abe K (2011) A practical method for apple cultivar identification and parent-offspring analysis using simple sequence repeat markers. *Euphytica* 177:135–150.

Moriya S, Iwanami H, Takahashi S, Kotoda N, Suzaki K, Abe K (2008) Evaluation and inheritance of crown gall resistance in apple rootstocks. *J Jpn Soc Hortic Sci* 77:236–241.

Moriya S, Iwanami H, Takahashi S, Kotoda N, Suzaki K, Yamamoto T, Abe K (2010) Genetic mapping of the crown gall resistance gene of the wild apple *Malus sieboldii. Tree Genet Genom* 6:195–203.

Moriya S, Kunihisa M, Okada K, Iwanami H, Iwata H, Minamikawa M, Katayose Y, Matsumoto T, Mori S, Sasaki H, Matsumoto T, Nishitani C, Terakami S, Yamamoto T, Abe K (2017a) Identification of QTLs for flesh mealiness in apple (*Malus × domestica* Borkh.). *Hortic J* 86:159–170.

Moriya S, Kunihisa M, Okada K, Shimizu T, Honda C, Yamamoto T, Muranty H, Denancé C, Katayose Y, Iwata H, Abe K (2017b) Allelic composition of *MdMYB1* drives red skin color intensity in apple (*Malus × domestica* Borkh.) and its application to breeding. *Euphytica* 213:78.

Moriya S, Okada K, Haji T, Yamamoto T, Abe K (2012a) Fine mapping of *Co*, a gene controlling columnar growth habit located on apple (*Malus × domestica* Borkh.) linkage group 10. *Plant Breed* 131:641–647.

Moriya S, Terakami S, Okada K, Shimizu T, Adachi Y, Katayose Y, Fujisawa H, Wu JZ, Kanamori H, Yamamoto T, Abe K (2019a) Identification of candidate genes responsible for the susceptibility of apple (*Malus × domestica* Borkh.) to Alternaria blotch. *BMC Plant Biol* 19:132.

Murata M, Noda I, Homma S (1995) Enzymatic browning of apples on the market: relationship between browning, polyphenol content, and polyphenol oxidase. *J Jpn Soc Food Sci Technol* 42:820–826.

Naing AT, Kim CK (2018) Roles of R2R3-MYB transcription factors in transcriptional regulation of anthocyanin biosynthesis in horticultural plants. *Plant Mol Biol* 98: 1–18.

Nishio S, Saito T, Terakami S, Takada N, Kato H, Itai A, Yamamoto T (2018) Identification of QTLs associated with conversion of sucrose to hexose in mature fruit of Japanese pear. *Plant Mol Biol Rep* 36:643–652.

Norelli JL, Wisniewski M, Fazio G, Burchard E, Gutierrez B, Levin E, Droby S (2017) Genotyping-by-sequencing markers facilitate the identification of quantitative trait loci controlling resistance to *Penicillium expansum* in *Malus sieversii. Plos One* 12:e0172949.

Nybom H, Afzadi MA, Sehic J, Hertog M (2013) DNA marker-assisted evaluation of fruit firmness at harvest and post-harvest fruit softening in a diverse apple germplasm. *Tree Genet Genom* 9:279–290.

Nybom H, Sehic J, Garkava-Gustavsson L (2008) Modern apple breeding is associated with a significant change in the allelic ratio of the ethylene production gene *Md-ACS1. J Hortic Sci Biotechnol* 83:673–677.

Nybom N (1959) On the inheritance of acidity in cultivated apples. *Hereditas* 45:332–350.

Ognjanov V, Vujanic-Varga D, Gasic K (1999) Breeding columnar apples in Novi Sad. *Acta Hortic* 484:207–209.

Okada K, Wada M, Moriya S, Katayose Y, Fujisawa H, Wu JZ, Kanamori H, Kurita K, Sasaki H, Fujii H, Terakami S, Iwanami H, Yamamoto T, Abe K (2016) Expression of a putative dioxygenase gene adjacent to an insertion mutation is involved in the short internodes of columnar apples (*Malus × domestica*). *J Plant Res* 129:1109–1126.

Okada K, Wada M, Takebayashi Y, Kojima M, Sakakibara H, Nakayasu M, Mizutani M, Nakajima M, Moriya S, Shimizu T, Abe K (2020) Columnar growth phenotype in apple results from gibberellin deficiency by ectopic expression of a dioxygenase gene. *Tree Physiol* 40:1205–1216.

Oraguzie N, Alspach P, Volz R, Whitworth C, Ranatunga C, Weskett R, Harker R (2009) Postharvest assessment of fruit quality parameters in apple using both instruments and an expert panel. *Postharvest Biol Technol* 52:279–287.

Oraguzie NC, Iwanami H, Soejima J, Harada T, Hall A (2004) Inheritance of the *Md-ACS1* gene and its relationship to fruit softening in apple (*Malus × domestica* Borkh.). *Theor Appl Genet* 108:1526–1533.

Oraguzie NC, Volz RK, Whitworth CJ, Bassett HCM, Hall AJ, Gardiner SE (2007) Influence of *Md-ACS1* allelotype germplasm collection on fruit and harvest season within an apple softening during cold air storage. *Postharvest Biol Technol* 44:212–219.

Orcheski B, Meng D, Bai Y, Fei ZJ, Cheng LL (2021) The transcriptomes of healthy and bitter pit-affected 'Honeycrisp' fruit reveal genes associated with disorder development and progression. *Tree Genet Genom* 17:37.

Orcheski B, Parker R, Brown S (2015) Pale green lethal disorder in apple (*Malus*) is caused by a mutation in the *PHYLLO* gene which is essential for phylloquinone (vitamin K_1) biosynthesis. *Tree Genet Genom* 11:131.

Otto D, Petersen R, Brauksiepe B, Braun P, Schmidt ER (2014) The columnar mutation ("*Co* gene") of apple (*Malus × domestica*) is associated with an integration of a Gypsy-like retrotransposon. *Mol Breed* 33:863–880.

Padmarasu S, Sargent DJ, Jaensch M, Kellerhals M, Tartarini S, Velasco R, Troggio M, Patocchi A (2014) Fine-mapping of the apple scab resistance locus *Rvi12* (*Vb*) derived from 'Hansen's baccata #2'. *Mol Breed* 34:2119–2129.

Padmarasu S, Sargent DJ, Patocchi A, Troggio M, Baldi P, Linsmith G, Poles L, Jänsch M, Kellerhals M, Tartarini S, Velasco R (2018) Identification of a leucine-rich repeat receptor-like serine/threonine-protein kinase as a candidate gene for *Rvi12* (*Vb*)-based apple scab resistance. *Mol Breed* 38:73.

Pagliarani G, Dapena E, Minarro M, Denancé C, Lespinasse Y, Rat-Morris E, Troggio M, Durel CE, Tartarini S (2016) Fine mapping of the rosy apple aphid resistance locus *Dp-fl* on linkage group 8 of the apple cultivar 'Murray'. *Tree Genet Genom* 12:56.

Parisi L, Lespinasse Y, Guillaumes J, Krüger J (1993) A new race of *Venturia inaequalis* virulent to apples with resistance due to the *Vf* gene. *Phytopathology* 83:533–537.

Parravicini G, Gessler C, Denancé C, Lasserre-Zuber P, Vergne E, Brisset MN, Patocchi A, Durel CE, Broggini GAL (2011) Identification of serine/threonine kinase and nucleotide-binding site-leucine-rich repeat (NBS-LRR) genes in the fire blight resistance quantitative trait locus of apple cultivar 'Evereste'. *Mol Plant Pathol* 12:493–505.

Patocchi A, Bigler B, Koller B, Kellerhals M, Gessler C (2004) *Vr2*: a new apple scab resistance gene. *Theor Appl Genet* 109:1087–1092.

Patocchi A, Walser M, Tartarini S, Broggini GAL, Gennari F, Sansavini S, Gessler C (2005) Identification by genome scanning approach (GSA) of a microsatellite tightly associated with the apple scab resistance gene *Vm*. *Genome* 48:630–636.

Patocchi A, Wehrli A, Dubuis PH, Auwerkerken A, Leida C, Cipriani G, Passey T, Staples M, Didelot F, Philion V, Peil A, Laszakovits H, Ruhmer T, Boeck K, Baniulis D, Strasser K, Vavra R, Guerra W, Masny S, Ruess F, Le Berre F, Nybom H, Tartarini S, Spornberger A, Pikunova A, Bus VGM (2020) Ten years of VINQUEST: first insight for breeding new apple cultivars with durable apple scab resistance. *Plant Dis* 104:2074–2081.

Peil A, Garcia-Libreros T, Richter K, Trognitz FC, Trognitz B, Hanke MV, Flachowsky H (2007) Strong evidence for a fire blight resistance gene of *Malus robusta* located on linkage group 3. *Plant Breed* 126:470–475.

Peil A, Hübert C, Wensing A, Horner M, Emeriewen OF, Richter K, Wöhner T, Chagné D, Orellana-Torrejon C, Saeed M, Troggio M, Stefani E, Gardiner SE, Hanke MV, Flachowsky H, Bus VGM (2019) Mapping of fire blight resistance in *Malus × robusta* 5 flowers following artificial inoculation. *BMC Plant Biol* 19:532.

Perazzolli M, Malacarne G, Baldo A, Righetti L, Bailey A, Fontana P, Velasco R, Malnoy M (2014) Characterization of resistance gene analogues (RGAs) in apple (*Malus × domestica* Borkh.) and their evolutionary history of the Rosaceae family. *PLoS One* 9:e83844.

Perchepied L, Chevreau E, Ravon E, Gaillard S, Pelletier S, Bahut M, Berthelot P, Cournol R, Schouten HJ, Vergne E (2021) Successful intergeneric transfer of a major apple scab resistance gene (*Rvi6*) from apple to pear and precise comparison of the downstream molecular mechanisms of this resistance in both species. *BMC Genomics* 22:843.

Petersen R, Djozgic H, Rieger B, Rapp S, Schmidt ER (2015) Columnar apple primary roots share some features of the columnar-specific gene expression profile of aerial plant parts as evidenced by RNA-Seq analysis. *BMC Plant Biol* 15:34.

Petkovsek MM, Stamper F, Veberic R (2007) Parameters of inner quality of the apple scab resistant and susceptible apple cultivars (*Malus domestica* Borkh.). *Sci Hortic* 114:37–44.
Pilet-Nayel ML, Moury B, Caffier V, Montarry J, Kerlan MC, Fournet S, Durel CE, Delourme R (2017) Quantitative resistance to plant pathogens in pyramiding strategies for durable crop protection. *Front Plant Sci* 8:1838.
Pirona R, Eduardo I, Pacheco I, Da Silva Linge C, Miculan M, Verde I, Tartarini S, Dondini L, Pea G, Bassi D, Rossini L (2013) Fine mapping and identification of a candidate gene for a major locus controlling maturity date in peach. *BMC Plant Biol* 13:166.
Potts SM, Khan MA, Han YP, Kushad MM, Korban SS (2014) Identification of quantitative trait loci (QTLs) for fruit quality traits in apple. *Plant Mol Biol Rep* 32:109–116.
Rikkerink HA, Hilario-Andrade EM, Dare AP, Gardiner SE, Yoon M, Bus VGM (2007) Resistance gene and uses thereof. New Zealand Patent Application 563032, filed 31 Oct 2007.
Robinson TL, Hoying SA, Reginato GH (2011) The tall spindle planting system: principles and performance. *Acta Hortic* 903:571–580.
Rowan DD, Hunt MB, Dimouro A, Alspach PA, Weskett R, Volz RK, Gardiner SE, Chagné D (2009) Profiling fruit volatiles in the progeny of a 'Royal Gala' × 'Granny Smith' apple (*Malus × domestica*) cross. *J Agric Food Chem* 57:7953–7961.
Rusholme-Pilcher RL, Celton JM, Gardiner SE, Tustin DS (2008) Genetic markers linked to the dwarfing trait of apple rootstock 'Malling 9'. *J Am Soc Hortic Sci* 133:100–106.
Rymenants M, van de Weg E, Auwerkerken A, De Wit I, Czech A, Nijland B, Heuven H, De Storme N, Keulemans W (2020) Detection of QTL for apple fruit acidity and sweetness using sensorial evaluation in multiple pedigreed full-sib families. *Tree Genet Genom* 16:71.
Saito K, Takeda K (1984) Genetic analysis of resistance to Alternaria blotch in apple. *Japan J Breed* 34:197–209.
Sampson DR, Cameron DF (1965) Inheritance of bronze foliage, extra petals and pendulous habit in ornamental crabapples. *Proc Am Soc Hortic Sci* 86:717–722.
Sassa H (2016) Molecular mechanism of the S-RNase-based gametophytic self-incompatibility in fruit trees of Rosaceae. *Breed Sci* 66:116–121.
Sato H, Otagaki S, Saelai P, Kondo S, Shiratake K, Matsumoto S (2017) Varietal differences in phenolic compounds metabolism of type 2 red-fleshed apples. *Sci Hortic* 219:1–9.
Sato T, Kudo T, Akada T, Wakasa Y, Niizeki M, Harada T (2004) Allelotype of a ripening-specific 1-amino cyclopropane-1-carboxylate synthase gene defines the rate of fruit drop in apple. *J Am Soc Hortic Sci* 129:32–36.
Savelyeva NN, Yushkov AN, Zemisov AS, Chivilev VV, Lyzhin AS (2018) Creating new varieties of columnar apple trees for the plantations of the intensive type. *Hortic Vitic* 5:16–22.
Schaffer RJ, Friel EN, Souleyre EJF, Bolitho K, Thodey K, Ledger S, Bowen JH, Ma JH, Nain B, Cohen D, Gleave AP, Crowhurst RN, Janssen BJ, Yao JL, Newcomb RD (2007) A genomics approach reveals that aroma production in apple is controlled by ethylene predominantly at the final step in each biosynthetic pathway. *Plant Physiol* 144:1899–1912.
Schifferstein HNJ, Verlegh PWJ (1996) The role of congruency and pleasantness in odor-induced taste enhancement. *Acta Psycol* 94:87–105.
Schlathölter I, Jänsch M, Flachowsky H, Broggini GAL, Hanke MV, Patocchi A (2018) Generation of advanced fire blight-resistant apple (*Malus × domestica*) selections of the fifth generation within 7 years of applying the early flowering approach. *Planta* 247:1475–1488.
Schöneberg A, Perren S, Felder B, Hollenstein R, Müller U, Szalatnay D, Hunziker K (2017) Feu bactérien - sensibilité des fruits à pépins. *Agroscope Fiche Technique* 732:1–4.
Schouten HJ, Brinkhuis J, van der Burgh A, Schaart JG, Groenwold R, Broggini GAL, Gessler C (2014) Cloning and functional characterization of the *Rvi15* (*Vr2*) gene for apple scab resistance. *Tree Genet Genom* 10:251–260.
Scorza R, Bassi D, Liverani A (2002) Genetic interactions of pillar (columnar), compact, and dwarf peach tree genotypes. *J Am Soc Hortic Sci* 127:254–261.
Sekido K, Hayashi Y, Yamada K, Shiratake K, Matsumoto S, Maejima T, Komatsu H (2010) Efficient breeding system for red-fleshed apple based on linkage with S3-RNase allele in 'Pink Pearl'. *HortScience* 45:534–537.
Shan W, Kuang J-F, Chen L, Xie H, Peng H-H, Xiao Y-Y, Li X-P, Chen W-X, He Q-G, Chen J-Y, Lu W-J (2012) Molecular characterization of banana NAC transcription factors and their interactions with ethylene signalling component EIL during fruit ripening. *J Exp Bot* 63:5171–5187.
Silfverberg-Dilworth E, Matasci CL, Van de Weg WE, Van Kaauwen MPW, Walser M, Kodde LP, Soglio V, Gianfranceschi L, Durel CE, Costa F, Yamamoto T, Koller B, Gessler C, Patocchi A (2006) Microsatellite markers spanning the apple (*Malus × domestica* Borkh.) genome. *Tree Genet Genom* 2:202–224.

Smolka A, Welander M, Olsson P, Holefors A, Zhu LH (2009) Involvement of the *ARRO-1* gene in adventitious root formation in apple. *Plant Sci* 177:710–715.

Soejima J, Yoshida Y, Haniuda T, Bessho H, Tsuchiya S, Masuda T, Komori S, Sanada T, Ito Y, Sadamori S, Kashimura Y (2010) New dwarfing apple rootstocks 'JM1', 'JM7' and 'JM8'. *Bull Nat Inst Fruit Tree Sci* 11:1–16.

Soejima J, Yoshida Y, Haniuda T, Bessho H, Tsuchiya S, Masuda T, Komori S, Sanada T, Ito Y, Sadamori S, Kashimura Y, Abe K, Kotoda N (2013) New semi- and extremely dwarfing apple rootstocks, 'JM2' and 'JM5'. *Bull NARO Inst Fruit Tree Sci* 16:19–36.

Soriano JM, Madduri M, Schaart JG, van der Burgh A, van Kaauwen MPW, Tomic L, Groenwold R, Velasco R, van de Weg E, Schouten HJ (2014) Fine mapping of the gene *Rvi18* (*V25*) for broad-spectrum resistance to apple scab, and development of a linked SSR marker suitable for marker-assisted breeding. *Mol Breed* 34:2021–2032.

Soufflet-Freslon V, Gianfranceschi L, Patocchi A, Durel CE (2008) Inheritance studies of apple scab resistance and identification of *Rvi14*, a new major gene that acts together with other broad-spectrum QTL. *Genome* 51:657–667.

Souleyre EJF, Chagné D, Chen XY, Tomes S, Turner RM, Wang MY, Maddumage R, Hunt MB, Winz RA, Wiedow C, Hamiaux C, Gardiner SE, Rowan DD, Atkinson RG (2014) The *AAT1* locus is critical for the biosynthesis of esters contributing to 'ripe apple' flavour in 'Royal Gala' and 'Granny Smith' apples. *Plant J* 78:903–915.

Souleyre EJF, Greenwood DR, Friel EN, Karunairetnam S, Newcomb RD (2005) An alcohol acyl transferase from apple (cv. Royal Gala), *MpAAT1*, produces esters involved in apple fruit flavor. *FEBS J* 272:3132–3144.

Stoeckli S, Mody K, Gessler C, Patocchi A, Jermini M, Dorn S (2008) QTL analysis for aphid resistance and growth traits in apple. *Tree Genet Genom* 4:833–847.

Sun LX, Bukovac MJ, Forsline PL, van Nocker S (2009) Natural variation in fruit abscission-related traits in apple (*Malus*). *Euphytica* 165:55–67.

Sun R, Li H, Zhang Q, Chen DM, Yang FQ, Zhao YB, Wang Y, Han YP, Zhang XZ, Han ZH (2014) Mapping for quantitative trait loci and major genes associated with fresh-cut browning in apple. *HortScience* 49:25–30.

Sunako R, Sakuraba W, Senda M, Akada S, Ishikawa R, Niizeki M, Harada T (1999) An allele of the ripening-specific 1-aminocyclopropane-1-carboxylic acid synthase gene (*ACS1*) in apple fruit with a long storage life. *Plant Physiol* 119:1297–1303.

Sundin GW (2014) Fire blight. In: Sutton TB, Aldwinckle, HS, Agnello, AM, Walgenbach, JF (eds) *Compendium of apple and pear diseases and pests*, second edition. APS Publications, St. Paul, MN, pp. 87–89.

Szankowski I, Waidmann S, Degenhardt J, Patocchi A, Paris R, Silfverberg-Dilworth E, Broggini G, Gessler C (2009) Highly scab-resistant transgenic apple lines achieved by introgression of *HcrVf2* controlled by different native promoter lengths. *Tree Genet Genom* 5:349–358.

Takos AM, Jaffe FW, Jacob SR, Bogs J, Robinson SP, Walker AR (2006) Light-induced expression of a *MYB* gene regulates anthocyanin biosynthesis in red apples. *Plant Physiol* 142:1216–1232.

Tan Y, Lv S, Liu X, Gao, T, Li T, Wang Y, Wu T, Zhang X, Han Y, Korban SS, Han Z (2017) Development of high-density interspecific genetic maps for the identification of QTLs conferring resistance to *Valsa ceratosperma* in apple. *Euphytica* 213:10.

Tanaka F, Okazaki K, Kashimura T, Ohwaki Y, Tatsuki M, Sawada A, Ito T, Miyazawa T (2016) Profiles and physiological mechanisms of sensory attributes and flavor components in watercored apple. *J Jpn Soc Food Sci Technol* 63:101–116.

Tazawa J, Oshino H, Kon T, Kasai S, Kudo T, Hatsuyama Y, Fukasawa-Akada T, Yamamoto T, Kunihisa M (2019) Genetic characterization of flesh browning trait in apple using the non-browning cultivar 'Aori 27'. *Tree Genet Genom* 15:49.

Teh SL, Kostick SA, Evans KM (2021) Genetics and breeding of apple scions. In: Korban SS (ed) *The apple genome*. Springer Nature, Switzerland, pp. 73–103.

Telias A, Kui LW, Stevenson DE, Cooney JM, Hellens RP, Allan AC, Hoover EE, Bradeen JM (2011) Apple skin patterning is associated with differential expression of *MYB10*. *BMC Plant Biol* 11:93.

Terakami S, Nishio S, Kato H, Takada N, Saito T, Yamamoto T (2021) Genetic mapping of the dominant gene controlling weeping habit in Japanese chestnut (*Castanea crenata* Sieb. et Zucc.). *Tree Genet Genom* 17:16.

Thapa R, Singh J, Gutierrez B, Arro J, Khan A (2021) Genome-wide association mapping identifies novel loci underlying fire blight resistance in apple. *Plant Genome* 14:e20087.

Tian YK, Wang C-H, Zhang J-S, James C, Dai H-Y (2005) Mapping *Co*, a gene controlling the columnar phenotype of apple, with molecular markers. *Euphytica* 145:181–188.
Tobutt KR (1985) Breeding columnar apples at east malling. *Acta Hortic* 159:63–68.
Tobutt KR (1988a) Columnar apple tree-maypole variety. US Plant Patent No 6, p. 184.
Tobutt KR (1988b) Columnar apple tree-telamon variety. US Plant Patent No 6, p. 224.
Tydeman HM (1933) Breeding experiments with "Paradise" apple rootstocks. *J Pomol Hortic Sci* 11:214–236.
Ubi BE (2004) External stimulation of anthocyanin biosynthesis in apple fruit. *Food Agric Environ* 2:65–70.
Umemura H, Otagaki S, Wada M, Kondo S, Matsumoto S (2013) Expression and functional analysis of a novel *MYB* gene, *MdMYB110a_JP*, responsible for red flesh, not skin color in apple fruit. *Planta* 238:65–76.
Urbanietz A, Dunemann F (2005) Isolation, identification and molecular characterization of physiological races of apple powdery mildew (*Podosphaera leucotricha*). *Plant Pathol* 54:125–133.
Urrestarazu J, Denancé C, Ravon E, Guyader A, Guisnel R, Feugey L, Poncet C, Lateur M, Houben P, Ordidge M, Fernández-Fernández F, Evans KM, Paprstein F, Sedlak J, Nybom H, Garkava-Gustavsson L, Miranda C, Gassmann J, Kellerhals M, Suprun I, Pikunova AV, Krasova NG, Torutaeva E, Dondini L, Tartarini S, Laurens F, Durel CE (2016) Analysis of the genetic diversity and structure across a wide range of germplasm reveals prominent gene flow in apple at the European level. *BMC Plant Biol* 16:130.
Urrestarazu J, Muranty H, Denancé C, Leforestier D, Ravon E, Guyader A, Guisnel R, Feugey L, Aubourg S, Celton JM, Daccord N, Dondini L, Gregori R, Lateur M, Houben P, Ordidge M, Paprstein F, Sedlak J, Nybom H, Garkava-Gustavsson L, Troggio M, Bianco L, Velasco R, Poncet C, Theron A, Moriya S, Bink M, Laurens F, Tartarini S, Durel CE (2017) Genome-wide association mapping of flowering and ripening periods in apple. *Front Plant Sci* 8:1923.
Vanblaere T, Szankowski I, Schaart J, Schouten H, Flachowsky H, Broggini GAL, Gessler C (2011) The development of a cisgenic apple plant. *J Biotechnol* 154:304–311.
Vanderplank JE (1968) *Disease resistance in plants*. Academic Press, New York.
Vanderzande S, Piaskowski JL, Luo F, Edge-Garza DA, Klipfel J, Schaller A, Martin S, Peace C (2018) Crossing the finish line: how to develop diagnostic DNA tests as breeding tools after QTL discovery. *J Hortic* 5:1.
Velasco R, Zharkikh A, Affourtit J, Dhingra A, Cestaro A, Kalyanaraman A, Fontana P, Bhatnagar SK, Troggio M, Pruss D, Salvi S, Pindo M, Baldi P, Castelletti S, Cavaiuolo M, Coppola G, Costa F, Cova V, Dal Ri A, Goremykin V, Komjanc M, Longhi S, Magnago P, Malacarne G, Malnoy M, Micheletti D, Moretto M, Perazzolli M, Si-Ammour A, Vezzulli S, Zini E, Eldredge G, Fitzgerald LM, Gutin N, Lanchbury J, Macalma T, Mitchell JT, Reid J, Wardell B, Kodira C, Chen Z, Desany B, Niazi F, Palmer M, Koepke T, Jiwan D, Schaeffer S, Krishnan V, Wu C, Chu VT, King ST, Vick J, Tao Q, Mraz A, Stormo A, Stormo K, Bogden R, Ederle D, Stella A, Vecchietti A, Kater MM, Masiero S, Lasserre P, Lespinasse Y, Allan AC, Bus V, Chagné D, Crowhurst RN, Gleave AP, Lavezzo E, Fawcett JA, Proost S, Rouze P, Sterck L, Toppo S, Lazzari B, Hellens RP, Durel CE, Gutin A, Bumgarner RE, Gardiner SE, Skolnick M, Egholm M, Van de Peer Y, Salamini F, Viola R (2010) The genome of the domesticated apple (*Malus × domestica* Borkh.). *Nat Genet* 42:833–839.
Verma S, Evans K, Guan Y, Luby JJ, Rosyara UR, Howard NP, Bassil N, Bink MCAM, van de Weg WE, Peace CP (2019) Two large-effect QTLs, *Ma* and *Ma3*, determine genetic potential for acidity in apple fruit: breeding insights from a multi-family study. *Tree Genet Genom* 15:18.
Vinatzer BA, Patocchi A, Gianfranceschi L, Tartarini S, Zhang H-B, Gessler C, Sansavini S (2001) Apple contains receptor-like genes homologous to the *Cladosporium fulvum* resistance gene family of tomato with a cluster of genes cosegregating with *Vf* apple scab resistance. *Mol Plant Microbe Interact* 14:508–515.
Vinatzer BA, Patocchi A, Tartarini S, Gianfranceschi L, Sansavini S, Gessler C (2004) Isolation of two microsatellite markers from BAC clones of the *Vf* scab resistance region and molecular characterization of scab-resistant accessions in *Malus* germplasm. *Plant Breed* 123:321–326.
Vinatzer BA, Zhang HB, Sansavini S (1998) Construction and characterization of a bacterial artificial chromosome library of aoole. *Theor Appl Genet* 97:1183–1190.
Visser T, Verhaegh JJ (1980) Resistance to poedery mildew (*Podosphaera leucotricha*) of apple seedlings growing under glasshouse and nursery conditions. In: Proceedings of the Eucarpia meeting of fruit tree breeding, Angers, 1979, pp. 111–120.
Vogt I, Wöhner T, Richter K, Flachowsky H, Sundin GW, Wensing A, Savory EA, Geider K, Day B, Hanke M-V, Peil A (2013a) Gene-for-gene relationship in the host-pathogen system *Malus × robusta* 5-*Erwinia amylovora*. *New Phytol* 197:1262–1275.
Vogt J, Schiller D, Ulrich D, Schwab W, Dunemann F (2013b) Identification of lipoxygenase (LOX) genes putatively involved in fruit flavour formation in apple (*Malus × domestica*). *Tree Genet Genom* 9:1493–1511.
Volk GM, Bramel P (2021) Apple genetic resources: diversity and conservation. In: Korban SS (ed) *The apple genome, compendium of plant genomes*. Springer Nature Switzerland, Cham, pp. 33–45.

Volz RK, Kumar S, Chagné D, Espley R, McGhie TK, Allan AC (2013) Genetic relationship between red flesh and fruit quality traits in apple. *Acta Hortic* 946:363–368.
Wada M, Iwanami H, Moriya S, Hanada T, Moriya-Tanaka Y, Honda C, Shimizu T, Abe K, Okada K (2018) A root-localized gene in normal apples is ectopically expressed in aerial parts of columnar apples. *Plant Growth Regul* 85:389–398.
Wakasa Y, Kudo H, Ishikawa R, Akada S, Senda M, Niizeki M, Harada T (2006) Low expression of an endopoly-galacturonase gene in apple fruit with long-term storage potential. *Postharvest Biol Technol* 39:193–198.
Wang LM, Yu BY, Zhao YN, Li YZ, Guo J, Zhu YD (2021) A putative 2OG-Fe(II) oxygenase's response to gibberellin deficiency is related to the internodal growth of columnar apples. *Acta Physiol Plant* 43:70.
Wang N, Jiang SH, Zhang ZY, Fang HC, Xu HF, Wang YC, Chen XS (2018) *Malus sieversii*: the origin, flavonoid synthesis mechanism, and breeding of red-skinned and red-fleshed apples. *Hortic Res* 5:70.
Watanabe D, Takahashi I, Jaroensanti-Tanaka N, Miyazaki S, Jiang K, Nakayasu M, Wada M, Asami T, Mizutani M, Okada K, Nakajima M (2021) The apple gene responsible for columnar tree shape reduces the abundance of biologically active gibberellin. *Plant J* 105:1026–1034.
Watkins CB, Erkan M, Nock JF, Iungerman KA, Beaudry RM, Moran RE (2005) Harvest date effects on maturity, quality, and storage disorders of 'Honeycrisp' apples. *HortScience* 40:164–169.
Way RD, Lamb RC, Pratt C, Cummins JN (1976) Pale green lethal gene in apple clones. *J Am Soc Hort Sci* 101:679–684.
Webster AD, Tobutt K (2001) Breeding and selection of new apple rootstocks at Horticulture Research International-East Malling. *Acta Hortic* 557:189–192.
Webster AD, Wertheim SJ (2003) Apple rootstock. In: Ferree DC, Warrington IJ (eds) *Apples: botany, production and uses*. CABI, Wallingford, pp. 91–124.
Werner DJ, Chaparro JX (2005) Genetic interactions of pillar and weeping peach genotypes. *HortScience* 40:18–20.
Wertheim SJ, Webster AD (2003) Propagation and nursery tree quality. In: Ferree DC, Warrington IJ (eds) *Apples: botany, production and uses*. CABI, Wallingford, pp. 125–151.
Wilcox AN, Angelo E (1936) Apple breeding studies I. Fruit color. *Proc Amer Soc Hortic Sci* 33:108–113.
Wöhner TW, Flachowsky H, Richter K, Garcia-Libreros T, Trognitz F, Hanke MV, Peil A (2014) QTL mapping of fire blight resistance in *Malus × robusta* 5 after inoculation with different strains of *Erwinia amylovora*. *Mol Breed* 34:217–230.
Wolters PJ, Schouten HJ, Velasco R, Si-Ammour A, Baldi P (2013) Evidence for regulation of columnar habit in apple by a putative 2OG-Fe(II) oxygenase. *New Phytol* 200:993–999.
Wood AR, Hernandez DG, Nalls MA, Yaghootkar H, Gibbs JR, Harries LW, Chong S, Moore M, Weedon MN, Guralnik JM, Bandinelli S, Murray A, Ferrucci L, Singleton AB, Melzer D, Frayling TM (2011) Allelic heterogeneity and more detailed analyses of known loci explain additional phenotypic variation and reveal complex patterns of association. *Hum Mol Genet* 20:4082–4092.
Würdig J, Flachowsky H, Saß A, Peil A, Hanke M (2015) Improving resistance of different apple cultivars using the *Rvi6* scab resistance gene in a cisgenic approach based on the Flp/FRT recombinase system. *Mol Breed* 35:95.
Xu K (2015) Arctic apples: a look back and forward. *New York Fruit Quartery* 23:21–24.
Xu KN, Wang AD, Brown S (2012) Genetic characterization of the *Ma* locus with pH and titratable acidity in apple. *Mol Breed* 30:899–912.
Xu ML, Huaracha E, Korban SS (2001) Development of sequence-characterized amplified regions (SCARs) from amplified fragment length polymorphism (AFLP) markers tightly linked to the *Vf* gene in apple. *Genome* 44:63–70.
Xu ML, Korban SS (2002) A cluster of four receptor-like genes resides in the *Vf* locus that confers resistance to apple scab disease. *Genetics* 162:1995–2006.
Yamanouchi H, Koyama A, Machii H, Takyu T, Muramatsu N (2009) Inheritance of a weeping character and the low frequency of rooting from cuttings of the mulberry variety 'Shidareguwa'. *Plant Breed* 128:321–323.
Yang F, Shen F, Liu Z, He M, Wang Y, Wu T, Xu X, Li W, Qiu C, Zhang X, Han Z (2021) Development and validation of functional markers from Dw1 and Dw2 loci to accurately predict apple rootstock dwarfing ability. *Res Square* doi:10.21203/rs.2.24669/v4.
Young M, Kuykendall LD, Martinez-Romero E, Kerr A, Sawada H (2001) A revision of *Rhizobium* Frank 1889, with an emended description of the genus, and the inclusion of all species of *Agrobacterium* Conn 1942 and *Allorhizobium undicola* de Lajudie et al. 1998 as new combinations: *Rhizobium radiobacter*, *R. rhizogenes*, *R. rubi*, *R. undicola* and *R. vitis*. *Int J Syst Evol Microbiol* 51:89–103.
Yu YG, Saghai Maroof MA, Buss GR (1996) Divergence allelomorphic relationship of a soybean virus resistance gene based on tightly linked DNA microsatellite and RFLP markers. *Thor Appl Genet* 92:64–69.

Yuan KJ, Wang CJ, Wang JH, Xin L, Zhou GF, Li LG, Shen GN (2014) Analysis of the MdMYB1 gene sequence and development of new molecular markers related to apple skin color and fruit-bearing traits. *Mol Genet Genom* 289:1257–1265.

Zhang J, Zhang QX, Cheng TR, Yang WR, Pan HT, Zhong JJ, Huang L, Liu EZ (2015) High-density genetic map construction and identification of a locus controlling weeping trait in an ornamental woody plant (*Prunus mume* Sieb. et Zucc). *DNA Res* 22:183–191.

Zhang LY, Hu J, Han XL, Li JJ, Gao Y, Richards CM, Zhang CX, Tian Y, Liu GM, Gul EA, Wang DJ, Tian Y, Yang CX, Meng MH, Yuan GP, Kang GD, Wu YL, Wang K, Zhang HT, Wang DP, Cong PH (2019a) A high-quality apple genome assembly reveals the association of a retrotransposon and red fruit colour. *Nat Commun* 10:1494.

Zhang Q, Karkee M, Tabb A (2019b) The use of agricultural robots in orchard management. In: Billingsley J (ed) *Robotics and automation for improving agriculture*. Burleigh Dodds Science Publishing, Cambridge, pp. 187–214.

Zhang Q, Ma BQ, Li H, Chang YS, Han YY, Li J, Wei GC, Zhao S, Khan MA, Zhou Y, Gu C, Zhang XZ, Han ZH, Korban SS, Li SH, Han YP (2012) Identification, characterization, and utilization of genome-wide simple sequence repeats to identify a QTL for acidity in apple. *BMC Genom* 13:537.

Zhang XJ, Wang LX, Chen XX, Liu YL, Meng R, Wang YJ, Zhao ZY (2014) A and *MdMYB1* allele-specific markers controlling apple (*Malus*×*domestica* Borkh.) skin color and suitability for marker-assisted selection. *Genet Mol Res* 13:9103–9114.

Zhang Y, Li P, Cheng L (2010) Developmental changes of carbohydrates, organic acids, amino acids, and phenolic compounds in 'Honeycrisp' apple flesh. *Food Chem* 123: 1013–1018.

Zhu YD, Zhang W, Li GC, Wang T (2007) Evaluation of inter-simple sequence repeat analysis for mapping the *Co* gene in apple (*Malus pumila* Mill.). *J Hortic Sci Biotechnol* 82:371–376.

Zhu YM, Barritt BH (2008) Md-ACS1 and Md-ACO1 genotyping of apple (*Malus* x *domestica* Borkh.) breeding parents and suitability for marker-assisted selection. *Tree Genet Genom* 4:555–562.

Zini E, Biasioli F, Gasperi F, Mott D, Aprea E, Mark TD, Patocchi A, Gessler C, Komjanc M (2005) QTL mapping of volatile compounds in ripe apples detected by proton transfer reaction-mass spectrometry. *Euphytica* 145:269–279.

3 Tools and Techniques for Allele Mining in Apple

Cheol Choi, Laura Dougherty, and Seunghyun Ban
Kyungpook National University

3.1 INTRODUCTION

Apples are a high-value commodity grown in temperate climate around the world. Over 86 million tons of apples were produced worldwide by 95 countries in 2020 (FAOSTAT, http://faostat.fao.org). Apple fruit are versatile, they can be eaten fresh or processed and some apple trees are used in ornamental landscaping for their aesthetic.

Apple (genus *Malus*), a member of the *Rosaceae* family, consist of over 30 species (Hancock et al., 2008). Both domesticated (*Malus domestica*) and wild apple (*Malus*) cultivars/species trees have different growth habits, disease resistance, and vigor, while fruits vary in taste, texture, color, size, and post-harvest storage qualities. While wild apples are not typically consumed, they are sources for genetic diversity and favorable traits that are introgressed into domesticated apples to create superior cultivars. Genetically, apple is a highly heterozygous outcrossing species with self-incompatibility. Conventional breeding of apples from initial cross, to seedling evaluation and cultivar release takes well over a decade (Hancock et al., 2008). Introgression from wild species can take over 50 years (Schouten et al., 2006). Apples have a long juvenile phase that can span years (typically 4–10 years) depending on genotype, environmental, and cultural practices (Visser, 1964). Coordination of resources (land, labor, supplies) for such an extended time is costly. The cost to breed a new apple variety is hundreds of thousands of dollars (Fenning & Gershenzon, 2002). Modern breeding efforts using marker-assisted selection screen progeny based on reliable genetic markers directly related to traits of interest, predicting phenotypes in the seedling stage which may not be visible for years saves resources by accelerating the removal of unwanted progeny (Wannemuehler et al., 2019).

The identification of allelic variations and the genes underlying traits of interest is essential for the expansion of molecular breeding. Segregating progeny, diverse germplasms, and somatic mutants provide backgrounds for trait exploration. Allele mining approaches to identify the genetic variation and candidate genes controlling those traits of interest can accelerate breeding efficiency (Kumar et al., 2010a). Here, we highlight techniques and tools used for allele mining in apples, key horticultural trait discoveries in apple and the application of allele mining for future breeding efforts.

3.2 APPLE REFERENCE GENOMES, TRANSCRIPTOMES, AND ONLINE TOOLS

The introduction and accessibility of Next-Generation Sequencing (NGS) and third-generation sequencing have ushered in a decade of *Malus* genome assemblies (Table 3.1) (Velasco et al., 2010; Li et al., 2016b; Daccord et al., 2017; Zhang et al., 2019; Chen et al., 2019; Broggini et al., 2020; Sun et al., 2020; Tian et al., 2022). Combinations of NGS short reads, third-generation sequencing long reads (>10 Kb), chromosome conformation capture (Hi-C) sequencing, optical mapping, and various assembly methods have unlocked the genetic code of apples. Advances in sequencing technologies and genome assemblies have produced phased genomes, revealing individual haplotypes. Current apple reference genomes from *Malus domestica*, *Malus baccata*, *Malus sieversii*, and *Malus sylvestris* provide insight into the genomic makeup and evolution of *Malus* (Peace et al., 2019; Duan et al., 2017; Sun et al., 2020) (Table 3.1).

 DOI: 10.1201/9781003386490-3

TABLE 3.1
Summary of Available Apple Reference Genomes

	Cultivar/Species	Mapped Genome Size for Haploid (Mb)	Predicted Number of Protein-Coding Genes	References
Malus domestica	Diploid 'Golden Delicious'	603.9	57,386	Velasco et al. (2010)
	Double haploid 'Golden Delicious'	643.2	42,140	Daccord et al. (2017)
	'Hanfu'	658.9	44,677	Zhang et al. (2019)
	'Gala'	652–688	45,199–45,352	Sun et al. (2020)
	'Galaxy Gala'	719.4	41,264	Broggini et al. (2020)
	'Royal Gala'	765	42,788	Tian et al. (2022)
	'Honeycrisp'	660–674	47,563–48,655	Khan et al. (2022)
Other *Malus* species	*Malus baccata*	655	46,144	Chen et al. (2019)
	Malus sieversii	667.7Mb	45,210	Sun et al. (2020)
	Malus sylvestris	661.1Mb	45,199	Sun et al. (2020)

3.2.1 The Diploid 'Golden Delicious' Genome

In 2010, the first domesticated apple (*Malus domestica*) draft genome was published in *Nature Genetics* (Velasco et al., 2010). The diploid cultivar 'Golden Delicious' (v1.0) was assembled using whole-genome shotgun sequencing approach. This genome assembly of 1,629 mega contigs made up 17 linkage groups (LGs) of 603.9 megabases (Mb), or 81.3% of the estimated genome while repetitive elements made up 98% of the unassembled genome (138.4 Mb). There were 57,386 predicted genes with 11,444 of those genes being apple specific and 4,021 transcription factors. In 2014, an updated version of the reference genome was released (v2.0). This version contains pseudo haplotype assemblies, derived from the initial contigs. In 2015, a third version of this reference genome was released (v3.0). This version involved the anchoring of contigs from 21 genetic maps (Bianco et al., 2014). In 2016, a hybrid de novo genome assembly of 'Golden Delicious' was released with greater coverage and assembly size of 632.4 Mb (Li et al., 2016b). This assembly contained 53,922 protein-coding genes and 2,765 non-coding RNA genes.

3.2.2 The Double Haploid 'Golden Delicious' Genome

In 2017, a double haploid 'Golden Delicious' genome (GDDH13) was published (Daccord et al., 2017). This genome assembled to 643.2 Mb with the total genome size estimated at 651 Mb. There are 42,140 genes in the genome with 1,965 non-protein-coding genes. The homozygosity of this GDDH13 genome allowed for higher genome resolution than the previous heterozygous 'Golden Delicious' diploid genome. This genome is widely used in genetic studies due to its extensive coverage and gene annotation.

3.2.3 Other Apple Genomes

Since the initial assemblies and releases of 'Golden Delicious' reference genomes, other *Malus domestica* genomes have been published. These include the 'Hanfu' apple, a Chinese cultivar with desirable bright red fruit skin, cold-resistance and long shelf-life (Zhao et al., 2009), 'Gala,' a top variety in the United States of America and 'Gala' sports: 'Galaxy Gala' and 'Royal Gala' (Sun et al., 2020; Broggini et al., 2020; Tian et al., 2022). Additionally important wild apple species

Malus baccata, *Malus sieversii*, and *Malus sylvestris* genomes have been released (Chen et al., 2019; Sun et al., 2020).

3.2.3.1 *M. domestica* Genomes

In 2019, the genome of cultivar 'Hanfu' ('Dongguang' × 'Fuji') was assembled from anther-derived trihaploid DNA (Zhang et al., 2019). This highly homozygous genome assembly, consisted of 17 chromosomes of 658.90 Mb, with a total genome size estimated as 708.54 Mb with transposable elements making up 393.88 Mb. Compared to the GDDH13 genome, this genome contained fewer gaps per chromosome. In total, 44,677 genes and 17 large inversions were identified.

A diploid 'Gala' phased reference genome was published in 2020 (Sun et al., 2020). The haploid genome was 652.4 Mb in size with 58.72% repetitive elements, similar to the Hanfu and GDDH13 genomes. The genome had 45,352 annotated genes. A draft reference genome of 'Galaxy Gala,' a sport mutant of 'Gala' has been completed (Broggini et al., 2020). This heterozygous genome contains 41,264 genes. Most recently, a tetra-haploid 'Royal Gala' genome (RGTH1) has been assembled from anther culture (Tian et al., 2022). This genome is 765 Mb and contains 42,788 genes. Recently, the 'Honeycrisp' phased reference genome was uploaded to bioRxiv (Khan et al., 2022). The size of the two assembled haplomes was 674 and 660 Mb, respectively, and 47,563 and 48,655 protein-coding genes were annotated from each haplome.

3.2.3.2 *Malus* Genomes

In 2019, the genome of the wild apple species *Malus baccata* was published (Chen et al., 2019). *Malus baccata* can hybridize with *Malus domestica,* and is an important apple species for rootstock, disease resistance, and cold tolerance. The assembled genome was 665 with 413 Mb of transposable elements. In total 46,144 coding genes were identified.

In 2020, phased diploid genome of *Malus sieversii* was published (Sun et al., 2020). The haploid genome assembly was 667.7 Mb with 59.36% of the genome containing repetitive elements. There were 45,210 annotated genes. Additionally in 2020, the phased diploid genome of *Malus sylvestris* was publish by the same group (Sun et al., 2020). The haploid version was 661.1 Mb with 59.31% repetitive sequences and 45,199 genes. *Malus sieversii* and *Malus sylvestris* are the main progenitor species of the domesticated apple (Duan et al., 2017; Sun et al., 2020).

3.2.4 Transcriptomes

The Genome Database for Rosaceae (GDR) *Malus × domestica* reference transcriptome V1 combines 2.1 billion RNA-sequencing reads from peer-reviewed data sets (Legay et al., 2015; Petersen et al., 2015; Xia et al., 2012;; Kim et al., 2012; Ke et al., 2014; Gusberti et al., 2013; Johnson & Zhu, 2015) along with 326,940 expressed sequence tags (ESTs). This reference transcriptome has 85,918 sequences which have been annotated for putative gene function, InterPro domains (Blum et al., 2021) and gene ontology (Conesa et al., 2005).

Additionally, an apple transcriptome was released based on RNA sequencing at 14 'Golden Delicious' fruit developmental stages (Bai et al., 2014). These RNA-seq reads were aligned to the 'Golden Delicious' genome v1.0 (Velasco et al., 2010) and also assembled de novo. In total, 71,178 genes were identified of which, 17,524 were novel transcripts.

3.2.5 Online Tools and Resources

The National Center for Biotechnology Information (NCBI) (http://www.ncbi.nlm.nih.gov) within the National Library of Medicine (NLM) at the National Institutes of Health (NIH) has many resources and tools for use (Benson et al., 2013; Sayers et al., 2021). Two important tools include GenBank® and basic local alignment search tool (BLAST). GenBank® is a nucleotide database of publically available sequences including genome assemblies and genes (Benson et al., 2013). BLAST is a powerful tool where input (nucleotide or protein) sequences are searched against the sequence databases to find similar sequences (Altschul et al., 1990). Within BLAST, sequences can be directly aligned and compared.

The GDR (https://www.rosaceae.org/) is an informative web-based resource to access publically available apple genomes and transcriptomes. (Jung et al., 2019). It contains genetic maps, diversity and single nucleotide polymorphism (SNP) array data, genetic markers, trait loci and breeding data (Evans et al., 2013). Individual genes and their functional annotations can also be searched. The site has an embedded JBrowse genome browser (Figure 3.1) (Buels et al., 2016) and BLAST function.

Genome data and epigenome data from the GDDH13 genome (Daccord et al., 2017) can be found on a site maintained by the Bucher lab (https://iris.angers.inra.fr/gddh13/). Similar to GDR, there is an embedded JBrowse viewer (Buels et al., 2016) that can be overlaid with gene models, transposable elements, epigenomics, SNP markers, mRNA contigs, and small RNAs mappings. Additionally, gene models from the original 'Golden Delicious' genome can be overlaid as well.

The RosBREED project (https://www.rosbreed.org/), involving US university scientists and international scientists, combines disease resistance with horticultural quality traits to develop new cultivars based on DNA tests and modern breeding methods. The International Rosaceae SNP Consortium created an apple 8K SNP array (Chagné et al., 2012).

High-quality Disease Resistant Apples for a Sustainable Agriculture (HIDRAS) (https://sites.unimi.it/camelot/hidras/index.php) is a database created by European scientists (Gianfranceschi & Soglio, 2003). It is a useful tool for breeders, providing molecular markers and linkage maps linked to fruit quality and pathogen resistance quantitative trait loci (QTLs). The site is no longer updated, but simple sequence repeat (SSR) markers, amplified sequences, and primer sequences are available

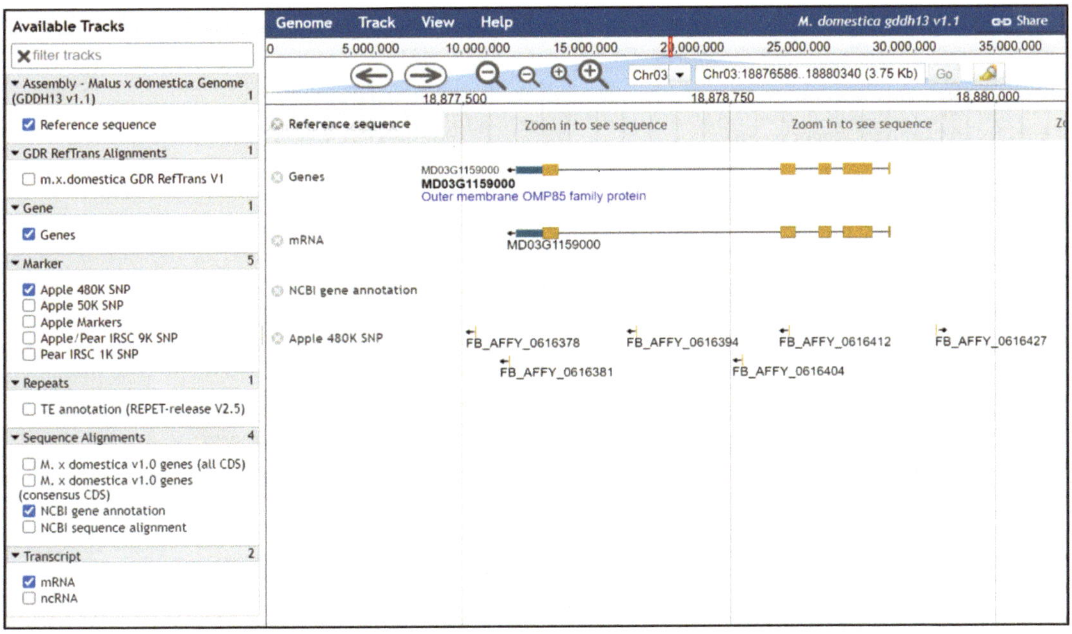

FIGURE 3.1 Screenshot of JBrowse genome browser on GDR site. On the right-hand side, different markers, alignments, and transcripts can be viewed. Genes are annotated with gene ID (MD03G1159000) and common gene name (outer membrane OMP85 family protein). The top right corner indicates what genome you are viewing and below that is the chromosome and chromosomal location.

from multiple papers (Guilford et al., 1997; Hokanson et al., 1998; Liebhard et al., 2002; Silfverberg-Dilworth et al., 2006; Celton et al., 2009; Wang et al., 2012).

3.3 ALLELE MINING METHODS IN APPLE

Allelic variations in genes underlying important horticultural traits in apple are of great interest to breeders. Genetic variations including insertions, deletions or polymorphisms within the coding sequence (CDS) of a gene, the 5′ and 3′ untranslated regions, introns or the promoter region can alter plants phenotypes (Kumar et al., 2010a). The phenotypes result from loss or gain of function mutations. In the CDS, insertions and deletions cause frame-shift mutations; however, insertions or deletions in multiples of three, add or remove codons (Roth, 1974). Insertions of transposable elements and large deletions are also sources of genetic change (Ban et al., 2022; Okada et al., 2016). SNPs or nucleotide substitutions can cause synonymous or non-synonymous changes in the CDS. Synonymous SNPs, sometimes referred to as silent mutations, do not change the codon/amino acid sequence. Non-synonymous SNPs cause codon changes, resulting in an amino acid substitution. Non-synonymous SNPs are classified as either missense (amino acid substitution) or nonsense (stop codon introduction) (Wasana, 2018). Another type of polymorphism is a SSR or microsatellite which are tandemly repeated nucleotide sequences (Powell et al., 1996).

There are different approaches for the identification of allelic variation that can be utilized depending on the trait of interest and material for the study. Such approaches include whole-genome resequencing, genotyping by sequencing (GBS), SNP arrays, genome-wide association studies (GWAS), SSR markers, linkage mapping, and RNA sequencing, among others. Often, different approaches complement each other and are used in studies together (Zhang et al., 2014; Xing et al., 2016). Study materials consist of bi-parental populations, sports and germplasm collections. For apple, there are several germplasm repositories that contain large collections of species and cultivars around the world (Kumar et al., 2010b; Gao et al., 2015; Liang et al., 2015; Watts et al., 2021; Venison et al., 2022). For example, the US Department of Agriculture *Malus* collection houses over 6,000 cultivars and 48 species from around the world (Dougherty et al., 2021).

3.3.1 Mapping Approaches

In apple, most traits are quantitative, meaning they have a measurable phenotypic variation owing to genetic and/or environmental influences. These traits are influenced by several genes of varying effects (Members of the Complex Trait, 2003). QTL analysis is a statistical method that combines phenotypic and genotypic data to explain the genetic basis of the trait and associate it to a genetic location (Myles & Wayne, 2008; Kenis et al., 2008). Linkage mapping is the association of genetic markers in a population to each other and is used to discovery recombination frequencies (Kulwal, 2018).

SSR, restriction fragment length polymorphisms (RFLP) (Williams et al., 1995), amplified fragment length polymorphism (AFLP) and SNP markers can all be used for QTL and trait mapping (Hokanson et al., 2001; Kenis & Keulemans, 2005; Lassois et al., 2016). Detailed linkage maps for apple cultivars can be found on HIDRAS, and numerous SSR markers have been developed for mapping (Liebhard et al., 2002; Silfverberg-Dilworth et al., 2006).

3.3.2 Variant Detection

3.3.2.1 Whole-Genome Resequencing

High-throughput sequencing reads can be mapped back to a reference genome to discover allelic variations. The success of whole-genome resequencing depends on the quality and completeness of the reference genome and compatibility of the sequenced sample to the reference genome. Some *Malus* species map poorly to the *Malus domestica* reference genomes because of differences in genome structure (Chen et al., 2019). This means that all variation cannot be detected. Whole-genome sequencing of individual cultivars, mapped back against the reference genome can be used to identify SNPs, or indels (Zhang et al., 2014; Xing et al., 2016). Bulk Sergeant Analysis (BSA) is a type of whole-genome resequencing where DNA from individual samples is bulked into two pools of contrasting phenotypes. The causal variations would be more abundant in the one pool over the other. BSA has been used to determine QTLs for weeping growth habit (Dougherty et al., 2018), acidity genes (Jia et al., 2018), and disease resistance (Shen et al., 2019).

3.3.2.2 Genotyping by Sequencing (GBS)

GBS is a cost effective way to obtain highly multiplexed reduced representation libraries (Elshire et al., 2011). Genomic DNA is digested with restriction enzymes, barcoded and sequenced. SNPs, rare variants, and structural variants can be identified for whole populations in a single experiment (Kumar et al., 2014). There are a variety of method types used for GBS (Scheben et al., 2017). Two types are restriction-site-associated DNA sequencing (RADseq) (Davey & Blaxter, 2011) and double-digest restriction-site-associated DNA sequencing (ddRADseq) (Peterson et al., 2012). Both RADseq and ddRADseq have been used to identify QTLs for fruit quality traits and drought stress (Sun et al., 2015; Wang et al., 2018). Additionally GBS has been used to identify SNPs related to apple fruit skin color (Gardner et al., 2014), examine the population structure and relatedness of the Pometum gene bank collection (Larsen et al., 2018), and genotype SNPs from 1,175 accessions in Canada's Apple Biodiversity Collection (Migicovsky et al., 2022).

3.3.2.3 SNP Arrays

SNP arrays are a type of DNA array where probes are primed with known SNPs and input samples hybridize, revealing their sequence for the particular SNP (LaFramboise, 2009). For apple, an initial array that consist of 237 SNPs was based on the 'Golden Delicious' v1.0 genome (Micheletti et al., 2011). An 8K SNP array by IRSC (International RosBreed SNP Consortium) based on the whole-genome sequencing data of 27 accessions was then produced (Chagne et al., 2012). A 20K Illumina Infinium array was developed from the FruitBreedomics consortium using 14 accessions and two double haploids from 'Golden Delicious' (Bianco et al., 2014). A few years later, an Affymetrix Apple480K array was developed based on a total of 67 resequenced data (Bianco et al., 2016).

The reason why the SNP array has been continuously developed in apple is that it can be easily used as a unified platform in the research community and has the advantage of being able to produce results in a relatively short time from many samples (Peace et al., 2019). However, the disadvantage is that allele mining for other gene regions not included in the SNP array cannot be performed, and since the accessions used for development were mainly *Malus domestica*, it may not be suitable for experiments targeting wild accessions.

3.3.2.4 RNA Sequencing

RNA sequencing is used to capture a snapshot of the transcriptome, analyze gene expression, facilitate the discovery of novel transcripts, identify alternative splicing, and detect of allele-specific and tissue-specific gene expression (Kukurba & Montgomery, 2015; Brown et al., 2017). Comparisons between differently expressed genes under a QTL of interest can help determine candidate genes underlying that QTL and SNPs can be identified in CDSs (Zhang et al., 2012b).

3.3.2.5 Genome-Wide Association Studies (GWAS)

GWAS is a type of SNP analysis to identify statistically significant variants associated with the trait of interest (Korte & Farlow, 2013). GWAS can be used with diverse samples and is a popular type of analysis in apple (Lee et al., 2017; Larsen et al., 2019; Cao et al., 2021; Thapa et al., 2021). Additionally, Pedigree-Based Analysis (PBA) for allele mining using multiple pedigreed full-sib families has also been actively used by RosBreed and FruitBreedomics (Guan et al., 2015; Howard et al., 2017).

3.3.3 Data Analysis Tools

WGS and GBS generate large data sets that must be analyzed. Numerous software packages and tools are available for data analysis. Command line-based bioinformatics pipelines are commonly used. Command line pipelines and packages are available, or custom scripts can be created in various programming languages. Some available open-source R packages include Bioconductor (https://www.bioconductor.org/) (Team, 2013) and R/qtl (Broman et al., 2003).

User-friendly bioinformatics software is also available for data analysis. QIAGEN CLC Genomics Workbench (https://digitalinsights.qiagen.com/) software has tools that can perform de novo assemblies, alignment to references, variant calling, and RNA-seq expression analysis, among others. It has been used frequently in apple research (Bai et al., 2014; Otto et al., 2014; Okada et al., 2016; Hargarten et al., 2018; Ban et al., 2022). Galaxy, an open-source software (https://usegalaxy.org), is another platform that can be used for assembly, mapping, variant calling, and other analysis (Afgan et al., 2018). Geneious by Dotmatics (https://www.geneious.com/) software provides similar tools as those listed above. Trait Analysis by aSSociation, Evolution and Linkage (TASSEL) is another open-source software to analyze traits associations, evolutionary patterns, and linkage disequilibrium (Bradbury et al., 2007).

3.4 ALLELE MINING FOR FRUIT QUALITY TRAITS

Apple fruit quality is perceived as combination of taste, texture, shape, and size of the fruit (Manalo, 1990). Quality can be further described as fruit flavor, sweetness, acidity, crispiness, firmness, and juiciness. The importance a consumer or a producer places on each quality attribute can vary (Harker et al., 2008; Yue et al., 2013).

3.4.1 Fruit Acidity

Fruit acidity directly influences consumers perceived fruit quality and preference (Harker et al., 2008; Zhang et al., 2010). Malic acid is predominant acid in mature apple fruit, though citric acid, fumaric acid and quinic acid are also present (Zhang et al., 2010). Malic acid content in fruit differs between wild and domesticated apples (Ma et al., 2015). On average, acidity level in wild species was 2.2 times higher compared to domesticated apples, with a boarder range of acidity; from 0.8 to 22.7 mg/g in wild species verses 0.5 to 18.9 mg/g in domesticated apples. When breeding apples, the acceptable acidity range for fresh market fruit is between 3 and 10mg/mL corresponding to a pH between 3.1 and 3.8 (Visser & Verhaegh, 1978).

Early acidity genetic studies in apple examined the ratio of acidity in segregating populations and determined that low acidity was determined by a single recessive gene designated as *Ma* (malic acid). The *Ma* allele represented medium and high acidity, and the *ma* allele represented low acidity (Visser & Verhaegh, 1978). Medium- and high-acidity genotypes exhibit dominance over low-acidity genotypes (Nybom, 1959; Brown & Harvey, 1971; Visser & Verhaegh, 1978).

Years later, the genetic location of *Ma* was mapped to LG16 through a QTLs study using a segregating bi-parental F_1 population with heterozygous (*Mama*) parents: 'Prima' and 'Fiesta' (Maliepaard et al., 1998). This mapping position was further confirmed by independent studies with different bi-parental F_1 populations (Kenis et al., 2008; Xu et al., 2012; Liebhard et al., 2003). Interestingly, the effect of the QTL *Ma* on fruit acidity differed depending on the parental types. For example, the *Ma* QTL explained 36% of fruit acidity variation in 'Fiesta' × 'Discovery' (Liebhard et al., 2003), 20%–34% in 'Telamon' × 'Braeburn' (Kenis et al., 2008), 42.3% in 'Royal Gala' × '*M. sieversii*: PI613988' (Xu et al., 2012), and 12.4% in 'Jiguan' × 'Wangshanhong' (Ma et al., 2016). This suggests the possibility of other gene involvement in acidity. Genetic fine mapping of the *Ma* QTL with SSR markers revealed two aluminum-activated malate transporter (ALMT) genes, *Ma1* and *Ma2*, regulating malate accumulation (Bai et al., 2012). Interestingly, a significant correlation was confirmed between TA and *Ma1* (R^2=0.4543, P=0.0021), but no significant correlation was found with *Ma2* (Bai et al., 2012). This result suggests that *Ma1*, not *Ma2*, is a strong candidate gene for regulating fruit acidity in apple.

Sequencing of *Ma1* alleles of 'G41' and 'Golden Delicious' revealed a SNP in the CDS at position 1455 changing a guanine 'G' to an adenine 'A.' This 'G' to 'A' SNP resulted in a nonsense mutation, introducing a premature stop codon, truncating 84 amino acids of the *Ma1* protein c-terminal end. Accessions homozygous for the premature stop codon (A/A) at position 1455 had significantly reduced acidity compared to heterozygous (G/A) and homozygous (G/G) accessions (Bai et al., 2012). Therefore, the high-acidity allele (*Ma1*) contains a 'G' at position 1455, while the low-acidity allele (*ma1*) contains an 'A' at position 1455. In recent study, when Ma1–1455G and Ma1–1455A proteins were expressed separately in *Xenopus laevis* oocytes and *Nicotiana benthamiana* cells, the Ma1–1455G protein showed significantly higher malate transport activity than the Ma1–1455A protein (Li et al., 2019a).

A CAPS marker was developed to target base 1455 in the *Ma1* gene. The marker amplified a 2 kB region of the *Ma1* gene. The PCR products were then digested with restriction enzyme BspHI (New England Bio Labs, Ispwich MA, USA) which cuts the PCR product if the low-acid allele is present (~1.7 kB and 249 bp fragments). This marker can be used to genotype and screen cultivars for high or low acidity (Bai et al., 2012).

In addition to the major QTL *Ma* on LG16, another QTL was identified on LG8 from diverse bi-parental F_1 populations. This LG8 QTL explained 13.5% of the acidity in 'Fiesta' × 'Discovery' (Liebhard et al., 2003), 13.5% in 'Jonathan' × 'Golden Delicious' (Zhang et al., 2012a), 8% in 'Telamon' × 'Braeburn' (Kenis et al., 2008), and 14.4% in 'Jiguan' × 'Wangshanhong' (Ma et al., 2016).

Fruit acidity related major QTLs located on LGs 8 and 16 were also identified through the genomic selection study using seven full-sib families (Kumar et al., 2012). Sun et al. (2015) identified 3,441 SNP markers by RADseq from a large (N=297) F_1 population ('Jonathan' × 'Golden Delicious'), and a malic acid QTL was identified on LG8, accounting for 39.1% acidity variance (Sun et al., 2015). This QTL region was aligned to LG8 at between positions 13.66 and 14.2 Mb. Candidate genes *MdMYB4* (R2R3-MYB transcription factor) and *MdME* (Malic enzyme gene) underlying the acidity QTL on LG8 were identified in this region by comparing the gene expression and fruit acidity changes during the fruit development in the high-acidity cultivar 'Jonathan' and the low-acidity cultivar 'Fuji' (Sun et al., 2015). Additional candidate genes *MdSAUR37*, *MdPP2CH*, and *MdMYB44* were identified on LG8 from QTL mapping and bulk segregant analysis sequencing (BSA-Seq) (Jia et al., 2018). Based on genetic variation of these three genes, the authors developed linked molecular markers and compared malate content to marker genotype using two F_1 populations and 92 *Malus* accessions. It is thought that apple fruit acidity is regulated by epistatic genetic control of these three genes. In addition to the major QTLs for acidity on LGs 8 and 16, minor QTLs were identified on LGs 2, 10, 13, 15, 17 (Kenis et al., 2008) and LG1 (4.3% TA variance explained), and LG6 (6.2% TA variance explained) (Xu et al., 2012).

The major QTLs on LGs 8 and 16 from these experiments on apple fruit acidity can largely explain lowered acidity level, but the large variation in medium and high-acidity groups remains largely unknown. Previous approaches, such as QTL mapping or GWAS, may lack the sensitivity needed for discovery of novel fruit acidity QTL, by only reconfirming the major QTLs on LGs 8 and 16, due to their large effects. The low-acidity genotypes at LGs 16 and 8 may prevent the detection of any significant minor QTLs that determine medium and high acidity. In an experiment using a F_1 segregating population from 'Royal Gala' × 'PI 613988' with no effect from the LG8 or *ma1/*

mal, LGs from Chr.4 and 6 were identified for high acidity (Ban & Xu, 2020). The combined effect of the QTLs on LGs 4 and 6 explained 50.7% of the malic acid variation in the *Mal/Mal* background. Further investigation is needed to identify candidate genes for LGs 4 and 6.

3.4.2 Fruit Color

Fruit skin color is an important factor in consumer preference (Iglesias et al., 2008). The red color of the fruit is determined by the accumulated anthocyanin content in the vacuole (Tsao et al., 2003), and it has been known to be regulated by both environmental and genetic factors. Many experiments have been conducted to find genetic factors of fruit skin color, and *MYB1* and *MYB10* genes have been identified as key genes through several allele mining techniques.

In the promoter region of MYB1, a SNP was identified and subsequently a dCAPS marker was developed to target the SNP. The marker can be used to predict red or green fruit skin color (Takos et al., 2016). Furthermore, overexpression of *MYB1* gene in Arabidopsis and cultured grape cells induced anthocyanin expression. However, more studies have been reported on the *MYB10* gene, which is known to explain more than 80% of phenotypic variance for fruit skin color in diverse *Malus* populations (Peace et al., 2019). Mapping of red foliage and red core color in 32 segregating population, *MdMYB10* gene was found in LG9, and the developed indel marker revealed the relationship between foliage and fruit color (Chagne et al., 2007). Furthermore, it was confirmed that anthocyanin expression was induced when the *MdMYB10* gene was transiently expressed in tobacco leaves and in the stable transformants of apple plants (Espley et al., 2007). Thus, red color is determined by the presence or absence of rearrangement of multiple repeats in the promoter region of the *MdMYB10* gene (Espley et al., 2009). It was also found that the degree of methylation of the *MdMYB10* gene promoter region in the yellow skin apple somatic mutation is an epigenetic factor of the mutation (El-Sharkawy et al., 2015). According to a whole-genome DNA methylation study, it was found that anthocyanin biosynthesis in apple fruit skin was induced on the methylation of flavonoid biosynthesis-related genes (*PAL*, *4CL*, *CYP98A*, *PER*, *CCoAOMT*, *CHS*, and *F3'H*) and *MYB10*, but hypermethylation from upstream region of bHLH74 inhibits the anthocyanin accumulation (Li et al., 2019b). Methylation of the promoter region of the *MdMYB1* gene also affects anthocyanin expression, and a study result was also published that the *MdAGO4s* gene binds to the *MdMYB1* promoter region and coordinates it (Jiang et al., 2020). In addition, the *ERF109* gene was also involved in light-induced anthocyanin biosynthesis, and it was also found that the long non-coding RNA MdLNC499 acts as a bridge between the transcription factors *MdWRKY1* and *MdERF109* (Ma et al., 2021).

3.4.3 Fruit Size

Fruit size is a particularly important factor in determining fruit quality and is known to be regulated by several genes or locus. According to several QTL studies, significantly associated regions have been found in almost all LGs, which means that fruit size is determined by complex regulatory pathway (Liebhard et al., 2003; Kenis et al., 2008; Devoghalaere et al., 2012). In addition to the QTL study, genetic factors involved in fruit size were investigated through comparative analysis using sports that showed a difference in fruit size from the original variety. 'Grand Gala,' which is a sport from 'Gala' and has a larger fruit size, increased cell production and cell size when compared to 'Gala' (Malladi & Hirst, 2010). Furthermore, cell cycle-related genes *MdCDKA1* and *MdCYCA2* have been highlighted as candidate genes for the fruit size difference based on their expression.

In another fruit size sport cultivar "Grand Longfeng," expression of auxin signaling-related genes was significantly different compared to its parent "Longfeng." The actual endogenous auxin level was also high in 'Grand Longfeng' compared to 'Longfeng.' Auxin biosynthesis-related genes *MdTAR1* and *MdYUCCA6* were proposed as candidate genes underlying the fruit size difference (Bu et al., 2020). Interestingly, the effect of auxin on apple fruit size has been previously reported

(Devoghalaere et al., 2012). When exogenous auxin was injected into developing apple fruit, cell expansion was initiated, which led to an increase in cell size.

Furthermore, auxin response factor (*ARF106*) gene was found in the fruit weight-related QTL region. The result that microRNA (miRNA) is involved in fruit size was also revealed by creating a transgenic apple line. When miRNA172 was overexpressed, the fruit size was reduced, and the expression level of miRNA172 was relatively low in varieties with large fruit sizes in the apple breeding population (Yao et al., 2015). Two more miRNA172 genes (*miRNA172g* and *miRNA172h*) involved in fruit size were discovered by a subsequent study, and these genes seem to have contributed to the increase in fruit size during specification prior to domestication (Duan et al., 2017). In addition, Duan et al. also revealed that cell division regulatory gene (MDP0000223854) and B-galactosidase genes (MDP0000921848 and MDP0000179821) were under the human selection.

3.4.4 Fruit Development and Maturity

For an in-depth understanding of traits such as fruit acidity, size, and color that affect the quality of fruit, it is difficult without an understanding of the fruit development process. This is because fruit quality at harvest time is the cumulative result of various changes during fruit development. Physiological changes during apple fruit development have been extensively studied and fruit development can be grouped into four partially overlapping stages: (1) The cell division stage; from fruit set to 4–6 weeks after full bloom. Fruit grows by cell division only during about first 1–2 weeks. (2) Cell expansion stage; from 2 weeks after full bloom until the fruit maturity. Once cell division end at around 4–6 weeks after full bloom, apple grows by cell expansion only. (3) Fruit maturation stage; Fruit ready to be harvest with a sharp decline of the starch and increase of sucrose. (4) Ripening stage; Apple fruit becomes sweeter and softer with ethylene production as it ripens (Lakso et al., 1995; Brookfield et al., 1997; Janssen et al., 2008; Eccher et al., 2014).

Major physiological changes in apple, such as cell growth, and carbohydrate metabolism, are known to be genetically regulated. Numerous studies have been conducted to identify candidate gene or allele that affect to fruit development and quality. For example, cell cycle transcripts (Janssen et al., 2008), and DNA and protein synthesis proteins (Li et al., 2016a) were significantly up-regulated during cell division stage, concurrent with high rates of cell division. Young fruit in the cell division stage exhibited eightfold greater transcriptional levels of photosynthesis, protein synthesis, cell proliferation, and differentiation genes when compared to ripening stage samples (Lee et al., 2007). Additionally, sucrose-phosphate synthesis (Li et al., 2012), and the starch degradation enzyme α-amylase (Janssen et al., 2008), peaked during the fruit maturity state, potentially implicated in the reduction of starch at maturation stage. However, recently ‘omics’ studies have focused on the early and ripening stages mostly, neglecting intermediate and maturation stages, limiting a holistic understanding of apple fruit development. For example, the mechanisms underpinning apple fruit abscission during the middle and late carbohydrate competition periods remain unclear (Abruzzese et al., 1995; Reyes et al., 2016). Although young fruit abscission has been linked to stress responses, hormones, and hormone biosynthesis genes, middle and late fruit drop has not been investigated (Ferrero et al., 2015).

As a regulator, hormone also modulates major physiological changes during fruit development (McAtee et al., 2013). Therefore, studies have been conducted to identify genes or alleles that control the production of hormones. Furthermore, crosstalk between hormones and their role during the transition of each developmental stage are also particularly important for deep understanding of fruit development. For example, a study of tomato has shown that early fruit development is associated with an increase of auxin, cytokinin, and gibberellin (Mariotti et al., 2011). In strawberry, gibberellins regulate the physiological processes during cell expansion and crosstalk with auxin, playing an important role in cell division and expansion (Csukasi et al., 2011). In apple, auxin level within fruit cortex increased during the cell division phase and decreased until the late cell expansion phase (Devoghalaere et al., 2012). Furthermore, ethylene, abscisic acid, auxin, and gibberellins proteins are up-regulated during the maturation phase (Li et al., 2016a). Not only in apple and but also in different fruit species, like peach, elevated levels of jasmonic acid were observed during the early fruit development stages (Ruiz et al., 2013). Early maturation somatic mutation from 'Fuji' has been identified, showing significantly different expression patterns in the genes encoding ethylene biosynthesis enzymes (*MdACS1, MDACO1*) and ethylene receptor proteins (*MdETR1, MdERS1, MdERS2*) (Wang et al., 2009b). However, a limited number of hormones and/or sampling point during the fruit development provide the restricted knowledge of the relationship between the hormone and transcriptome changes. Therefore, global transcriptome and hormone change analysis is required to understand a comprehensive and parallel evaluation of the transcriptome and plant hormonal changes that regulate major physiological events, such as cell division, cell expansion, starch accumulation, and starch decline.

According to a recent study using Gala and its somatic mutant Autumn Gala (maturation date delayed by 4 weeks), it was found that there is a significant difference in maturation date according to the genotype of *MdActin7* (MD06G1127300: actin encoding gene) on chromosome 6 (Ban et al., 2022). More specifically, heterozygous (*MdACT7/Mdact7*) accessions have shown significantly later maturation date than the homozygous genotype (*MdACT7/MdACT7*) accessions. In addition, the gene *NAC18.1* (MD03G1222600), NAC-NOR transcription factor known to regulate fruit maturation and ripening in tomato (Giovannoni et al., 2005), on chromosome 3 has been proven to be a major QTL of fruit maturation in apple (Migicovsky et al., 2021). According to QTL studies, significant associations have been found in several chromosomes (Table 3.2), indicating that the regulation of maturity is a complex action of several genes (Liebhard et al., 2003; Kenis et al., 2008; Morimoto et al., 2013; Chagné et al., 2014; Kunihisa et al., 2014; Migicovsky et al., 2016; Urrestarazu et al., 2017; Migicovsky et al., 2021; Ban et al., 2022).

3.4.5 Post-Harvest Storage

Apples are climacteric fruits that are ripened after harvesting. Ethylene is produced in the core of apples at the beginning of ripening after maturation and then respiration increases, and ripening is accelerated from this point (DeEll et al., 2001; Giovannoni, 2001; Johnston et al., 2002). During the ripening process, the fruit naturally loses its firmness, and post-harvest softening is one of the main reasons stored apples lose commercial value (Johnston et al., 2002; Harker et al., 2008). Therefore, the study of the post-harvest of apples is mainly about how to maintain fruit marketability and supply products continuously throughout the year. For this purpose, various storage and processing

TABLE 3.2
List of References and Locations of Maturation-Related QTLs in Apple

References	Linkage Group or Location of Maturation Date-Related Genes
Liebhard et al. (2003)	LG3
Kenis et al. (2008)	LGs 3, 9, 10, 16
Morimoto et al. (2013)	LG9
Kunihisa et al. (2014)	LGs 3, 10, 15, 16
Chagné et al. (2014)	LGs 2, 3, 5, 6, 7, 9, 10, 11, 14, 15, 16
Urrestarazu et al. (2017)	LGs 3, 10, 16
Migicovsky et al. (2016, 2021)	LG3
Ban et al. (2022)	LG6

techniques have been developed and used. For example, 1-methyl-cyclo-propene (1-MCP) treatment has been used frequently in apple. 1-MCP preferentially attaches to the ethylene receptor, thereby inhibiting ethylene-induced ripening (Fan et al., 1999; Watkins et al., 2000; Greene, 2009). However, 1-MCP treatment and/or controlled atmosphere (CA) storage can cause storage injury like flesh browning (Burmeister & Dilley, 1995; Watkins et al., 2005; Fawbush et al., 2008), and the facilities and costs required for 1-MCP treatment and CA storage itself are significant (McCormick et al., 2012). For this reason, development of new varieties with good fruit quality after harvest without 1-MCP treatment or CA storage, and the search for genes related to post-harvest fruit quality have been undergone.

QTL regions associated with fruit firmness and flesh texture were found in various LGs (L01, L03, L08, L10, L15, and L16) (King et al., 2000; Maliepaard et al., 2001; Seymour et al., 2002). Apple shelf-life and post-harvest storage are closely related to ethylene production. Ethylene biosynthesis-related genes, 1-aminocyclopropane-1-carboxylate synthase (ACS) and 1-aminocyclopropane-1-carboxylate oxidase (ACO), were identified in LGs 15 and 10, respectively (Costa et al., 2005). In particular, the *MdACS1* (MD15G1302200) gene found in LG15 showed a positive correlation with ethylene production, whereas the *MdACS3a* (Md15G1203500) gene exhibited a negative correlation (Sunako et al., 1999; Harada et al., 2000; Wang et al., 2009a). Transgenic apples in which the *MdACS1* gene was silenced significantly reduced ethylene production and exhibited excellent shelf-life (Dandekar et al., 2004). *MdERF3* and *MdERF118* genes involved in the ethylene response were also found to be involved in fruit firmness and crispness (Wu et al., 2021).

In addition to ethylene biosynthesis-related genes, cell wall-degrading enzymes are directly involved in fruit softening in the reduction of fruit firmness (Brummell, 2006). High expression levels of *MdAF3*, a gene encoding α-L-arabinofuranosidase (a-AFase), were associated with mealiness symptoms and high a-AFase activity (Nobile et al., 2011). In addition, when *MdPG1*, which is an endo-polygalacturonase 1 (PG1) gene, was down-regulated, firmer apples could be produced compared to the control (Atkinson et al., 2012).

3.5 TREE QUALITY TRAITS

3.5.1 Disease Resistance

3.5.1.1 Fire Blight

Fire blight, caused by the gram-negative bacterium *Erwinia amylovora*, is a devastating disease in apple that can destroy orchards in a single growing season (Thomson, 2000). The most effective way to manage fire blight outbreaks is with the antibiotic streptomycin (Norelli et al., 2003); however, streptomycin-resistant *E. amylovora* has been discovered (Russo et al., 2008). Natural resistance to fire blight is a well-known trait of wild *Malus* species. To further understand natural resistance in apple, genetic studies have been conducted to determine the genetic regions controlling fire blight resistance. A cross between *'Malus fusca'* (Resistant) × 'Ida Red' (susceptible) was made, and the resulting F_1 progenies were inoculated with fire blight and scored based on lesion length (Emeriewen et al., 2014). A QTL on LG10 was identified (*Mfu10*) by DArT, and SSR markers on chromosome 10 were highly correlated to resistance and explained 65.6% of the variation. Additional QTLs for fire blight resistance have been discovered on LG3 of *'Malus robusta 5'* (80% variation explained) (Flachowsky et al., 2007), LG12 of *'Malus floribunda 821'* (40% variation explained) (Durel et al., 2009), LG12 of ornamental crabapple 'Evereste' (53% variation explained) (Durel et al., 2009), and LG7 of *Malus domestica* 'Fiesta' (40% variation explained) (Calenge et al., 2005, Khan et al., 2006).

3.5.2 Architecture

3.5.2.1 Weeping Architecture

Weeping apple trees, characterized by downward growth, are desired for ornamental landscaping. Studies have concluded that the weeping growth habit is controlled by a single dominant allele called *Weeping* (*W*) based on inheritance studies (Sampson & Cameron, 1965; Brown, 1992). Two genetic markers (GD147, $CS44_{1150}$) have been linked to the weeping trait (Just, 2001). To map *W* in apple, three F_1 populations segregating for weeping growth habit were utilized. Each population used was generated from crosses with a weeping growth habit parent ('Lousia,' 'Cheal's Weeping,' and 'NY-100')×standard growth habit parent ('NY-051,' 'Evereste,' and 'NY-011') (Dougherty et al., 2018). Two sets of DNA (weeping and standard) were pooled from 'Cheal's Weeping'×'Evereste' F_1 progeny for sequencing. Analysis of weeping pool-specific variants and variants shared between weeping and standard pool with allele frequency differences of 50% were used for mapping, identifying four QTLs associated with weeping growth habit. The major QTL (*W*) was mapped to chromosome 13, and minor QTLs *W2, W3* and *W4* were mapped to chromosomes 10, 16, and 5, respectively (Dougherty et al., 2018). SSR markers were used to fine map *W* and *W2* in the three F_1 populations, and marker Ch13–8547 was tightly linked to the *W*. RNA sequencing was completed with weeping and standard F_1 progeny to look for potential candidate genes underlying *W* and *W2*. Analysis of differentially expressed genes within *W* and *W2* revealed three and five possible candidate genes, respectively.

3.5.3 Other Tree Traits

3.5.3.1 Bud Break

As perennials, apple trees enter a state of dormancy during the cold season. The cease of active growth protects the meristems from unfavorable climate conditions (Miotto et al., 2019).

Apple trees have two growth cycles: a growth phase and a dormancy phase. An apple tree enters a state of dormancy by various circumstances, and in the case of bud dormancy, it is divided into three main categories (Lang et al., 1987): (1) paradormancy, when the growth of the lateral bud is suppressed by apical dominance; (2) endodormancy, which is caused by environmental factors such as low temperatures and short day length. To resume growing, low temperatures above a certain level must be accumulated known as chilling requirement. (3) In ecodormancy, dormancy is still maintained by unfavorable environmental factors such as cold temperature, and growth is initiated only when a sufficient amount of heat is accumulated (known as heat requirement). However, due to climate change, flowering time tends to be earlier in the spring, when low-temperature damage can damage or kill blooming buds. These lost buds can lead to lower fruit set and yield, directly causing economic damages the farms.

Various efforts to understand dormancy have been made in apple. According to the QTL study, several cell cycle-related genes were found in LG9 from two segregating F_1 populations (Celton et al., 2011). QTL related to the initiation of vegetative budbreak was also found in LG9 (van Dyk et al., 2010), and a major flowering gene *AGL24* was found in LG9 as well (Allard et al., 2016). Another flowering gene *FT* was found in LG12, and MADS-box (*DAM*) genes associated with dormancy were found in LGs 8 and 15 (Allard et al., 2016). Dormancy- and flowering-related genes were also identified by transcriptome analysis. The *flowering locus C-like* gene and the *MADS AFFECTING FLOWERING* gene were significantly up-regulated during dormancy release (Porto et al., 2015). On the other hand, the *MdDAM* genes *MdMADS063* and *MdMADS125* and the *MdFLC-like* genes *MdMADS135* and *MdMADS136* showed a dramatic decrease in expression during dormancy release, suggesting that they act as flowering repressors (Kumar et al., 2016a). DAM genes also showed a tissue-specific expression pattern, with low expression levels in vegetative organs and high levels in dormant buds and seeds (Porto et al., 2016). Not only flowering-related genes, but also lipid-related gene (Saito et al., 2017), carbon storage-related gene (Falavigna et al., 2018), and various phytochrome-related genes (Takeuchi et al., 2018; Kumar et al., 2017) were associated with dormancy and showed differences in expression levels. The actual effect of these genes on dormancy and flowering was verified through the transgenic line. Apple transgenic tree silencing *MdDAM1* did not enter dormancy during winter, suggesting that *MdDAM1* is a key regulator of bud dormancy (Moser et al., 2020). In addition, overexpression of genes *MdDAMb* and *MdSVPa* led to the delayed budbreak in transgenic apple tree (Wu et al., 2017). Overexpression of *MdTFL1*, TERMINAL FLOWER 1 (TFL1) homologous gene, in Arabidopsis led to the delayed flowering compared to the wild type (Kotoda & Wada, 2005). Furthermore, overexpression of FLOWERING LOCUS T (FT)-like genes *MdFT1* and *MdFT2* in Arabidopsis also led to the earlier flowering compared to wild type (Kotoda et al., 2010). To further understand the mechanisms of flowering, a nucleoporin gene, *MdNup54* which was observed to have high expression in flower buds and flowers was analyzed (Zhang et al., 2022). When *MdNuc54* was overexpressed in *Arabidopsis,* those plants flowered significantly earlier than the wild type. When overexpressing the heat-shock protein *MdHSP70* in *Arabidopsis*, the plants also had earlier flowering times than the wild type. Additionally, using a yeast-two-hybrid assay, they found *MdNup54* interacts with heat-shock protein *MdHSP70* suggesting they may mediate nuclear and cytoplasmic transport to regulate apple flowering (Zhang et al., 2022). There are also studies that demonstrate budbreak after dormancy is mediated by epigenetic regulation by DNA methylation (Chen et al., 2022; Kumar et al., 2016b).

3.5.3.2 Parthenocarpic Apple Fruit

Parthenocarpy is the development of fruit without fertilization. Parthenocarpic fruits are sterile and seedless (Picarella & Mazzucato, 2019). An advantage to parthenocarpy is that the fruit develops without the need for pollinators, which have been declining in recent years (Council, 2007). Known parthenocarpic domesticated apples include ‘Spencer Seedless,’ ‘Wellington Bloomless,’ and ‘Rae Ime.’ These cultivars produce flowers without petals or stamens, but with two sets of sepals and two sets of carpels (Yao et al., 2001).

To gain insight on parthenocarpy in apple, F_1 progeny from a ‘Wijcik’ (non-parthenocarpic)×‘Wellington Bloomless’ (parthenocarpic) cross were assessed for parthenocarpy phenotype (Tobutt, 1994). None of the F_1 progeny exhibited parthenocarpic traits, so another set of crosses were made with four F_1 progeny (non-parthenocarpic)×‘Spencer Seedless’ (parthenocarpic). The resulting four F_1 populations segregated for parthenocarpy and non-parthenocarpy at a 1:1 ratio, suggesting parthenocarpy is inherited by a single recessive gene (Tobutt, 1994).

To identify the candidate gene for parthenocarpy, researchers noticed that an Arabidopsis mutant *pistillata* (*pi*) (Goto & Meyerowitz, 1994) had a very similar apetalous phenotype to the parthenocarpic apples and investigated the *Pistillata* homolog (*MdPI*) in ‘Rae Ime’ and ‘Granny Smith’ (Yao et al., 2001). *MdPI* is a MADS-box transcription factor. Expression analysis revealed that the *MdPI* gene was not expressed in ‘Rae Ime,’ similar to what was observed in the Arabidopsis mutant, but was expressed in ‘Granny Smith.’ Cloning and sequencing of the ‘Rae Ime’ *MdPI* gene revealed a retrotransposon insertion in the 4th intron which abolished gene function. Cloning and sequencing of ‘Spencer Seedless’ and ‘Wellington Bloomless’ *MdPI* genes both had retrotransposon insertions in the 6th intron (Yao et al., 2001).

The independent retrotransposon insertions in the 4th and 6th introns of *MdPI* for ‘Rae Ime,’ ‘Spencer Seedless,’ and ‘Wellington Bloomless’ are causal for the apetalous and parthenocarpic phenotypes. Future breeding efforts can genetically target *MdPI* for knockdown expression or loss of function mutations, leading to the development of more seedless cultivars. The recessive trait can also be bred into new cultivars with selection of heterozygous *MdPI* parents.

3.6 FUTURE APPLICATIONS

Substantial progress has been made in the assembly of apple reference genomes over the last decade. As sequencing platforms continue to improve, apple reference genomes will continue to improve as well, with fewer gaps and assembly errors. The first apple pangenome consisting of three

Malus species, including *M. domestica* (43 cultivars), *M. sieversii* (37 cultivars), and *M. sylvestris* (11 cultivars), was released in 2020 (Sun et al., 2020). Considering that various traits, especially disease resistance, are known to exist in various wild accessions, the pangenome should be expanded upon to include other *Malus* species and genetic diversity. The addition of more *Malus* species will unlock further genetic diversity underlying apple traits. As climate changes continue to intensify, abiotic traits will be of great interest. Drought, flooding and heat tolerance are emerging as important tree traits (Jangea et al., 2017) and more emphasis is being placed on traits related to dormancy. The Brazilian apple germplasm has already screened their germplasm for genetic variability to look for cultivars with low chilling requirements to adapt to climate change (Mariano et al., 2019).

The identification of allelic variations in apple traits presents an excellent opportunity for genome editing for the improvement of established cultivars (Songstad et al., 2017). Agrobacterium-based transformations and tissue culture propagations are routinely preformed in apple (Bhatti & Jha, 2010). Functional characterization of candidate genes can be achieved with transformations. The suppression or overexpression of select genes may give desirable phenotypes. The Artic Apple® was developed to suppress a polyphenol oxidase (*PPO*) enzyme involved in fruit flesh browning, and has been cleared for market, leading the way for potentially more genetically edited apples in the future (Stowe & Dhingra, 2021). CRISPR-Cas9 gene editing has also successfully been carried out in apple along with transgene free editing (Malnoy et al., 2016; Nishitani et al., 2016; Pompili et al., 2020). Allele mining is highly important and necessary to determine genetic variations underlying traits of interest that can be developed into markers for breeding purposes and/or targeted in genome editing.

3.7 CONCLUSION

The apple research community has produced a wealth of resources for genetic studies in apple. Genetic studies to uncover allelic variations underlying key horticultural traits in apple, will allow breeders to generate superior cultivars. The general techniques of allele mining have been applied to numerous genetic studies in apple over the years and have been proven to uncover causal genetic variations in key quality traits. As techniques and technologies continue to advance, apple researchers will be able to answer more questions and gain fundamental understandings of how allelic variations affect phenotypes in apple.

REFERENCES

Abruzzese A, Mignani I, Cocucci SM, 1995. Nutritional-status in apples and June drop. *J Am Soc Hortic Sci* **120**, 71–4.

Afgan E, Baker D, Batut B, et al., 2018. The Galaxy platform for accessible, reproducible and collaborative biomedical analyses: 2018 update. *Nucleic Acids Res* **46**, W537–W44.

Allard A, Bink MC, Martinez S, et al., 2016. Detecting QTLs and putative candidate genes involved in budbreak and flowering time in an apple multiparental population. *J Exp Bot* **67**, 2875–88.

Altschul SF, Gish W, Miller W, Myers EW, Lipman DJ, 1990. Basic local alignment search tool. *J Mol Biol* **215**, 403–10.

Atkinson RG, Sutherland PW, Johnston SL, et al., 2012. Down-regulation of POLYGALACTURONASE1 alters firmness, tensile strength and water loss in apple (*Malus × domestica*) fruit. BMC Plant Biol **12**, 1–13.

Bai Y, Dougherty L, Li M, Fazio G, Cheng L, Xu K, 2012. A natural mutation-led truncation in one of the two aluminum-activated malate transporter-like genes at the Ma locus is associated with low fruit acidity in apple. *Mol Genet Genom* **287**, 663–78.

Bai Y, Dougherty L, Xu K, 2014. Towards an improved apple reference transcriptome using RNA-seq. *Mol Genet Genom* **289**, 427–38.

Ban S, El-Sharkawy I, Zhao J, Fei Z, Xu K, 2022. An apple somatic mutation of delayed fruit maturation date is primarily caused by a retrotransposon insertion associated large deletion. *Plant J* **106**, 1609–25.

Ban S, Xu K, 2020. Identification of two QTLs associated with high fruit acidity in apple using pooled genome sequencing analysis. *Hortic Res* **7**, 171.

Benson DA, Cavanaugh M, Clark K, et al., 2013. GenBank. *Nucleic Acids Res* **41**, D36–42.
Bhatti S, Jha G, 2010. Current trends and future prospects of biotechnological interventions through tissue culture in apple. Plant Cell Rep **29**, 1215–25.
Bianco L, Cestaro A, Linsmith G, et al., 2016. Development and validation of the Axiom((R)) Apple480K SNP genotyping array. *Plant J* **86**, 62–74.
Bianco L, Cestaro A, Sargent DJ, et al., 2014. Development and validation of a 20K single nucleotide polymorphism (SNP) whole genome genotyping array for apple (Malus×domestica Borkh). *PLoS One* **9**, 10e110377.
Blum M, Chang H-Y, Chuguransky S, et al., 2021. The InterPro protein families and domains database: 20 years on. *Nucleic Acids Res* **49**, D344–54.
Bradbury PJ, Zhang Z, Kroon DE, Casstevens TM, Ramdoss Y, Buckler ES, 2007. TASSEL: software for association mapping of complex traits in diverse samples. *Bioinformatics* **23**, 2633–5.
Broggini Ga L, Schlathölter I, Russo G, et al., 2020. Chromosome-scale de novo diploid assembly of the apple cultivar 'Gala Galaxy'. *bioRxiv*. doi:10.1101/2020.04.25.058891.
Broman KW, Wu H, Sen S, Churchill GA, 2003. R/qtl: QTL mapping in experimental crosses. *Bioinformatics* **19**, 889–90.
Brookfield P, Murphy P, Harker R, Macrae E, 1997. Starch degradation and starch pattern indices; interpretation and relationship to maturity. *Postharv Biol Technol* **11**, 23–30.
Brown AG, Harvey DM, 1971. Nature and inheritance of sweetness and acidity in cultivated apple. *Euphytica* **20**, 68–80.
Brown JWS, Calixto CPG, Zhang R, 2017. High-quality reference transcript datasets hold the key to transcript-specific RNA-sequencing analysis in plants. *New Phytol* **213**, 525–30.
Brown SK, 1992. Genetics of apple. Plant Breed Rev **9**, 333–66.
Brummell DA, 2006. Cell wall disassembly in ripening fruit. *Funct Plant Biol* **33**, 103–19.
Bu H, Yu W, Yuan H, Yue P, Wei Y, Wang A, 2020. Endogenous auxin content contributes to larger size of apple fruit. Front Plant Sci **11**, 592540.
Buels R, Yao E, Diesh CM, et al., 2016. JBrowse: a dynamic web platform for genome visualization and analysis. *Genome Biol* **17**, 66.
Burmeister DM, Dilley DR, 1995. A scald-like controlled-atmosphere storage disorder of empire apples - a chilling injury-induced by Co2. *Postharvest Biol Technol* **6**, 1–7.
Calenge F, Drouet D, Denancé C, et al., 2005. Identification of a major QTL together with several minor additive or epistatic QTLs for resistance to fire blight in apple in two related progenies. *Theor Appl Genet* **111**, 128–35.
Cao F, Li Z, Jiang L, et al., 2021. Genome-wide association study (GWAS) of leaf wax components of apple. *Stress Biol* **1,** 13.
Celton JM, Martinez S, Jammes MJ, et al., 2011. Deciphering the genetic determinism of bud phenology in apple progenies: a new insight into chilling and heat requirement effects on flowering dates and positional candidate genes. *New Phytol* **192**, 378–92.
Celton JM, Tustin DS, Chagné D, Gardiner SE, 2009. Construction of a dense genetic linkage map for apple rootstocks using SSRs developed from *Malus* ESTs and *Pyrus* genomic sequences. *Tree Genet Genomes* **5**, 93–107.
Chagne D, Carlisle CM, Blond C, et al., 2007. Mapping a candidate gene (MdMYB10) for red flesh and foliage colour in apple. *BMC Genomics* **8**, 212.
Chagne D, Crowhurst RN, Troggio M, et al., 2012. Genome-wide SNP detection, validation, and development of an 8K SNP array for apple. PLoS One 7, e31745.
Chagné D, Dayatilake D, Diack R, et al., 2014. Genetic and environmental control of fruit maturation, dry matter and firmness in apple (*Malus × domestica* Borkh.). *Hortic Res* **1**, 1–12.
Chen W, Tamada Y, Yamane H, et al., 2022. H3K4me3 plays a key role in establishing permissive chromatin states during bud dormancy and bud break in apple. *Plant J* **111**, 1015–31.
Chen X, Li S, Zhang D, et al., 2019. Sequencing of a wild apple (*Malus baccata*) genome unravels the differences between cultivated and wild apple species regarding disease resistance and cold tolerance. *Genes Genomes Genet* **9**, 2051–60.
Conesa A, Götz S, García-Gómez JM, Terol J, Talón M, Robles M, 2005. Blast2GO: a universal tool for annotation, visualization and analysis in functional genomics research. *Bioinformatics* **21**, 3674–6.
Costa F, Stella S, Van De Weg WE, et al., 2005. Role of the genes Md-ACO1 and Md-ACS1 in ethylene production and shelf life of apple (*Malus domestica* Borkh). *Euphytica* **141**, 181–90.
Council NR, 2007. *Status of Pollinators in North America*. National Academies Press.

Csukasi F, Osorio S, Gutierrez JR, et al., 2011. Gibberellin biosynthesis and signalling during development of the strawberry receptacle. *New Phytol* **191**, 376–90.
Daccord N, Celton J-M, Linsmith G, et al., 2017. High-quality de novo assembly of the apple genome and methylome dynamics of early fruit development. *Nat Genet* **49**, 1099–106.
Dandekar AM, Teo G, Defilippi BG, et al., 2004. Effect of down-regulation of ethylene biosynthesis on fruit flavor complex in apple fruit. Transgen Res **13**, 373–84.
Davey JW, Blaxter ML, 2011. RADSeq: next-generation population genetics. *Brief Funct Genom* **9**, 416–23.
Deell JR, Khanizadeh S, Saad F, Ferree DC, 2001. Factors affecting apple fruit firmness - a review. *J Am Pomol Soc* **55**, 8–27.
Devoghalaere F, Doucen T, Guitton B, et al., 2012. A genomics approach to understanding the role of auxin in apple (Malus×domestica) fruit size control. BMC Plant Biol **12**, 1–15.
Dougherty L, Singh R, Brown S, Dardick C, Xu K, 2018. Exploring DNA variant segregation types in pooled genome sequencing enables effective mapping of weeping trait in Malus. *J Exp Bot* **69**, 1499–516.
Dougherty L, Wallis A, Cox K, Zhong G-Y, Gutierrez B, 2021. Phenotypic evaluation of fire blight outbreak in the USDA Malus collection. *Agronomy* **11**, 144.
Duan N, Bai Y, Sun H, et al., 2017. Genome re-sequencing reveals the history of apple and supports a two-stage model for fruit enlargement. *Nat Commun* **8**, 249.
Durel C-E, Denance C, Brisset M-N, 2009. Two distinct major QTL for resistance to fire blight co-localize on linkage group 12 in apple genotypes 'Evereste' and Malus floribunda clone 821. *Genome* **52**, 139–47.
Eccher G, Ferrero S, Populin F, Colombo L, Botton A, 2014. Apple (Malus domestica L. Borkh) as an emerging model for fruit development. *Plant Biosyst* **148**, 157–68.
El-Sharkawy I, Liang D, Xu K, 2015. Transcriptome analysis of an apple (Malus×domestica) yellow fruit somatic mutation identifies a gene network module highly associated with anthocyanin and epigenetic regulation. *J Exp Bot* **66**, 7359–76.
Elshire RJ, Glaubitz JC, Sun Q, et al., 2011. A robust, simple genotyping-by-sequencing (GBS) approach for high diversity species. PLoS One 6, e19379.
Emeriewen O, Richter K, Kilian A, et al., 2014. Identification of a major quantitative trait locus for resistance to fire blight in the wild apple species *Malus fusca. Mol Breed* **34**, 407–19.
Espley RV, Brendolise C, Chagne D, et al., 2009. Multiple repeats of a promoter segment causes transcription factor autoregulation in red apples. *Plant Cell* **21**, 168–83.
Espley RV, Hellens RP, Putterill J, Stevenson DE, Kutty-Amma S, Allan AC, 2007. Red colouration in apple fruit is due to the activity of the MYB transcription factor, MdMYB10. *Plant J* **49**, 414–27.
Evans K, Jung S, Lee T, et al., 2013. Addition of a breeding database in the Genome Database for Rosaceae. Database 2013, bat078.
Falavigna VDS, Porto DD, Miotto YE, et al., 2018. Evolutionary diversification of galactinol synthases in Rosaceae: adaptive roles of galactinol and raffinose during apple bud dormancy. *J Exp Bot* **69**, 1247–59.
Fan XT, Blankenship SM, Mattheis JP, 1999. 1-Methylcyclopropene inhibits apple ripening. *J Am Soc Hortic Sci* **124**, 690–5.
Fawbush F, Nock JF, Watkins CB, 2008. External carbon dioxide injury and 1-methylcyclopropene. *Postharv Biol Technol* **48**, 92–8.
Fenning TM, Gershenzon J, 2002. Where will the wood come from? Plantation forests and the role of biotechnology. *Trends Biotechnol* **20**, 291–6.
Ferrero S, Carretero-Paulet L, Mendes MA, et al., 2015. Transcriptomic signatures in seeds of apple (Malus domestica L. Borkh) during fruitlet abscission. PLoS One 10, e0120503.
Flachowsky H, Richter K, Garcia-Libreros T, et al., 2007. Confirmation of the fire blight QTL of Malus×robusta 5 on linkage group 3. *Plant Breed* **126**, 297–303.
Gao Y, Liu F, Wang K, et al., 2015. Genetic diversity of Malus cultivars and wild relatives in the Chinese National Repository of Apple Germplasm Resources. *Tree Genet Genomes* **11**, 106.
Gardner KM, Brown P, Cooke TF, et al., 2014. Fast and cost-effective genetic mapping in apple using next-generation sequencing. *Genes Genomes Genet* **4**, 1681–7.
Gianfranceschi L, Soglio V, 2003. The European project HiDRAS: innovative multidisciplinary approaches to breeding high quality disease resistant apples. *Acta Hortic* **663**, 327–30.
Giovannoni J, 2001. Molecular biology of fruit maturation and ripening. *Annu Rev Plant Physiol Plant Mol Biol* **52**, 725–49.
Giovannoni J, Tanksley S, Vrebalov J, Noensie F, 2005. NOR gene compositions and methods for use thereof. Google Patent.

Goto K, Meyerowitz EM, 1994. Function and regulation of the Arabidopsis floral homeotic gene PISTILLATA. *Genes Dev* **8**, 1548–60.

Greene DW, 2009. Effect of sprayable 1-MCP (harvista) on preharvest drop and fruit quality of ‘McIntosh’ apples. HortScience 44, 1037.

Guan YZ, Peace C, Rudell D, Verma S, Evans K, 2015. QTLs detected for individual sugars and soluble solids content in apple. *Mol Breed* **35**, 135.

Guilford P, Prakash S, Zhu JM, et al., 1997. Microsatellites in *Malus × domestica* (apple): abundance, polymorphism and cultivar identification. *Theor Appl Genet* **94**, 249–54.

Gusberti M, Gessler C, Broggini GA, 2013. RNA-Seq analysis reveals candidate genes for ontogenic resistance in Malus-Venturia pathosystem. PLoS One 8, e78457.

Hancock J, Luby J, Brown S, Lobos G, 2008. Apples. In: Hancock JF (ed) *Temperate Fruit Crop Breeding.* Springer, Dordrecht. pp. 1–38.

Harada T, Sunako T, Wakasa Y, Soejima J, Satoh T, Niizeki M, 2000. An allele of the 1-aminocyclopropane-1-carboxylate synthase gene (Md-ACS1) accounts for the low level of ethylene production in climacteric fruits of some apple cultivars. *Theor Appl Genet* **101**, 742–6.

Hargarten H, Waliullah S, Kalcsits L, Honaas LA, 2018. Leveraging transcriptome data for enhanced gene expression analysis in apple. *J Am Soc Hort Sci* **143**, 333–46.

Harker FR, Kupferman EM, Marin AB, Gunson FA, Triggs CM, 2008. Eating quality standards for apples based on consumer preferences. *Postharvest Biol Technol* **50**, 70–8.

Hokanson SC, Lamboy WF, Szewc-Mcfadden AK, Mcferson JR, 2001. Microsatellite (SSR) variation in a collection of *Malus* (apple) species and hybrids. *Euphytica* **118**, 281–94.

Hokanson SC, Szewc-Mcfadden AK, Lamboy WF, Mcferson JR, 1998. Microsatellite (SSR) markers reveal genetic identities, genetic diversity and relationships in a *Malus × domestica* Borkh. core subset collection. *Theor Appl Genet* **97**, 671–83.

Howard NP, Van De Weg E, Bedford DS, et al., 2017. Elucidation of the ‘Honeycrisp’ pedigree through haplotype analysis with a multi-family integrated SNP linkage map and a large apple (Malus × domestica) pedigree-connected SNP data set. Hortic Res **4**, 17003.

Iglesias I, Echeverría G, Soria Y, 2008. Differences in fruit colour development, anthocyanin content, fruit quality and consumer acceptability of eight ‘Gala’ apple strains. *Sci Hortic* **119**, 32–40.

Jangra S, Mishra A, Kamboj D, Yadav NR, Yadav RC, 2017. Engineering abiotic stress tolerance traits for mitigating climate change. In: Gahlawat SK, Salar RK, Siwach P, Duhan JS, Kumar S, Kaur P (eds) *Plant Biotechnology: Recent Advancements and Developments.* Springer Singapore, Singapore, pp. 59–73.

Janssen BJ, Thodey K, Schaffer RJ, et al., 2008. Global gene expression analysis of apple fruit development from the floral bud to ripe fruit. BMC Plant Biol **8**, 16.

Jia DJ, Shen F, Wang Y, et al., 2018. Apple fruit acidity is genetically diversified by natural variations in three hierarchical epistatic genes: MdSAUR37, MdPP2CH and MdALMTII. *Plant J* **95**, 427–43.

Jiang S, Wang N, Chen M, et al., 2020. Methylation of MdMYB1 locus mediated by RdDM pathway regulates anthocyanin biosynthesis in apple. *Plant Biotechnol J* **18**, 1736–48.

Johnson FT, Zhu Y, 2015. Transcriptome changes in apple peel tissues during CO_2 injury symptom development under controlled atmosphere storage regimens. *Hortic Res* **2**, 15061.

Johnston JW, Hewett EW, Hertog MLATM, 2002. Postharvest softening of apple (*Malus domestica*) fruit: a review. *NZ J Crop Hortic Sci* **30**, 145–60.

Jung S, Lee T, Cheng CH, et al., 2019. 15 years of GDR: new data and functionality in the Genome Database for Rosaceae. *Nucleic Acids Res* **47**, D1137–45.

Just BJ, 2001. Molecular markers for weeping plant habit and powdery mildew (*Podosphaera Leucotricha*) resistance from the ornamental crabapple ‘Red Jade’. Cornell University, August.

Ke X, Yin Z, Song N, et al., 2014. Transcriptome profiling to identify genes involved in pathogenicity of Valsa mali on apple tree. *Fungal Genet Biol* **68**, 31–8.

Kenis K, Keulemans J, 2005. Genetic linkage maps of two apple cultivars (*Malus × domestica* Borkh.) based on AFLP and microsatellite markers. *Mol Breed* **15**, 205–19.

Kenis K, Keulemans J, Davey MW, 2008. Identification and stability of QTLs for fruit quality traits in apple. *Tree Genet Genomes* **4**, 647–61.

Khan A, Carey SB, Serrano A, et al. 2022. A phased, chromosome-scale genome of ‘Honeycrisp’ apple (*Malus domestic*a). *bioRxiv*, doi:10.1101/2022.08.24.505160.

Khan MA, Duffy B, Gessler C, Patocchi A, 2006. QTL mapping of fire blight resistance in apple. *Mol Breed* **17**, 299–306.

Kim J, Park JH, Lim CJ, et al., 2012. Small RNA and transcriptome deep sequencing proffers insight into floral gene regulation in Rosa cultivars. *BMC Genomics* **13**, 657.

King GJ, Maliepaard C, Lynn JR, et al., 2000. Quantitative genetic analysis and comparison of physical and sensory descriptors relating to fruit flesh firmness in apple (*Malus pumila* Mill.). *Theor Appl Genet* **100**, 1074–84.

Korte A, Farlow A, 2013. The advantages and limitations of trait analysis with GWAS: a review. *Plant Methods* **9**, 29.

Kotoda N, Hayashi H, Suzuki M, et al., 2010. Molecular characterization of FLOWERING LOCUS T-like genes of apple (Malus×domestica Borkh.). *Plant Cell Physiol* **51**, 561–75.

Kotoda N, Wada M, 2005. MdTFL1, a TFL1-like gene of apple, retards the transition from the vegetative to reproductive phase in transgenic Arabidopsis. *Plant Sci* **168**, 95–104.

Kukurba KR, Montgomery SB, 2015. RNA sequencing and analysis. Cold Spring Harb Prot **2015**, 951–969.

Kulwal PL, 2018. Trait mapping approaches through linkage mapping in plants. In: Varshney RK, Pandey MK, Chitikineni A (eds) *Plant Genetics and Molecular Biology*. Springer International Publishing, Cham, pp. 53–82.

Kumar G, Arya P, Gupta K, Randhawa V, Acharya V, Singh AK, 2016a. Comparative phylogenetic analysis and transcriptional profiling of MADS-box gene family identified DAM and FLC-like genes in apple (*Malus × domestica*). *Sci Rep* **6**, 20695.

Kumar G, Gupta K, Pathania S, et al., 2017. Chilling affects phytohormone and post-embryonic development pathways during bud break and fruit set in apple (*Malus domestica* Borkh.). *Sci Rep* **7**, 42593.

Kumar G, Rattan UK, Singh AK, 2016b. Chilling-mediated DNA methylation changes during dormancy and its release reveal the importance of epigenetic regulation during winter dormancy in apple (Malus×domestica Borkh.). *PLoS One* **11**, e0149934.

Kumar GR, Sakthivel K, Sundaram RM, et al., 2010a. Allele mining in crops: prospects and potentials. *Biotechnol Adv* **28**, 451–61.

Kumar S, Chagne D, Bink MCAM, Volz RK, Whitworth C, Carlisle C, 2012. Genomic selection for fruit quality traits in apple (*Malus × domestica* Borkh.). PLoS One 7, e36674.

Kumar S, Volz RK, Alspach PA, Bus VGM, 2010b. Development of a recurrent apple breeding programme in New Zealand: a synthesis of results, and a proposed revised breeding strategy. *Euphytica* **173**, 207–22.

Kumar S, Volz RK, Chagné D, Gardiner S, 2014. Breeding for apple (*Malus × domestica* Borkh.) fruit quality traits in the genomics era. In: Tuberosa R, Graner A, Frison E (eds) *Genomics of Plant Genetic Resources: Volume 2. Crop Productivity, Food Security and Nutritional Quality*. Springer, Netherlands, Dordrecht, pp. 387–416.

Kunihisa M, Moriya S, Abe K, et al., 2014. Identification of QTLs for fruit quality traits in Japanese apples: QTLs for early ripening are tightly related to preharvest fruit drop. *Breed Sci* **64**, 240–51.

Laframboise T, 2009. Single nucleotide polymorphism arrays: a decade of biological, computational and technological advances. *Nucleic Acids Res* **37**, 4181–93.

Lakso AN, Grappadelli LC, Barnard J, Goffinet MC, 1995. An expolinear model of the growth-pattern of the apple fruit. *J Hortic Sci* **70**, 389–94.

Lang GA, Early JD, Martin GC, Darnell RL, 1987. Endodormancy, paradormancy, and ecodormancy - physiological terminology and classification for dormancy research. *HortScience* **22**, 371–7.

Larsen B, Gardner K, Pedersen C, et al., 2018. Population structure, relatedness and ploidy levels in an apple gene bank revealed through genotyping-by-sequencing. PLoS One 13, e0201889.

Larsen B, Migicovsky Z, Jeppesen AA, et al., 2019. Genome-wide association studies in apple reveal loci for aroma volatiles, sugar composition, and harvest date. *Plant Genome* **12**, 2.

Lassois L, Denancé C, Ravon E, et al., 2016. Genetic diversity, population structure, parentage analysis, and construction of core collections in the French apple germplasm based on ssr markers. Plant Mol Biol Rep **34**, 827–44.

Lee SJ, Ban SH, Kim GH, Kwon SI, Kim JH, Choi C, 2017. Identification of potential gene-associated major traits using GBS-GWAS for Korean apple germplasm collections. *Plant Breed* **136**, 977–86.

Lee YP, Yu GH, Seo YS, et al., 2007. Microarray analysis of apple gene expression engaged in early fruit development. *Plant Cell Reps* **26**, 917–26.

Legay S, Guerriero G, Deleruelle A, et al., 2015. Apple russeting as seen through the RNA-seq lens: strong alterations in the exocarp cell wall. *Plant Mol Biol* **88**, 21–40.

Li CL, Dougherty L, Coluccio AE, et al., 2019a. Apple ALMT9 requires a conserved C-terminal domain for malate transport underlying fruit acidity. *Plant Physiol* **182**, 992–1006.

Li MJ, Feng FJ, Cheng LL, 2012. Expression patterns of genes involved in sugar metabolism and accumulation during apple fruit development. PLoS One 7, e33055.

Li MJ, Li DX, Feng FJ, Zhang S, Ma FW, Cheng LL, 2016a. Proteomic analysis reveals dynamic regulation of fruit development and sugar and acid accumulation in apple. *J Exp Bot* **67**, 5145–57.

Li WF, Ning GX, Mao J, Guo ZG, Zhou Q, Chen BH, 2019b. Whole-genome DNA methylation patterns and complex associations with gene expression associated with anthocyanin biosynthesis in apple fruit skin. *Planta* **250**, 1833–47.

Li X, Kui L, Zhang J, et al., 2016b. Improved hybrid de novo genome assembly of domesticated apple (*Malus × domestica*). *Gigascience* **5**, 35.

Liang W, Dondini L, De Franceschi P, Paris R, Sansavini S, Tartarini S, 2015. Genetic diversity, population structure and construction of a core collection of apple cultivars from Italian germplasm. Plant Mol Biol Rep **33**, 458–73.

Liebhard R, Gianfranceschi L, Koller B, et al., 2002. Development and characterisation of 140 new microsatellites in apple (*Malus × domestica* Borkh.). *Mol Breed* **10**, 217–41.

Liebhard R, Kellerhals M, Pfammatter W, Jertmini M, Gessler C, 2003. Mapping quantitative physiological traits in apple (*Malus × domestica* Borkh.). *Plant Mol Biol* **52**, 511–26.

Ma B, Chen J, Zheng H, et al., 2015. Comparative assessment of sugar and malic acid composition in cultivated and wild apples. *Food Chem* **172**, 86–91.

Ma BQ, Zhao S, Wu BH, et al., 2016. Construction of a high density linkage map and its application in the identification of QTLs for soluble sugar and organic acid components in apple. *Tree Genet Genomes* **12**, 1.

Ma H, Yang T, Li Y, et al., 2021. The long noncoding RNA MdLNC499 bridges MdWRKY1 and MdERF109 function to regulate early-stage light-induced anthocyanin accumulation in apple fruit. *Plant Cell* **33**, 3309–3330.

Maliepaard C, Alston FH, Van Arkel G, et al., 1998. Aligning male and female linkage maps of apple (Malus pumila Mill.) using multi-allelic markers. *Theor Appl Genet* **97**, 60–73.

Maliepaard C, Sillanpaa MJ, Van Ooijen JW, Jansen RC, Arjas E, 2001. Bayesian versus frequentist analysis of multiple quantitative trait loci with an application to an outbred apple cross. *Theor Appl Genet* **103**, 1243–53.

Malladi A, Hirst PM, 2010. Increase in fruit size of a spontaneous mutant of ‘Gala’ apple (*Malus × domestica* Borkh.) is facilitated by altered cell production and enhanced cell size. *J Exp Bot* **61**, 3003–13.

Malnoy M, Viola R, Jung M-H, et al., 2016. DNA-free genetically edited grapevine and apple protoplast using CRISPR/Cas9 ribonucleoproteins. Front Plant Sci 7, 1904.

Manalo AB, 1990. Assessing the importance of apple attributes: an agricultural application of conjoint analysis. *Northeastern J Agric Resour Econ* **19**, 118–24.

Mariano LC, Zchonski FL, Da Silva CM, Da-Silva PR, 2019. Genetic variability in a Brazilian apple germplasm collection with low chilling requirements. PeerJ 6, e6265.

Mariotti L, Picciarelli P, Lombardi L, Ceccarelli N, 2011. Fruit-set and early fruit growth in tomato are associated with increases in indoleacetic acid, cytokinin, and bioactive gibberellin contents. *J Plant Growth Regul* **30**, 405–15.

Mcatee P, Karim S, Schaffer R, David K, 2013. A dynamic interplay between phytohormones is required for fruit development, maturation, and ripening. *Front Plant Sci* **4**, 79.

Mccormick R, Neuwald DA, Streif J, 2012. Commercial apple CA storage temperature regimes with 1-MCP (SmartFresh™): benefits and risks. *Acta Hortic* **934**, 263–70.

Members of the Complex Trait Consortium, 2003. The nature and identification of quantitative trait loci: a community’s view. *Nat Rev Genet* **4**, 911–6.

Micheletti D, Troggio M, Zharkikh A, et al., 2011. Genetic diversity of the genus *Malus* and implications for linkage mapping with SNPs. *Tree Genet Genomes* **7**, 857–68.

Migicovsky Z, Douglas GM, Myles S, 2022. Genotyping-by-sequencing of Canada’s apple biodiversity collection. Front Genet **13**, 2177.

Migicovsky Z, Gardner KM, Money D, et al., 2016. Genome to phenome mapping in apple using historical data. *Plant Genome* **9**, 1–15.

Migicovsky Z, Yeats TH, Watts S, et al., 2021. Apple ripening is controlled by a NAC transcription factor. *Front Genet* **12**, 671300.

Miotto YE, Tessele C, Czermainski ABC, et al., 2019. Spring is coming: genetic analyses of the bud break date locus reveal candidate genes from the cold perception pathway to dormancy release in apple (*Malus × domestica* Borkh.). *Front Plant Sci* **10**, 33.

Morimoto T, Hiramatsu Y, Banno K, 2013. A major QTL controlling earliness of fruit maturity linked to the red leaf/red flesh trait in apple cv. ‘Maypole’. *J Jpn Soc Hortic Sci* **82**, 97–105.

Moser M, Asquini E, Miolli GV, et al., 2020. The MADS-box gene MdDAM1 controls growth cessation and bud dormancy in apple. Front Plant Sci 11, 1003.

Myles C, Wayne M, 2008. Quantitative trait locus (QTL) analysis. *Nat Edn* **1**(1), 208.

Nishitani C, Hirai N, Komori S, et al., 2016. Efficient genome editing in apple using a CRISPR/Cas9 system. Sci Rep **6**, 1–8.

Nobile PM, Wattebled F, Quecini V, Girardi CL, Lormeau M, Laurens F, 2011. Identification of a novel alpha-L-arabinofuranosidase gene associated with mealiness in apple. *J Exp Bot* **62**, 4309–21.

Norelli JL, Jones AL, Aldwinckle HS, 2003. Fire blight management in the twenty-first century: using new technologies that enhance host resistance in apple. Plant Dis **87**, 756–65.

Nybom N, 1959. On the inheritance of acidity in cultivated apples. *Hereditas* **45**, 332–50.

Okada K, Wada M, Moriya S, et al., 2016. Expression of a putative dioxygenase gene adjacent to an insertion mutation is involved in the short internodes of columnar apples (*Malus × domestica*). *J Plant Res* **129**, 1109–26.

Otto D, Petersen R, Brauksiepe B, Braun P, Schmidt ER, 2014. The columnar mutation ("Co gene") of apple (Malus × domestica) is associated with an integration of a Gypsy-like retrotransposon. Mol Breed **33**, 863–80.

Peace CP, Bianco L, Troggio M, et al., 2019. Apple whole genome sequences: recent advances and new prospects. *Hortic Res* **6**, 59.

Petersen R, Djozgic H, Rieger B, Rapp S, Schmidt ER, 2015. Columnar apple primary roots share some features of the columnar-specific gene expression profile of aerial plant parts as evidenced by RNA-Seq analysis. BMC Plant Biol **15**, 34.

Peterson BK, Weber JN, Kay EH, Fisher HS, Hoekstra HE, 2012. Double digest RADseq: an inexpensive method for de novo SNP discovery and genotyping in model and non-model species. PLoS One 7, e37135.

Picarella ME, Mazzucato A, 2019. The occurrence of seedlessness in higher plants; insights on roles and mechanisms of parthenocarpy. Front Plant Sci 9, 1997.

Pompili V, Dalla Costa L, Piazza S, Pindo M, Malnoy M, 2020. Reduced fire blight susceptibility in apple cultivars using a high-efficiency CRISPR/Cas9-FLP/FRT-based gene editing system. *Plant Biotechnol J* **18**, 845–58.

Porto DD, Bruneau M, Perini P, et al., 2015. Transcription profiling of the chilling requirement for bud break in apples: a putative role for FLC-like genes. *J Exp Bot* **66**, 2659–72.

Porto DD, Da Silveira Falavigna V, Arenhart RA, et al., 2016. Structural genomics and transcriptional characterization of the Dormancy-Associated MADS-box genes during bud dormancy progression in apple. Tree Genet Genomes 12, 46.

Powell W, Machray GC, Provan J, 1996. Polymorphism revealed by simple sequence repeats. *Trends Plant Sci* **1**, 215–22.

Reyes F, Dejong T, Franceschi P, Tagliavini M, Gianelle D, 2016. Maximum growth potential and periods of resource limitation in apple tree. Front Plant Sci **7**, 233.

Roth JR, 1974. Frameshift mutations. *Annu Rev Gene* **8**, 319–46.

Ruiz KB, Trainotti L, Bonghi C, Ziosi V, Costa G, Torrigiani P, 2013. Early methyl jasmonate application to peach delays fruit/seed development by altering the expression of multiple hormone-related genes. *J Plant Growth Regul* **32**, 852–64.

Russo NL, Burr TJ, Breth DI, Aldwinckle HS, 2008. Isolation of streptomycin-resistant isolates of Erwinia amylovora in New York. *Plant Dis* **92**, 714–8.

Saito T, Wang S, Ohkawa K, et al., 2017. Lipid droplet-associated gene expression and chromatin remodelling in LIPASE 5′-upstream region from beginning- to mid-endodormant bud in 'Fuji' apple. *Plant Mol Biol* **95**, 441–9.

Sampson D, Cameron D, 1965. Inheritance of bronze foliage extra petals and pendulous habit in ornamental crabapples 1965. *Proc Amer Soc Horti Sci* **86**, 717–22.

Sayers EW, Beck J, Bolton EE, et al., 2021. Database resources of the National Center for Biotechnology Information. *Nucleic Acids Res* **49**, D10–7.

Scheben A, Batley J, Edwards D, 2017. Genotyping-by-sequencing approaches to characterize crop genomes: choosing the right tool for the right application. *Plant Biotechnol J* **15**, 149–61.

Schouten HJ, Krens FA, Jacobsen E, 2006. Cisgenic plants are similar to traditionally bred plants: international regulations for genetically modified organisms should be altered to exempt cisgenesis. EMBO Rep **7**, 750–3.

Seymour GB, Manning K, Eriksson EM, Popovich AH, King GJ, 2002. Genetic identification and genomic organization of factors affecting fruit texture. *J Exp Bot* **53**, 2065–71.

Shen F, Huang Z, Zhang B, et al., 2019. Mapping gene markers for apple fruit ring rot disease resistance using a multi-omics approach. *Genes Genom Genet* **9**, 1663–78.

Silfverberg-Dilworth E, Matasci CL, Van De Weg WE, et al., 2006. Microsatellite markers spanning the apple (Malus × domestica Borkh.) genome. *Tree Genet Genomes* **2**, 202–24.
Songstad DD, Petolino JF, Voytas DF, Reichert NA, 2017. Genome editing of plants. *Crit Rev Plant Sci* **36**, 1–23.
Stowe E, Dhingra A, 2021. Development of the arctic(r) apple. *Plant Breed Rev* **44**, 273–96.
Sun R, Chang YS, Yang FQ, et al., 2015. A dense SNP genetic map constructed using restriction site-associated DNA sequencing enables detection of QTLs controlling apple fruit quality. BMC Genomics 16, 747.
Sun X, Jiao C, Schwaninger H, et al., 2020. Phased diploid genome assemblies and pan-genomes provide insights into the genetic history of apple domestication. *Nat Genet* **52**, 1423–32.
Sunako R, Sakuraba W, Senda M, et al., 1999. An allele of the ripening-specific 1-aminocyclopropane-1-carboxylic acid synthase gene (ACS1) in apple fruit with a long storage life. Plant Physiol **119**, 1297–303.
Takeuchi T, Matsushita MC, Nishiyama S, Yamane H, Banno K, Tao R, 2018. RNA-sequencing analysis identifies genes associated with chilling-mediated endodormancy release in apple. *J Am Soc Hortic Sci* **143**, 194–206.
Takos AM, Jaffe FW, Jacob SR, Bogs J, Robinson SP, Walker AR, 2006. Light-induced expression of a MYB gene regulates anthocyanin biosynthesis in red apples. Plant Physiol **142**, 1216–32.
Team RC, 2013. R: a language and environment for statistical computing. R Foundation for Statistical Computing, Vienna, Austria. URL https://www.R-project.org/
Thapa R, Singh J, Gutierrez B, Arro J, Khan A, 2021. Genome-wide association mapping identifies novel loci underlying fire blight resistance in apple. Plant Genome **14**, e20087.
Thomson SV, 2000. Epidemiology of fire blight. In: Vanneste JL (ed) *Fire Blight: The Disease and Its Causative Agent, Erwinia amylovora*. CAB international, Wallingford. 9-36.
Tian Y, Thrimawithana A, Ding T, et al., 2022. Transposon insertions regulate genome-wide allele-specific expression and underpin flower colour variations in apple (*Malus* spp.). *Plant Biotechnol J* **20**, 1285–97.
Tobutt KR, 1994. Combining apetalous parthenocarpy with columnar growth habit in apple. *Euphytica* **77**, 51–4.
Tsao R, Yang R, Young JC, Zhu H, 2003. Polyphenolic profiles in eight apple cultivars using high-performance liquid chromatography (HPLC). *J Agric Food Chem* **51**, 6347–53.
Urrestarazu J, Muranty H, Denancé C, et al., 2017. Genome-wide association mapping of flowering and ripening periods in apple. *Front Plant Sci* **8**, 1–19.
Van Dyk MM, Soeker MK, Labuschagne IF, Rees DJG, 2010. Identification of a major QTL for time of initial vegetative budbreak in apple (*Malus × domestica* Borkh.). *Tree Genet Genomes* **6**, 489–502.
Velasco R, Zharkikh A, Affourtit J, et al., 2010. The genome of the domesticated apple (Malus × domestica Borkh.). *Nat Genet* **42**, 833–9.
Venison EP, Litthauer S, Laws P, et al., 2022. Microsatellite markers as a tool for active germplasm management and bridging the gap between national and local collections of apple. *Genet Resour Crop Evol* **69**, 1817–32.
Visser T, 1964. Juvenile phase and growth of apple and pear seedlings. *Euphytica* **13**, 119–29.
Visser T, Verhaegh JJ, 1978. Inheritance and selection of some fruit characters of apple. 1. Inheritance of low and high acidity. *Euphytica* **27**, 753–60.
Wang A, Aldwinckle H, Forsline P, et al., 2012. EST contig-based SSR linkage maps for Malus × domestica cv Royal Gala and an apple scab resistant accession of M. sieversii, the progenitor species of domestic apple. *Mol Breed* **29**, 379–97.
Wang A, Yamakake J, Kudo H, et al., 2009a. Null mutation of the MdACS3 gene, coding for a ripening-specific 1-aminocyclopropane-1-carboxylate synthase, leads to long shelf life in apple fruit. Plant Physiol **151**, 391–9.
Wang AD, Tan DM, Tatsuki M, et al., 2009b. Molecular mechanism of distinct ripening profiles in 'Fuji' apple fruit and its early maturing sports. *Postharvest Biol Technol* **52**, 38–43.
Wang H, Zhao S, Mao K, et al., 2018. Mapping QTLs for water-use efficiency reveals the potential candidate genes involved in regulating the trait in apple under drought stress. BMC Plant Biol **18**, 136.
Wannemuehler SD, Luby JJ, Yue C, Bedford DS, Gallardo RK, Mccracken VA, 2019. A cost-benefit analysis of DNA informed apple breeding. *HortScience* **54**, 1998–2004.
Wasana S, 2018. Polymorphisms: genetic diversity and disease susceptibility. In: Liu Y, (ed). *Genetic Diversity and Disease Susceptibility*. IntechOpen. https://doi.org/10.5772/intechopen.76728
Watkins CB, Erkan M, Nock JE, Iungerman KA, Beaudry RM, Moran RE, 2005. Harvest date effects on maturity, quality, and storage disorders of 'Honeycrisp' apples. *HortScience* **40**, 164–9.
Watkins CB, Nock JF, Whitaker BD, 2000. Responses of early, mid and late season apple cultivars to postharvest application of 1-methylcyclopropene (1-MCP) under air and controlled atmosphere storage conditions. Postharvest Biol Technol **19**, 17–32.

Watts S, Migicovsky Z, Mcclure KA, et al., 2021. Quantifying apple diversity: a phenomic characterization of Canada's apple biodiversity collection. *Plants People Planet* **3**, 747–60.

Williams JGK, Hanafey MK, Rafalski JA, Tingey SV, 1995. Genetic analysis using random amplified polymorphic DNA markers. In: Wu R (ed) *Recombinant DNA Methodology II.* Academic Press, Boston, pp. 849–84.

Wu B, Shen F, Wang X, et al., 2021. Role of MdERF3 and MdERF118 natural variations in apple flesh firmness/crispness retainability and development of QTL-based genomics-assisted prediction. *Plant Biotechnol J* **19**, 1022–37.

Wu R, Tomes S, Karunairetnam S, et al., 2017. SVP-like MADS box genes control dormancy and budbreak in apple. *Front Plant Sci* **8**, 477.

Xia R, Zhu H, An YQ, Beers EP, Liu Z, 2012. Apple miRNAs and tasiRNAs with novel regulatory networks. *Genome Biol* **13**, R47.

Xing L, Zhang D, Song X, et al., 2016. Genome-wide sequence variation identification and floral-associated trait comparisons based on the re-sequencing of the 'Nagafu No. 2' and 'Qinguan' varieties of apple (Malus domestica Borkh.). Front Plant Sci 7, 908.

Xu K, Wang A, Brown S, 2012. Genetic characterization of the Ma locus with pH and titratable acidity in apple. *Mol Breed* **30**, 899–912.

Yao J-L, Dong Y-H, Morris Ba M, 2001. Parthenocarpic apple fruit production conferred by transposon insertion mutations in a MADS-box transcription factor. *Proc Natl Acad Sci* **98**, 1306–11.

Yao JL, Xu J, Cornille A, et al., 2015. A microRNA allele that emerged prior to apple domestication may underlie fruit size evolution. *Plant J* **84**, 417–27.

Yue C, Gallardo RK, Luby J, et al., 2013. An investigation of U.S. apple producers' trait prioritization-evidence from audience surveys. *HortScience* **48**, 1378–84.

Zhang C, Zhang X, Cheng B, et al., 2022. MdNup54 interactions with MdHSP70 involved in flowering in apple. *Front Plant Sci* **13**, 903808.

Zhang L, Hu J, Han X, et al., 2019. A high-quality apple genome assembly reveals the association of a retrotransposon and red fruit colour. Nat Commun 10, 1494.

Zhang Q, Ma BQ, Li H, et al., 2012a. Identification, characterization, and utilization of genome-wide simple sequence repeats to identify a QTL for acidity in apple. BMC Genomics 13, 537.

Zhang S, Chen W, Xin L, et al., 2014. Genomic variants of genes associated with three horticultural traits in apple revealed by genome re-sequencing. *Hortic Res* **1**, 14045.

Zhang Y, Zhu J, Dai H, 2012b. Characterization of transcriptional differences between columnar and standard apple trees using RNA-Seq. *Plant Mol Biol Rep* **30**, 957–65.

Zhang YZ, Li PM, Cheng LL, 2010. Developmental changes of carbohydrates, organic acids, amino acids, and phenolic compounds in 'Honeycrisp' apple flesh. *Food Chem* **123**, 1013–8.

Zhao D, Liu G, Lv D, Qin S, Ma H, Wang H, 2009. Evaluation of the specific characters of Hanfu apple cultivar. *J Fruit Sci* **26**, 6–12.

4 Allele Mining in *Citrus*

Prospects and Potentials for Functional Genomics, Evolutional Studies, and Genomics-Assisted Breeding

Tokurou Shimizu and Hiroyoshi Iwata
Institute of Fruit Tree and Tea Science, NARO

4.1 INTRODUCTION

Citrus is a woody perennial fruit tree produced in a wide range of tropical to temperate regions. Fresh fruit and juice are the major products, but processed products, including essential oil and canned fruit, are also distributed worldwide (Hodgson 1967; Swingle and Reece 1967). The Rutaceae family, including genera *Citrus*, *Poncirus*, and *Fortunella*, show a wide diversity in fruit traits (size, shape, the thickness of rind, the color of rind or flesh, aroma, soluble solid content (SSC), acidity or else; Figure 4.1), inflorescence, disease resistance, yield, and alternate bearing (Swingle and Reece 1967). Such diversities are observed even within a population derived from a single cross (Scora 1975; Combrink et al. 2013, 2015). Despite the diversity observed within the genus *Citrus*, international trade is dominated by particular varieties such as sweet orange, lemon, lime, grapefruit, and mandarins include Satsuma, Clementine, and ponkan (Intergovernmental Group on Citrus Fruits 2016). Other varieties, such as yuzu and pummelo, a sort of minor mandarin varieties with natural hybrids including tangor and tangelo, and also *Citrus*-related genera kumquat (*Fortunella* spp.), trifoliate orange (*Poncirus trifoliata*), and finger limes are produced and distributed locally. The citrus genome is a rich resource of numerous alleles that deliver such diversity.

In *Citrus*, the mining of polymorphisms and alleles has been conducted to identify genes responsible for the variation of traits among varieties and reveal phylogenetic relationships within the genus *Citrus*. Identifying genes and alleles associated with a particular trait will allow improvement of the trait through breeding, evaluation of the trait by molecular tools, and the development of high-precision control techniques. The first step in functional genomics studies, which evaluates the involvements of genes in traits, is the isolation of the genes of interest. Before publishing the first citrus reference genome, the primary approach involved isolating citrus orthologs by referring to the function of genes that had been investigated in detail in other plants (Talon and Gmitter 2008; Gmitter et al. 2012). For molecular phylogenetic analysis, it is essential to find and evaluate a sufficient number of polymorphisms to cover the entire genome. Although random markers were initially used, they were replaced by highly accurate codominant markers as partial genome information became available (Talon and Gmitter 2008; Gmitter et al. 2012; Shimizu et al. 2020). Different approaches have been adopted for individual studies in the past for mining polymorphisms and alleles in citrus. Following the release of the reference genome, searching polymorphisms and alleles throughout the genome shifted to the mainstream approach to examine the function of genes inferred from genomic regions that are suggested to be associated with specific traits. In this review,

DOI: 10.1201/9781003386490-4

FIGURE 4.1 Diversity of fruit characteristics observed among 22 citrus varieties.

we will introduce past efforts in allele mining of citrus, focusing on gene identification and polymorphism detection, identifying alleles associated with traits, and applications for understanding the genealogical correlations of indigenous varieties and the domestication process. This review also presents applications of genome-wide DNA marker genotypes combined with genomic information to citrus breeding and discuss promising future efforts for fast and accurate mining of alleles associated with complex traits.

4.2 TOOLS AND RESOURCES FOR ALLELE DISCOVERY IN CITRUS

4.2.1 Prerequisite for Allele Mining in Citrus

Locating a genomic region associated with the trait is the first step for identifying causative genes and alleles involved in a trait of interest. Even if the causal gene is not identified, detecting alleles associated with the trait of interest will enable the early selection of breeding seedlings. DNA marker analysis allowed dissecting the complex traits by locating associated genomic regions for a trait of interest by linkage mapping and quantitative trait locus (QTL) analysis (Shimizu et al. 2020). Those analyses enabled us to understand phenotypic variation observed among citrus varieties by means of genetic differences through estimating causative genes and genetic variations that lie behind the trait diversity. Next-generation sequencing (NGS) technologies and large-scale genotyping platforms accelerated validating varieties by genome-wide DNA marker genotype data, and it expanded our knowledge of the phylogenetic relationship and the origin of *Citrus*. Amino acid sequences of the deduced genes of citrus show considerable homology to those of the model plant *Arabidopsis thaliana* (Shimizu et al. 2017). These technologies, theories, and informative resources developed and accumulated in annual crops and model plants facilitated the study of citrus.

In genus *Citrus*, the large tree size, long juvenile period, and high genome heterozygosity restricted the development and phenotyping of a hybrid population for linkage analysis and hampered the introduction of new technology into the study. Applications developed in annual crops for high-density linkage mapping and high-throughput analysis are also available effectively in *Citrus*, but some methods are hard to use. As examples, TILLING (Targeting Induced Local Lesions IN Genomes) and Eco-TILLING (Ecotype TILLING) have been used as valuable tools for allele mining and functional gene analysis in the study of annual crops (McCallum et al. 2000; Barkley and Wang 2008; Kurowska et al. 2011). TILLING detects point mutation within a target region by screening a pooled DNA set prepared from EMS-mutagenized plants with gel electrophoresis or by NGS analysis (Till et al. 2006; Kumar et al. 2017). Though TILLING anticipates the population is homozygosed by selfing ethyl methanesulfonate (EMS)-treated seeds (McCallum et al. 2000), this process is hard to apply in fruit trees, such as citrus, because it will change the whole traits drastically, even if some means enable to avoid constraints for the long juvenility, available orchard area, and phenotyping. Another approach, Eco-TILLING, detects natural genetic variation among many accessions (Till et al. 2006; Barkley and Wang 2008). While Eco-TIILING avoids using a population derived from selfing, the extra bands that are produced due to natural polymorphism in a highly heterozygous genome hinder detecting a target variation associated with a trait (Barkley and Wang 2008), and no successful application of TILLING or Eco-TILLING to fruit trees has been reported to date. In the next section, we briefly summarize the tools and resources that have been used for allele mining so far in citrus.

4.2.2 Tools and Genome Resources for Allele Mining

Until the 1990s, random amplified polymorphic DNA (RAPD), restriction fragment length polymorphism (RFLP), inter-simple sequence repeat (ISSR), amplified fragment length polymorphism (AFLP), and isozyme were the mainstream detection methods for the genotyping study in *Citrus* (Gmitter et al. 2012; Shimizu et al. 2020). Early attempts of citrus genome study collected expressed sequence tags (ESTs) and partial genome sequences since the 21st century motivated the development of codominant PCR-RFLPs (a.k.a. cleaved amplified polymorphic sequences = CAPS), simple sequence repeat (SSR), and single nucleotide polymorphism (SNP) markers (Gmitter et al. 2012; Shimizu et al. 2020). In 2006, the first high-density genotyping array for citrus consisting of 30,171 probe sets for the measurement of transcript expression and 5,023 SNP probe sets was developed as an Affymetrix GeneChip® (Roose et al. 2006). Using this GeneChip®, transcriptome analysis of the organic acid metabolism of lemon fruit suggested the association of H^+-ATPase AHA10 proton pump with citric acid accumulation (Aprile et al. 2011). The first reference genome sequencing efforts of *Citrus* released doubled haploid sweet-orange genome in 2013 (Xu et al. 2013), followed by the haploid Clementine genome sequence in 2014 (Wu et al. 2014). The genomes of other major citrus varieties have subsequently been released for Satsuma (Shimizu et al. 2017) and lemon (Guardo et al. 2021). Also, a highly accurate sweet-orange genome was released in 2021 (Wang et al. 2021). The published reference genome sequence accelerated the genome-wide DNA marker development and studies using large genome-wide genotype data. Linkage analysis and QTL analysis mapped at least 17 traits, such as disease or nematode resistance, environment response, dwarfing, fruit acidity, apomixis, fruit quality, and aroma volatiles on linkage maps (Shimizu et al. 2020). Likewise, genome-wide association analysis (GWAS) and genomic prediction studies demonstrated the identification of the genomic region linked to fruit traits and accurate estimation of the traits from the marker genotype data (Gois et al. 2016; Minamikawa et al. 2017).

Analysis of among-variety polymorphisms of genome-wide DNA markers disclosed their genealogy and ancestor estimation, and extended our knowledge of the domestication process of the genus *Citrus*. Those achievements, as an example, facilitated a comprehensive study that allowed tracing the inheritance of a haplotype block that locates the male sterility by linkage mapping analysis that referred to the inferred citrus genealogy (Goto et al. 2018).

4.2.3 Resources for Functional Genome Analysis

Molecular cloning and functional analysis of a gene with known function in other plants were the primary approaches to accumulate nucleotide sequence resources of citrus and verify the involvement of it in a physiological process in citrus. Those early functional genome studies examined the role of genes that known functions especially revealed in model plants in fruit traits, floral organogenesis, and disease resistance in citrus. Initial attempts for gene isolation identified genes for sugar metabolism (Komatsu et al. 1996, 2002), organic acid metabolism (Sadka et al. 2000), biosynthesis of anthocyanins (Moriguchi et al. 1999; Lo Piero et al. 2006), carotenoids (Kato et al. 2004; Fanciullino et al. 2007), and limonoids (Kita et al. 2000) then revealed expression profiles of the identified genes. Those studies focused on expression profiling analysis, but some studies also evaluated polymorphisms among their homologous genes (Komatsu et al. 1996, 2002; Moriguchi et al. 1999) and genetic diversity among varieties (Lo Piero et al. 2006; Fanciullino et al. 2007).

Juvenility, flowering, and fruit set are factors that severely influence the stable production of fruit in fruit trees, including citrus. Therefore, gene isolation and functional analysis for flower development have long been primary research subjects. Citrus orthologs *TFL* (Pillitteri et al. 2004) and *FT* (Hisada et al. 1997; Endo et al. 2005), which are the homeotic genes involved in the ABC model of floral genesis in the model plant, were isolated and demonstrated to induce precocious flowering in citrus transformants (Pillitteri et al. 2004; Endo et al. 2005). Other homeotic genes involved in floral genesis including *FT/TFL1* homolog (*CuMFT1*) (Nishikawa et al. 2008), *LEAFY* and *APETALLA* 1 (Pillitteri et al. 2004), *MADS43* (Hou et al. 2014; Ye et al. 2021), *AGL9* (Ye et al. 2021), and *SVP/AGL24* (Sun et al. 2016) have been isolated.

Plant hormones, gibberellins, in particular, serve to induce fruit set, growth, and ripening (Talon et al. 1990). In citrus, the transient increase of gibberellin concentration in the floral organs during flowering is known to be closely related to parthenocarpy (Talón et al. 1990, 1992). The gibberellin 20-oxidase gene (*GA20ox*) that catalyzes the biosynthesis of gibberellins was first isolated from Carrizo citrange in 2001 (Vidal et al. 2001). Vidal and colleagues confirmed the biological activity of the isolated gene in transgenic tobacco that showed plant height elongation, larger inflorescences, and longer peduncles than those of control plants by ectopic expression of the introduced gene (Vidal et al. 2001). Another study in Satsuma isolated two *GA20ox* orthologs and demonstrated their biological activity that induced elongated inflorescences in transgenic *Arabidopsis* (Kotoda et al. 2015). They also isolated genes catalyzing gibberellin inactivation, gibberellin 2-oxidase (*GA2ox*), from Satsuma in 2017 and demonstrated their biological activity that induced dwarfing by transgenic expression analysis using *Arabidopsis* (Kotoda et al. 2017). Several *GA20ox* and *GA2ox* orthologs have been identified further through the survey of deduced genes in the draft genome of Satsuma. In addition, the survey of the deduced gene identified the candidate genes of the gibberellin 3-oxidase (*GA3ox*) that catalyze the conversion of GA precursors to bioactive GAs (Shimizu et al. 2017).

Those studies aimed to evaluate the functions of the isolated genes, and genetic diversity or phylogenetic relationships with other varieties were not always examined within the isolated genes or their adjacent genomic region. Furthermore, the number of genes isolated and evaluated in detail remains a few, and the associations of alleles and interspecific trait variations are not fully elucidated. It is not easy to overcome these situations, but the resequencing analysis using the citrus reference genome, which will be discussed later, facilitated the collection of genome-wide polymorphism data of individual cultivars (Wang et al. 2021). It enabled an approach to identify candidate genes from large-scale allele mining by referring to the inferred gene annotation.

4.3 GENOME-WIDE SEQUENCING FOR GENE FAMILY DISCOVERY

Studies of whole-genome sequencing analysis estimated 24,533–29,445 genes in the citrus genomes (Xu et al. 2013; Wu et al. 2014; Shimizu et al. 2017). Because the number of genes whose functions have been analyzed in citrus is limited, using well-established functional gene databases of model plants with rich evidence facilitated the functional genomics study in citrus. Comparative analysis of the deduced protein-coding sequences showed that they were well conserved among Satsuma, haploid Clementine, sweet orange, pummelo, citron, and close relative *Atalantia* (Shimizu et al. 2017). Their amino acid sequences showed relatively high homology with *Arabidopsis* and demonstrated the usefulness of a model plant as a reference for the functional genome studies in citrus. Degenerated PCR targeting conserved regions is a common technique to isolate orthologous genes from genomic DNA and complementary DNA (cDNA). This method is simple and suitable for isolating groups of genes with high sequence similarity.

On the other hand, the success or failure of degenerated PCR depends on the primer design, and care must be taken to isolate groups of genes that are not highly similar. Another approach, *in silico* mining, enables a comprehensive discovery of genes predicted from the genome sequence by specifying multiple common criteria of a specified gene family. This emerging method has the advantage of rapidly identifying candidate gene clusters. It allows the quick launch of polymorphism detection among samples and expression analysis by using the corresponding genomic sequence and also promotes related analyses of the target genes. As genomic information becomes accumulated, this effort is now a widely adopted approach. This section describes the major gene groups inferred from the whole-genome sequence to date.

4.3.1 Allelic Regulation of Carotenoid Composition

Citrus fruits, especially mandarins and oranges, contain high levels of carotenoids in the peel and pulp, and their content affects the product's retail price. The health benefits of carotenoids have recently been recognized (Ma et al. 2020; Saini et al. 2022), and breeding carotenoid-rich new varieties have now been addressed (Nonaka et al. 2012; Raveh et al. 2020). Citrus varieties show wide variation in the contents and compositions of carotenoid components, and previous studies demonstrated their genetic regulation (Fanciullino et al. 2006; Alquézar et al. 2008; Nonaka et al. 2012; Ma et al. 2013). Gene expression analysis among sweet orange, which predominantly accumulates violaxanthin isomers in the juice sacs, Satsuma, which accumulates predominantly β-cryptoxanthin, and lemon, which accumulates low levels of carotenoids have revealed a variety-specific regulation of gene expression in the carotenoid biosynthesis pathway (Kato 2012; Ikoma et al. 2016). Fanciullino and colleagues (2007) evaluated the genetic polymorphism among seven carotenoid biosynthesis genes *Psy*, *Pds*, *Zds*, *Lcyb*, *Lcy-e*, *Hy-b*, and *Zep* in 25 varieties. They showed that the allelic diversity observed at the locus of both lycopene cyclase genes, *Lcy-b* and *Lcy-e1*, is associated with interspecific diversity in carotenoid accumulation in citrus. Another study by Sugiyama and colleagues (2010) detected four alleles in the zeaxanthin epoxidase (*ZEP*) gene region, and the expression levels of transcripts derived from each allele showed differences among varieties that differed in their carotenoid composition. Those observations shed light on the role of allele-specific gene expression in the variation of carotenoid composition. A deletion assay study of the 5′ upstream region of lycopene β-cyclases identified an enhancer element that confers significant promoter activity at 574-513 bp of the promoter region (Lu et al. 2016). Their study showed differences in copy number of the enhancer element (three copies in sweet orange, heterozygous one/two copies in pummelo, and one/three in grapefruit) and suggested these differences would affect the carotenoid composition in these varieties. In particular, a 1,677 bp insertion sequence was identified in the mandarin genome, suggesting an association with *LCYb* gene expression, and also proposed the use of the enhancer element region to distinguish different species.

With the aim of developing β-cryptoxanthin-rich new scions than Satsuma, the potential alleles of genes in the carotenoid metabolic pathway were mined from 13 parental varieties used in citrus breeding, and then identified seven *PSY* alleles, seven *HYb* alleles, 11 *ZEP* alleles, five *NCED* alleles, and four alleles for the eQTL (expression quantitative trait) that control the transcription levels of PDS and ZDS among the ancestral varieties (Fujii et al. 2021). Association analysis between carotenoid composition and gene expression for 263 breeding pedigrees demonstrated *PSY-a*, and

ZEP-e alleles at *PSY* and *ZEP* loci had strong positive effects on increasing the total carotenoid content, including β-cryptoxanthin and violaxanthin, in juice sac tissues. These results indicated that the allele combination of each gene affects the carotenoid composition of individual varieties and suggested the possibility of developing selection methods for increasing the carotenoid content in citrus fruit.

4.3.2 Cytochrome P450 Gene Family (CYPs)

Plant cytochrome P450s (*CYP*s) is a group of heme-thiolate enzymes involved in fatty acid biosynthesis and secondary metabolites, including plant hormones and ultraviolet protections (anthocyanins and flavonoids), alkaloids, tannins, terpenoids, carotenoids, and other various compounds (Mizutani and Ohta 2010). In plants, *CYP*s comprise a large gene family. Mittapelli et al. identified 296, 272, and 202 *CYP* genes in the haploid Clementine genome, doubled haploid sweet-orange genome, and diploid sweet-orange genome, respectively (Mittapelli et al. 2014). The deduced *CYP* genes represented nine clans (*CYP51, 71, 72, 74, 85, 86, 97, 710,* and *711*) consisting of 43 families. Among them, the whole set of A-type cytochrome P450 genes belonged to 18 families in a single clan (*CYP71*), and another whole set of non-A-type cytochrome P450 genes belonged to 25 families in eight clans (*CYP51, 72, 74, 85, 86, 97, 710,* and *711*). Likewise, another genome study of *CYP* genes in the Satsuma draft genome deduced 260 *CYP* genes belonged to 46 families in nine clans (Shimizu et al. 2017). A homology search of *CYP714A* or *CYP714C* genes that catalyze the 13-hydroxylation process of gibberellins suggested in the analysis of model plants estimated their citrus orthologs in the Satsuma genome. The *CYP* genes identified in these three citrus genomes coincided well, but more *CYP* gene families were inferred in the Satsuma genome than in Clementine and sweet-orange genomes. Meanwhile, the deduced *CYP* genes in the Clementine genome suggested two out-grouped genes (*Ccl-CYP706A16* and *Ccl-CYP706B1*), but they were unable to be amplified with sweet-orange genomic DNA by PCR and suggested variations of their genomes. Although further analysis of individual *CYP* genes is still awaited, these studies revealed that the *CYP* genes deduced in the three citrus genomes show variations despite their interspecific similarities. Additional studies will provide insights into the role of this interesting group of genes with diverse functions in the genus *Citrus* by verifying the interspecific polymorphisms of the *CYP* genes and surrounding sequences and their copy number variations among the varieties.

4.3.3 Isoprenoid Pathway

Isoprenoids are a diverse family of compounds synthesized from C_5 precursors isopentenyl diphosphate (IPP) and dimethylallyl diphosphate (DMAPP) (Pulido et al. 2012). In plants, isoprenoids are synthesized by two independent pathways, the mevalonate (MVA) pathway in the cytoplasm and the 2-*C*-methyl-D-erythritol 4-phosphate (MEP) pathway in the plastid, and it plays a vital role in providing substrates for the biosynthesis of pigments and plant hormones, such as abscisic acid, gibberellins, cytokinins, and terpenoids as aroma components (Pulido et al. 2012; Vranová et al. 2013). The isoprenoid pathway is highly regulated in many organisms, from microbes to animals to plants, producing tens of thousands of such a variety of end products from a single simple compound (Vranová et al. 2013; Kumari et al. 2013). Citrus is a rich source of secondary metabolites, and understanding the isoprenoid synthesis pathway will help to elucidate and utilize a wide range of physiological processes that covers the biosynthesis of pigments and aroma compounds and plant hormone-mediated physiology of setting, growth, and maturation. On the other hand, although there is a report (Xu et al. 2009) that evaluated the isoprenoid biosynthesis gene cluster in transcriptome analysis of citrus fruit, there is no report focused on this gene cluster in citrus. The genome analysis of Satsuma deduced citrus orthologs in each step of the isoprenoid pathway; eight orthologs for six genes in the MVA pathway, ten orthologs for seven genes in the MEP pathway, and 11 orthologs for four genes in the pathway to synthesis geranylgeranyl diphosphate from isopentenyl diphosphate

(Shimizu et al. 2017). The estimated diversity of each gene was generally the same or slightly less than that of the model plant *Arabidopsis thaliana*, suggesting that the duplication of genes in the isoprenoid biosynthetic pathway and the acquisition of new functions are not a cause resulting in the synthesis of a wide variety of terpenoid compounds in citrus. The results may imply that there are no apparent differences between the citrus and other plants regarding the regulation of the isoprenoid pathway of supplying substrates for various compounds. On the other hand, no studies in citrus that mine the alleles of each gene of the isoprenoid pathway and examined the allelic regulation of gene expression on supplying isoprenoid compounds via MEV or MEP pathways. Future studies are anticipated to evaluate each gene of the isoprenoid pathway focusing on supplying substrates for downstream pathways to terpenoid, gibberellin, and carotenoid biosynthesis.

4.3.4 Terpenoid Biosynthesis

Terpenoids are a wide range of compounds consisting of more than 25,000 members classified into monoterpenes (C_{10}) and sesquiterpenes (C_{15}) that contain two and three isoprene units, and diterpenes (C_{20}) and triterpenes (C_{30}) contain two and three terpene units (Croteau et al. 2000; Zwenger and Basu 2008). Monoterpenes and sesquiterpenes are the majority of volatile components, and in citrus, the terpenoids accumulate primarily in the oil glands in the rind, a source of essential oil (Tranchida et al. 2012; Jing et al. 2015). The essential oil prepared from the citrus fruit is widely distributed as a fragrance material (González-Mas et al. 2019). The aroma of each variety is determined by its aroma composition, and the content of each aroma component is strictly controlled genetically for each variety. Therefore, efforts have been made to understand the terpenoid biosynthesis pathway to modify each terpenoid for the purpose of developing new scions with excellent aroma (Rouseff et al. 1994; Miyazaki et al. 2011, 2012). The initial step of the transformation of the prenyl diphosphate precursors, geranyl diphosphate, farnesyl diphosphate, and geranylgeranyl diphosphate supplied by isoprenoid pathway to various terpenoid classes are catalyzed by the respective monoterpene (C_{10}), sesquiterpene (C_{15}), and diterpene synthases (C_{20}) (Trapp and Croteau 2001). Terpenoid compounds are synthesized by terpenoid synthase (TPS) genes under temporal and spatial regulations in close coordination with the MEV and MEP pathways (Trapp and Croteau 2001; Nagegowda 2010; Singh and Sharma 2015).

Attempts at isolating TPS genes from the *Arabidopsis* genome resulted in the isolation of more than 30 TPS genes, which were classified into three classes based on their nucleotide sequences and exon-intron structures (Trapp and Croteau 2001). Meanwhile, an *in silico* survey of TPSs for the sweet-orange genome identified 95 loci from which 55 encode for putative functional TPSs (Alquézar et al. 2017). Detailed analysis of the deduced TPS genes identified putative full-length genes corresponding to the angiosperm TPS family, *TPS-a*, *TPS-b*, *TPS-c*, *TPS-e/f*, and *TPS-g* and showed that they are one of the largest TPS family characterized. Functional analysis of the expressed proteins of the seven isolated genes confirmed that they have a catalytic activity to synthesize sesquiterpenes. The deduced TPS genes were discovered in six of the nine chromosomes of the citrus and showed to locate closely within particular narrow regions of the chromosome. Yu and colleagues (2017) demonstrated in the QTL analysis of aroma components using the 'Fortune' × 'Murcott' population and found that the QTLs of individual monoterpene and sesquiterpene content were clustered in particular regions in the genome. This result is reminiscent of the unequal distribution of TPSs in the citrus genome (Alquézar et al. 2017) and could be a clue to explain the genetic background of the resemblance of aroma between parents and offspring, as observed in the sweet-orange aroma with the fruit of Kiyomi tangor (Nishiura et al. 1983). On the other hand, most of the detected QTLs for terpenes also overlapped with QTLs for the chemicals classified as alcohol, aldehyde, ester, and ketone. Although the influence of each QTL on the regulation of aroma composition requires further study, if the differences in TPS gene expression can explain terpenoid content and aroma impression, DNA markers to detect alleles in the TPS gene among varieties would provide clues to modifying aroma properties.

4.3.5 WRKY Transcription Factor Family

WRKY is a family of plant transcription factor (TF) genes participating in growth, development, metabolism, and response to environmental cues (Phukan et al. 2016; Chen et al. 2019). In citrus, studies reported identifying WRKY binding motif sequences in the 5′ upstream region in the D-limonene synthase gene (Nishikawa et al. 2009) and the *ZEP* gene (Sugiyama et al. 2010). Ayadi et al. (2016) identified 100 WRKY TFs from the published genomes of Clementine and sweet orange (Xu et al. 2013; Wu et al. 2014) and also from the partial genome sequence of Satsuma (Shimizu et al. 2017). The following study by Ayadi et al. (2019) identified 51 WRKY TFs from the Satsuma genome and corresponded them to the previously identified citrus WRKY TFs. They also demonstrated that the identified WRKY TFs are distributed throughout the citrus genome and suggested an association between citrus WRKY and fruit acidification, essential oil biosynthesis, and abiotic/biotic stress tolerance by expression profiling. By extending the results obtained so far, it is anticipated to unravel the hidden role of each WRKY gene on the physiological process and their regulation through their expression profiling and allelic variations among varieties.

4.3.6 MADS-Box Gene Family

MADS-box TFs are important regulatory factors for plant organogenesis, development, and differentiation (Smaczniak et al. 2012). In 2014, a genome-wide survey study identified 84 MADS-box genes from the haploid Clementine genome and 52 from the sweet-orange genome (Hou et al. 2014). The identified 136 MADS-box genes were categorized into five groups. An analysis of the identified genes with rice and *Arabidopsis* genomes showed a lower percentage of MADS-box gene duplication in the Clementine mandarin and sweet-orange genomes than in *Arabidopsis*. Furthermore, another analysis of the identified 136 MADS-box genes found 128 new putative microsatellites within flanking sequences. Sixteens of the putative microsatellite were selected and examined for validation and transferability test using 11 citrus species. Among 16 SSRs, 15 were identified as valid loci and were highly transferable among citrus species. Their study demonstrated the usefulness of an approach identifying polymorphisms within the MADS-box genes and their flanking region for developing DNA markers to distinguish diverse citrus varieties and species. The called alleles will be helpful for future applications to find valid alleles tightly associated with an important trait. This approach first requires finding the role of a particular MADS-box gene in a trait of interest. Among the MADS-box genes found in the genome-wide survey, *CiMADS43* was reported to interact with *CiAGL9*, indicating that these two genes are closely related to seasonal citrus flowering and leaf development (Ye et al. 2021). The same group also isolated a MADS-box gene *PtAGL24* from trifoliate orange (*Poncirus trifoliata* L. Raf.) and showed it as a homolog of known AGAMOUS-LIKE 24 (*AGL24*).

Expression analysis indicated that *PtAGL24* was expressed in most organs, with higher expression in mature flowers (Sun et al. 2016). Ectopic expression of *PtAGL24* in wild-type *Arabidopsis* promoted early flowering and suggested an essential role in the process of floral transition and diverse functions in development. While the functions of MADS-box genes are diverse, focusing on the association between the interspecific polymorphism and expression levels of individual genes will help us to understand the differences in traits among varieties at the molecular level and to advance our understanding of those traits at the molecular level.

4.4 ALLELE MINING IN CITRUS

This section describes traits for which the causative gene has been identified, or the candidate gene locus has been refined to a narrow genome region. As mentioned in the previous section, many constraints of citrus in identifying causal genes make it almost impossible to apply methods that are standard techniques in model plants, such as separating candidate genes by backcrossing, mutation search by TILLING/eco-TILLING, and functional verification of candidate genes by the transformation of citrus plants. Instead, different approaches to citrus have been adopted, including evaluating spontaneous mutants such as sports, linkage mapping using single-cross populations, and comparative analysis of a genomic region for the candidate gene among multiple varieties. Each study has been accomplished based on many earlier studies conducted long. Although there are still many constraints in the gene mining of citrus, the availability of reference genome sequences has contributed to identifying and understanding the function of candidate genes by clarifying the genome sequence and gene locations of the candidate region and also by allowing a quick survey of polymorphisms among varieties in that region.

4.4.1 *Ruby* Gene Determines Anthocyanin Accumulation in Blood Orange

Butelli and colleagues identified *Ruby* gene to be responsible for accumulating anthocyanins in blood oranges (Butelli et al. 2012). Blood oranges are high in anthocyanins, red to purple pigments, which have been suggested to have health benefit (de Pascual-Teresa and Sanchez-Ballesta 2008; de Pascual-Teresa et al. 2010; Cappellini et al. 2021). On the other hand, it is known that color development depends on the environment temperature during the ripening period and storage conditions, resulting in difficulties in cultivation management and quality control (Sicilia et al. 2020). In an early study on blood orange, Lo Piero and colleagues isolated cDNAs of 4-gene phenylalanine ammonia-lyase (PAL), chalcone synthase (CHS), dihydroflavonol 4-reductase (DFR), and UDP-glucose flavonoid glucosyl transferase (UFGT) involved in anthocyanin accumulation and showed that anthocyanin accumulation in blood orange is regulated at the gene expression level (Lo Piero et al. 2005). Licciardello and colleagues evaluated genes showing differential expression between blood orange and blonde orange using the suppression subtractive hybridization (SSH) assay and identified genes involved in the anthocyanin pathway, such as PAL, CHS, DFR, anthocyanidin synthase (ANS), and UFGT, and showed their higher expression in blood orange (Licciardello et al. 2008).

Butelli and colleagues (2012) isolated the blood orange *R2R3* Myb ortholog cDNA (*Ruby*), referring that *R2R3 Myb* gene regulates anthocyanin accumulation in plants (Kobayashi 2004). They demonstrated the expression of the isolated *Ruby* gene in blood orange, but the expression of the *Ruby* gene was not found in blond orange. They also demonstrated that introducing the *Ruby* gene into tobacco and expressing it under the CaMV 35S promoter increased purple-red pigmentation. Furthermore, coexpression assay with the bHLH (basic helix-loop-helix) TF, known to interact with anthocyanin-type *Myb*, increased expression levels. Comparing the sequences around *Ruby* gene between blood orange and blond orange, a retrotransposon insertion was found in the 5′ upstream region of the blood orange Moro and named Tcs1 (Figure 4.2). Comparative gene expression profiling showed the high expression of *Ruby* in blood orange, but not in blond orange, pummelo, and mandarin due to defects in promoter function. Contrarily, in Moro, the inserted LTR of Tcs1 functions as a promoter, synthesizing transcripts from *Ruby* and activating anthocyanin biosynthesis that results in pigmentation. Another blood orange of different origin found in China, Jingxian, also shows an independent retrotransposon insertion in the upstream region of the *Ruby* gene and confirmed it to activate the *Ruby* gene similar to Moro and cause pigmentation (Figure 4.2).

Recently, a study reported that methylation of the promoter is involved in the induction of *Ruby* gene expression by low temperatures (Sicilia et al. 2020). While numerous strains are known in citrus that were found as sports or selected from nucellar seedlings, identifying the mutated gene or loci in the somatic cells is generally challenging. The approach shown in the study of *Ruby* gene is that finding candidate genes through expression profiling and metabolome analysis and examining their association to specify the causative gene based on genome analysis will facilitate identifying the causative mutated gene.

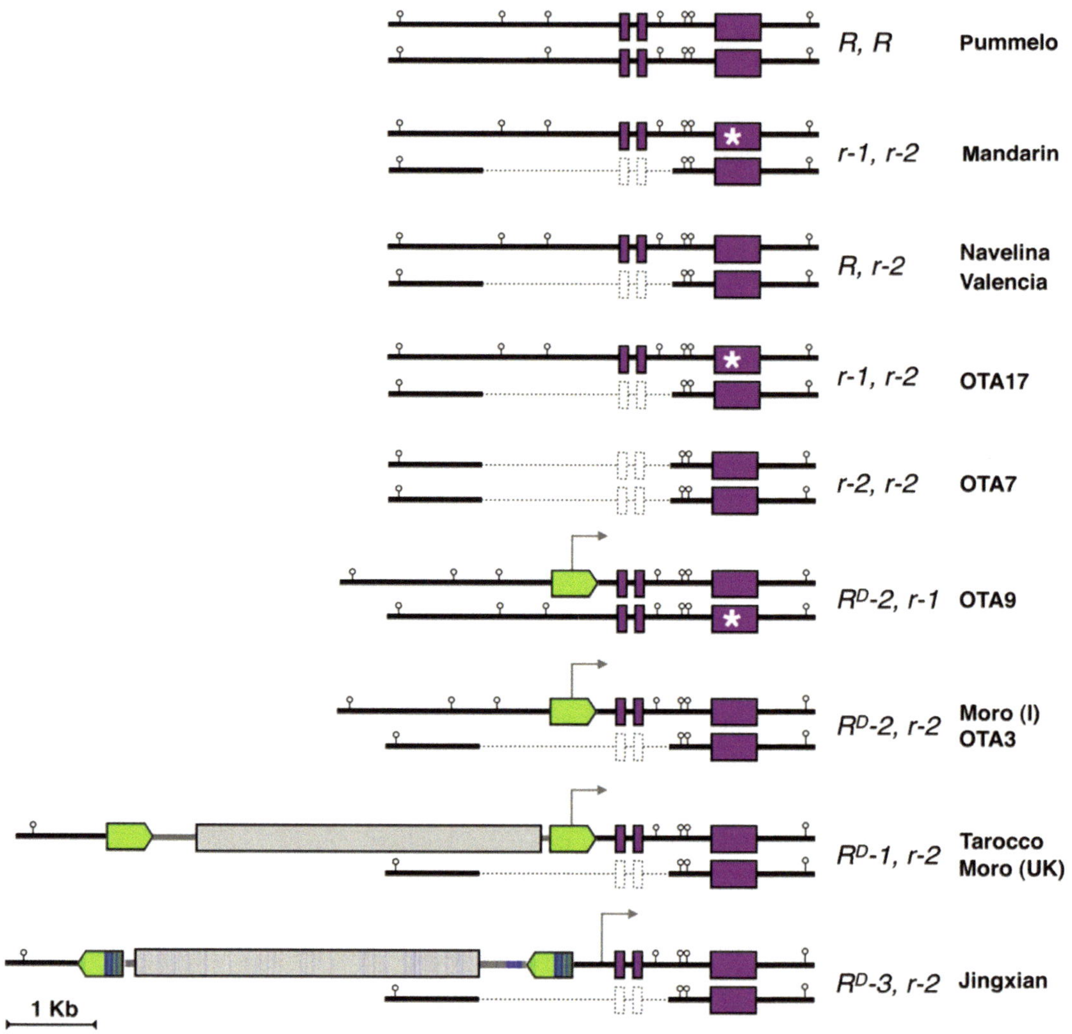

FIGURE 4.2 Maps of structures of the *Ruby* locus in the different citrus species, orange accessions, and hybrids (Butelli et al. 2012).

4.4.2 Polyembryony

Polyembryony is a type of apomixis, and citrus is classified as sporophytic apomixis, in which a sexually derived embryo and multiple nucellar embryos coexist in the same seed (Hand and Koltunow 2014; Zhang et al. 2018). Considerable numbers of citrus varieties are known to form polyembryonic seeds in which a single sexual embryo and asexual embryos coexist in the same seed (Ueno et al. 1967; Frost and Soost 1968; Kepiro and Roose 2007). Major commercial varieties that include sweet orange, grapefruit, Satsuma, and lemon are polyembryonic, while Kishu mandarin and most of pummelo varieties are monoembryonic (Ueno et al. 1967). Due to the high heterogeneity of citrus genome, monoembryonic varieties are incapable of clonal propagation by seed, but polyembryony provides a choice to propagate scion and rootstock by seed other than grafting (Frost and Soost 1968; Kepiro and Roose 2007; Zhang et al. 2018). Understanding the clonal propagation system by seed has been anticipated in all annual crops for agricultural production values. Early studies suggested that a single principal dominant gene regulates the occurrence of citrus polyembryony (Parlevliet and Cameron 1959; Iwamasa et al. 1967).

Based on these studies, initial studies intended to identify the loci for the polyembryony in citrus used molecular markers to carry out linkage analysis (Nakano et al. 2008; Kepiro and Roose 2009). A subsequent study constructed a 47 cM long linkage map and showed the association of this region with the polyembryony by Bulked Segregation Analysis (BSA) and linkage analysis using a population that segregated polyembryony (Nakano et al. 2008). According to the haplotype analysis with the constructed linkage map, they proposed a gene responsible for determining polyembryony, *P*, and suggested that Satsuma and sweet orange possess this *P* gene heterozygously (Nakano et al. 2008). Their subsequent study further isolated and sequenced three bacterial artificial chromosome (BAC) clones corresponding to the polyembryonic *P* loci inferred from the linkage map and then reconstructed a contiguous 380 kb nucleotide sequence containing 70 open reading frames (ORFs) (Nakano et al. 2012). They surveyed SNPs within this region for 35 polyembryonic and 18 monoembryonic varieties and strains, then showed that genotypes found within this region show strong associations with embryony and suggested this locus harbors a common region that regulates polyembryony of citrus. The arrangement of the predicted ORFs showed high synteny to the genome of chromosome 4 in *Vitis vinifera* and linkage group VI in *Populus trichocarpa*, suggesting that this region is well conserved among plants.

Wang et al. (2017) sequenced the entire genome of 100 primitive, wild and cultivated citrus accessions and narrowed the polyembryony locus to an 80 kb region containing 11 candidate genes from association analysis. Further expression profiling analysis in this 80 kb region found a gene *CitRWP* that expressed a higher level in ovules of polyembryonic varieties among 11 candidate genes and identified the insertion of a miniature inverted-repeat transposable element in its promoter region that cosegregated with polyembryony (see Xu et al. (2021) also for further details).

The development of DNA markers to detect insertional variation around *CitRWP* as polymorphisms among varieties is anticipated to promote early selection for polyembryony. Introducing the polyembryonic trait into rootstock varieties offers the advantage of producing large quantities of stable quality rootstock. While introducing the monoembryonic trait into scion varieties disables clonal propagation, it will work to prevent unauthorized replication of a patented variety via seed and helps protect the breeder's rights.

4.4.3 Allele Mining for Self/Cross-Incompatibility

Self-incompatibility (SI) system of citrus has been of interest in obtaining sufficient hybrid seeds and breeding seedless varieties in breeding programs (Kim et al. 2021; Yamamoto et al. 2006; Ngo et al. 2011). Like Rosaceae plants and Brassicaceae plants, SI has been known in several citrus varieties, including Hyuganatsu and Clementine (Frost and Soost 1968; Zhang et al. 2018; Yamamoto et al. 2006). Based on hybridization tests and phenotypic evaluations, SI in citrus has been assumed to be a gametophytic SI system, and attempts have been made to infer *S* alleles and *S* genotypes (Yamamoto et al. 2006; Ngo et al. 2011; Kim et al. 2021). Although the study of citrus SI is still ongoing and the whole picture is not yet fully understood, rapid progress has been made in isolating associated genes and their functional analysis by referring to genome information.

Zhou and colleagues (2018) demonstrated that the S_4 and S_5 alleles of citrus are derived from pummelo through the analysis of allele frequencies of native citrus accessions and also showed that all accessions with S_4 alleles have their origin in Japan except Kunenbo. The pollen *S* gene in citrus is supposed to be an *S*-RNase, a member of the T2 RNase family (Liang et al. 2017; Honsho et al. 2021). Early transcriptome analysis identified an aspartic-acid-rich protein as the candidate gene to determine SI in Clementine and its self-compatible strain 'Comune' (Caruso et al. 2012). Recently, another attempt identified *S*-RNase genes that expressed pistil-specific by *in silico* analysis of the sweet-orange genome (Liang et al. 2017). In addition, homology searches against the sweet-orange and Clementine genomes identified 16 *S*-RNase genes (Liang et al. 2017). The isolated *S*-RNase protein inhibited pollen tube elongation in specific varieties, indicating that the isolated

S-RNase is involved in self-recognition. The same group reported that *S* ribonucleases (*S*-RNases) and *S*-locus F-boxes (SLFs) as *S* determinants function as the SI system in citrus (Liang et al. 2020). Developing DNA markers to discriminate each *S* allele will facilitate further study of the SI system and breeding.

4.4.4 Allele Mining for Breeding Alternaria Brown Spot (ABS) Resistant Scion

ABS is a fungal disease caused by *Alternaria alternata* (Fr.) Keissl. that gives symptoms on young leaves and young shoots appear as brown to black spots surrounded by a yellow halo (Catara et al. 2020). ABS affects leaves, twigs, and fruits and has a large economic impact on tangerines and mandarin cultivation (Cuenca et al. 2018; Catara et al. 2020). Citrus varieties show variations in susceptibility to ABS, and Clementine, Willowleaf mandarin, and Satsuma are resistant, whereas Dancy and 'Fortune' mandarins, 'Orlando', 'Minneola' and 'Nova' tangelos, and the 'Murcott' tangor are susceptible (Catara et al. 2020). Breeding programs aimed to introduce ABS resistance into scion and rootstock have been carried out by several groups, and marker-assisted selection (MAS) techniques were required for efficient selection of the candidate seedlings (Machado et al. 2011; Cuenca et al. 2018). Linkage analysis carried out by different research groups independently inferred that one or a few recessive genes regulate ABS resistance (*ABSr*) (Dalkilic et al. 2005; Gulsen et al. 2010; Campos et al. 2017; Catara et al. 2020). Subsequently, Cuenca and colleagues located ABSr to a 3.3 Mb genomic region near the centromere of chromosome III by BSA and targeted genetic mapping by half tetrad analysis using SSR and SNP markers (Cuenca et al. 2013). Then they further refined the *ABSr* loci to within the 366 kb range by fine mapping with nine SNPs selected from 22 SNPs mined by sequence analysis of a 9.7 kb region in the 3.3 Mb identified genomic region (Cuenca et al. 2016). This 366 kb region harbored nine genes associated with resistance to pathogens, including eight resistance (*R*) gene homologs, and three of these genes were regarded as primary candidates for ABS resistance. An identified SNP closest to the *ABSr* loci (SNP08) revealed a very high association with the resistant/susceptible phenotype in 40 citrus accessions and converted to a KASP (Kompetitive allele-specific PCR) marker for MAS. Recently, they additionally developed two different DNA marker systems for calling the alleles of SNP08 with high-resolution melting (HRM) analysis and also with a CAPS marker coupled with *Bcc*I restriction enzyme digestion (Arlotta et al. 2020). They demonstrated the availability of the developed marker systems for replacing the KASP marker in the MAS of ABS resistance in a breeding program. The developed CAPS system is advantageous to discard ABS susceptible hybrids with a simple protocol. In contrast, the HRM system has the benefit of examining additional allele information nearby SNP08 in a single assay. These three marker systems provide flexible choices for the MAS depending on the purpose of MAS and the available equipment.

4.5 STUDIES AIMING TO IDENTIFY VARIETIES, PEDIGREE ANALYSIS, AND DOMESTICATION

4.5.1 Classification and Genealogy Study Using SSR and SNP Markers

Understanding the processes of the diversification and domestication of the genus *Citrus* is a fundamental theme that has long awaited unraveling. Early investigations successfully inferred phylogenetic relationships and ancestral species among citrus varieties using initial marker systems (isozymes, RAPD, SCAR, and RFLP) (Scora 1975; Nicolosi 2007). Barkley and colleagues revealed the genetic diversity and population structure of citrus varieties using SSR markers (Barkley et al. 2006). Curk and colleagues revealed the process of lime and lemons that emerged from the hybridization of ancestral papeda (*C. micrantha*), citron (*C. medica*), mandarin (*C. reticulata*), and pummelo (*C. maxima*) by evaluating 133 citrus accessions using carefully designed and verified 123 SNP markers (Curk et al. 2016).

In citrus, it has been reported that both SSR and SNP markers could be decreased transferability depending on the target varieties (Luro et al. 2008; Ollitrault et al. 2012). PCR amplification of less conserved genomic regions among varieties could result in insufficient amplification from either chromosome of a particular plant, resulting in a null allele to that chromosome, even if it is a heterozygous locus, thereby showing a false genotype and resulting in reduced transferability. Because miscalled genotypes can lead to misinterpretation in assessing genetic diversity and relationship estimation of varieties, careful verification must be done for the marker genotype data. In particular, detecting and eliminating miscalled marker genotypes is indispensable for estimating phylogenetic relationships and kinships, especially for estimating the parents-offspring trio.

Shimizu and colleagues selected 123 SSR markers that were anticipated to be reliable by eliminating marker genotype data that showed false genotypes by evaluating them with 122 known trio combinations (Shimizu et al. 2016). From the analysis using genotypic data of the selected SSR markers, they inferred parental varieties for 22 indigenous varieties and alleged parents for 45 varieties. Then they inferred parental varieties for 22 indigenous varieties and alleged parents for 45 varieties by analyzing the genotype data obtained with the selected SSR markers. In another study, they confirmed the trio relationships of Satsuma or Clementine and their parental varieties by verifying that the genotypes called from whole-genome sequencing in these trios were inherited in a Mendelian manner (Shimizu et al. 2017). Both SSR and SNP markers are codominant and highly discriminative, but the parents of triploid citrus cultivars were inferred by taking advantage of the SSR marker to detect two or more alleles (Nakano et al. 2019).

On the other hand, Nagano and colleagues demonstrated the genetic relationship of 35 species of the subfamily Aurantioideae that covers the subdivision of Clauseneae that consists of genera Micromelinae, Clauseninae, and Merrillinae and the subdivision of Citreae that consists of genera Triphasiinae, Citrinae, and Balsamocitrinae with genome-wide marker genotype data obtained by RAD-Seq analysis (Nagano et al. 2018). Currently, the use of genome-wide polymorphisms, mentioned in the next section, is becoming the primary approach for the analysis of phylogenetic evaluation and evolutionary study in citrus. In contrast, the simplicity of analysis, low cost, and more accessible data handling make using a small number of reliable codominant markers still highly valuable. Continuous study to develop simple and flexible DNA markers is still beneficial to meet broad needs for comprehensive studies.

4.5.2 Domestication Process Study Based on *Ruby* Gene

As mentioned in the previous section, various retrotransposon insertion mutations were found in the *Ruby* gene region that confers anthocyanin accumulation in blood oranges, and it provides an excellent clue to investigate the phylogenetic study in citrus. Detailed analysis of sequences and retrotransposon insertions around the *Ruby* gene identified insertion/deletions (INDELs), mutations on TATA-box, CAAT-box, STOP codon, and frameshift mutations across major citrus varieties in addition to Tcs1/Tcs2 insertions and translocations in the vicinity of the *Ruby* gene (Butelli et al. 2017). The identified variations affected the expression of the *Ruby* gene and revealed that the *Ruby* gene plays an essential role in anthocyanin accumulation in the leaf and flower organs, especially in citron and lemon. Phylogenetic relationships among basic species and their interspecific hybrids were examined regarding their allelic constitution according to the *Ruby* gene, and their neighboring variations confirmed the derivation of commercial varieties during domestication. Another evaluation based on the *Ruby* gene sequence elucidated the relationships within the subgenus Papeda, and allowed a new genetic classification of mandarins. The phylogenetic analysis focusing on structural variations around specific genes will provide alternative insight from the analyses based on comprehensive genome-wide marker genotype data.

4.5.3 Understanding the Domestication Process of the Genus *Citrus*

Advances in whole-genome sequence analysis of diverse citrus species and the reduced cost for resequencing analysis have led to a comprehensive understanding of the domestication process that produced current varieties from ancestral species. The study of the domestication process first estimated the ancestral species and then classified alleles called from genome-wide marker genotypes into individual ancestral species. The identified alleles became a valuable tool in confirming the interspecific admixture of the ancestral species' genomes in modern citrus varieties. From the whole-genome sequencing analysis, Xu and colleagues indicated the origin of sweet orange, a back-crossed hybrid between pummelo and mandarin (Xu et al. 2013). Wu and colleagues revealed that cultivated types derived from two progenitor species, ancestral pummelo (*C. maxima*) and ancestral mandarin (*C. reticulata*), from the sequencing analysis of haploid Clementine, mandarin, pummelo, sweet-orange, and sour-orange genomes (Wu et al. 2014). Genome analysis also revealed that sour orange is an F_1 hybrid of the ancestral species.

Furthermore, ponkan mandarin is widely assumed to represent *C. reticulata*, but their study showed it has substantial admixture from *C. maxima*. Besides the progress of identifying the ancestor species and revealing the admixture of their genome in modern varieties, the time and place of origin of primitive citrus have been a matter of subject to debate. Wu and colleagues estimated the center of origin – the triangle formed by northeastern India, northern Myanmar, and northwestern Yunnan with whole-genome sequence analysis (Wu et al. 2018). They further estimated 12 ancestral species, including four already known ancestral citrus species (pummelo, mandarin, citron, and papeda) by genome analysis of diverse citrus accessions and proposed an evolution process of the genus *Citrus*. Recently, they revealed the origin of tachibana and shiikwasha by collecting native citrus accession from Asian regions, especially in Japan, and sequencing their genomes (Wu et al. 2021). The insights into the process of producing major varieties and their relationship to ancestral species from these studies will facilitate the breeding of citrus by achieving accurate early selection methods.

4.6 CONCLUDING REMARKS AND FUTURE PROSPECTS

Genes responsible for various traits, such as fruit characteristics, morphogenesis, reproductive physiology, juvenility, and disease resistance, have been found despite many limitations, and alleles responsible for among-variety variations have been identified. The availability of several reference genomes in citrus and the reduced cost of NGS analysis allows for the generation of large amounts of genotypic information in a short period of time (Shimizu et al. 2020). Such progress accelerated the construction of high-density linkage maps and the survey of interspecific polymorphisms of genotypes/alleles in citrus and enabled the MAS of breeding seedlings using DNA markers designed based on the identified alleles. Verifying the associations among detected alleles and traits by referring to the genome information and the gene information allowed us to understand the molecular mechanism of a particular genotype or alleles that cause among-variety phenotypic variations. On the other hand, conventional linkage analysis and QTL analysis are limited to analyzing traits controlled by one or a few genes, which is an obstacle to the analysis of complex traits involving many genes. The high heterozygosity of citrus also requires a complex process for mining causal genes, along with the limitation that it is not easy to develop a crossed population.

GWAS is an approach to detect loci and alleles that show statistical associations with phenotypic variations in the trait of interest by assessing linkage disequilibrium within genome-wide polymorphisms among wide varieties and lines, and is beneficial in avoiding the development of a crossed population only for the purpose of genetic analysis (Shimizu 2020; Shimizu et al. 2020). A citrus study of GWAS that evaluated 17 fruit traits using 1,841 SNPs detected genomic regions above a significance threshold for at least 10 fruit traits (Minamikawa et al. 2017). Genome-wide prediction (a.k.a. genomic prediction) of the breeding and/or genetic values of traits that are intricately controlled by many genes is another approach that allows the selection of individuals based on genome-wide marker genotypes without having to estimate the locations and effects of causal loci (Gois et al. 2016; Minamikawa et al. 2017). Minamikawa and colleagues evaluated genome-wide prediction for 17 fruit traits by developing prediction models based on 787 citrus accessions and hybrids and demonstrated that seven traits (fruit weight, fruit hardness, color of pericarp, color of flesh, firmness of locule membrane, SSC, and easiness of peeling) showed relatively high prediction accuracy ($r > 0.6$) (Minamikawa et al. 2017). Gois and colleagues evaluated the prediction of 10 traits by genome-wide prediction for 180 hybrids of 'Pera' sweet orange (*Citrus sinensis* Osbeck) and 'Murcott' tangor (*C. sinensis* Osbeck × *C. reticulata* Blanco) populations using 5,287 DArT_seq genotype data (Gois et al. 2016). As a result, although it was examined in a single population, they showed the prediction accuracy of the traits was from 0.53 to 0.64 and suggested that selection based on the genome-wide prediction is superior to phenotypic selection in terms of genetic gain per unit of time. Thus, the use of genome-wide marker genotype data becomes common in citrus breeding as well as in genealogy evaluation and evolutionary study. These studies also showed that the enrichment of the number of markers provided and the number of individuals with trait information in both GWAS and genome-wide prediction effectively improves the detection of the associated genomic region and prediction accuracy (Gois et al. 2016; Minamikawa et al. 2017). High-density genotyping arrays and Skim-sequencing approaches (Kumar et al. 2021), such as NGS-based Genotyping-by-Sequencing (GBS) (Elshire et al. 2011) and RAD-Seq (Davey and Blaxter 2010), have been used to quickly obtain large-scale polymorphic data from a large number of individuals. These methods are already being used for breeding and genome-wide genotyping purposes in citrus (Shimizu 2020; Shimizu et al. 2020). Expanding the use of low-cost and rapid genotyping methods and increasing the number of SNPs and samples to be tested will lead to identifying novel alleles strongly associated with important traits through the future analysis of large-scale marker genotype data.

Efforts to refine trait evaluation protocols are also essential to meet the raised accuracy and higher throughput demand for improving breeding efficiency. Trait evaluation, especially in the field, is still performed almost manually by breeders and drastic advancements are indispensable, but it has many constraints (Iwata et al. 2016). In an attempt to evaluate fruit quality traits by analyzing the cross-sectional images of citrus fruit using machine learning approaches, good correlations were demonstrated between the trait values extracted from the images and the manually evaluated values of fruit traits (Minamikawa et al. 2022). Furthermore, this approach revealed previously unrecognized correlations between traits. The use of high-throughput trait evaluation will enrich trait data and contribute to identifying new genomic regions and alleles involved in individual traits, as well as detect novel alleles involved in multiple traits. This will bring us a new understanding of the functions of those novel alleles.

REFERENCES

Alquézar B, Rodrigo MJ, Zacarías L (2008) Carotenoid biosynthesis and their regulation in citrus fruits. *Tree For Sci Biotechnol* 2(1):23–35.

Alquézar B, Rodríguez A, de la Peña M, Peña L (2017) Genomic analysis of terpene synthase family and functional characterization of seven sesquiterpene synthases from *Citrus sinensis*. *Front Plant Sci* 8:1481.

Aprile A, Federici C, Close TJ, De Bellis L, Cattivelli L, et al. (2011) Expression of the H +-ATPase AHA10 proton pump is associated with citric acid accumulation in lemon juice sac cells. *Funct Integr Genom* 11(4):551–563.

Arlotta C, Ciacciulli A, Strano MC, Cafaro V, Salonia F, et al. (2020) Disease resistant citrus breeding using newly developed high resolution melting and CAPS protocols for Alternaria brown spot marker assisted selection. *Agronomy* 10(9):1368.

Ayadi M, Hanana M, Kharrat N, Merchaoui H, Marzoug RB, et al. (2016) The WRKY transcription factor family in Citrus: valuable and useful candidate genes for citrus breeding. *Appl Biochem Biotechnol* 180(3):516–543.

Ayadi M, Mzid R, Ben Ayed R, Kharrat N, Merchaoui H, et al. (2019) Genome-wide identification and molecular characterization of Citrus unshiu WRKY transcription factors in Satsuma mandarin: clues for putative involvement in cell growth, fruit ripening, and stress response. *TURK J Agric For* 43(2):209–231.

Barkley N, Wang M (2008) Application of TILLING and EcoTILLING as reverse genetic approaches to elucidate the function of genes in plants and animals. *Curr Genom* 9(4):212–226.
Barkley NA, Roose ML, Krueger RR, Federici CT (2006) Assessing genetic diversity and population structure in a citrus germplasm collection utilizing simple sequence repeat markers (SSRs). *Theor Appl Genet* 112(8):1519–1531.
Butelli E, Garcia-Lor A, Licciardello C, Las Casas G, Hill L, et al. (2017) Changes in anthocyanin production during domestication of *Citrus*. *Plant Physiol* 173(4):2225–2242.
Butelli E, Licciardello C, Zhang Y, Liu J, Mackay S, et al. (2012) Retrotransposons control fruit-specific, cold-dependent accumulation of anthocyanins in blood oranges. *Plant Cell* 24(3):1242–1255.
Campos NA, Panis B, Carpentier SC (2017) Somatic embryogenesis in coffee: the evolution of biotechnology and the integration of omics technologies offer great opportunities. *Front Plant Sci* 8:1460.
Cappellini F, Marinelli A, Toccaceli M, Tonelli C, Petroni K (2021) Anthocyanins: from mechanisms of regulation in plants to health benefits in foods. *Front Plant Sci* 12:748049.
Caruso M, Merelo P, Distefano G, La Malfa S, Lo Piero AR, et al. (2012) Comparative transcriptome analysis of stylar canal cells identifies novel candidate genes implicated in the self-incompatibility response of *Citrus clementina*. *BMC Plant Biol* 12(1):20.
Catara V, Suming D, Sarris PF (2020) Molecular mechanisms for resistance to biotic stresses. In: Gentile A, Malfa SL, Deng Z (eds) *The Citrus Genome*. Springer, Cham, pp. 281–294.
Chen X, Li C, Wang H, Guo Z (2019) WRKY transcription factors: evolution, binding, and action. *Phytopathol Res* 1(1):13.
Combrink NK, Bijzet Z, Sippel AD, Booyse M, Labuschagne MT (2015) Genotypic variation of rind colour in citrus tangor Kiyomi families. *Acta Hortic* 1065:439–448.
Combrink NK, Labuschagne MT, Bijzet Z (2013) Variation of fruit size and shape in Kiyomi tangor families. *Sci Hortic (Amsterdam)* 162:357–364.
Croteau R, Kutchan TM, Lewis NG (2000) Natural products (secondary metabolites). In: Buchanan BB, Gruissem W, Jones RL (eds) *Biochemistry Molecular Biology of Plants*. American Society of Plant Physiologists, Rock Ville, pp. 1250–1318.
Cuenca J, Aleza P, Garcia-Lor A, Ollitrault P, Navarro L (2016) Fine mapping for identification of citrus Alternaria brown spot candidate resistance genes and development of new SNP markers for marker-assisted selection. *Front Plant Sci* 7:1948.
Cuenca J, Aleza P, Vicent A, Brunel D, Ollitrault P, et al. (2013) Genetically based location from triploid populations and gene ontology of a 3.3-Mb genome region linked to Alternaria brown spot resistance in citrus reveal clusters of resistance genes. *PLoS One* 8(10):e76755.
Cuenca J, Garcia-Lor A, Navarro L, Aleza P (2018) Citrus genetics and breeding. In: Al-Khayri JM, Mohan Jain S, Johnson DV (eds) *Advances in Plant Breeding Strategies: Fruits*. Springer International Publishing, Cham, pp. 403–436.
Curk F, Ollitrault F, Garcia-Lor A, Luro F, Navarro L, et al. (2016) Phylogenetic origin of limes and lemons revealed by cytoplasmic and nuclear markers. *Ann Bot* 117(4):565–583.
Dalkilic Z, Timmer LW, Gmitter Jr FG (2005) Linkage of an Alternaria disease resistance gene in mandarin hybrids with RAPD fragments. *J Am Soc Hortic Sci* 130(2):191–195.
Davey JW, Blaxter ML (2010) RADSeq: next-generation population genetics. *Brief Funct Genom* 9(5–6):416–423.
de Pascual-Teresa S, Moreno DA, García-Viguera C (2010) Flavanols and anthocyanins in cardiovascular health: a review of current evidence. *Int J Mol Sci* 11(4):1679–1703.
de Pascual-Teresa S, Sanchez-Ballesta MT (2008) Anthocyanins: from plant to health. *Phytochem Rev* 7(2):281–299.
Elshire RJ, Glaubitz JC, Sun Q, Poland JA, Kawamoto K, et al. (2011) A robust, simple genotyping-by-sequencing (GBS) approach for high diversity species. *PLoS One* 6(5):e19379.
Endo T, Shimada T, Fujii H, Kobayashi Y, Araki T, et al. (2005) Ectopic expression of an *FT* homolog from *Citrus* confers an early flowering phenotype on trifoliate orange (*Poncirus trifoliata* L. Raf.). *Transgen Res* 14(5):703–712.
Fanciullino AL, Dhuique-Mayer C, Luro F, Casanova J, Morillon R, et al. (2006) Carotenoid diversity in cultivated citrus is highly influenced by genetic factors. *J Agric Food Chem* 54(12):4397–4406.
Fanciullino A-L, Dhuique-Mayer C, Luro F, Morillon R, Ollitrault P (2007) Carotenoid biosynthetic pathway in the *Citrus* genus: number of copies and phylogenetic diversity of seven genes. *J Agric Food Chem* 55(18):7405–7417.
Frost HB, Soost RK (1968) Seed reproduction: development of gametes and embryos. In: Reuther W, Batchelor LD, Webber HJ (eds) *The Citrus Industry, Volume II*, Revised Ed. University of California Press, Berkeley, pp. 290–324.

Fujii H, Nonaka K, Minamikawa MF, Endo T, Sugiyama A, et al. (2021) Allelic composition of carotenoid metabolic genes in 13 founders influences carotenoid composition in juice sac tissues of fruits among Japanese citrus breeding population. *PLoS One* 16(2):e0246468.

Gmitter FG, Chen C, Machado MA, Souza AA, Ollitrault P, et al. (2012) Citrus genomics. *Tree Genet Genom* 8(3):611–626.

Gois IB, Borém A, Cristofani-Yaly M, De Resende MDV, Azevedo CF, et al. (2016) Genome wide selection in citrus breeding. *Genet Mol Res* 15(4):1–14.

González-Mas MC, Rambla JL, López-Gresa MP, Blázquez MA, Granell A (2019) Volatile compounds in *Citrus* essential oils: a comprehensive review. *Front Plant Sci* 10:12.

Goto S, Yoshioka T, Ohta S, Kita M, Hamada H, et al. (2018) QTL mapping of male sterility and transmission pattern in progeny of Satsuma mandarin. *PLoS One* 13(7):e0200844.

Guardo M Di, Moretto M, Moser M, Catalano C, Troggio M, et al. (2021) The haplotype-resolved reference genome of lemon (*Citrus limon* L. Burm f.). *Tree Genet Genom* 17(6):46.

Gulsen O, Uzun A, Canan I, Seday U, Canihos E (2010) A new citrus linkage map based on SRAP, SSR, ISSR, POGP, RGA and RAPD markers. *Euphytica* 173(2):265–277.

Hand ML, Koltunow AMG (2014) The genetic control of apomixis: asexual seed formation. *Genetics* 197(2):441–450.

Hisada S, Akihama T, Endo T, Moriguchi T, Omura M (1997) Expressed sequence tags of citrus fruit during rapid cell development phase. *J Am Soc Hortic Sci* 122(6):808–812.

Hodgson RW (1967) Horticultural varieties of *Citrus*. In: Reuther W, Webber HJ, Batchelor LD (eds) *The Citrus Industry*. University of California Press, Berkeley, pp. 431–591.

Honsho C, Ushijima K, Anraku M, Ishimura S, Yu Q, et al. (2021) Association of T2/S-RNase with self-incompatibility of Japanese citrus accessions examined by transcriptomic, phylogenetic, and genetic approaches. *Front Plant Sci* 12:638321.

Hou XJ, Liu SR, Khan MRG, Hu CG, Zhang JZ (2014) Genome-wide identification, classification, expression profiling, and SSR marker development of the MADS-box gene family in *Citrus*. *Plant Mol Biol Report* 32(1):28–41.

Ikoma Y, Matsumoto H, Kato M (2016) Diversity in the carotenoid profiles and the expression of genes related to carotenoid accumulation among citrus genotypes. *Breed Sci* 66(1):139–147.

Intergovernmental Group on Citrus Fruits F (2016) *Citrus Fruit Statistics 2015*. Food and Agriculture Organization of the United Nations, Rome.

Iwamasa M, Ueno I, Nishiura M (1967) Inheritance of nucellar embryony in *Citrus*. *Bull Fruit Tree Res Stn E* 2:9–24.

Iwata H, Minamikawa MF, Kajiya-Kanegae H, Ishimori M, Hayashi T (2016) Genomics-assisted breeding in fruit trees. *Breed Sci* 66(1):100–115.

Jing L, Lei Z, Zhang G, Pilon AC, Huhman D V., et al. (2015) Metabolite profiles of essential oils in citrus peels and their taxonomic implications. *Metabolomics* 11(4):952–963.

Kato M (2012) Mechanism of carotenoid accumulation in citrus fruit. *J Jpn Soc Hortic Sci* 81(3):219–233.

Kato M, Ikoma Y, Matsumoto H, Sugiura M, Hyodo H, et al. (2004) Accumulation of carotenoids and expression of carotenoid biosynthetic genes during maturation in citrus fruit. *Plant Physiol* 134(2):824–837.

Kepiro JL, Roose ML (2007) Nucellar embryony. In: Khan IA (ed) *Citrus Genetics, Breeding and Biotechnology*. CAB International, Wallingford, pp. 141–149.

Kepiro JL, Roose ML (2009) AFLP markers closely linked to a major gene essential for nucellar embryony (apomixis) in *Citrus maxima* × *Poncirus trifoliata*. *Tree Genet Genom* 6(1):1–11.

Kim JH, Sato M, Wakana A, Takamatsu F, Sakai K, et al. (2021) Genotyping of *Citrus* accessions with S_9 and/or S_{10} alleles for self-incompatibility and their allelic distribution. *Hortic Sci Technol* 39(6):807–822.

Kita M, Hirata Y, Moriguchi T, Endo-Inagaki T, Matsumoto R, et al. (2000) Molecular cloning and characterization of a novel gene encoding limonoid UDP-glucosyltransferase in *Citrus*. *FEBS Lett* 469(2–3):173–178.

Kobayashi S (2004) Retrotransposon-induced mutations in grape skin color. *Science* 304(5673):982–982.

Komatsu A, Moriguchi T, Koyama K, Omura M, Akihama T (2002) Analysis of sucrose synthase genes in citrus suggests different roles and phylogenetic relationships. *J Exp Bot* 53(366):61–71.

Komatsu A, Takanokura Y, Omura M, Akihama T (1996) Cloning and molecular analysis of cDNAs encoding three sucrose phosphate synthase isoforms from a citrus fruit (*Citrus unshiu* Marc.). *Mol Gen Genet* 252(3):346–351.

Kotoda N, Matsuo S, Honda I, Yano K, Shimizu T (2015) Isolation and functional analysis of two gibberellin 20-oxidase genes from Satsuma mandarin (Citrus unshiu Marc.). *Hortic J* 85(2):128–140.

Kotoda N, Matsuo S, Honda I, Yano K, Shimizu T (2017) Gibberellin 2-oxidase genes from Satsuma mandarin (Citrus unshiu Marc.) caused late flowering and dwarfism in transgenic Arabidopsis. *Hortic J* 85(2):128–140.

Kumar APK, McKeown PC, Boualem A, Ryder P, Brychkova G, et al. (2017) TILLING by sequencing (TbyS) for targeted genome mutagenesis in crops. *Mol Breed* 37(2):14.
Kumar P, Choudhary M, Jat BS, Kumar B, Singh V, et al. (2021) Skim sequencing: an advanced NGS technology for crop improvement. *J Genet* 100(2):38.
Kumari S, Priya P, Misra G, Yadav G (2013) Structural and biochemical perspectives in plant isoprenoid biosynthesis. *Phytochem Rev* 12(2):255–291.
Kurowska M, Daszkowska-Golec A, Gruszka D, Marzec M, Szurman M, et al. (2011) TILLING - a shortcut in functional genomics. *J Appl Genet* 52(4):371–390.
Liang M, Cao Z, Zhu A, Liu Y, Tao M, et al. (2020) Evolution of self-compatibility by a mutant S_m-RNase in citrus. *Nat Plants* 6(2):131–142.
Liang M, Yang W, Su S, Fu L, Yi H, et al. (2017) Genome-wide identification and functional analysis of S-RNase involved in the self-incompatibility of citrus. *Mol Genet Genom* 292(2):325–341.
Licciardello C, Russo MP, Vale' G, Recupero RG (2008) Identification of differentially expressed genes in the flesh of blood and common oranges. *Tree Genet Genom* 4(2):315–331.
Lo Piero AR, Puglisi I, Petrone G (2006) Gene characterization, analysis of expression and in vitro synthesis of dihydroflavonol 4-reductase from [*Citrus sinensis* (L.) Osbeck]. *Phytochemistry* 67(7):684–695.
Lo Piero AR, Puglisi I, Rapisarda P, Petrone G (2005) Anthocyanins accumulation and related gene expression in red orange fruit induced by low temperature storage. *J Agric Food Chem* 53(23):9083–9088.
Lu S, Zhang Y, Zheng X, Zhu K, Xu Q, et al. (2016) Isolation and functional characterization of a lycopene β-cyclase gene promoter from citrus. *Front Plant Sci* 7:1367.
Luro FL, Costantino G, Terol J, Argout X, Allario T, et al. (2008) Transferability of the EST-SSRs developed on Nules clementine (*Citrus clementina* Hort ex Tan) to other Citrus species and their effectiveness for genetic mapping. *BMC Genomics* 9:287.
Ma G, Zhang L, Matsuta A, Matsutani K, Yamawaki K, et al. (2013) Enzymatic formation of β-citraurin from β-cryptoxanthin and zeaxanthin by carotenoid cleavage dioxygenase4 in the flavedo of citrus fruit. *Plant Physiol* 163(2):682–695.
Ma G, Zhang L, Sugiura M, Kato M (2020) Citrus and health. In: Talon M, Caruso M, Gmitter FG (eds) *The Genus Citrus*. Elsevier, Amsterdam, pp. 495–511.
Machado MA, Cristofani-Yaly M, Bastianel M (2011) Breeding, genetic and genomic of citrus for disease resistance. *Rev Bras Frutic* 33(1):158–172.
McCallum CM, Comai L, Greene EA, Henikoff S (2000) Targeting induced local lesions IN genomes (TILLING) for plant functional genomics. *Plant Physiol* 123(2):439–442.
Minamikawa MF, Nonaka K, Hamada H, Shimizu T, Iwata H (2022) Dissecting breeders' sense *via* explainable machine learning approach: application to fruit peelability and hardness in Citrus. *Front Plant Sci* 13:832749.
Minamikawa MF, Nonaka K, Kaminuma E, Kajiya-Kanegae H, Onogi A, et al. (2017) Genome-wide association study and genomic prediction in citrus: potential of genomics-assisted breeding for fruit quality traits. *Sci Rep* 7(1):4721.
Mittapelli SR, Maryada SK, Khareedu VR, Vudem DR (2014) Structural organization, classification and phylogenetic relationship of cytochrome P450 genes in *Citrus clementina* and *Citrus sinensis*. *Tree Genet Genom* 10(2):399–409.
Miyazaki T, Plotto A, Baldwin EA, Reyes-De-Corcuera JI, Gmitter FG (2012) Aroma characterization of tangerine hybrids by gas-chromatography-olfactometry and sensory evaluation. *J Sci Food Agric* 92(4):727–735.
Miyazaki T, Plotto A, Goodner K, Gmitter FG (2011) Distribution of aroma volatile compounds in tangerine hybrids and proposed inheritance. *J Sci Food Agric* 91(3):449–460.
Mizutani M, Ohta D (2010) Diversification of P450 genes during land plant evolution. *Annu Rev Plant Biol* 61(1):291–315.
Moriguchi T, Kita M, Tomono Y, Endo-Inagaki T, Omura M (1999) One type of chalcone synthase gene expressed during embryogenesis regulates the flavonoid accumulation in citrus cell cultures. *Plant Cell Physiol* 40(6):651–655.
Nagano Y, Mimura T, Kotoda N, Matsumoto R, Nagano AJ, et al. (2018) Phylogenetic relationships of Aurantioideae (Rutaceae) based on RAD-Seq. *Tree Genet Genom* 14(1):6.
Nagegowda DA (2010) Plant volatile terpenoid metabolism: biosynthetic genes, transcriptional regulation and subcellular compartmentation. *FEBS Lett* 584(14):2965–2973.
Nakano M, Shimada T, Endo T, Fujii H, Nesumi H, et al. (2012) Characterization of genomic sequence showing strong association with polyembryony among diverse *Citrus* species and cultivars, and its synteny with *Vitis* and *Populus*. *Plant Sci* 183:131–142.

Nakano M, Shimizu T, Kuniga T, Nesumi H, Omura M (2008) Mapping and haplotyping of the flanking region of the polyembryony locus in *Citrus unshiu* Marcow. *J Jpn Soc Hortic Sci* 77(2):109–114.

Nakano M, Shimizu T, Sugawa S, Kaneyoshi J, Fujii H, et al. (2019) Determining the parental combinations of the triploid acid citrus cultivars 'Yellow Bell' and 'Tahiti lime' using DNA marker analyses. *Sci Hortic (Amsterdam)* 246(August 2018):893–897.

Ngo BX, Kim J-H, Wakana A, Isshiki S, Mori T (2011) Estimation of self-incompatibility genotypes of citrus cultivars with Got-3 allozyme markers. *J Jpn Soc Hortic Sci* 80(3):284–294.

Nicolosi E (2007) Origin and taxonomy. In: Khan IA (ed) *Citrus Genetics, Breeding and Biotechnology*. CABI, CAB International, Oxfordshire, pp. 19–43.

Nishikawa F, Endo T, Shimada T, Fujii H, Shimizu T, et al. (2008) Isolation and characterization of a Citrus *FT/TFL1* homologue (*CuMFT1*), which shows quantitatively preferential expression in *Citrus* seeds. *J Jpn Soc Hortic Sci* 77(1):38–46.

Nishikawa F, Endo T, Shimada T, Fujii H, Shimizu T, et al. (2009) Characterization of the 5′ flanking region of the citrus *d*-limonene synthase gene, which shows a quantitatively preferential expression in peel. *J Jpn Soc Hortic Sci* 78(1):84–89.

Nishiura M, Shichijo T, Ueno I, Iwamasa M, Kihara T, et al. (1983) New citrus cultivar "Kiyomi" tangor. *Bull Fruit Tree Res Station Ser B Okitsu* 10:1–9.

Nonaka K, Kita M, Ikoma Y, Fukamachi H, Imai A, et al. (2012) Genetic differences and environmental variations in carotenoid contents of fruit flesh in parental population used in citrus breeding in Japan. *J Am Soc Hortic Sci* 137(4):243–249.

Ollitrault P, Terol J, Garcia-Lor A, Bérard A, Chauveau A, et al. (2012) SNP mining in *C. clementina* BAC end sequences; transferability in the *Citrus* genus (Rutaceae), phylogenetic inferences and perspectives for genetic mapping. *BMC Genomics* 13(1):13.

Parlevliet JE, Cameron JW (1959) Evidence on the inheritance of nucellar embryony in *Citrus*. *Proc Am Soc Hort Sci* 74:252–260.

Phukan UJ, Jeena GS, Shukla RK (2016) WRKY transcription factors: molecular regulation and stress responses in plants. *Front Plant Sci* 7:760.

Pillitteri L, Lovatt C, Walling L (2004) Isolation and characterization of a *TERMINAL FLOWER* homolog and its correlation with juvenility in citrus. *Plant Physiol* 135(July):1540–1551.

Pulido P, Perello C, Rodriguez-Concepcion M (2012) New insights into plant isoprenoid metabolism. *Mol Plant* 5(5):964–967.

Raveh E, Goldenberg L, Porat R, Carmi N, Gentile A, et al. (2020) Conventional breeding of cultivated citrus varieties. In: Gentile A, Malfa SL, Deng Z (eds) *The Citrus Genome, Compendium of Plant Genomes*. Springer Nature, Cham, pp. 33–48.

Roose M, Wanamaker S, Matthew L, Timothy C, Davies C, et al. (2006) A GeneChip(r) for citrus. In: Plant and Animal Genome XIV Conference. San Diego, CA, p. 82.

Rouseff R, Gmitter F, Grosser J (1994) Citrus breeding and flavour. In: Piggott JR, Paterson A (eds) *Understanding Natural Flavors*. Springer, New York, pp. 113–127.

Sadka A, Dahan E, Or E, Cohen L (2000) $NADP^+$-isocitrate dehydrogenase gene expression and isozyme activity during citrus fruit development. *Plant Sci* 158(1–2):173–181.

Saini RK, Ranjit A, Sharma K, Prasad P, Shang X, et al. (2022) Bioactive compounds of citrus fruits: a review of composition and health benefits of carotenoids, flavonoids, limonoids, and terpenes. *Antioxidants* 11(2):239.

Scora RW (1975) On the history and origin of citrus. *Bull Torrey Bot Club* 102(6):369–375.

Shimizu T (2020) Genomic breeding. In: Talon M, Caruso M, Gmitter FG (eds) *The Genus Citrus*. Elsevier, Duxford, pp. 149–169.

Shimizu T, Aka Kacar Y, Cristofani-Yaly M, Curtolo M, Machado MA (2020) Markers, maps, and marker-assisted selection. In: Gentile A, Malfa SL, Deng Z (eds) *The Citrus Genome*. Springer, Cham, pp. 107–139.

Shimizu T, Kitajima A, Nonaka K, Yoshioka T, Ohta S, et al. (2016) Hybrid origins of citrus varieties inferred from DNA marker analysis of nuclear and organelle genomes. *PLoS One* 11(11):e0166969.

Shimizu T, Tanizawa Y, Mochizuki T, Nagasaki H, Yoshioka T, et al. (2017) Draft sequencing of the heterozygous diploid genome of Satsuma (*Citrus unshiu* Marc.) using a hybrid assembly approach. *Front Genet* 8:180.

Sicilia A, Scialò E, Puglisi I, Lo Piero AR (2020) Anthocyanin biosynthesis and DNA methylation dynamics in sweet orange fruit [*Citrus sinensis* L. (Osbeck)] under cold stress. *J Agric Food Chem* 68(26):7024–7031.

Singh B, Sharma RA (2015) Plant terpenes: defense responses, phylogenetic analysis, regulation and clinical applications. *3 Biotech* 5(2):129–151.

Smaczniak C, Immink RGH, Angenent GC, Kaufmann K (2012) Developmental and evolutionary diversity of plant MADS-domain factors: insights from recent studies. *Development* 139(17):3081–3098.
Sugiyama A, Ikoma Y, Fujii H, Shimada T, Endo T, et al. (2010) Structure and expression levels of alleles of *Citrus* zeaxanthin epoxidase genes. *J Jpn Soc Hortic Sci* 79(3):263–274.
Sun LM, Zhang JZ, Hu CG (2016) Characterization and expression analysis of PtAGL24, a SHORT VEGETATIVE PHASE/AGAMOUS-LIKE 24 (SVP/AGL24)-type MADS-box gene from trifoliate orange (*Poncirus trifoliata* L. Raf.). *Front Plant Sci* 7:823.
Swingle WT, Reece PC (1967) The botany of *Citrus* and its wild relatives. In: Reuther W, Webber HJ, Batchelor LD (eds) *The Citrus Industry*. University of California Press, Berkeley, pp. 190–430.
Talon M, Gmitter FG (2008) Citrus genomics. *Int J Plant Genom* 2008:528361.
Talón M, Hedden P, Primo-Millo E (1990) Gibberellins in Citrus sinensis: a comparison between seeded and seedless varieties. *J Plant Growth Regul* 9(1–4):201–206.
Talon M, Zacarias L, Primo-Millo E (1992) Gibberellins and parthenocarpic ability in developing ovaries of seedless mandarins. *Plant Physiol* 99(4):1575–1581.
Talon M, Zacarias L, Primo-Millo E (1990) Hormonal changes associated with fruit set and development in mandarins differing in their parthenocarpic ability. *Physiol Plant* 79(2):400–406.
Till BJ, Zerr T, Comai L, Henikoff S (2006) A protocol for TILLING and Ecotilling in plants and animals. *Nat Protoc* 1(5):2465–2477.
Tranchida PQ, Bonaccorsi I, Dugo P, Mondello L, Dugo G (2012) Analysis of Citrus essential oils: state of the art and future perspectives. A review. *Flav Fragr J* 27(2):98–123.
Trapp SC, Croteau RB (2001) Genomic organization of plant terpene synthases and molecular evolutionary implications. *Genetics* 158(2):811–832.
Ueno I, Iwamasa M, Nishiura M (1967) Embryo number of various varieties of *Citrus* and it's relatives. *Bull Hort Res Sta Jpn Ser B* 7:11–21.
Vidal AM, Gisbert C, Talón M, Primo-Millo E, López-Díaz I, et al. (2001) The ectopic overexpression of a citrus gibberellin 20-oxidase enhances the non-13-hydroxylation pathway of gibberellin biosynthesis and induces an extremely elongated phenotype in tobacco. *Physiol Plant* 112(2):251–260.
Vranová E, Coman D, Gruissem W (2013) Network analysis of the MVA and MEP pathways for isoprenoid synthesis. *Annu Rev Plant Biol* 64:665–700.
Wang L, Huang Y, Liu Z, He J, Jiang X, et al. (2021) Somatic variations led to the selection of acidic and acidless orange cultivars. *Nat Plants* 7(7):954–965.
Wang X, Xu Y, Zhang S, Cao L, Huang Y, et al. (2017) Genomic analyses of primitive, wild and cultivated citrus provide insights into asexual reproduction. *Nat Genet* 49(5):765–772.
Wu GA, Prochnik S, Jenkins J, Salse J, Hellsten U, et al. (2014) Sequencing of diverse mandarin, pummelo and orange genomes reveals complex history of admixture during citrus domestication. *Nat Biotechnol* 32:656–662.
Wu GA, Sugimoto C, Kinjo H, Azama C, Mitsube F, et al. (2021) Diversification of mandarin citrus by hybrid speciation and apomixis. *Nat Commun* 12(1):1–10.
Wu GA, Terol J, Ibanez V, López-García A, Pérez-Román E, et al. (2018) Genomics of the origin and evolution of *Citrus*. *Nature* 554(7692):311–316.
Xu Q, Chen L-L, Ruan X, Chen D, Zhu A, et al. (2013) The draft genome of sweet orange (*Citrus sinensis*). *Nat Genet* 45(1):59–66.
Xu Q, Yu K, Zhu A, Ye J, Liu Q, et al. (2009) Comparative transcripts profiling reveals new insight into molecular processes regulating lycopene accumulation in a sweet orange (*Citrus sinensis*) red-flesh mutant. *BMC Genomics* 10:540.
Xu Y, Jia H, Wu X, Koltunow AM, Deng X, et al. (2021) Regulation of nucellar embryony, a mode of sporophytic apomixis in *Citrus* resembling somatic embryogenesis. *Curr Opin Plant Biol* 59:101984.
Yamamoto M, Kubo T, Tominaga S (2006) Self-and cross-incompatibility of various citrus accessions. *J Jpn Soc Hortic Sci* 75(5):372–378.
Ye L-X, Zhang J-X, Hou X-J, Qiu M-Q, Wang W-F, et al. (2021) A MADS-box gene CiMADS43 is involved in citrus flowering and leaf development through interaction with CiAGL9. *Int J Mol Sci* 22(10):5205.
Yu Y, Bai J, Chen C, Plotto A, Yu Q, et al. (2017) Identification of QTLs controlling aroma volatiles using a 'Fortune' × 'Murcott' (*Citrus reticulata*) population. *BMC Genomics* 18(1):646.
Zhang S, Liang M, Wang N, Xu Q, Deng X, et al. (2018) Reproduction in woody perennial Citrus: an update on nucellar embryony and self-incompatibility. *Plant Reprod* 31(1):43–57.
Zhou X-H, Kim J-H, Wakana A, Sakai K, Kajiwara K, et al. (2018) Distribution and evolution of citrus with S_4 and/or S_5 gene alleles for self-incompatibility with special focus on the origin of Satsuma mandarin (*Citrus unshiu* Marc.; $S_f S_4$). *Genet Resour Crop E* 65(3):1013–1033.
Zwenger S, Basu C (2008) Plant terpenoids : applications and future potentials. *Biotechnol Mol Biol Rev* 3(1):1–7.

5 Allele Mining for Complex Traits in Grapevine

Anuradha Upadhyay, Roshni R. Samarth, Nandni Goyal, and Kashmir Singh

5.1 INTRODUCTION

Grapevine (*Vitis vinifera* L.), a fruit crop of high economic value, is grown worldwide under different climatic and soil conditions. Grapes are used for fresh consumption, processed into wine and raisins, and also used for extraction of seed oil, pigments, and secondary metabolites. Thus, the breeding objectives vary depending upon the end use. The focus is on bunch and berry appearance for table grapes, while pulp content and berry thickness are traits of interest for raisin grape varieties. For wine varieties emphasis is given to aroma and secondary metabolites constituents of juice. These traits are complex in nature, and most often genetics of these traits is not well understood in grape. Identification of grape genotypes with superior traits is of paramount importance for varietal improvement. Domestication and selection process results in changes in genetic makeup of plants. These variations need to be analyzed in depth for the identification of superior genotypes. Allele mining is one such approach that allows identification and isolation of specific variations of genes (alleles) that are associated with desirable traits. These traits can include resistance to diseases, improved fruit quality, and increased yield. Allele mining is typically done through genome sequencing, where the entire genetic makeup is analyzed to identify specific variations of genes that are associated with specific traits. The identified superior allele can be used in breeding programs to create new grape varieties that have improved characteristics. Allele mining can also be used for the identification of haplotypes and development of markers closely linked to the traits for their use in molecular breeding.

5.2 SKIN COLOR

Berry skin color is an important quality trait. It affects the quality of processed products like wine and raisins, and influences the market value of table grapes due to bunch appearance. Due to domestication and selection process, large variation is observed in berry skin color and grape genotypes with wide range of colors ranging from the green/yellow to the dark blue tones exist. The color type and intensity depend on the amount and composition of anthocyanin. Delphinidin, petunidin, cyanidin, malvidin, and peonidin glycoside derivatives are the major anthocyanins in colored grapes. Anthocyanin synthesis pathway involves several enzymes and is part of a complex phenylalanine pathway. Numerous studies have pointed out the significant role of Myb transcription factor, a regulator of anthocyanin pathway, in skin color development (Kobayashi et al. 2002), and variation in *VvMybA*1 gene is found associated with skin color variation in grape. While colored grapes contain functional *VvMybA* genes, white varieties are found to have non-functional *VvMybA* genes. Insertion of retro element Gret1 in promoter region of *VvMybA1* resulted in non-functional allele *VvMybA1a* (Kobayashi et al. 2004). The alleles *VvMybA1b* and *VvMybA1c* lack Gret1, and are dominant over non-functional alleles (Azuma et al. 2007). Walker et al. (2007) found that the berry color locus comprises two genes, *VvMybA1* and *VvMybA2*, and either gene can regulate color in the grape berry. The white berry allele of *VvMybA2* is inactivated by two non-conservative mutations.

DOI: 10.1201/9781003386490-5

A detailed analysis of allele variation in *VvMybA1* in 200 accessions of *V. vinifera* identified association of insertion of the retroelement Gret1 in the promoter region of *VvMybA*1 with white skin color. In red- and pink-colored genotypes, additional polymorphism was observed including variation due to excision of retroelement (This et al. 2007). Thus, variation in single gene resulted in allelic series association with skin color variation in grapes. The presence of heterozygous partially excised Gret1 retrotransposon in the promoter region of the MYBA1 anthocyanin regulator resulted in mild red-skinned variant (Ferreira et al. 2019). Later on, five polymorphisms in a cluster of Myb genes containing four genes, viz. *VvMybA1*, *VvMybA2*, *VvMybA3*, and *VvMybA4*, were observed to account for 84% of the variation in skin color (Fournier-Level et al. 2009). These clusters were identified in the QTL region for skin color. These polymorphisms either resulted in structural changes in the MYB proteins or altered the promoter regions of *VvMybA*s. In *Vitis labrusca*, functional haplotypes HapE2 and HapE2 consisting of *VlMybA2* and *VlMybA1–3* determine skin color and anthocyanin content correlated with the type of haplotype (Azuma et al. 2011). Analysis of wild grapevine (*Vitis vinifer*a subsp. *sylvestris* and subsp. *sativa*) revealed that variation in the two transcriptional regulators generated a novel allele series via length polymorphisms in *VvMybA*1 and a point mutation in *VvMybA*2, which is lacking in cultivated grapevine (Carrasco et al. 2015). Several haplotypes were identified which varied between wild genotypes of western Mediterranean region and eastern regions. Teinturier grape cultivars have red-fleshed berries due to ectopic anthocyanin formation and have economic importance as wines of these varieties can be used for blending for color enhancement. The molecular analysis of color locus of these varieties revealed that insertion of 408 bp repetitive DNA in promoter region of *VvMybA1* (grapevine color enhancer, GCE) is responsible for the red-fleshed berry phenotypes of teinturier cultivars (Röckel et al. 2020). Allelic variation generated due to difference in repeat numbers resulted in varieties with varied anthocyanin content.

Though the major determinant of skin color is *VvMybA1*, *VvMybA2* is shown to regulate the anthocyanin synthesis. The regulatory mechanism of two allele of *VvMybA2*, viz. *VvMybA2r* and *VvMyb2w* has been analyzed. The overexpression of *VvMybA2r* allele in tobacco up-regulates the expression of anthocyanin biosynthetic genes, whereas ectopic expression of *VvMybA2w* inactivates the expression of anthocyanin biosynthetic genes due to CA dinucleotide deletion which shortened the C-terminal transactivation region and subsequent disruption of the transcriptional activation activity of *VvMybA2w* (Jiu et al. 2021). Allele *VvMybA2r* positively regulated anthocyanin biosynthesis by forming the *VvMybA*2r-*VvMYCA1*-*VvWDR1* complex. Analysis of *Myb* haplotype in 213 grape genotypes showed that dark-colored genotypes contain HapC-N and HapE2, whereas haplotypes HapB and HapC-Rs were prevalent in red-colored genotypes.

5.3 BUNCH ARCHITECTURE-RELATED TRAITS

5.3.1 Sex Determination

All the wild *Vitis* species are dioecious having either male or female flowers, whereas cultivated varieties belonging to *Vitis vinifera* are hermaphrodite and self-pollinating. The change in mating system took place during domestication (This et al. 2006). The identification of sex-specific alleles allows selection of desirable hermaphrodite haplotype seedlings, thus improving the breeding efficacy and saving the labor and space required for progeny maintenance.

The hermaphrodite alleles derive from male alleles of wild grapevines, with successive recombination events (Picq et al. 2014). Transition from dioecy to hermaphroditism is considered to be the key step of grape domestication. The SEX-DETERMINING REGION (*SDR*) in *Vitis* is located on chromosome 2 (Riaz et al. 2006; Battilana et al. 2013). Based on the analysis of three Simple Sequence Repeat (SSR) markers around sex locus, in a population of 132 *V. sylvestris* accessions and 171 *V. vinifera* cultivars, one SSR haplotype accounted for 66% of the hermaphrodite individuals (Battilana et al. 2013). In same study, size variants of the VVIB23 microsatellite sequence were found to be associated with the sex alleles M, H, and f, and thus useful for determining the flower type of grape germplasm. Massonnet et al. (2020) performed complete genome sequence of eight grape genotypes with different flower types and identified 20 haplotypes for *Vitis SDR* (5 H, 12 F, and 3 M haplotypes), of which six haplotypes were linked to the flower sex. A mutation in *VviINP1* gene resulted in male sterility and M allele for *VviYABBY3* was found associated with female sterility. Also, the expression level of M and H alleles of *VviAPT3*, gene involved in cytokinin synthesis, influenced sex determination (Massonnet et al. 2020). *VviAPRT3* is specifically expressed in the carpel primordial of male plants and has role in the abortion of pistil structures (Coito et al. 2017). In *Vitis amurensis*, a molecular marker developed based on 22-bp InDel discriminated female alleles from male or hermaphrodite allele (Men et al. 2021). The allele variation for gene *VviFSEX* is found to be associated with dioecy. The female plants display a single allele for this gene, whereas two alleles are detected in male and hermaphrodite plants (Coito et al. 2017). *VviPLATZ1*, a transcription factor gene in *SDR*, is identified as candidate female flower morphology factor (Iocco-Corena et al. 2021). Loss of function of this gene controls female flower morphology in *Vitis*.

5.3.2 Flowering

Environment has profound effect on bud burst and flowering time. Thus, the information on the interaction of genotype or allele combination with environment will be useful for the selection of grape genotypes better suited for a particular climatic condition. QTL regions controlling flowering time have been identified on different chromosomes, and the role of several probable candidate genes for flowering has been examined. TERMINAL FLOWER 1 (*TFL1*) is involved in the transcription repression of floral meristem. In grape, this gene is found to delay flowering and leads to ramose inflorescence structure. Single dominant mutation in this gene resulted in extreme cluster proliferation and delayed anthesis somatic variant of Carignan (Fernandez et al. 2010). This somatic variant has reiterated reproductive meristem (*RRM*). The sequencing of *VvTFL1A* gene, including its promoter, in 140 grape genotypes revealed higher variation in promoter and intron sequences as compared to exons (Fernandez et al. 2014). Detailed analysis of haplotype network in this gene identified three haplogroups which were consistent with geographical origin and use and were associated with flowering time, cluster width, and berry size.

Kamal et al. (2019) identified allele combinations of candidate genes correlated with early or late flowering phenotypes. Early flowering phenotype was found associated with specific allele combinations of flowering time candidate genes in QTL regions on five different chromosomes. Alleles of *VvHUA2a* and *VvGAI1* correlated with late flowering were identified. In Arabidopsis, *HUA2* is known to regulate the expression of the floral gene AGAMOUS (*AG*) and *FLC*. Mutants of *VvGAI1* exhibit insensitivity to GA, and expression of *VvFL*, a floral development gene, exhibits correlation with inflorescence in these mutants (Boss and Thomas 2002). Similarly, early flowering-related alleles for genes *VvTM6* and *VvWNK6* were identified. *VvTM6* is a MADS-box B-class floral identity gene affecting the petal development, whereas *VvWNK6* is involved in circadian rhythm.

5.3.3 Bunch Compactness

Development and growth of inflorescences, pollination efficiency, fruit set, and berry growth determine the size and shape of the grape bunch. The bunch architecture influences the commercial value of table grape. Also, disease management is easier in loose bunches as compared to compact bunches. Several indices have been proposed to estimate bunch compactness (Tello and Ibáñez 2014). In recent years, much emphasis has been given to understand the genetic basis of bunch architecture-related traits, and several candidate genes and determinant allele have been identified. A comprehensive approach identified a set of SNPs associated with bunch compactness and associated traits (Tello et al. 2016). Grimplet et al. (2019) analyzed clones of Tempranillo at phenotypic, hormonal, and genomic levels and detected allelic variation in several genes with probable effect on bunch architecture. A transcription factor *VvGRF4* has been identified as major gene affecting inflorescence architecture. Mutations in the binding site of regulatory miRNA resulted in prevention of downregulation of *VvGRF4* in loose-clustered clones of Pinot Noir (Rossmann et al. 2020). Two identified mutated alleles functioned in heterozygous state and resulted in increased cell numbers in pedicle of loose clusters. This role of *VvGRF4* in increasing pedicle length was confirmed by Richter et al. (2020) in a larger set of loose and compact clones of Pinot Noir. Besides *VvGRF4*, 14 other genes showed differential expression in loose and compact clones (Richter et al. 2020).

Allele variation in *VviUCC1*, a gene encoding for uclacyanin type 1 cell wall protein, showed strong association with bunch compact-related trait (Tello et al. 2016). Sequence analysis of *VviUCC1* gene in 114 genotypes revealed close association of a series of SNPs with cluster compactness and rachis trait variation. Two of these SNPs were detected in *cis*-transcriptional regulatory elements indicating phenotype effect due to regulation of expression (Tello et al. 2020).

5.3.4 Berry Size

Large amount of morphological variation is observed in nature for berry size due to domestication and selection pressure for various purpose. *VvNAC26* gene is reported to be involved in early flower and berry development in the grapevine (Tello et al. 2015). The SNP and INDEL polymorphisms were detected in *VvNAC26* gene promoter and coding sequences, and 26 haplotypes were observed. Twenty-six haplotypes were clustered in two major haplogroups correlated with berry size. A set of polymorphisms explained 8%–21.7% trait. A haplotype involving only three polymorphisms explained up to 26.6% variation in berry size indicating involvement of *VvNAC26* in determining the berry size. This gene influences hormonal pathways and interacts with *VvMADS9*, a MADS-box gene to regulate grape fruit and seed development (Zhang et al. 2021). A set of SNPs and InDel markers associated with berry size, was identified by (Muñoz-Espinoza et al. 2020), which were validated in a segregating population as well as grape genotypes of a different genetic background. Fernandez et al. (2014) identified an InDel polymorphism in *VvTFL1A* gene, which explained 25% variation in berry weight in a population of 140 grape genotypes.

5.3.5 Berry Firmness

Berry firmness is a prime factor for quality and marketing of produce. Based on QTL mapping, Correa et al. (2016) identified ten genes involved in cation/calcium exchange, a potential cellulose synthase, a xylosyltransferase (*XTH*) and putative invertase as candidate genes for berry firmness. Differential expression of *XTH* and other cell wall modification genes like *PE*, *PL*, *PG*, *β-GAL*, *GATL*, *WAK*, and *EXP* was observed in two grape genotypes, Red Globe and Muscat Hamburg differing in berry firmness (Ma et al. 2020; Wang et al. 2020). Besides cell wall synthesis genes, abscisic-aldehyde oxidase-like and endoglucanase 3 genes (Jiang et al. 2020) and 23 transcription factors (Ma et al. 2020) were found associated with berry firmness. These genes showed higher level of expression in grape genotype with firm berries.

5.3.6 Seed Development

In table grapes, seedlessness is one of the important quality traits. Bold-sized fruits with no or little seeds improve marketability and suitability for food processing industry (Varoquaux et al. 2000). Thus, the understanding of genetic basis of seedlessness is important for successful development of table grape varieties with seedless and bold berry. *SEED DEVELOPMENT INHIBITOR* (*SDI*) locus is known for stenospermocarpy seed abortion in *Vitis vinifera* cultivars. Royo et al. (2018) identified *sdi* mutation (323 kb region) on chromosome 18 through mapping of two Crimson Seedless-derived populations. The *sdi*-linked seed abortion was identified through comparative RNA sequencing of seed traces of seedless and seeded. RNA sequence analysis enabled to identify the detrimental nucleotide changes in two genes at *sdi* locus. The resequencing of two genes exhibited a point mutation resulting in substitution of arginine to leucine in *VviAGL11* gene responsible for stenospermocarpy. B3 domain transcription factor genes have important and diverse functions in vegetative and reproductive growth. Their role is well studied in other crops, but not extensively deliberated in grapevines. Fifty B3 domain transcription factor genes are identified in grapes (Ahmad et al. 2019). Many B3 genes were expressed differentially in early developing seeds from seeded and seedless cultivars, indicating their role in seed development. Homeobox transcription factor has multiple functional roles during growth and development process of plant. Expression profiling of this class of transcription factor during ovule development in seeded and seedless cultivars suggested a potential role in ovule abortion associated with seedlessness (Li et al. 2017). *VvHB58* gene (encoding a homeodomain-leucine zipper (*HD-Zip I*) transcription factor) was found to control fruit size and seed number (Ren et al. 2019). The expression of this gene in transgenic tomato resulted in modifications of hormone signaling pathways. Results indicated that the DNA methylation levels during the seed development stages may lead to seed abortion. Similarly, mutation in transcription factor gene *VviINO* resulted in outer integument formation of ovule and seed formation (di Rienzo et al. 2021). The stenospermocarpic seedlessness in grapes is controlled by single dominant and multiple recessive genes. Genome resequencing and population genomics resulted in the identification of three candidate genes (*Vitvi01g00455*, *Vitvi08g01528*, and *Vitvi18g01237*) exhibiting association with recessive regulatory loci for seedlessness in grape. SNP haplotype around these genes clustered seedless and seeded genotypes in different groups (Kim et al. 2020).

5.3.7 Berry Ripening

Berry development in grape involves through three distinct phases and displays a double sigmoid growth curve (Coombe and McCarthy 2000). During berry development, two periods of growth are separated by a lag phase of slow expansion. The first phase is characterized by the accumulation of organic acids in the vacuoles and synthesis of precursors of phenolic compounds. This phase is followed by a lag phase, and the stage after the lag phase is known as veraison which is characterized by accumulation of sugars, berry softening, and color development in colored varieties. After veraison, glucose and fructose accumulate, whereas levels of organic acids decrease. Thus, berry ripening is a complex phenomenon and is regulated by several hormones and signaling pathways (Fortes et al. 2015). Transcriptome analysis of berry ripening revealed modulation of several cellular and metabolic pathways during different stages of ripening (Balic et al. 2018; Ma and Yang 2019; Li et al. 2022). Attempts have been made to identify the key regulators of these processes. Though the grape is non-climacteric fruit, ethylene is shown to have a role in berry ripening. Gene 1-aminocyclopropane-1-carboxylic acid oxidase (*ACO*) is the important gene of ethylene biosynthesis pathway. Muñoz-Robredo et al. (2013) demonstrated enhanced accumulation of transcripts of three *VvACO* during ripening process. The external application of ethylene leads to few folds accumulation of *VvACOs1* transcripts at veraison. D'Incà et al. (2021) identified *VviCARPO* (Controlled Adjustment of Ripening and maturation of Plant Organs), a NAC transcription factor as a key regulator of grapevine maturation. Induction of this TF preceded the expression of ripening-associated genes in grape. *VviCARPO* upregulated genes *VviSGR1* and *VviMYBA1* resulting in induced chlorophyll degradation and anthocyanin accumulation.

5.3.8 Aroma

The aroma, *i.e.*, the olfactory component of flavor, has significant relevance to wine industry. The aroma is comprised of various groups of compounds which are synthesized by different metabolic pathways. The major compound categories contributing to aroma mono- and sesquiterpenes, methoxypyrazines, furan derivatives, lipoxygenase pathway products, and phenylpropanoid pathway products (Lin et al. 2019b). α-Terpineol is an important monoterpene found in aromatic varieties. Gene α-terpineol synthase (*VviTer*) is involved in its synthesis in grape berries (Martin and

Bohlmann 2004). Yang et al. (2017) performed gene association analysis in 61 grape genotypes and observed correlation of two SNPs in coding region of the gene *VviTer* with higher levels of α-terpineol.

Sesquiterpene synthases (largest subfamily of grapevine terpene synthase (*VviTPS*) genes) are well-known volatile metabolites for flavor and aroma of wines. In grape, a large cluster of sesquiterpene synthases is present on chromosome 18. Comparative sequence analysis of multiple genotypes and grapevine reference genome identified two gene, viz. *VviTPS04* and *VviTPS10* as paralogues and/or allelic variants (Smit et al. 2021). Detailed analysis identified genotype-specific structural variations and polymorphisms which affected the gene function. One of the compounds 'Rotundone' is known to be responsible for peppery aroma in Shiraz and have been reported in many spices. The characterization of sesquiterpene synthase, *VvGuaS*, from the developing berries elucidated that it produces α-guaiene, a precursor of rotundone (Drew et al. 2016). A novel allele '*VvGuaS*' of *VvTPS24* gene was identified, and the resulted enzyme showed 99.5% identity with previously reported *VvPNSeInt* (produces various selinene-type sesquiterpenes). Difference of 6 amino acid residue (out of 561) was identified in the enzymes, and amino acids T414 and V530 were located within the active site of *VvGuaS* of substrate 'farnesyl pyrophosphate.' The mutation of these amino acids resulted into conversion of one enzyme to other. Similarly, PHOSPHOLIPID HYDROPEROXIDE GLUTATHIONE PEROXIDASE is identified as a candidate gene for grassy smelling volatile (*E*)-2-hexenal based on its presence in close vicinity of peak marker in QTL region on chromosome 2 (Alahakoon et al. 2022).

Among the grape and wine aromas, muscat have been admired largely. The key compound contributing to muscat flavor is monoterpene linalool. 1-Deoxy-D-xylulose 5-phosphate synthase (*VviDXS*) gene was identified as the gene responsible for muscat flavor and is co-localized with QTL for monoterpene content on chromosome 5 (Battilana et al. 2009; Duchêne et al. 2017). The superior allele of this gene in cultivar Gewurztraminer was required for high level of terpenol (Duchêne et al. 2017). A candidate gene association analysis in 148 grape varieties identified K284N mutation which was present in heterozygous state in majority of the muscat cultivars (Emanuelli et al. 2010). A SNP in the coding region of *VvDXS* was reported to result in "gain of function" mutation which is involved in muscat flavor (Morcia et al. 2021).

The green bell pepper aroma in some wine varieties is due to the presence of 3-isobutyl-2-methoxypyrazine (*IBMP*). Enzymes *O*-methyltransferases (*OMTs*) catalyze the final *O*-methylation of the nonvolatile 3-alkyl-2-hydroxypyrazine precursors (Dunlevy et al. 2013). Among the four genes, a superior allele of gene *VviOMT3* with efficient catalytic ability was detected in Cabernet Sauvignon. Association study in 91 *V. vinifera* cultivars found that at least one copy of Cabernet Sauvignon *VviOMT3* allele is required for high levels of methoxypyrazine (Dunlevy et al. 2013).

5.4 ABIOTIC STRESSES

Grapes are cultivated worldwide under a wide range of environmental conditions from temperate to tropical regions. Growing under such diverse conditions, vines experience various stresses and develop adaption mechanism for survival. Cramer (2010) reviewed the whole plant and molecular responses toward the abiotic stresses. The frequent responses for water deficit, salinity, and chilling were found to be associated with hormone metabolism (especially ABA), photosynthesis, protein synthesis, signaling, and cellular protection. Genes related to various stresses were mapped and had shown that the ABA is a primary regulator for abiotic stress tolerance. ABA is known to be involved in regulation of many processes and stress response. ABA affects transporters for water and ions in roots (Rattanakon et al. 2016). ABA was also found to affect genes for ABA degradation, conjugation, and signaling as well as many genes encoding transcription factors (Rattanakon et al. 2016). Four ABA receptor PYR/PYL/RCAR (*PYL*) genes (*VvPYL1*, *VvPYL4*, *VvPYL7*, and *VvPYL8*) were found to respond toward stress like cold, salt, or polyethylene glycol under the regulation of *cis*-regulatory elements (Zhao et al. 2020). Haddad and Japelaghi (2015) reported response of grape peroxiredoxin PrxII C (*VvPrxII C*) gene to different abiotic stresses. Peroxidase acts as a reductant and decreases hydrogen peroxide and its alkyl form. The analysis of gene sequence indicated the presence of several *cis*-acting elements in the promotor region. The expression analysis of Vv*PrxIIC* gene under various stresses (abiotic, oxidative, and anti-oxidants application) indicated the role of putative regulatory elements of the promoter.

The secondary metabolite 'resveratrol' has a significant contribution for stress tolerance. Analysis of 95 grapevine accessions including cultivated and wild grape accessions from China, with SSR markers identified eight markers encompassing 19 alleles, significantly associated with resveratrol content (Zheng et al. 2017). Three positive (for allele 232) and one negative (for allele 236) linkages were reported at Sh5 locus on chromosome 16. In grape variety Campbell Early, differential expression was observed for flavonoid and stilbene synthesis genes at different elevated temperatures (Kim et al. 2018). Gradual induction of *resveratrol synthase* (*STS1*) gene was recorded at all temperature and reduced at 48 h, and higher level of expression was observed at 35°C as compared to 25°C. Similar pattern of expression was observed for stilbene synthase genes (*STS11*, *STS12*, and *STS13*); however, expression decreased with temperature increase.

5.5 DISEASE RESISTANCE

Diverse disease-resistance genes are identified in grapes using allele-mining approaches, which provide resistance against bacterial, viral, and fungal pathogens. The exploitation of allele mining in exploring disease-resistance genes in grapes against different pathogens is described below.

5.5.1 Powdery Mildew Resistance

Fungal pathogens majorly affect crop productivity in grapes. One of the major fungal diseases is powdery mildew caused by *Erysiphe necator* which is affecting grapevines worldwide. Various resistance loci have been mined from cultivated as well as wild relatives against this pathogen. For instance, *Run1* was the first locus identified from the wild variety *Muscadinia rotundifolia* and introgressed to *V. vinifera*, and further different markers that cosegregate with *Run1* were also identified (Pauquet et al. 2001; Donald et al. 2002). Similarly, the *Ren1* locus was also present as a single dominant allele in 'Kishmish Vatkana' with its closest markers (VMC9H4-2, VMCNG4E10-1, and UDV-020) (Hoffmann et al. 2008). More exploration of disease-resistant genes in different varieties revealed the presence of *Ren1* also in 'Dzhandzhal kara'. Further *Ren1* gene exhibited co-segregation with *NBS-LRR* (nucleotide-binding site-leucine rich repeat) cluster in both 'Kishmish Vatkana' and 'Dzhandzhal kara' (Coleman et al. 2009). Further exploration of PM resistance in *V. vinifera* subsp. *sylvestris* DVIT3351.27 and O34-16 accessions revealed the presence of an allele of the same marker co-segregating with the *Ren1* locus (Riaz et al. 2013). Also, more diversity was observed in the allelic haplotypes of the SSR markers linked with this locus in *V. vinifera* subsp. *sylvestris* (DVIT3351.27). Further, SNP found in the SSR marker co-segregating with this locus was common to both wild and cultivated relatives which provides evidence for the presence of powdery mildew fungus in Asia for longer times (Riaz et al. 2020). Many other R loci (*Run2.1*, *Run2.2*, *Ren2*, *Ren3*, *Ren4*, *Ren5*, *Ren6*, and *Ren7*) were also identified and mapped for improvement of disease resistance in grapes against PM (Dalbó et al. 2001; Welter et al. 2007; Ramming et al. 2011; Riaz et al. 2011; Blanc et al. 2012; Pap et al. 2016). Recently, different structural variations are identified in haplotypes of *Run1.2* and *Run2.2* using haplotype phasing in *Muscadinia rotundifolia* and further constitutively expressing *NBS-LRR* genes associated with these loci were explored that can be used as novel resources for the production of PM resistant variety using genetic engineering or breeding techniques (Massonnet et al. 2022). Of all, *Ren4* provides rapid non-race specific resistance and seedlessness to the plant. Markers assisted with this locus are also known to play important role in the tracking of this gene (Mahanil et al. 2012). However, some of these loci activate rapid defense responses, while others provide resistance at later stages to the plant (Massonnet et al. 2022).

5.5.2 Downy Mildew Resistance

Another prominent fungal disease of the grape plant is downy mildew (DM) caused by *Plasmopara viticola*. Many hazardous fungicides are extensively applied to avoid yield losses. These fungicides are not only labor intensive or costly but also have destructive effects on the environment. Hence, it's very necessary to mine disease-resistance locus for the development of superior and resistant varieties of grapes. Several genomic loci are identified that enhance fungal resistance in crop plants; for instance, *Rpv1* (Resistance to *Plasmopara viticola* 1) and *Rpv2* were identified in *Muscadinia rotundifolia* Michaux (Merdinoglu et al. 2003). Similarly, the *Rpv3* locus was identified in *V. rupestris* and is majorly associated with hypersensitive responses as it enhances necrosis of infected tissue (Welter et al. 2007; Bellin et al. 2009; Eisenmann et al. 2019; Possamai et al. 2020), and several haplotypes of *Rpv3* were identified in different species of grapes (*V. rupestris*, *V. lincecumii*, *V. riparia*, and *V. labrusca*) (Di Gaspero et al. 2012). Further, *Rpv3-3* was identified as a major haplotype associated with disease resistance (Vezzulli et al. 2019). However, additional mining of variation in the *Rpv3-1* haplotype in wild and cultivated relatives confirmed gene duplication and transposable element movement as major sources of variation in this haplotype (Foria et al. 2020). Additionally, genomic characterization and identification of two major segregating SNP markers further confirmed its crucial role in providing DM resistance to the plant (Wairich et al. 2021). Many other loci against were identified and characterized, for instance, *Rpv4*, *Rpv5*, *Rpv6*, *Rpv7*, *Rpv8*, *Rpv9*, and *Rpv10* up to *Rpv28* in different species (Welter et al. 2007; Bellin et al. 2009; Marguerit et al. 2009; Blasi et al. 2011; Moreira et al. 2011; Schwander et al. 2012; Venuti et al. 2013; Ochssner et al. 2016; Lin et al. 2019a; Sapkota et al. 2019; Bhattarai et al. 2021). In recent years, three loci (*Rpv29*, *Rpv30*, and *Rpv31*) are identified against *Plasmopara viticola* in grapes through a GWAS approach (Sargolzaei et al. 2020). These identified loci can be used for the generation of homozygous resistant lines like; homozygous lines of grapes were developed in *Rpv1* and *Rpv3* loci that exhibited resistance against DM infection (Sánchez-Mora et al. 2017). Some of these loci can be used as candidates for marker-assisted breeding, and the plants containing pyramided *Rpv1* and *Rpv3.1* loci were more resistant than the plant containing only one locus (Saifert et al. 2018). These identified loci can be used as a potential source for the production of DM-resistant varieties.

5.5.3 *Phomopsis* Cane and Leaf Spot Resistance

Many other fungal diseases can affect plant survival; for instance, *Phomopsis* cane and leaf spot disease is caused by the *Diaporthe ampelina* (syn=*Phomopsis viticola*), a hemibiotrophic fungus. Its infection majorly causes lesion development on fruit clusters. Two resistance loci (*Rda1* and *Rda2*) are identified against this pathogen in *V. cinerea* B9 and 'Horizon.' Further mapping confirmed the presence of *Rda1* on chromosome 15 of *V. cinerea* B9 and Rda2 on chromosome 7 of 'Horizon'. Additionally, QTL mapping of 'Chardonnay'×*V. cinerea* B9 progeny revealed the association of 2 *NB-LRR* (*VIT_15s0046g02730* and *VIT_15s0046g02800*) with *Rda1* loci (Barba et al. 2018).

5.5.4 Black Rot Resistance

Guignardia Bidwell (hemibiotrophic fungus) is the causal agent of black rot disease in grapes. Most of the European grapevine cultivars and PM as well as DM-resistant varieties are susceptible to the black rot disease; hence, it is necessary to identify the resistance locus against this disease to produce block rot-resistant varieties of grapes. First, QTL mapping of grapevine rootstock 'Börner' (*V. riparia* Gm183×*V. cinerea* Arnold) revealed the presence of two QTL loci (*Rgb1* and *Rgb2*), which provide resistance against this pathogen. Further detailed mapping of *Rgb1* predicted the linked markers of this locus (Rex et al. 2014). Interestingly, both loci *Rgb1* and *Rgb2* were present on linkage groups 14 and 16, and also exhibited overlapping with QTLs of PM resistance on these linkage groups (Dalbó et al. 2000). Also, mapping of GF.GA-47-42×'Villard blanc' population revealed the presence of resistance-conferring

QTL on linkage group 14 similar to *Rgb1*. Further two *Rgb1-linked* markers (Gf14-41 and Gf14-42) were also identified (Hausmann et al. 2017). These markers can be used for marker-assisted breeding approaches. However, in a recent study, these two loci in 'Börner' were confirmed to be originated from *Vitis riparia* (Weber et al. 2022). Additionally, a high-resistance locus on chromosome 14 was confirmed with its specifically associated SNP marker in segregating offsprings of 'Merzling' (*V. vinifera* and *V. rupestris*×*V. aestivalis* var. *lincecumii*) and 'Teroldego' (*V. vinifera*) (Bettinelli et al. 2022). These identified loci can be used as major resources for the introduction of black rot resistance in PM or DM-resistant varieties for the production of superior varieties of grapes.

5.5.5 Pierce's Disease Resistance

Allele mining provides resistance against bacterial pathogens that severely affect many crop species. One of the major bacteria is *Xylella fastidiosa* (XF), which affects the yield of many important fruit crops and causes Pierce's disease in grapevines. Most of the *V. vinifera* varieties show susceptibility toward this pathogen; hence, it is necessary to identify resistant resources through allele/gene mining. Krivanek et al. (2006) identified and mapped QTL locus against XF as *PDR1* (Pierce's disease resistance 1) on chromosome 14. Further, allele mining of this locus was done, and *PdR1a* and *PdR1b* alleles were identified from line b43-17 of *V. arizonica/candicans* and *PDR1c* allele from the line of b40-14 *V. arizonica* (Riaz et al. 2008). Walker et al. (2014) succeeded in the generation of 97% *V. vinifera* plant with *PDR1b* allele taken from *V. arizonica* line b43-17 using backcrossing. Further, physical mapping of the *PDR1b* allele led to the identification of five different ORFs (Open Reading Frames). Transfer formation of susceptible varieties with two of these ORFs exhibited a limited role in providing resistance against this pathogen (Agüero et al. 2022). Another region providing strong resistance against XF was identified as *PDR2* on chromosome 8 from line b42-26 of *V. arizonica*. However, many ongoing studies are focusing on marker-assisted breeding of these identified loci and the identification of more loci. Novel accessions (T03–16, ANU67, and b41-13) for resistance against XF were also explored. Moreover, these accessions exhibit resistance patterns that do not resemble the resistance provided by *PDR1-linked* markers, which indicated that this resistance comes from other locations in the genome (Riaz et al. 2018). In addition to this, recent reports also identified and mapped resistant loci from different accessions that can be further used in marker-assisted breeding approaches (Huerta-Acosta et al. 2023).

Overall, many resistance loci are identified, mapped, and confirmed in various species against various pathogens. However, many more resistance-providing alleles of genes are yet to be mined for various diseases; for instance, markers linked with resistance against anthracnose caused by *Elsinoe ampelina* in grapes are identified, but specific loci are yet to be identified (Kim et al. 2008; Shetty et al. 2015). Grapevine trunk diseases also cause major destructive effects on viticulture; hence, resistant gene mining in cultivated species of different regions or wild relatives and further, their introgression into the varieties that are already resistant to other pathogens is a very important step for proper sustainability of wine production (Mondello et al. 2018).

REFERENCES

Agüero CB, Riaz S, Tenscher AC, Bistué C, Walker MA. Molecular and functional characterization of two RGA type genes in the *PdR*1b locus for Pierce's disease resistance in *Vitis arizonica/candicans*. *Plant Cell Tissue Organ Cult.* 2022;151(3):497–510. doi:10.1007/s11240-022-02366-6.

Ahmad B, Zhang S, Yao J, Rahman MU, Hanif M, et al. Genomic organization of the B3-domain transcription factor family in grapevine (*Vitis vinifera* L.) and expression during seed development in seedless and seeded cultivars. *Int J Mol Sci.* 2019;20(18):4553. doi:10.3390/ijms20184553.

Alahakoon D, Fennell A, Helget Z, Bates T, Karn A, et al. Berry anthocyanin, acid, and volatile trait analyses in a grapevine-interspecific F2 population using an integrated GBS and rhAmpSeq genetic map. *Plants.* 2022;11(5):696. doi:10.3390/plants11050696.

Azuma A, Kobayashi S, Yakushiji H, Yamada M, Mitani N, et al. *VvMybA*1 genotype determines grape skin color. *Vitis*. 2007;46:154–5. doi:10.5073/vitis.2007.46.154-155.

Azuma A, Udo Y, Sato A, Mitani N, Kono A, et al. Haplotype composition at the color locus is a major genetic determinant of skin color variation in Vitis×labruscana grapes. *Theor Appl Genet*. 2011;122(7):1427–38. doi:10.1007/s00122-011-1542-7.

Balic I, Vizoso P, Nilo-Poyanco R, Sanhueza D, Olmedo P, et al. Transcriptome analysis during ripening of table grape berry cv. Thompson Seedless. *PLoS One*. 2018;13(1):e0190087. doi:10.1371/journal.pone.0190087.

Barba P, Lillis J, Luce RS, Travadon R, Osier M, et al. Two dominant loci determine resistance to *Phomopsis* cane lesions in F1 families of hybrid grapevines. *Theor Appl Genet*. 2018;131(5):1173–89. doi:10.1007/s00122-018-3070-1.

Battilana J, Costantini L, Emanuelli F, Sevini F, Segala C, et al. The 1-deoxy-D-xylulose 5-phosphate synthase gene co-localizes with a major QTL affecting monoterpene content in grapevine. *Theor Appl Genet*. 2009;118(4):653–69. doi:10.1007/s00122-008-0927-8.

Battilana J, Lorenzi S, Moreira FM, Moreno-Sanz P, Failla O, et al. Linkage mapping and molecular diversity at the flower sex locus in wild and cultivated grapevine reveal a prominent SSR haplotype in hermaphrodite plants. *Mol Biotechnol*. 2013;54(3):1031–7. doi:10.1007/s12033-013-9657-5.

Bellin D, Peressotti E, Merdinoglu D, Wiedemann-Merdinoglu S, Adam-Blondon AF, et al. Resistance to *Plasmopara viticola* in grapevine 'Bianca' is controlled by a major dominant gene causing localised necrosis at the infection site. *Theor Appl Genet*. 2009;120(1):163–76. doi:10.1007/s00122-009-1167-2.

Bettinelli P, Nicolini D, Giovannini O, Costantini L, Stefanini M, et al. Towards marker-assisted breeding for black rot resistance: High-density linkage mapping and identification of a major QTL in the cultivar 'Merzling' (*V. rupestris*×*V. aestivalis* var. lincecumii). In: Töpfer R, editor. *XIII International Symposium on Grapevine Breeding and Genetics, Pioneering Wines (PIWIs) - Innovation and Tradition*. Germany: Institute for Grapevine Breeding Geilweilerhof I Siebeldingen, Julius-Kühn-Institut, Bundesforschungsinstitut für Kulturpflanzen; 2022. p. 45.

Bhattarai G, Fennell A, Londo JP, Coleman C, Kovacs LG. A novel grape downy mildew resistance locus from *Vitis rupestris*. *Am J Enol Vitic*. 2021;72(1):12–20. doi:10.5344/ajev.2020.20030.

Blanc S, Wiedemann-Merdinoglu S, Dumas V, Mestre P, Merdinoglu D. A reference genetic map of *Muscadinia rotundifolia* and identification of *Ren*5, a new major locus for resistance to grapevine powdery mildew. *Theor Appl Genet*. 2012;125(8):1663–75. doi:10.1007/s00122-012-1942-3.

Blasi P, Blanc S, Wiedemann-Merdinoglu S, Prado E, Rühl E, et al. Construction of a reference linkage map of *Vitis amurensis* and genetic mapping of *Rpv*8, a locus conferring resistance to grapevine downy mildew. *Theor Appl Genet*. 2011;123(1):43–53. doi:10.1007/s00122-011-1565-0.

Boss PK, Thomas MR. Association of dwarfism and floral induction with a grape 'green revolution' mutation. *Nature*. 2002;416(6883):847–50. doi:10.1038/416847a.

Carrasco D, De Lorenzis G, Maghradze D, Revilla E, Bellido A, et al. Allelic variation in the *VvMybA*1 and *VvMybA*2 domestication genes in natural grapevine populations (*Vitis vinifera* subsp. *sylvestris*). *Plant Syst E*. 2015;301(6):1613–24. doi:10.1007/s00606-014-1181-y.

Coito JL, Ramos MJ, Cunha J, Silva HG, Amâncio S, et al. *VviAPRT3* and *VviFSEX*: Two genes involved in sex specification able to distinguish different flower types in *Vitis*. *Front Plant Sci*. 2017;8:98. doi:10.3389/fpls.2017.00098.

Coleman C, Copetti D, Cipriani G, Hoffmann S, Kozma P, et al. The powdery mildew resistance gene *REN*1 co-segregates with an *NBS-LRR* gene cluster in two Central Asian grapevines. *BMC Genetics*. 2009;10(1):89. doi:10.1186/1471-2156-10-89.

Coombe BG, McCarthy MG. Dynamics of grape berry growth and physiology of ripening. *Aust J Grape Wine Res*. 2000;6(2):131–5. doi:10.1111/j.1755-0238.2000.tb00171.x.

Correa J, Mamani M, Muñoz-Espinoza C, González-Agüero M, Defilippi BG, et al. New stable QTLs for berry firmness in table grapes. *Am J Enol Vitic*. 2016;67(2):212–7. doi:10.5344/ajev.2015.15049.

Cramer GR. Abiotic stress and plant responses from the whole vine to the genes. *Aust J Grape Wine Res*. 2010;16(s1):86–93. doi:10.1111/j.1755-0238.2009.00058.x.

D'Incà E, Foresti C, Orduña L, Amato A, Vandelle E, et al. Initiation of organ maturation and fruit ripening in grapevine is controlled by the CARPO-NAC transcription factor. *bioRxiv - Plant Biol*. 2021. doi:10.1101/2021.11.13.468481.

Dalbó MA, Weeden NF, Reisch BI. QTL analysis of disease resistance in interspecific hybrid grapes. *Acta Hortic*. 2000;528:217–22. doi:10.17660/ActaHortic.2000.528.29.

Dalbó MA, Ye GN, Weeden NF, Wilcox WF, Reisch BI. Marker-assisted selection for powdery mildew resistance in grapes. *J Amer Soc Hort Sci.* 2001;126(1):83–9. doi:10.21273/jashs.126.1.83.

Di Gaspero G, Copetti D, Coleman C, Castellarin SD, Eibach R, et al. Selective sweep at the *Rpv*3 locus during grapevine breeding for downy mildew resistance. *Theor Appl Genet.* 2012;124(2):277–86. doi:10.1007/s00122-011-1703-8.

di Rienzo V, Imanifard Z, Mascio I, Gasser CS, Skinner DJ, et al. Functional conservation of the grapevine candidate gene *INNER NO OUTER* for ovule development and seed formation. *Hort Res.* 2021;8(1):29. doi:10.1038/s41438-021-00467-5.

Donald TM, Pellerone F, Adam-Blondon AF, Bouquet A, Thomas MR, et al. Identification of resistance gene analogs linked to a powdery mildew resistance locus in grapevine. *Theor Appl Genet.* 2002;104(4):610–8. doi:10.1007/s00122-001-0768-1.

Drew DP, Andersen TB, Sweetman C, Møller BL, Ford C, et al. Two key polymorphisms in a newly discovered allele of the *Vitis vinifera TPS24* gene are responsible for the production of the rotundone precursor α-guaiene. *J Exp Bot.* 2016;67(3):799–808. doi:10.1093/jxb/erv491.

Duchêne E, Butterlin G, Claudel P, Dumas V, Jaegli N, et al. Genetic determinism of the 'Muscat' flavour in grapevine (*Vitis vinifera* L.) cultivars. *Acta Hortic.* 2017;1157:87–92. doi:10.17660/ActaHortic.2017.1157.14.

Dunlevy JD, Dennis EG, Soole KL, Perkins MV, Davies C, et al. A methyltransferase essential for the methoxypyrazine-derived flavour of wine. *Plant J.* 2013;75(4):606–17. doi:10.1111/tpj.12224.

Eisenmann B, Czemmel S, Ziegler T, Buchholz G, Kortekamp A, et al. *Rpv*3-1 mediated resistance to grapevine downy mildew is associated with specific host transcriptional responses and the accumulation of stilbenes. *BMC Plant Biol.* 2019;19(1):343. doi:10.1186/s12870-019-1935-3.

Emanuelli F, Battilana J, Costantini L, Le Cunff L, Boursiquot JM, et al. A candidate gene association study on muscat flavor in grapevine (*Vitis vinifera* L.). *BMC Plant Biol.* 2010;10:241. doi:10.1186/1471-2229-10-241.

Fernandez L, Le Cunff L, Tello J, Lacombe T, Boursiquot JM, et al. Haplotype diversity of *VvTFL1A* gene and association with cluster traits in grapevine (*V. vinifera*). *BMC Plant Biol.* 2014;14:209. doi:10.1186/s12870-014-0209-3.

Fernandez L, Torregrosa L, Segura V, Bouquet A, Martinez-Zapater JM. Transposon-induced gene activation as a mechanism generating cluster shape somatic variation in grapevine. *Plant J.* 2010;61(4):545–57. doi:10.1111/j.1365-313X.2009.04090.x.

Ferreira V, Matus JT, Pinto-Carnide O, Carrasco D, Arroyo-García R, et al. Genetic analysis of a white-to-red berry skin color reversion and its transcriptomic and metabolic consequences in grapevine (*Vitis vinifera* cv. 'Moscatel Galego'). *BMC Genom.* 2019;20(1):952. doi:10.1186/s12864-019-6237-5.

Foria S, Copetti D, Eisenmann B, Magris G, Vidotto M, et al. Gene duplication and transposition of mobile elements drive evolution of the *Rpv3* resistance locus in grapevine. *Plant J.* 2020;101(3):529–42. doi:10.1111/tpj.14551.

Fortes AM, Teixeira RT, Agudelo-Romero P. Complex interplay of hormonal signals during grape berry ripening. *Molecules (Basel, Switzerland).* 2015;20(5):9326–43. doi:10.3390/molecules20059326.

Fournier-Level A, Le Cunff L, Gómez C, Doligez A, Ageorges A, et al. Quantitative genetic bases of anthocyanin variation in grape (*Vitis vinifera* L. ssp. *sativa*) berry: A quantitative trait locus to quantitative trait nucleotide integrated study. *Genetics.* 2009;183:1127–39. doi:10.1534/genetics.109.103929.

Grimplet J, Ibáñez S, Baroja E, Tello J, Ibáñez J. Phenotypic, hormonal, and genomic variation among *Vitis vinifera* clones with different cluster compactness and reproductive performance. *Front Plant Sci.* 2019;9:1917. doi:10.3389/fpls.2018.01917.

Haddad R, Japelaghi RH. Isolation of grape peroxiredoxin gene responding to abiotic stresses. *Russ J Plant Physiol.* 2015;62(6):856–65. doi:10.1134/S1021443715050088.

Hausmann L, Rex F, Töpfer R. Evaluation and genetic analysis of grapevine black rot resistances. *Acta Hortic.* 2017;1188:285–90. doi:10.17660/ActaHortic.2017.1188.37.

Hoffmann S, Di Gaspero G, Kovács L, Howard S, Kiss E, et al. Resistance to *Erysiphe necator* in the grapevine 'Kishmish vatkana' is controlled by a single locus through restriction of hyphal growth. *Theor Appl Genet.* 2008;116(3):427–38. doi:10.1007/s00122-007-0680-4.

Huerta-Acosta KG, Riaz S, Tenscher A, Walker MA. Genetic characterization of Pierce's disease resistance in a *Vitis arizonica/monticola* wild grapevine. *Am J Enol Vitic.* 2023;74(1):0740003. doi:10.5344/ajev.2022.22021.

Iocco-Corena P, Chaïb J, Torregrosa L, Mackenzie D, Thomas MR, et al. *VviPLATZ*1 is a major factor that controls female flower morphology determination in grapevine. *Nat Commun.* 2021;12(1):6995. doi:10.1038/s41467-021-27259-8.

Jiang J, Fan X, Zhang Y, Tang X, Li X, et al. Construction of a high-density genetic map and mapping of firmness in grapes (*Vitis vinifera* L.) based on whole-genome resequencing. *Int J Mol Sci.* 2020;21(3):797. doi:10.3390/ijms21030797.

Jiu S, Guan L, Leng X, Zhang K, Haider MS, et al. The role of *VvMybA2r* and *VvMybA2w* alleles of the *MYBA2* locus in the regulation of anthocyanin biosynthesis for molecular breeding of grape (*Vitis* spp.) skin coloration. *Plant Biotechnol J.* 2021;19(6):1216–39. doi:10.1111/pbi.13543.

Kamal N, Ochßner I, Schwandner A, Viehöver P, Hausmann L, et al. Characterization of genes and alleles involved in the control of flowering time in grapevine. *PLoS One.* 2019;14(7):e0214703. doi:10.1371/journal.pone.0214703.

Kim GH, Yun HK, Choi CS, Park JH, Jung YJ, et al. Identification of AFLP and RAPD markers linked to anthracnose resistance in grapes and their conversion to SCAR markers. *Plant Breed.* 2008;127(4):418–23. doi:10.1111/j.1439-0523.2008.01488.x.

Kim MS, Hur YY, Kim JH, Jeong SC. Genome resequencing, improvement of variant calling, and population genomic analyses provide insights into the seedlessness in the genus *Vitis*. *G3.* 2020;10(9):3365–77. doi:10.1534/g3.120.401521.

Kim SA, Oh S-K, Ahn SY, Yun HK. Expression of flavonoid and stilbene synthesis genes in grape berries is affected by high temperature. *Hortic Sci Technol.* 2018;36(5):607–18. doi:10.12972/kjhst.20180061.

Kobayashi S, Goto-Yamamoto N, Hirochika H. Retrotransposon-induced mutations in grape skin color. *Science.* 2004;304(5673):982. doi:10.1126/science.1095011.

Kobayashi S, Ishimaru M, Hiraoka K, Honda C. *Myb*-related genes of the Kyoho grape (*Vitis labruscana*) regulate anthocyanin biosynthesis. *Planta.* 2002;215(6):924–33. doi:10.1007/s00425-002-0830-5.

Krivanek AF, Riaz S, Walker MA. Identification and molecular mapping of *PdR*1, a primary resistance gene to Pierce's disease in *Vitis*. *Theor Appl Genet.* 2006;112(6):1125–31. doi:10.1007/s00122-006-0214-5.

Li Y, Zhu Y, Yao J, Zhang S, Wang L, et al. Genome-wide identification and expression analyses of the homeobox transcription factor family during ovule development in seedless and seeded grapes. *Sci Rep.* 2017;7(1):12638. doi:10.1038/s41598-017-12988-y.

Li Y-M, Sun X-R, You J-L, Li X-J, Li B, et al. Transcriptional analysis of the early ripening of 'Summer Black' grape in response to abscisic acid treatment. *Sci Hortic.* 2022;299:111054. doi:10.1016/j.scienta.2022.111054.

Lin H, Leng H, Guo Y, Kondo S, Zhao Y, et al. QTLs and candidate genes for downy mildew resistance conferred by interspecific grape (*V. vinifera* L. × *V. amurensis* Rupr.) crossing. *Sci Hortic.* 2019a;244:200–7. doi:10.1016/j.scienta.2018.09.045.

Lin J, Massonnet M, Cantu D. The genetic basis of grape and wine aroma. *Hort Res.* 2019b;6(1):81. doi:10.1038/s41438-019-0163-1.

Ma L, Sun L, Guo Y, Lin H, Liu Z, et al. Transcriptome analysis of table grapes (*Vitis vinifera* L.) identified a gene network module associated with berry firmness. *PLoS One.* 2020;15(8):e0237526. doi:10.1371/journal.pone.0237526.

Ma Q, Yang J. Transcriptome profiling and identification of the functional genes involved in berry development and ripening in *Vitis vinifera*. *Gene.* 2019;680:84–96. doi:10.1016/j.gene.2018.09.033.

Mahanil S, Ramming D, Cadle-Davidson M, Owens C, Garris A, et al. Development of marker sets useful in the early selection of *Ren*4 powdery mildew resistance and seedlessness for table and raisin grape breeding. *Theor Appl Genet.* 2012;124(1):23–33. doi:10.1007/s00122-011-1684-7.

Marguerit E, Boury C, Manicki A, Donnart M, Butterlin G, et al. Genetic dissection of sex determinism, inflorescence morphology and downy mildew resistance in grapevine. *Theor Appl Genet.* 2009;118(7):1261–78. doi:10.1007/s00122-009-0979-4.

Martin DM, Bohlmann J. Identification of *Vitis vinifera* (-)-alpha-terpineol synthase by *in silico* screening of full-length cDNA ESTs and functional characterization of recombinant terpene synthase. *Phytochemistry.* 2004;65(9):1223–9. doi:10.1016/j.phytochem.2004.03.018.

Massonnet M, Cochetel N, Minio A, Vondras AM, Lin J, et al. The genetic basis of sex determination in grapes. *Nat Commun.* 2020;11(1):2902. doi:10.1038/s41467-020-16700-z.

Massonnet M, Vondras AM, Cochetel N, Riaz S, Pap D, et al. Haplotype-resolved powdery mildew resistance loci reveal the impact of heterozygous structural variation on NLR genes in *Muscadinia rotundifolia*. *G3.* 2022;12(8). doi:10.1093/g3journal/jkac148.

Men Y, Li JR, Shen HL, Yang YM, Fan ST, et al. *VaAPRT3* gene is associated with sex determination in *Vitis amurensis*. *Front Genet.* 2021;12:727260. doi:10.3389/fgene.2021.727260.

Merdinoglu D, Wiedeman-Merdinoglu S, Coste P, Dumas V, Haetty S, et al. Genetic analysis of downy mildew resistance derived from *Muscadinia rotundifolia*. *Acta Hortic.* 2003;603:451–6. doi:10.17660/ActaHortic.2003.603.57.

Mondello V, Songy A, Battiston E, Pinto C, Coppin C, et al. Grapevine trunk diseases: A review of fifteen years of trials for their control with chemicals and biocontrol agents. *Plant Disease*. 2018;102(7):1189–217. doi:10.1094/pdis-08-17-1181-fe.

Morcia C, Tumino G, Raimondi S, Schneider A, Terzi V. Muscat flavor in grapevine: A digital PCR assay to track allelic variation in *VvDXS* gene. *Genes*. 2021;12(5):747. doi:10.3390/genes12050747.

Moreira FM, Madini A, Marino R, Zulini L, Stefanini M, et al. Genetic linkage maps of two interspecific grape crosses (*Vitis* spp.) used to localize quantitative trait loci for downy mildew resistance. *Tree Genet Genom*. 2011;7(1):153–67. doi:10.1007/s11295-010-0322-x.

Muñoz-Espinoza C, Di Genova A, Sánchez A, Correa J, Espinoza A, et al. Identification of SNPs and InDels associated with berry size in table grapes integrating genetic and transcriptomic approaches. *BMC Plant Biol*. 2020;20(1):365. doi:10.1186/s12870-020-02564-4.

Muñoz-Robredo P, Gudenschwager O, Chervin C, Campos-Vargas R, González-Agüero M, et al. Study on differential expression of 1-aminocyclopropane-1-carboxylic acid oxidase genes in table grape cv. Thompson Seedless. *Postharvest Biol Technol*. 2013;76:163–9. doi:10.1016/j.postharvbio.2012.10.006.

Ochssner I, Hausmann L, Toepfer R. *Rpv*14, a new genetic source for *Plasmopara viticola* resistance conferred by *Vitis cinerea*. *Vitis*. 2016;55:79–81. doi:10.5073/vitis.2016.55.79-81.

Pap D, Riaz S, Dry IB, Jermakow A, Tenscher AC, et al. Identification of two novel powdery mildew resistance loci, *Ren*6 and *Ren*7, from the wild Chinese grape species Vitis piasezkii. *BMC Plant Biol*. 2016;16(1):170. doi:10.1186/s12870-016-0855-8.

Pauquet J, Bouquet A, This P, Adam-Blondon AF. Establishment of a local map of AFLP markers around the powdery mildew resistance gene *Run*1 in grapevine and assessment of their usefulness for marker assisted selection. *Theor Appl Genet*. 2001;103(8):1201–10. doi:10.1007/s001220100664.

Picq S, Santoni S, Lacombe T, Latreille M, Weber A, et al. A small XY chromosomal region explains sex determination in wild dioecious *V. vinifera* and the reversal to hermaphroditism in domesticated grapevines. *BMC Plant Biol*. 2014;14:229. doi:10.1186/s12870-014-0229-z.

Possamai T, Migliaro D, Gardiman M, Velasco R, De Nardi B. *Rpv* mediated defense responses in grapevine offspring resistant to *Plasmopara viticola*. *Plants*. 2020;9(6):781. doi:10.3390/plants9060781.

Ramming DW, Gabler F, Smilanick J, Cadle-Davidson M, Barba P, et al. A single dominant locus, *Ren*4, confers rapid non-race-specific resistance to grapevine powdery mildew. *Phytopathology*. 2011;101(4):502–8. doi:10.1094/phyto-09-10-0237.

Rattanakon S, Ghan R, Gambetta GA, Deluc LG, Schlauch KA, et al. Abscisic acid transcriptomic signaling varies with grapevine organ. *BMC Plant Biol*. 2016;16(1):72. doi:10.1186/s12870-016-0763-y.

Ren F, Ren C, Zhang Z, Duan W, Lecourieux D, et al. Efficiency optimization of CRISPR/Cas9-mediated targeted mutagenesis in grape. *Front Plant Sci*. 2019;10:612. doi:10.3389/fpls.2019.00612.

Rex F, Fechter I, Hausmann L, Töpfer R. QTL mapping of black rot (*Guignardia bidwellii*) resistance in the grapevine rootstock 'Börner' (*V. riparia* Gm183 × *V. cinerea* Arnold). *Theor Appl Genet*. 2014;127(7):1667–77. doi:10.1007/s00122-014-2329-4.

Riaz S, Boursiquot J-M, Dangl GS, Lacombe T, Laucou V, et al. Identification of mildew resistance in wild and cultivated Central Asian grape germplasm. *BMC Plant Biol*. 2013;13(1):149. doi:10.1186/1471-2229-13-149.

Riaz S, Huerta-Acosta K, Tenscher AC, Walker MA. Genetic characterization of *Vitis* germplasm collected from the southwestern US and Mexico to expedite Pierce's disease-resistance breeding. *Theor Appl Genet*. 2018;131(7):1589–602. doi:10.1007/s00122-018-3100-z.

Riaz S, Krivanek AF, Xu K, Walker MA. Refined mapping of the Pierce's disease resistance locus, *PdR*1, and sex on an extended genetic map of *Vitis rupestris* × *V. arizonica*. *Theor Appl Genet*. 2006;113(7):1317–29. doi:10.1007/s00122-006-0385-0.

Riaz S, Menéndez CM, Tenscher A, Pap D, Walker MA. Genetic mapping and survey of powdery mildew resistance in the wild Central Asian ancestor of cultivated grapevines in Central Asia. *Hort Res*. 2020;7(1):104. doi:10.1038/s41438-020-0335-z.

Riaz S, Tenscher AC, Ramming DW, Walker MA. Using a limited mapping strategy to identify major QTLs for resistance to grapevine powdery mildew (*Erysiphe necator*) and their use in marker-assisted breeding. *Theor Appl Genet*. 2011;122(6):1059–73. doi:10.1007/s00122-010-1511-6.

Riaz S, Tenscher AC, Rubin J, Graziani R, Pao SS, et al. Fine-scale genetic mapping of two Pierce's disease resistance loci and a major segregation distortion region on chromosome 14 of grape. *Theor Appl Genet*. 2008;117(5):671–81. doi:10.1007/s00122-008-0802-7.

Richter R, Rossmann S, Gabriel D, Töpfer R, Theres K, et al. Differential expression of transcription factor- and further growth-related genes correlates with contrasting cluster architecture in *Vitis vinifera* 'Pinot Noir' and *Vitis* spp. genotypes. *Theor Appl Genet*. 2020;133(12):3249–72. doi:10.1007/s00122-020-03667-0.

Röckel F, Moock C, Braun U, Schwander F, Cousins P, et al. Color intensity of the red-fleshed berry phenotype of *Vitis vinifera* teinturier grapes varies due to a 408 bp duplication in the promoter of *VvMybA1*. *Genes*. 2020;11(8):891. doi:10.3390/genes11080891.

Rossmann S, Richter R, Sun H, Schneeberger K, Töpfer R, et al. Mutations in the miR396 binding site of the growth-regulating factor gene *VvGRF4* modulate inflorescence architecture in grapevine. *Plant J*. 2020;101(5):1234–48. doi:10.1111/tpj.14588.

Royo C, Torres-Pérez R, Mauri N, Diestro N, Cabezas JA, et al. The major origin of seedless grapes is associated with a missense mutation in the MADS-Box gene *VviAGL11*. *Plant Physiol*. 2018;177(3):1234–53. doi:10.1104/pp.18.00259.

Saifert L, Sánchez-Mora FD, Assumpção WT, Zanghelini JA, Giacometti R, et al. Marker-assisted pyramiding of resistance loci to grape downy mildew. *Pesquisa Agropecuária Brasileira*. 2018;53:602–10. doi:10.1590/S0100-204X2018000500009.

Sánchez-Mora FD, Saifert L, Zanghelini J, Assumpção WT, Guginski-Piva CA, et al. Behavior of grape breeding lines with distinct resistance alleles to downy mildew (*Plasmopara viticola*). *Crop Breed Appl Biotechnol*. 2017;17:141–9. doi:10.1590/1984-70332017v17n2a21.

Sapkota S, Chen LL, Yang S, Hyma KE, Cadle-Davidson L, et al. Construction of a high-density linkage map and QTL detection of downy mildew resistance in *Vitis aestivalis*-derived 'Norton'. *Theor Appl Genet*. 2019;132(1):137–47. doi:10.1007/s00122-018-3203-6.

Sargolzaei M, Maddalena G, Bitsadze N, Maghradze D, Bianco PA, et al. *Rpv*29, *Rpv*30 and *Rpv*31: Three novel genomic loci associated with resistance to *Plasmopara viticola* in *Vitis vinifera*. *Front Plant Sci*. 2020;11:562432. doi:10.3389/fpls.2020.562432.

Schwander F, Eibach R, Fechter I, Hausmann L, Zyprian E, et al. *Rpv*10: A new locus from the Asian *Vitis* gene pool for pyramiding downy mildew resistance loci in grapevine. *Theor Appl Genet*. 2012;124(1):163–76. doi:10.1007/s00122-011-1695-4.

Shetty D, Sawant I, Narkar S, Ghule S. Screening of grape vine genotypes to identify sources of resistance to anthracnose disease and identifying biochemical marker associated with resistance. *Indian Phytopath*. 2015;68:424–31.

Smit SJ, Vivier MA, Young PR. Seeing the forest through the (phylogenetic) trees: Functional characterisation of grapevine terpene synthase (*VviTPS*) paralogues and orthologues. *Plants*. 2021;10(8):1520. doi:10.3390/plants10081520.

Tello J, Ibáñez J. Evaluation of indexes for the quantitative and objective estimation of grapevine bunch compactness. *Vitis*. 2014;53:9–16. doi:10.5073/vitis.2014.53.9-16.

Tello J, Torres-Pérez R, Flutre T, Grimplet J, Ibáñez J. *VviUCC1* nucleotide diversity, linkage disequilibrium and association with rachis architecture traits in grapevine. *Genes*. 2020;11(6):598. doi:10.3390/genes11060598.

Tello J, Torres-Pérez R, Grimplet J, Carbonell-Bejerano P, Martínez-Zapater JM, et al. Polymorphisms and minihaplotypes in the *VvNAC26* gene associate with berry size variation in grapevine. *BMC Plant Biol*. 2015;15(1):253. doi:10.1186/s12870-015-0622-2.

Tello J, Torres-Pérez R, Grimplet J, Ibáñez J. Association analysis of grapevine bunch traits using a comprehensive approach. *Theor Appl Genet*. 2016;129(2):227–42. doi:10.1007/s00122-015-2623-9.

This P, Lacombe T, Cadle-Davidson M, Owens CL. Wine grape (*Vitis vinifera* L.) color associates with allelic variation in the domestication gene *VvMybA1*. *Theor Appl Genet*. 2007;114(4):723–30. doi:10.1007/s00122-006-0472-2.

This P, Lacombe T, Thomas MR. Historical origins and genetic diversity of wine grapes. *Trends Genet TIG*. 2006;22(9):511–9. doi:10.1016/j.tig.2006.07.008.

Varoquaux F, Blanvillain R, Delseny M, Gallois P. Less is better: New approaches for seedless fruit production. *Trends Biotechnol*. 2000;18(6):233–42. doi:10.1016/s0167-7799(00)01448-7.

Venuti S, Copetti D, Foria S, Falginella L, Hoffmann S, et al. Historical introgression of the downy mildew resistance gene *Rpv*12 from the Asian species *Vitis amurensis* into grapevine varieties. *PLoS One*. 2013;8(4):e61228. doi:10.1371/journal.pone.0061228.

Vezzulli S, Malacarne G, Masuero D, Vecchione A, Dolzani C, et al. The *Rpv*3-3 haplotype and stilbenoid induction mediate downy mildew resistance in a grapevine interspecific population. *Front Plant Sci*. 2019;10. doi:10.3389/fpls.2019.00234.

Wairich A, Malabarba J, Buffon V, Porto DD, Togawa R, et al. Molecular characterization of the Rpv3 locus towards the development of KASP markers for downy mildew resistance in grapevine (Vitis spp.). *Euphytica*. 2021;218(1):5. doi:10.1007/s10681-021-02952-3.

Walker AR, Lee E, Bogs J, McDavid DA, Thomas MR, et al. White grapes arose through the mutation of two similar and adjacent regulatory genes. *Plant J*. 2007;49(5):772–85. doi:10.1111/j.1365-313X.2006.02997.x.

Walker MA, Riaz S, Tenscher A. Optimizing the breeding of Pierce's disease resistant winegrapes with marker-assisted selection. *Acta Hortic*. 2014;1046:139–43. doi:10.17660/ActaHortic.2014.1046.17.

Wang H, Yan A, Sun L, Zhang G, Wang X, et al. Novel stable QTLs identification for berry quality traits based on high-density genetic linkage map construction in table grape. *BMC Plant Biol*. 2020;20(1):411. doi:10.1186/s12870-020-02630-x.

Weber P, Rex F, Töpfer R, Hausmann L. Characterization of the black rot resistance loci (*Rgb*1 and *Rgb*2) of 'Börner' and development of associated markers suitable for marker-assisted selection in grapevine breeding. In: Töpfer R, editor. *XIII International Symposium on Grapevine Breeding and Genetics, Pioneering Wines (PIWIs) - Innovation and Tradition. Institute for Grapevine Breeding Geilweilerhof I Siebeldingen*. Germany: Julius-Kühn-Institut, Bundesforschungsinstitut für Kulturpflanzen; 2022. p. 164.

Welter LJ, Göktürk-Baydar N, Akkurt M, Maul E, Eibach R, et al. Genetic mapping and localization of quantitative trait loci affecting fungal disease resistance and leaf morphology in grapevine (*Vitis vinifera* L). *Mol Breed*. 2007;20(4):359–74. doi:10.1007/s11032-007-9097-7.

Yang X, Guo Y, Zhu J, Ma N, Sun T, et al. Associations between the α-terpineol synthase gene and α-terpineol content in different grapevine varieties. *Biotechnol Biotechnol Equip*. 2017;31(6):1100–5. doi:10.1080/13102818.2017.1364978.

Zhang S, Dong R, Wang Y, Li X, Ji M, et al. NAC domain gene *VvNAC26* interacts with *VvMADS9* and influences seed and fruit development. *Plant Physiol Biochem*. 2021;164:63–72. doi:10.1016/j.plaphy.2021.04.031.

Zhao Y, Qi G, Ren F, Wang Y, Wang P, et al. Analysis of *PYL* genes and their potential relevance to stress tolerance and berry ripening in grape. *J Amer Soc Hort Sci*. 2020;145(5):308–17. doi:10.21273/jashs04942-20.

Zheng X, Shi J, Yu Y, Shen Y, Tan B, et al. Exploration of elite stilbene synthase alleles for resveratrol concentration in wild Chinese *Vitis* spp. and *Vitis* cultivars. *Front Plant Sci*. 2017;8:487. doi:10.3389/fpls.2017.00487.

6 Allele Mining and Marker Identification for Germplasm Characterization and Breeding in Berries

Birutė Frercks, Rytis Rugienius, Ingrida Mažeikienė, Jurgita Vinskienė, and Gunars Lacis

6.1 INTRODUCTION

Strawberries (*Fragaria* spp.) are one of the most important and widely cultivated berry crop. Most strawberry production is supplied by China, followed by the USA, Mexico, Turkey, Egypt, and Spain. Although strawberries are now widely grown all over the world, they are a relatively young hybrid species, originating about 300 years ago. The diploid strawberry (*Fragaria vesca*), also known as wild strawberry, has a small genome ($2n = 2x = 14$, 240 Mbp), easily reproduces vegetatively and has relatively short, 12–16 weeks, development cycle from seed to seed. These advantages make the diploid strawberry a more attractive model than the commercial octoploid strawberry (*Fragaria* × *ananassa*) ($2n = 8x = 56$) and other plants of the Rosaceae family for functional genomics research. *Fragaria vesca* is one of four ancestors of cultivated *Fragaria* × *ananassa* and is the sub-genomic donor among *F. viridis*, *F. iinumae*, and *F. nipponica*. The fruits of cultivated *Fragaria* × *ananassa* exhibit a large size, intense color, and varied aromas and flavors compared with diploid varieties. Strawberry breeding programs are mainly focused on agronomic traits of fruits such as firmness, color, disease, and pest resistance, while little attention has been paid to the nutritional quality of the fruit. Strawberries (and other berries) have interesting and valuable nutritional properties and a very high antioxidant capacity, mainly due to their high content of (poly)phenolic compounds, and biosynthesis of metabolites with beneficial effects on human health have been reported. Strawberries provide a unique opportunity to study and exploit higher levels of ploidy and allopolyploid composition, possible ancestors, dominance of sub-genomes and the influence of ploidy on adaptability and berry quality.

The domestication story of *Vaccinium* spp. (blue- and cranberry) compared to other berries is very short, reaching about 100 years. Within the *Vaccinium* genus, cranberry (*Vaccinium macrocarpon* Ait.) and blueberry (*Vaccinium* L.) have been successfully domesticated and are considered the second most important soft fruit, after strawberry. High-bush blueberry and cranberry are distributed over the world. Ploidy levels across *Vaccinium* spp. vary: high-bush blueberry (*Vaccinium corymbosum* L.) is an autotetraploid ($2n = 4x = 48$), while cranberry (*V. macrocarpon*) is diploid ($2n = 2x = 24$); however, both are self-fertile crops. The fruits of blue- and cranberry contain a variety of vitamins and organic acids as well as abundant proanthocyanidins, anthocyanins, and flavonoids and have a cancer-preventing and anti-inflammatory effect. Despite the economic importance and high medicinal value of blueberries and cranberries, molecular research has been limited until recently. Just with the introduction of the technique of next-generation sequencing (NGS), the exponential increase of genetic data began. As an effective tool for analyzing the diversity of germplasm

 DOI: 10.1201/9781003386490-6

resources, molecular marker technology is currently the most suitable method for determining genetic relationships and investigating the origin and evolution of species. Recent achievements in blueberry and cranberry genomics have resulted to development of large number of informative simple sequence repeat (SSR) and single nucleotide polymorphism (SNP) markers, multiple high-density linkage maps with quantitative trait loci (QTLs) for different traits.

Currants and gooseberries (*Ribes* spp.) are important berry crops in the temperate zone of the Northern Hemisphere, mainly in Eurasia. They belong to the genus *Ribes* (family *Grossulariaceae*, order *Saxifragales*), which contains more than 150 species of spiny and non-spiny shrubs. Blackcurrant (*R. nigrum* L., *R. nigrum* var. *pauciflorum*, *R. nigrum* var. *sibiricum*), red currant (*R. rubrum* L., *R. spicatum* E. Robson, *R. petraeum* Wulfen), and gooseberry (*R. uva-crispa* L.) are mostly cultivated species groups. Other species of *Ribes* (*R. americanum* Mill., *R. dikuscha* Fisch. ex Turcz., *R. aureum* Pursh, and others) are also used in smaller quantities, as well as for decorative purposes and as donors of traits important for breeding. A common feature of all *Ribes* species is ploidy – they are all diploid with the basic chromosome number $x=8$ (Brennan 2008). Other characteristic features of the genus are cytological uniformity and lack of crossbreeding barriers. These characteristics have made it possible to create a number of interspecific hybrids, especially for blackcurrants, even more so – most modern blackcurrant cultivars are complex multi-stage interspecific hybrids. The largest and most successful blackcurrant breeding programs are in Scotland, Poland, Lithuania, Latvia, Norway, Russia, and New Zealand. Correspondingly, the largest and the most diverse germplasm collections are also available in these places. They are involved in variable uses and research. Unfortunately, the extensive involvement of the best hybrids in most breeding programs and the extensive exchange of plant material between breeding programs raise the risk of a decrease in overall genetic diversity. The leveling of the biological properties of advanced plant material can also create a lack of traits' donors necessary for the mitigation of arising challenges. Red currant and gooseberry breeding is less common and occurs fragmentarily and in smaller quantities. Considering the biological properties of these crops and the targeted breeding activities of humans, their further research and application without gene mining approaches are hard to imagine.

6.2 GENOME CONTENT

6.2.1 Genome Content of Strawberries

Strawberries are one of the main berry crops around the world. According to statistical data provided by the Food and Agriculture Organization (FAO), strawberries in 2020 occupied an area of 384,668 ha (FAOSTAT 2022). The harvest reached 8,861,381 tons. Strawberries are grown on all continents (except Antarctica) and in different climate zones. It is interesting to note that although strawberry areas have not changed much in the last 20 years, the production compared to 2001 increased almost twice. This shows that strawberry cultivation has become much more intensive. About 37.6% of the world's strawberry is produced in China, where strawberry production has tripled over the past 20 years. Other important strawberry growers are the USA, Mexico, Turkey, Egypt, and Spain. Strawberries are not only visually appealing, aromatic, and delicious, but also rich in beneficial nutrients such as vitamin C, phenolic compounds, and microelements. They are characterized by their anti-inflammatory and even chemoprotective properties (Fierascu et al. 2020).

Although strawberries are now very widespread all over the world, they are a relatively young hybrid species that appeared about 300 years ago. It is almost unique that the origin of this relatively new hybrid species, which arose from the accidental crossing of two octoploid species grown in European botanical gardens, can be traced quite precisely (Hancock 1999). Strawberry, on the other hand, is an allopolyploid (octoploid) species, and its genome history can be quite complex. With current methods of genome analysis and bioinformatics, it is possible to trace with some probability the circumstances of the origin of this species from wild species' ancestors, and even the chronology of its formation, and in which countries of the world its ancestors lived.

The sequencing and analysis of the genome of the related diploid species *F. vesca* are very helpful in examining the genome of *F.* × *ananassa. F. vesca* has been repeatedly proposed as a model for *Rosaceae* research, based on the facts that *F. vesca* is a diploid perennial species with a small, fully sequenced genome, it can be propagated by seed or clonally via stolons, and the cycle from seed to seed is relatively short, only 12–16 weeks. Furthermore, *F. vesca*, as the maternal ancestor of *F.* × *ananassa*, shares essential sequence identities with this economically important fruit crop (Hilmarsson et al. 2017; Rousseau-Gueutin et al. 2009). Unlike *Arabidopsis thaliana*, *F. vesca* is perennial and bears fruits that has high nutritional value. The wide geographic range of *F. vesca* from subtropical to arctic regions increases its potential as a model for studying adaptive traits. A lot of information for genome studies is available after the sequencing of *F. vesca* and the analysis and publishing of the data (Shulaev et al. 2011). In further *F. vesca* genome studies, the number of identified and described protein-coding and other gene sequences increased consistently with the use of increasingly sophisticated research and bioinformatics methods. According to last updated annotation of *F. vesca* Hawaii-4 genome (version v4.0.a2, Li et al. 2019b), in total 34,007 protein-coding genes with 98.1% complete Benchmarking Universal Single-Copy Orthologs (BUSCOs, Seppey et al. 2019) were obtained. Several dense linkage maps were created, and about 30% of genes were found to be alternatively spliced (AS) isoforms (Darwish et al. 2015; Edger et al. 2018; Li et al. 2019b, 2018; Tennessen et al. 2014). Recently two more genotypes of *F. vesca* were sequenced: CFRA 2339 with 30,349 genes (Alger et al. 2021) 'Yellow wonder' with 34,007 genes (Joldersma et al. 2022). Hawaii-4 and 'Yellow wonder' have yellow fruits, while CFRA 2339 produces red fruit and is runner less.

Hirakawa et al. (2014) reported the first draft genome of octoploid garden strawberry (*F.* × *ananassa* cv. 'Reikou'). Gene prediction was performed *ab initio* using Augustus 2.7. The total number of predicted genes in *F.* × *ananassa* (FANhybrid_r1.2) was 45,377. In the same study, the genomes of several *Fragaria* species were sequenced and assembled, including *F. orientalis*, *F. iinume*, *F. nipponica*, and *F. bucharica*. A near-complete chromosome-wide assembly for *F.* × *ananassa* cv. 'Camarosa' was generated using Illumina, 10X Genomics, and PacBio long reads (Edger et al. 2019). The authors annotated 108,087 protein-coding genes together with 30,703 genes encoding long non-coding RNAs, with a BUSCO completeness score of up to 99.17%. Transposable element related sequences represent ~36% of the total genome assembly, with LTR-RTs being the most abundant (~28%). Plastid and mitochondrial genomes were also assembled, annotated, and checked for completeness. In the same year, a *F.* × *ananassa* reference transcriptome was generated using PacBio sequencing (Yuan et al. 2019). The authors used long-read sequencing (LRS) technology and RNA-seq analysis to investigate polyploid transcriptome and differentially expressed transcriptome diversity and complexity at four stages of strawberry fruit development. It was demonstrated active roles of auxin, ABA pathway, and anthocyanin biosynthesis gene families in the regulation of fruit ripening. Liu et al. (2021a) presented improved annotation of the *F.* × *ananassa* genome, v1.0.a2, with 108,447 gene models and 97.85% complete BUSCOs. Isoforms for 11,044 genes and adds 5′ and 3′ UTR information for a large proportion of protein-coding genes was presented also. In the Genome Database for Rosaceae (GDR, https://www.rosaceae.org/species/fragaria/fragaria_spp; Jung et al. 2019) is presented data about sequencing of two more *F.* × *ananassa* genotypes: 'FL15.89-25' breeding accession from University of Florida and cv 'Royal Royce' from the University of California, Davis, as well as data about sequencing of other *Fragaria* species.

The complete sequencing of the genome of *F.* × *ananassa* led to the identification of diploid progenitors that were long unknown and have become the subject of intense debate in recent years (Li et al. 2021). Based on interspecific hybridization data, cytogenetic and molecular studies, it was found that strawberries have formed from two to three diploid clades (sub-genomes) and their diploid ancestors are *F. vesca*, *F. iinumae,* and several other related diploid species from East Asia (Njuguna et al. 2013; Rousseau-Gueutin et al. 2009; Sun et al. 2021; Tennessen et al. 2014; Yang and Davis 2017). Determining exactly which current diploid species were the ancestors of the current octoploids is not easy and depends on the system of phylogenetic analysis that can be different.

Interspecific hybridization during the formation of *Fragaria* species, genetic recombination in newly formed intermediate polyploids, incomplete lineage sorting, and the fact that some of the ancestral species are now extinct may be relevant for such determination. Therefore, phylogenetic methods can lead to erroneous conclusions (Feng et al. 2021). Edger et al. (2019) presented a phylogenetic analysis of *F.* × *ananassa* and hypothesized that a hexaploid ancestor, a hybrid whose genome included ancestors of *F. nipponica*, *F. viridis*, and *F. iinumae*, crossed from Asia to North America and hybridized with the native *F. vesca* ssp. *bracteata*, about 1.1 million years ago. *F. moschata* is proposed as one of the intermediate species. However, Liston et al. (2020) reanalyzed the four sub-genomes in a phylogenomic context and supported the view that the ancestors of *F.* × *ananassa* included *F. vesca* and *F. iinumae*, but denied the contribution of *F. viridis*, *F. nipponica*, and *F. moschata*. In response, a reanalysis of chromosome-wide *F. Iinumae* genome transcriptome data and a comparative genomic analysis using PhyDSby Edger et al. (2020) confirmed the previous model for the origin of octoploid strawberries. The third group sequenced and assembled the genomes of three wild diploid species, *F. nilgerrensis*, *F. nubicola*, and *F. viridis* (Feng et al. 2021). After aligning these three genomes with the previously sequenced genomes of *F. vesca* and *F. iinumae* using sppIDer (Langdon et al. 2018), it was confirmed that *F. vesca* and *F. iinumae*, but not others, are progenitors of cultivated garden strawberries. A third of the reads were assigned to *F. vesca* and *F. iinumae*, whereas only 12%, 11%, and 11% of the reads were assigned to *F. viridis*, *F. nilgerrensis* and *F. nubicola*, respectively (Feng et al. 2021). It is interesting to note that some sub-genomes dominate over other sub-genomes in *F.* × *ananassa*. Tennessen et al. (2014) claim that due to introgression caused by homoploid hybridization, homeologous exchanges, and inter-chromosomal rearrangements, *F. iinumae* and similar sub-genomes became closer or more like *F. vesca*. On the other hand, among diploid species in general, a complex pattern of gene flow has been observed between *Fragaria* species such as *F. iinumae*, *F. nilgerensis*, *F. nubicola* as well as *F. viridis* and *F. vesca*. Edger et al. (2019) found that dominant *F. vesca* chromosomes contain 20.2% more protein-coding genes and 14.2% more incRNA genes, and 19.5% less transposon sequences than the chromosomes of other sub-genomes. Transposable elements close to genes are also less abundant in the *F. vesca* sub-genome, as well as there are more tandem gene repeats in it, which suggests that *F. vesca* genes are more frequently transcribed and have a greater importance for adaptation than other sub-genomes. Most of resistance (*R*) genes determining disease resistance were also found in the *F. vesca* sub-genome. Even 32.6% of nucleotide binding site-leucin rich repeat (*NBS-LRR*) genes encoded in three other sub-genomes originate from homoeologous exchanges with the *F. vesca* sub-genome. A total of 426 *NBS-LRR* genes were identified, among which the majority (195) were genes encoding an N-terminal coiled-coil (*CC*), 79 genes encoding toll interleukin 1 receptor (*TIR*), and 24 genes encoding resistance to powdery mildew 8 (*RPW8*) domains (Edger et al. 2019). The analysis showed that certain metabolic pathways, including those responsible for strawberry flavor, color, and aroma, are largely controlled by the dominant sub-genome. For example, *F. vesca* homeologs in octoploid strawberry are responsible for the biosynthesis of 88.8% of anthocyanins, the metabolites responsible for the red pigments in ripe strawberry fruits, as well as 89.2% biosynthesis of geranyl acetate, a terpene associated with fruit aroma; and 95.3% of fructose biosynthesis associated with sweetness (Edger et al. 2019). Barbey et al. (2019) observed even higher levels of *F. vesca* sub-genome dominance, but gene expression was not always higher. Nevertheless, other genomes apparently give for the allopolyploid certain advantages in adaptation. *Fragaria* octoploids spread in new areas not occupied by other *Fragaria* species – in the dry mountain areas of South America and wetter meadow areas of North America, where the growth conditions were very different. The higher plasticity characteristic for polyploids allows them to conquer new ecological niches with new environmental conditions (Wei et al. 2019). Furthermore, the octoploid fern has some of the genes obtained from other diploid species, possible ancestors of *F.* × *ananassa*. These are mainly transcription factors, *R* genes, protein kinases, and genes related to flowering time and fruit quality, including color, taste, texture, and aroma (Feng et al. 2021).

The multiple genomes of *Fragaria* also provide a unique opportunity to study variation in genomic features (e.g., gene content) at the genus level. Genome-wide intra- and interspecies comparisons can reveal shared (core) or species-specific (invariant) gene content that has been gained or lost since divergence from the most recent common ancestor (Qiao et al. 2021). Many wild species of *Fragaria*, due to positive selection, have several genes related to the adaptation of plants of those species to specific growing conditions. For example, *F. daltoniana*, which has adapted to grow in mountains over a relatively large area (>3,200 m) in the southwestern regions of China, has many genes such as the "ubiquitin-dependent ERAD pathway" related to reproduction and response to abiotic stresses such as drought, cold, and heat (Qiao et al. 2021). *F. nilgerensis*, which grows in southern and north-eastern China, has similar adaptations. Its genome is 50 Mb larger than that of *F. vesca*, but this increase is only due to the higher number of transposons. In this species, a significant increase in genes involved in the interaction between plants and pathogens, biosynthesis of brassinosteroids and phenylpropanoids, metabolism of starch and sucrose, cyanoamin acids, and plant hormone signaling was found (Zhang et al. 2020a). These data are important for the development of new adaptive strawberry varieties. The berries of some *Fragaria* species are red, some are white. It is known to depend on the formation and accumulation of the natural pigment's anthocyanins in the berries, which are determined by the activity of the *R2R3-MYB* transcription factor and the *FaMYB1* and *FaMYB10* genes. By studying the detailed structure of these genes, various mutations and their expression features, a lot of important knowledge about the synthesis of anthocyanins in strawberries was obtained. Anthocyanins are natural pigments and antioxidants whose quantity and modification of composition in strawberries and other berries can increase their nutritional and medicinal value (Qiao et al. 2021; Starkevi~ et al. 2020). Many valuable characteristics, such as cold resistance, disease resistance, firm fruit, high aroma, different fruit colors, etc., are present in wild *Fragaria* species that can be used for cultivated strawberries (Hancock 1999; Zhang et al. 2020b). For example, five ploidy interspecific hybrids (5*x*, 6*x*, 8*x*, 9*x*, and 10*x*) with strong cold resistance were obtained in interspecific crosses by Luo et al. (2018). Majority of the hybrids exhibited stronger cold resistance than lower cold resistant parent. For instance, 80.0% and 83.3% hybrids, respectively, from 'Toyonoka' × *F. viridis* and 'Chandler' × *F. orientalis* revealed superior cold resistance than the lower cold resistant parents 'Toyonoka' and 'Chandler'. Those hybrids are valuable resources for strawberry cultivar improvement. Strawberry breeders developing new strawberry varieties focus on improving stress resistance, fruit quality, productivity, and disease resistance. Some desirable traits of lower ploidy species have been introduced into cultivated strawberry by pollination through unreduced gametes or by chromosome doubling with colchicine treatment (Bors and Sullivan 2005; Noguchi et al. 2002; Sangiacomo and Sullivan 1994).

6.2.2 Genome Content of Blueberries and Cranberries

Vaccinium L. genus contains 450 species including wild and culturally economically important berries (Colle et al. 2019; Edger et al. 2022). These berries, both wild and cultivated, have great benefits for human health. The main commercially grown crops of family *Ericaceae* were domesticated in the 20th century. There are American or 'large' cranberry *Vaccinium macrocarpon* Ait. and blueberry of section *Cyanococcus*, including lowbush (*V. angustifolium L.*), northern high-bush (*V. corymbosum* L.), and southern high-bush blueberry (*V. darrowii*). Bioactive compounds (mainly various phenolic compounds and ascorbic acid) from blueberries and cranberries have many health benefits, like cancer-preventing, anti-inflammatory against urinary tract infections, antimicrobial activities, and prevention of dental and gum diseases, stomach ulcers, lowering cholesterol, and biofilm formation (Michalska and Łysiak 2015; Skrovankova et al. 2015; Karlsons et al. 2018).

High-bush blueberry is an economically high-value crop and distributed everywhere in the world over the past decades (Colle et al. 2019; Bhatt and Debnath 2021). The production of blueberries in the world doubled during the last decade and currently stands at just under 1 million tonnes. The leading countries in decreasing order are the USA, Canada, Chile, and Peru (FAOSTAT 2022).

The 'large' cranberry is native to North America and recently domesticated where the production area stretches from the USA, Canada to Europe and from Chile to New Zealand (Vorsa and Zalapa 2020). Production of 'large' cranberry slightly lower (about 663,000 tons), but the same leading countries, except for Peru, and the fourth leader is Turkey (FAOSTAT 2022).

Breeding programs of *Vaccinium* spp. going about 100 years (blueberry earlier, cranberry later) and have made significant contributions to improving resistance to biotic and abiotic factors of berry plants. The genetic resistance to phytoplasma-incited disease was primary goal in breeding programs of *Vaccinium* spp. in 1929 (Kawash et al. 2022). The selection focused on releasing cultivars with disease resistance, higher fruit quality, reduced chilling hours in warmer regions, increased cold hardiness in colder regions, and higher yield under high pH, temperature, and radiative stress (Lobos and Hancock 2015; Kawash et al. 2022). More than 50 cultivars are listed for growers at this time of blueberry and cranberry, and the range changes every year. The breeding process using traditional approaches is a long and tedious, takes several decades to new cultivar release. Many genetic molecular studies in blueberry and cranberry have been oriented to genetic fingerprinting of cultivars, genetic mapping, and developing tools for marker-assisted selection (MAS) (Zalapa et al. 2015). These comprehensive genetic studies *Vaccinium* spp. performed in recent years open the way to accelerated and targeted selection. MAS capacity in *Vaccinium* breeding programs enables breeders to select and pyramid valuable traits (Zdepski et al. 2016; Die et al. 2020; Colle et al. 2019; Fong et al. 2021; Kawash et al 2022).

The gene pool of blueberry and cranberry is different in their ploidy (Edger et al. 2022). The lowbush blueberry (*V. angustifolium*) and high-bush blueberries (*V. corymbosum*) are a tetraploid ($2n=4x=48$) (Bruederle and Vorsa 1994; Bhatt and Debnath 2021; Redpath et al. 2022). By sequence counting, it was shown the genome of cultivated tetraploid high-bush blueberry (*V. corymbosum*) consists of 48 pseudomolecules with ~1.68 Gb of assembled sequence, and an average of 32,140 protein-coding genes per haplotype (Colle et al. 2019). Southern blueberry *V. darrowii* is diploid ($2n=2x=24$), and the genome was assembled at the chromosome level recently (Cui et al. 2022). Genome size of *V. darrowii* according to data of flow cytometry is about 560.5 and 555.38 Mb according to K-mer frequency analysis of Illumina short-read sequencing data. Protein-coding genes 41,815 of *V. darrowii* were annotated (Cui et al. 2022). The 'large' cranberry *V. macrocarpon* is diploid ($2n=2x=24$) and self-fertile, and at the same time, the wild relative *V. oxycoccos* genome occurs as diploid, tetraploid, and hexaploidy levels (Kawash et al. 2022). *V. macrocarpon* genome size is slightly smaller in size compared to blueberry, and it is 470 Mb according to flow cytometry data (Zdepski et al. 2016). Genome size of *V. macrocarpon* according to NGS data was 484 Mb, and genome was assembled in 124 contigs and 48,647 protein-coding genes were annotated (Kawash et al. 2022).

By hybridization of low- and high-bush blueberries, the new half-high blueberry cultivars were developed (Edger et al. 2022). Another gene pool for the development of blueberry cultivars with higher adaption to environment power are wild species of *Vaccinium* genus. Interspecific hybrids among *V. corymbosum*, *V. angustifolium*, *V. darrowii*, *V. ashei*, *V. tenellum*, *V. constablei*, and *V. elliottii* have been created (Lobos and Hancock 2015). *V. darrowii* is a subtropical wild species, a native of the South-eastern USA, and was used as a source of reduced chilling requirements, adaptation to hot and humid summers (Chavez and Lyrene 2009) to introduce low-chill southern high-bush cultivars (Hancock et al. 2008). Cultivars of *V. darrowii* and hybrid plants derived from crosses between *V. corymbosum* and *V. darrowii* have been developed for cultivation in warm southern and western regions of North America. *V. elliottii* is another wild source for blueberry breeding. It is diploid, and its hybridization with tetraploid high-bush blueberry cultivars (*V. corymbosum*) resulted in tetraploid vigorous and fertile hybrids (Norden et al. 2020). Fertile intersectional hybrids were obtained by crossing species from different subgenera *V. meridionale* (section *Pyxothamnus*) and high-bush blueberry (*V. corymbosum* section *Cyanococcus*) (Ehlenfeldt and Luteyn 2021). Another technique to obtain polyploid plants is the induction of polyploidy *in vitro* plants with mutagens like colchicine and recently was introduced for blueberries (Marangelli et al. 2022).

The gene pool of cranberries is more limited compared to blueberries as only cultivated American cranberry (*V. macrocarpon*) and wild cranberry (*V. oxycoccos*) belong to the section *Oxycoccos* (Edger et al. 2022). It has been established that the genes of *V. oxycoccos* preserved in duplicate, including genes related to protection against pests and diseases, and therefore this wild genotype remains a donor of resistance genes for the improvement of the *V. macrocarpon* genome (Kawash et al. 2022). Cranberries are a young crop with a shallow breeding tradition, and many varieties were considered slightly improved wild genotypes (Zdepski et al. 2016). Nevertheless, based on the genetic data presented by Kawash et al. (2022), who evaluated and compared the genomes of four varieties of cranberry with wild type, it was shown that the unique variety of heterozygous SNPs distinguishes *V. macrocarpon* from *V. oxycoccos* well. It turned out that not all varieties of *V. macrocarpon* have lost valuable genes, such as genes that are responsible for the synthesis of phytoalexins for pest control or improve growth through nitrogen metabolism. Ehlenfeldt et al. (2018) reported on crossbreeding of interspecific taxonomic sections and genotypes of various ploidies, noting that there are several successful and promising variants, but the combination of *V. meridionale* and *V. macrocarpon* (the possibility of triploid production) is unsuccessful, according to the authors' opinion (Ehlenfeldt et al. 2018).

6.2.3 Genome Content of Currants and Gooseberries

Currants and gooseberries belonging to the genus *Ribes* are not among the world's major fruit and berry crops. However, they play an important role, especially considering the high nutritional value of the obtained berries and their processing products. They provide consumers with the necessary vitamins, antioxidants, and other health-beneficial compounds (Cortez et al. 2019). According to the statistics provided by the FAO, in 2020, currants and gooseberries occupied 155,000 ha in the world, and 771,000 tons of harvest were obtained (FAOSTAT 2022). It should be mentioned that nearly all cultivation – 99% of the entire amount, both in terms of area and yield – takes place in Europe, with the Russian Federation, Poland, Ukraine, Lithuania, and the UK being the main producers.

The *Ribes* genus contains more than 150 species in temperate zones from Eurasia to North America, Central and South America mountain regions, and North Africa (Brennan 2008). Of all this wealth of species, 10–12 are involved in cultivating or creating cultivars, which is a lot compared to other cultivated fruit and berry crops. The dominant species groups in cultivation are blackcurrants (species and varieties *R. nigrum* L., *R. nigrum* var. *pauciflorum*, *R. nigrum* var. *sibiricum*, *R. americanum* Mill., *R. dikuscha* Fisch. ex Turcz.), red currants (species *R. rubrum* L., *R. spicatum* E. Robson, *R. petraeum* Wulfen), gooseberries (species *R. uva-crispa* L.), and golden currants (*R. aureum* Pursh). Another feature of using species of the genus *Ribes* is the extensive interspecies crossings. Most of today's blackcurrant cultivars are complex multi-stage interspecies crossings, the identification of cultivar composing species is relatively complex, and in some instances, it complicates research and breeding (Lācis et al. 2022). An important setback that should also be mentioned is the content of phenolic compounds and tannins in the leaves of *Ribes* plants which may inhibit PCR and other molecular biology research methods (Pikunova et al. 2019). This can be solved by selecting appropriate isolation methods and implementing DNA purification steps.

The next factor limiting the genetic research of *Ribes* plants could be the genome size. The available information shows that the genome size (DNA amount 1C (Mbp)) in the *Ribes* genus is determined for *R. glutinosum* -534.10 (Hanson et al. 2001), *R. odoratum* -833.00 (Zhang et al. 2013), *R. oxyacanthoides* -872.20 (Bai et al. 2012), *R. missouriense* -886.90 (Bai et al. 2012), *R. uva-crispa* -921.20, *R. petraeum* -940.80, *R. rubrum* -950.60, and *R. alpinum* -989.80 (Chiche et al. 2003). Although the analyzed *Ribes* species are defined as diploid, their genome size is comparable to higher ploidy fruit species, such as domestic plums or sour cherries (respectively, 607.60 and 896.70 DNA amount 1C (Mbp)). An equivalent genome size was also found in *Vaccinium* species, with an average value of 658.23 (Pellicer and Leitch 2020). Considering these circumstances and crops' relatively minor economic impact, no complete genome sequence has been published for any of the *Ribes* species (ncbi.nlm.nih.gov).

Despite the reasons mentioned above, there is a need for genomic information, and therefore research is being initiated based on the study of transcriptomes, identification of involved genes and *de novo* sequencing. In one such study, total RNA sequencing was performed using 311 genotypes from an F_1 full-sib population (Russell et al. 2011). In this case, 7,000 new SNPs and 3,000 expressed sequence tag (EST)-SSR markers were identified, which were later verified on 33 breeding lines, 15 commercial cultivars and five relative wild species. Obtaining genomic information was of fundamental importance in the further study of the heredity of traits and significantly improved QTL mapping activities. Obtained genomic information also indicated a high proportion of novel genes that control various putative functions in different genome sites.

In connection with the impact of climate change on the cultivation of currants and gooseberries, it is important to understand the phenology of development and the involvement of control genes. A critical development phase is dormancy release, the genetic control studied by analyzing gene expression in leaf buds (Hedley et al. 2010). As a result, three genes co-localize with previously identified QTLs for these traits (Brennan et al. 2008), confirming them. The analysis of transcriptome has also been used to study the genetic control of flower buds and flower development in *R. diacanthum* research (Zhou et al. 2019). High-throughput sequencing from six stages of flower development made it possible to identify candidate genes; 2,785 were differentially expressed, the majority were male or female-biased, and a relationship with MADS-box genes was detected.

The population established by Hedley and colleagues (2010) was also used in a gene expression micro assay study related to blackcurrant fruit development (Jarret et al. 2013). By performing lipid content phenotyping and comparing six stages of fruit development, transcripts related to lipid metabolism were analyzed, indicating relationships between transcript abundance and plastid function and their changes during fruit development. Phenotyping of sugars, organic acids, polyphenols, and anthocyanins was performed using the same six stages of fruit development in the blackcurrant cultivar 'Ben Finlay' (Jarret et al. 2018). The study identified changes in the expression of genes controlling metabolites related to fruit development, and also showed the relationship of transcript or metabolite profile with the fruit development stage. By sequencing the cDNA of two genes that control the synthesis of cyanidin- and delphinidin-based anthocyanins and analyzing their expression in blackcurrants, it was possible to identify a positive association with the level of anthocyanin and soluble carbohydrates during maturation (Li et al. 2019a). The obtained gene sequences showed homology with other plant species.

The cultivation of currants and gooseberries is significantly affected by various environmental stresses, both abiotic and biotic. A plant's response to environmental stress is determined by appropriate heredity mechanisms, the identification and understanding of which can ensure the creation of more suitable new varieties. One of the factors that a grower can face is drought stress. By conducting a study with the blackcurrant variety 'Ben Gairn' with and without irrigation, the effect of drought stress on the expression of specific genes was demonstrated: genes controlling plant hormones (e.g. gibberellin, auxin), ethylene-forming enzyme, metabolic responses, cell wall, and cell cycle genes (Čerekovič et al. 2015). Constant resistance to pathogen attacks is even more important, their mitigation in the reduced background of pesticide use. The general response of plants to pests is also important, which is ensured by pathogenesis-related genes (*PRs*) and coronatine-insensitive 1 (*COI1*) defense-response pathways. Their study in samples of different *Ribes* species (*R. americanum*, *R. dikusha*, *R. glandulosum*, *R. hudsonianum*, *R. nigrum*, *R. pauciflorum*, *R. sanguineum*, *R. uva-crispa*) was carried out to identify *PR*s and *COI1* homologs in *Ribes* (Juškytė et al. 2022). As a result, the homology of the studied genes with other plant species was proven, and specific molecular markers were developed for identification in *Ribes* plants. Pathways related to resistance to a specific pathogen (Blackcurrant Reversion Virus, BRV) were identified by analyzing transcripts in the resistant blackcurrant cultivar 'Aldoniai' using BRV inoculation experiments (Mažeikienė et al. 2022). More than 150,000 transcripts were obtained, of which 31% were unigenes. As a result, it was established that phenylpropanoid, cutin, suberin, and wax biosynthesis pathways were activated in infected plants, and the involved candidate genes were identified. The obtained information will allow a better understanding of resistance reactions against BRV and form the basis for selecting resistant cultivars.

6.3 ALLELE MINING

6.3.1 Characterization of Allelic Diversity in Strawberries

Currently, molecular markers are widely used in the investigation of genetic diversity and genotype identification of varieties, as well as in the creation of genetic linkage maps for diploid and octoploid strawberry species, cultivars, and accessions, for identification of resistance, and quality trait locus (QTL) (Rui et al. 2018). Applications of molecular marker techniques in strawberry breeding and production were reviewed by Kunihisa (2011), Whitaker (2011), and Rui et al. (2018). Different types of molecular markers including SSR, inter-simple sequence repeat (ISSR), amplified fragment length polymorphism (AFLP), random amplified polymorphic DNA (RAPD), and sequence-characterized amplified region (SCAR) have been used for mapping genomes of diploid *Fragaria vesca* and octoploid *Fragaria × ananassa* plants (Mohamed et al. 2014). Few strawberry marker databases are available at this moment. *F. vesca* and *F. × ananassa* EST-derived SSR markers and *F. × ananassa* transcriptome-derived SSR markers (FATS) were developed by Isobe et al. (2013) and placed to strawberry genome and resource database entry (https://strawberry-garden.kazusa.or.jp) created by Japan Kazusa DNA Research Institute. Many different molecular markers, genetic maps, and QTLs are available at the GDR (https://www.rosaceae.org/ species/ fragaria/ fragaria_spp).

6.3.1.1 Characterization of Allelic Diversity of Strawberries by Molecular Markers

High repeatability and a large number of polymorphic products are the main benefits of AFLP markers over RAPDs. Despite this, the AFLP method has not gained much popularity in strawberry genetic studies.

*Pst*I-based AFLP was successfully used for Polish strawberry cultivars and breeding lines identification (Tyrka et al. 2002). Peng et al. (2015) used AFLP markers to assess the genetic diversity of Chinese (Tibet) *Fragaria ananassa* Duch. Results showed high genetic diversity in the studied strawberry population. The average polymorphism was~59%. In Lithuanian strawberry accessions, polymorphic loci were amplified by 79% of AFLP primer combinations (Rugienius et al. 2015). The most informative was the EcoRI-ACC/MseI-CAC primer combination. The authors remarked that PCR-based SSR and AFLP methods can effectively be used for the identification of strawberry cultivars.

RAPD markers are well distributed throughout the genome and are expressed as dominant traits. In strawberry, these DNA markers were used for cultivar identification or to distinguish wild species. Harrison et al.'s (2000) study of *Fragaria* spp. from the northern USA showed very low genetic variability among wild populations. Three RAPD primers, OPJ04, OPJ05, and OPJ06 primers were the best in discriminating Sothern American (Argentina) strawberry cultivars (García et al. 2002). Five RAPD primers were used for the evaluation of the genetic variation of Indonesian strawberry cultivars (Aristya et al. 2019). Three primers gave a high polymorphic DNA percent (from 53.33% to 88.89%). Some specific bands identified as *Fragaria* spp. markers were detected with all primers. Rigotti (2002) used RAPD and PCR techniques for the identification of genetic markers unique to *B. cinerea* Pers.: Fr. in *F. × ananassa* Duch. (Rigotti 2002).

Sequence-characterized amplified regions (SCARs) markers can be developed from ISSR, RAPD, or AFLP profiles. Compared to other genetic markers, they are simpler, sequence-specific, more reliable, and reproducible. Albani et al. (2004) reported about three ISSR-developed SCAR markers related to the seasonal flowering locus (*SLF*) gene (Albani et al. 2004). Luk'yanchuk et al. (2018) used AFLP-derived SCAR markers for red stele root rot (*Rpf1* gene) and anthracnose (*Rca2* gene) resistance genetic screening in wild species of genus *Fragaria* L. A PCR product (240 bp) corresponding to *Rca2* gene was detected in the 'Laetitia' variety. Meanwhile, amplicons corresponding to *Rpf1* gene (285 bp) were not detected in the analyzed strawberry genotypes (Luk'yanchuk et al. 2018). Later, the same markers (STSRca2_240 and SCARR1A) were used for testing Russian and foreign-breeding strawberry (*Fragaria* L.) varieties (Lyzhin and Luk'yanchuk 2021). Mesifurane (marker FaOMT-SI/NO) and γ-decalactone content in fruits (marker FaFAD1-F/R) were analyzed, too. Anthracnose-resistant foreign ('Elianny' and 'Troubadour') and Russian ('Borovitskaya' and 'Sudarushka') varieties were detected. They all carried the *Rca2* allele. 'Bylinnaya' and *F. virginiana* ssp. *platypetala* had red stele root rot resistant gene *Rpf1*. PCR amplicons of fruit aroma genes (*FaOMT* and *FaFAD1*) were identified in some strawberry species (*F. virginiana* ssp. *platypetala*, *F. orientalis*, and *F. moschata*) and many other analyzed genotypes.

Microsatellites usually are used as primers to amplify mainly the ISSR sequences of different sizes, and the analysis is quick and simple (Marwal and Gaur 2020; Reddy et al. 2002). These markers are highly polymorphic and, compared to RAPD, have high reproducibility. Corrêa et al. (2021) used ISSR markers to evaluate the variability of strawberry cultivars, developed by the Brazilian UNICENTRO's strawberry breeding program (USBP). Out of nine, two markers (UBC 807 and UBC 834) generated the largest number (20) of loci. The average number of loci per primer was 16. It was reported that the genetic similarity between three-way hybrids and most commercial cultivars and heirloom cultivars was only 30%.

The genetic diversity of Canadian strawberry cultivars and hybrids was analyzed by Debnath et al. (2008). 17 ISSR primers generated polymorphic bands with an average of 15 bands per primer. Genetically, all studied strawberry cultivars and lines were very similar (from 63% to 77%) (Debnath et al. 2008). Nunes et al.'s (2014) study showed that octoploid strawberry hybrids, developed by the breeding program in Brasil (Agricultural Research Agency of Minas Gerais), were very diverse. Two cultivars, 'Dover' and 'Oso Grande', were offered as the best for further strawberry breeding. SNP is the most common type of sequence polymorphism. SNP-based markers are used in population and molecular genetics, and provide valuable genetic information due to their abundance and universality in the genome. Meanwhile, they are included in plant breeding programs to improve crop plants and protect germplasm (Marwal and Gaur 2020; Morgil et al. 2020). The first high-throughput genotyping platform for *F.* × *ananassa* was developed by Bassil et al. (2015). The Affymetrix IStraw90 Axiom SNP array is a valuable tool for genetic investigations such as the creation of high-density linkage maps, the identification of QTLs, and genome-wide association studies, thereby establishing a foundation for marker-assisted breeding in octoploid strawberry (Bassil et al. 2015). This high-density SNP array was used for fruit flavonoids metabolic QTLs (mQTLs) and color-related QTLs detection in 'Capitola' × 'CF1116' pseudo progenies by Labadie et al. (2020a). An SNP marker related to the anthocyanidin reductase (*ANR*) gene underlying the major pelargonidin-3-glucoside mQTL was detected.

First cleaved amplified polymorphic sequence (CAP) markers were used by Kunihisa et al. (2003) for the identification of imported Japanese strawberry cultivars from fruits. CAP and derived CAP (dCAP) markers were applied for the *Fragaria vesca* and *Fragaria bucharica* reference mapping population by Ruiz-Rojas et al. (2010). These SNP-derived techniques mainly are used to create/disrupt recognition sites of a restriction enzyme. One SNP was discovered in the current study on average per 60 bp between two strawberry species (Ruiz-Rojas et al. 2010).

6.3.1.2 Characterization of Strawberries Allelic Diversity by Microsatellite Markers

The most used markers in strawberry genotyping are SSRs or microsatellites. Microsatellites are codominant, highly polymorphic, and transferable molecular markers. SSR markers can be genomic (span non-coding regions of the genome), chloroplast, or EST-SSRs. EST-SSRs are located in the genic transcribed regions and have higher transferability to related species. Most microsatellite primers were developed from *F.* × *ananassa*, *F. vesca*, *F. viridis*, and *F. viriginiana* (Wambui 2010). SSR markers were used for generation genome maps for octoploid (*F.* × *ananassa* and *F. virginiana*) and diploid (*F. vesca*, *F. viridis*, and *F. bucharica*) *Fragaria* spp. (Mohamed et al. 2014). Sargent et al. (2009) developed a molecular linkage map for the cultivated strawberry from progenies of *F.* × *ananassa* population 'Redgauntlet' × 'Hapil' (Sargent et al. 2009). It was composed mainly of SSR (71) and gene-specific (5) markers. Two years later, Sargent et al. (2011) created new polymorphic SSR markers from previously unmapped areas of the diploid *F. vesca* L. 'Hawaii 4' genome and thus contributed to the development of the genetic linkage maps in strawberry (Sargent et al. 2011). A year later, Sargent et al. (2012) used mostly transferrable sequence-characterized SSR and gene-specific markers to update and modify the linkage map for the cultivated strawberry (*Fragaria* × *ananassa*) (Sargent et al. 2012). Van Dijk et al. (2014) generated an extremely detailed genetic SSR linkage map from a cross between 'Holiday' × 'Korona' cultivars (van Dijk et al. 2014).

Generating functional SSR markers is a time-consuming process. First, the initial identification of microsatellite loci, primer selection, and polymorphism evaluation should be performed. In addition, some prior knowledge of the genome of the target species is required. To avoid this, methods based on the transfer of SSR markers between species or genera have been implemented. A total of 6125 EST-SSR markers were developed for *Fragaria* species and used for providing a linkage framework (Mohamed et al. 2014). For *Fragaria* species, markers were created and used to create a linking architecture. *Rubus ellipticus* EST-SSR markers were used for cross-transferability analysis in *Fragaria* × *ananassa* Duch. cultivars (Sharma et al. 2021). In total, 94% of amplified markers were polymorphic. Contig 218 had the most regions of diversity and the highest polymorphism information content (PIC) value 0.68. Thus, it was the most suitable for transferability. EST-SSR markers were used for global (Eurasia, South and North America, Japan, and Island) *F. vesca* population structure and genetic diversity analysis by Hilmarsson et al. (2017). Eighty-two percent of the amplified markers were polymorphic. The average allele number per locus was not high (4.46). The FVES0109 marker generated the largest number of alleles (16) and the highest PIC value (0.78). In the Eurasian strawberry collection, the highest number of alleles (1.98) was detected. Meanwhile, in Japanese strawberry genotypes, it was the lowest (1.08). Structure analysis showed that strawberries from Eurasia and America were strongly separated. In the Eurasian group, Icelandic strawberry genotypes created a separate cluster.

SSR primers were used for old northern and southern European and northern American strawberry cultivars identification and germplasm genetic structure and diversity evaluation (Horvath et al. 2011). Genetic structure analysis divided analyzed strawberry population into four subpopulations: old and modern European and American, old European, American and modern North European, and American and modern south European. Very high genetic variability (92%) was detected within these four groups. In addition, diversity among old and modern European cultivars was not detected, showing that old European germplasm is very conservative. High polymorphism information content (PIC) value (0.6 and higher) was recorded with MCAD FAC_001, UFFa09E12, UFFa11G07, and ChFaM036 microsatellite markers (Horvath et al. 2011).

High polymorphism was detected by Hadonou et al. (2004) in microsatellite loci of *F. vesca* accessions. The most informative loci for discriminating strawberry genotypes were EMFv004, EMFv003, and EMFv006. The average number of alleles per locus was 4.9. Furthermore, a high transferability of primers to *Fragaria* species and *F.* × *ananassa* was detected, too (Hadonou et al. 2004).

*Fa*1A-6 and *Fa*2A-6 SSR markers were the most informative in distinguishing Japanese commercial cultivars (Shimomura and Hirashima 2006). Govan et al. (2008) announced that SSR primers were very useful for genotyping octopoid strawberry cultivars and species accessions.

Chloroplast SSR markers first were used by Wambui (2010). Diversity in *Fragaria* species collection at the US Department of Agriculture was evaluated. Twenty-eight haplotypes were amplified with four primer pairs: ccmp2, ccmp5, ccmp6, and ccmp7. ccmp6 marker generated the largest number of alleles (10) and the highest genetic diversity (0.78). Furthermore, the transferability of SSR markers developed from *F. ananassa*, *F. vesca*, *F. bucharica*, and *F. viridis* in *Fragaria* species was tested (Wambui 2010).

The first research on the use of SSR markers to map the genetic links between Iranian strawberry genotypes and generate genetic fingerprints was done by Hosseini et al. (2013). Scientists used SSR (16) and EST-SSR (4) markers for genetic diversity analysis of local and exotic Iranian strawberry varieties. All studied markers were polymorphic, and the average number of alleles per locus was 5.9. The most informative locus was UAFv7648 which produced 16 alleles (PIC value 0.87). Meanwhile, the less informative were FAC-008, UAFv8936, and UFFa16H07 loci (Hosseini et al. 2013).

The genetic diversity of Lithuanian strawberry cultivars was assessed by 11 SSR markers (Rugienius et al. 2015). The largest number (21) of alleles was observed at the EMFv-104 locus, and the average number of alleles per locus was high (16.5). The most informative primer pairs for distinguishing strawberry genotypes were ARSFL-009 and CFACT-152. However, the average genetic diversity among primer pairs was low (0.34), indicating that non-coding regions in the strawberry genome are very conservative. Honjo et al. (2011) identified a high polymorphism in Japanese strawberry (*F.* × *ananassa*, *Fragaria* species) cultivars and accessions. The largest allele number per locus (34) was detected in the EMFv104 locus, too (Honjo et al. 2011). One new primer, FxaHGA02P13, was announced as very informative for discriminating analyzed strawberry genotypes. However, a low number of alleles (15) per locus was amplified using EMFv104 primer for genotyping strawberry cultivars grown in the USA (Florida) (Brunings et al. 2010).

SSR markers were successfully applied for population structure analysis in commercially grown octoploid strawberry from Korea, Germany, the USA, the UK, and Japan (Yoon et al. 2012). The largest number of alleles (10) was detected per locus CFACT110, and a high number (~49%) of specific alleles were detected. Phylogenetic analysis showed that German accessions were genetically very diverse (Yoon et al. 2012).

DNA fingerprints of Mexican strawberry varieties were obtained with nine SSR markers. EMFn181 and EMFv104 markers generated the highest number of alleles (11 and 12, respectively) (Tlatelpa et al. 2019). It was noticed that CP0615, CPLE7 (Mexico), and 'Festival' (Florida, USA) varieties were genetically different from other investigated strawberries.

DNA barcode markers were used for genetic diversity evaluation in wild and cultivated Pakistani strawberry (Qarni et al. 2022). Two markers, *ITS2* (ribosomal DNA region) and *rbclC* (cpDNA region), were the most suitable for the identification of *Fragaria* species.

It was shown that SSR markers can be successfully used for the identification of Korean ever-bearing and June-bearing strawberry cultivars (Kim et al. 2019). FxaHGA02P13 and EMFv104 markers were the most informative and generated 21 and 19 alleles per locus. The authors noticed that ever-bearing and June-bearing strawberry cultivars could not be well distinguished because of the low number of markers and populations analyzed in the study.

Genotyping by SSR molecular markers was effectively used to assess the genetic differences or similarities among cultivated and wild *Fragaria* spp. in the US strawberry genotypes (Biswas et al. 2019). An average of 3.43 alleles per locus was detected by 35 markers. The most polymorphic was the FG1a/b locus.

6.3.1.3 Allelic Mining and Quantitative Trait Loci (QTLs) Mapping of Agronomically Important Traits

The appearance of molecular marker technology offers breeders powerful new tools for identifying complex quantitative traits (early/late fruit harvesting, modulation of the runnering/flowering trade-off, disease resistance). In addition, compared to conventional breeding techniques, DNA marker technology enables breeders to boost productivity, decrease expenses, and shorten the time required to produce new varieties and hybrids.

The use of molecular markers closely connected to QTLs of desirable features (disease resistance, flowering time, fruit color, sterility) may facilitate the development of strawberry cultivars with economically significant traits. Table 6.1 presents the genome-wide identification of genes, transcription factors, and QTLs relevant to strawberry resistance to abiotic and biotic stresses, ascorbic acid and anthocyanin accumulation, sugar transport, carbohydrate metabolism, signal transduction, tissue development, hormone response, fruit development and ripening, and fruit color and flavor regulation.

One of the most agriculturally desirable features in strawberry selection is day-neutrality or ever-bearing strawberry cultivars. Gor et al. (2016) reported about putative molecular marker FaP2E11,

TABLE 6.1
Genome-Wide Identification of Protein-Coding Genes, Transcription Regulators, and QTLs in *F.* × *ananassa*

Gene	Product	Response	References
FaG6PDH1–19	Glucose-6-phosphate dehydrogenase	Carbohydrate metabolism, multiple abiotic stress, and hormonal response	Lei et al. (2022)
FaSUT	Major facilitator superfamily	Sugar transporter, sugar accumulation in berries	Liu et al. (2020)
FaNAC	Transcription factors	Development and ripening of the fruit	Moyano et al. (2018)
FaPFRU	Quantitative trait loci	Flowering habit and fruit quality	Verma et al. (2017)
FaAP2	Transcription factors	Fruit color and aroma regulation	Sheng et al. (2021)
FaWRKY	Transcription factors	Transcription factors, adaptation, monoculture cropping tolerance	Chen and Liu (2019)
FaMAPK	Mitogen-activated protein kinases	Upregulation increases the content of endogenous ABA, sucrose, and anthocyanin in strawberry fruits; regulates ripening	Li et al. (2022a)
FaGMP	GDP-D-mannose pyrophosphorylase	Ascorbin acid metabolism	Lin et al. (2021)
FaAKR23	Aldo-keto reductases	Ascorbic acid accumulation, fruit ripening	Liu et al. (2022)
FaGST	Glutathione *S*-transferase	Anthocyanin accumulation	Lin et al. (2020)
R2R3-FaMYB	Transcription factors	Anthocyanin biosynthesis, differential gene expression	Liu et al. (2021c)
FaPP2C	The protein phosphatase 2C	Signal transduction, organ development, tissue-specific response, and abiotic stress-tolerance	Haider et al. (2019)
FaRALF	Rapid alkalinization factors	Tissue development, fungal infection, and hormone response	Zhang et al. (2020a)
FaZAT10	C_2H_2-type zinc finger proteins	Induced by drought, salt, low temperature, ABA, and MeJA treatments	Li et al. (2022b)
R (NLR)	Resistance genes	Disease resistance	Barbey et al. (2019)
FaTLP	Thaumatin like-proteins	Defense against pathogenic fungi	Zhang et al. (2022)
FaCDPK	Calcium-dependent protein kinases	Response to abiotic stress and ABA treatment	Crizel et al. (2020)
FW1–5	Resistance gene	Resistance to fusarium wilt	Pincot et al. (2022)
FaRPc6(7)C(D)	Quantitative trait loci	Resistance to crown rot disease	Nellist et al. (2019)

which is related to day-neutrality in strawberry (Gor et al. 2016). *Fragaria × ananassa* Duch. cultivars (American and Australian) and new breeding lines (European and Japanese) were analyzed. In the later study, Gor et al. (2017) analyzed markers linked to genes controlling strawberry aroma and six putative markers were detected. Four of them were specific for chloroplast. Besides, scientists noticed that the nuclear-specific marker FaP1D7 was associated with the synthesis of methyl butanoate in strawberry fruits (Gor et al. 2017).

Hawkins et al. (2016) sequenced and compared genomes of diploid strawberry accessions, 'Hawaii 4', 'Rügen', and 'Yellow Wonder'. An SNP in the *FveMYB10* gene, responsible for the yellow color of *F. vesca* fruits, was identified (Hawkins et al. 2016). Salinas et al.'s (2019) study demonstrated that a set of *Fragaria × ananassa* selections and cultivars had an allele in so-called *Fragaria* resistance to *C. acutatum* locus 1 (*FaRCa1*) and were resistant to *C. acutatum* isolates from Florida.

The purpose of Lerceteau-Köhler et al.'s (2002) study was to detect putative QTLs associated with strawberry resistance to anthracnose, and few AFLP markers were found linked to the *Rca2* gene (Lerceteau-Köhler et al. 2002). Cockerton et al. (2018) reported that five stable and multiple transient QTLs were associated with powdery mildew resistance in mapping populations of *Fragaria × ananassa.*

Perrotte et al. (2016) were interested in the quick and effective breeding of perpetual flowering cultivars of strawberry via MAS. The scientists used a selective mapping strategy to identify QTL locus *FaPFRU* flanking markers, and a homologous to FLOWERING LOCUS T (*FT*) gene candidate (*FaFT*) was detected.

Gaston et al. (2013) reported the identification of a major QTL (*FaPFRU*), controlling perpetual flowering and runnering in *F. × ananassa* Duch. genotypes. Salinas et al. (2017) identified accessions with the *FaPFRU* locus and offered an easy single-marker, Bx215, diagnostic test for selecting *F. × ananassa* cultivars with the *FaPFRU* locus.

Another important goal for strawberry breeding is the development of the sensory and nutritional quality of strawberry fruits (Labadie et al. 2020b). Labadie et al. (2020b) studied the fruit flavonoid and antioxidant content in 'Capitola' and line CF1116 and their progenitors. Flavonoid metabolic, color, and antioxidant-related QTLs were mapped onto male and female linkage maps.

In Barbey et al.'s (2020) research, expression of QTL (eQTL) genetic markers linked to varied mature receptacle transcript accumulation in *F. × ananassa* genotypes was detected. A total of 286 eQTL-associated genes (e.g., pectin, phenylpropanoid metabolism, fruit-based transcription factors, carotenoids, and nutrient-associated transcripts) were discovered.

6.3.2 Characterization of Allelic Diversity in *Vaccinium* spp. Berries

Despite the economic importance and medicinally high-value of both blue- and cranberries, the molecular research was limited until recently (Schlautman et al. 2017b; Vorsa and Zalapa 2020; Edger et al. 2022). Exponential increase of genetic data started with the introduction of the NGS technique (Vorsa and Zalapa 2020; Edger et al. 2022). To date, the Genome Database for Vaccinium (GDV, https://www.vaccinium.org) is public available, and it contains molecular markers, genes, genomes, and genetic maps for *Vaccinium* spp.

6.3.2.1 Characterization of Blueberries and Cranberries Allelic Diversity by Dominant Molecular Markers

For identification of *Vaccinium* spp. genotypes, diverse molecular markers were used. The genetic diversity and identification of *Vaccinium* spp. cultivars were evaluated with dominant markers, like RAPD (Albert et al. 2003; Novy et al. 1994; Novy and Vorsa 1996; Stewart and Excoffier 1996, Qu and Hancock 1997; Debnath 2005; Giongo et al. 2006; Debnath 2007; Quililongo et al. 2013; Carvalho et al. 2014; Cho et al. 2017; Gawroñski et al. 2017), AFLP (Albert et al. 2003; Inostroza-Blancheteau et al. 2011), ISSR (Garriga et al. 2013; Carvalho et al. 2014; An et al. 2015; Gawroñski et al. 2017; Carvalho et al. 2018), and the dominant or codominant SCAR (Cho et al. 2017; Polashock and Vorsa 2002) (Table 6.2). Some of RAPD markers were suitable for evaluation

TABLE 6.2
Dominant Molecular Markers Used for *Vaccinium* spp. Genetic Diversity Studies

Marker Locus	Marker Type	Country, References	No. of Genotypes	*Vaccinium* spp.
OPA-01	RAPD	Canada, Debnath (2005)	13	*V. macrocarpon*
		Canada, Debnath (2007)	43	
OPA-02	RAPD	USA, Novy et al. (1994)	22	*V. macrocarpon*
OPA-03	RAPD	Canada, Debnath (2005)	13	*V. macrocarpon*
OPA-04	RAPD	USA, Novy et al. (1994)	22	*V. macrocarpon*
		USA, Novy and Vorsa (1996)	123	
		USA, Stewart and Excoffier (1996)	6	
OPA-05	RAPD	USA, Novy et al. (1994)	22	*V. macrocarpon*
		Italy, Giongo et al. (2006)	38	*V. corymbosum*
OPA-06	RAPD	Canada, Debnath (2005)	13	*V. macrocarpon*
OPA-07	RAPD	USA, Novy et al. (1994)	22	*V. macrocarpon*
		USA, Novy and Vorsa (1996)	123	*V. corymbosum*
		USA, Stewart and Excoffier (1996)	6	
		Italy, Giongo et al. (2006)	38	
OPA-08	RAPD	USA, Novy and Vorsa (1996)	123	*V. macrocarpon*
OPA-09	RAPD	Canada, Debnath (2007)	43	*V. macrocarpon*
		USA, Novy and Vorsa (1996)	123	*V. corymbosum*
		USA, Stewart and Excoffier (1996)	6	
		South Korea, Cho et al. (2017)	45	
OPA-10	RAPD	Canada, Debnath (2007)	43	*V. macrocarpon*
		USA, Novy and Vorsa (1996)	123	
OPA-11	RAPD	Canada, Debnath (2007)	43	*V. macrocarpon*
		USA, Novy et al. (1994)	22	
		USA, Novy and Vorsa (1996)	123	
OPA-12	RAPD	USA, Novy and Vorsa (1996)	123	*V. macrocarpon*
		Italy, Giongo et al. (2006)	38	*V. corymbosum*
OPA-13	RAPD	Canada, Debnath (2007)	43	*V. macrocarpon*
		USA, Novy and Vorsa (1996)	123	*V. corymbosum*
		USA, Stewart and Excoffier (1996)	6	
		Italy, Giongo et al. (2006)	38	
OPA-14	RAPD	Canada, Debnath (2007)	43	*V. macrocarpon*
		USA, Novy and Vorsa (1996)	123	
OPA-15	RAPD	Canada, Debnath (2007)	43	*V. macrocarpon*
		Canada, Debnath (2005)	13	*V. corymbosum*
		South Korea, Cho et al. (2017)	45	
OPA-16	RAPD	Canada, Debnath (2007)	43	*V. macrocarpon*
		Canada, Debnath (2005)	13	*V. corymbosum*
		South Korea, Cho et al. (2017)	45	
OPA-17	RAPD	Canada, Debnath (2007)	43	*V. macrocarpon*
		USA, Novy and Vorsa (1996)	123	
OPA-18	RAPD	Canada, Debnath (2007)	43	*V. macrocarpon*
		Canada, Debnath (2005)	13	
		USA, Stewart and Excoffier (1996)	6	
OPA-19	RAPD	Canada, Debnath (2005)	13	*V. macrocarpon*
		South Korea, Cho et al. (2017)	45	*V. corymbosum*
OPA-20	RAPD	Canada, Debnath (2007)	43	*V. macrocarpon*
		Canada, Debnath (2005)	13	*V. corymbosum*
		Italy, Giongo et al. (2006)	38	
OPB-01	RAPD	Italy, Giongo et al. (2006)	38	*V. corymbosum*

(*Continued*)

TABLE 6.2 (*Continued*)
Dominant Molecular Markers Used for *Vaccinium* spp. Genetic Diversity Studies

Marker Locus	Marker Type	Country, References	No. of Genotypes	*Vaccinium* spp.
OPB-04	RAPD	USA, Stewart and Excoffier (1996)	6	*V. macrocarpon*
OPB-05	RAPD	Italy, Giongo et al. (2006)	38	*V. corymbosum*
OPB-08	RAPD	South Korea, Cho et al. (2017)	45	*V. corymbosum*
OPB-18	RAPD	USA, Stewart and Excoffier (1996)	6	*V. macrocarpon*
OPC-06	RAPD	Portugal, Carvalho et al. (2014)	10	*V. corymbosum*
OPC-07	RAPD	Italy, Giongo et al. (2006)	38	*V. corymbosum*
OPC-12	RAPD	Portugal, Carvalho et al. (2014)	10	*V. corymbosum*
OPC-20	RAPD	South Korea, Cho et al. (2017)	45	*V. corymbosum*
OPE-01	RAPD	Portugal, Carvalho et al. (2014)	10	*V. corymbosum*
OPE-05	RAPD	Portugal, Carvalho et al. (2014)	10	*V. corymbosum*
OPE-12	RAPD	South Korea, Cho et al. (2017)	45	*V. corymbosum*
OPE-17	RAPD	South Korea, Cho et al. (2017)	45	*V. corymbosum*
OPF-07	RAPD	South Korea, Cho et al. (2017)	45	*V. corymbosum*
OPF-13	RAPD	South Korea, Cho et al. (2017)	45	*V. corymbosum*
OPG-09	RAPD	South Korea, Cho et al. (2017)	45	*V. corymbosum*
OPG-10	RAPD	South Korea, Cho et al. (2017)	45	*V. corymbosum*
OPH-20	RAPD	South Korea, Cho et al. (2017)	45	*V. corymbosum*
OPI-07	RAPD	Portugal, Carvalho et al. (2014)	10	*V. corymbosum*
OPK-17	RAPD	South Korea, Cho et al. (2017)	45	*V. corymbosum*
OPK-19	RAPD	South Korea, Cho et al. (2017)	45	*V. corymbosum*
OPL-02	RAPD	South Korea, Cho et al. (2017)	45	*V. corymbosum*
OPL-12	RAPD	South Korea, Cho et al. (2017)	45	*V. corymbosum*
OPM-13	RAPD	South Korea, Cho et al. (2017)	45	*V. corymbosum*
OPM-18	RAPD	South Korea, Cho et al. (2017)	45	*V. corymbosum*
OPN-03	RAPD	South Korea, Cho et al. (2017)	45	*V. corymbosum*
OPN-04	RAPD	Portugal, Carvalho et al. (2014)	10	*V. corymbosum*
OPN-07	RAPD	South Korea, Cho et al. (2017)	45	*V. corymbosum*
OPN-19	RAPD	South Korea, Cho et al. (2017)	45	*V. corymbosum*
OPO-04	RAPD	Portugal, Carvalho et al. (2014)	10	*V. corymbosum*
OPO-10	RAPD	Portugal, Carvalho et al. (2014)	10	*V. corymbosum*
OPQ-04	RAPD	South Korea, Cho et al. (2017)	45	*V. corymbosum*
OPR-09	RAPD	South Korea, Cho et al. (2017)	45	*V. corymbosum*
OPR-12	RAPD	South Korea, Cho et al. (2017)	45	*V. corymbosum*
OPR-16	RAPD	South Korea, Cho et al. (2017)	45	*V. corymbosum*
		Portugal, Carvalho et al. (2014)	10	
OPR-19	RAPD	Portugal, Carvalho et al. (2014)	10	*V. corymbosum*
OPU-02	RAPD	South Korea, Cho et al. (2017)	45	*V. corymbosum*
OPU-07	RAPD	South Korea, Cho et al. (2017)	45	*V. corymbosum*
OPW-11	RAPD	South Korea, Cho et al. (2017)	45	*V. corymbosum*
OPY-11	RAPD	South Korea, Cho et al. (2017)	45	*V. corymbosum*
OPY-13	RAPD	South Korea, Cho et al. (2017)	45	*V. corymbosum*
OPY-17	RAPD	South Korea, Cho et al. (2017)	45	*V. corymbosum*
UBC168	RAPD	South Korea, Cho et al. (2017)	45	*V. corymbosum*
UBC190	RAPD	South Korea, Cho et al. (2017)	45	*V. corymbosum*

(*Continued*)

TABLE 6.2 (*Continued*)
Dominant Molecular Markers Used for *Vaccinium* spp. Genetic Diversity Studies

Marker Locus	Marker Type	Country, References	No. of Genotypes	*Vaccinium* spp.
UBC203	RAPD	Canada, Debnath (2007)	43	*V. macrocarpon*
		Canada, Debnath (2005)	13	
UBC222	RAPD	Canada, Debnath (2007)	43	*V. macrocarpon*
		Canada, Debnath (2005)	13	
UBC239	RAPD	Canada, Debnath (2007)	43	*V. macrocarpon*
		Canada, Debnath (2005)	13	
UBC256	RAPD	South Korea, Cho et al. (2017)	45	*V. corymbosum*
UBC268	RAPD	South Korea, Cho et al. (2017)	45	*V. corymbosum*
UBC297	RAPD	South Korea, Cho et al. (2017)	45	*V. corymbosum*
UBC320	RAPD	South Korea, Cho et al. (2017)	45	*V. corymbosum*
UBC333	RAPD	South Korea, Cho et al. (2017)	45	*V. corymbosum*
UBC402	RAPD	USA, Novy and Vorsa (1996)	123	*V. macrocarpon*
UBC424	RAPD	South Korea, Cho et al. (2017)	45	*V. corymbosum*
UBC436	RAPD	USA, Novy and Vorsa (1996)	123	*V. macrocarpon*
UBC471	RAPD	USA, Novy and Vorsa (1996)	123	*V. macrocarpon*
UBC472	RAPD	USA, Novy and Vorsa (1996)	123	*V. macrocarpon*
UBC474	RAPD	USA, Novy and Vorsa (1996)	123	*V. macrocarpon*
UBC478	RAPD	USA, Novy and Vorsa (1996)	123	*V. macrocarpon*
UBC482	RAPD	USA, Novy and Vorsa (1996)	123	*V. macrocarpon*
UBC489	RAPD	USA, Novy and Vorsa (1996)	123	*V. macrocarpon*
UBC493	RAPD	USA, Novy and Vorsa (1996)	123	*V. macrocarpon*
UBC500	RAPD	USA, Novy and Vorsa (1996)	123	*V. macrocarpon*
UBC504	RAPD	South Korea, Cho et al. (2017)	45	*V. corymbosum*
UBC531	RAPD	South Korea, Cho et al. (2017)	45	*V. corymbosum*
UBC550	RAPD	South Korea, Cho et al. (2017)	45	*V. corymbosum*
G1[a]	RAPD	Poland, Gawroński et al. (2017)	19	*V. corymbosum*
G2	RAPD	Poland, Gawroński et al. (2017)	19	*V. corymbosum*
G3	RAPD	Poland, Gawroński et al. (2017)	19	*V. corymbosum*
G4	RAPD	Poland, Gawroński et al. (2017)	19	*V. corymbosum*
G5	RAPD	Poland, Gawroński et al. (2017)	19	*V. corymbosum*
G6	RAPD	Poland, Gawroński et al. (2017)	19	*V. corymbosum*
G7	RAPD	Poland, Gawroński et al. (2017)	19	*V. corymbosum*
G8	RAPD	Poland, Gawroński et al. (2017)	19	*V. corymbosum*
G9	RAPD	Poland, Gawroński et al. (2017)	19	*V. corymbosum*
UBC801	ISSR	Canada, An et al. (2015)	5	*V. macrocarpon*
UBC807	ISSR	Portugal, Carvalho et al. (2014)	10	*V. corymbosum*
		Portugal, Carvalho et al. (2018)	16	
UBC808	ISSR	Canada, An et al. (2015)	5	*V. macrocarpon*
UBC809	ISSR	Canada, An et al. (2015)	5	*V. macrocarpon*
UBC810	ISSR	Canada, An et al. (2015)	5	*V. macrocarpon*
UBC816	ISSR	Canada, An et al. (2015)	5	*V. macrocarpon*
UBC817	ISSR	Canada, An et al. (2015)	5	*V. macrocarpon*
UBC818	ISSR	Chile, Garriga et al. (2013)	13	*V. corymbosum*
UBC821	ISSR	Portugal, Carvalho et al. (2014)	10	*V. corymbosum*
		Portugal, Carvalho et al. (2018)	16	

(*Continued*)

TABLE 6.2 (*Continued*)
Dominant Molecular Markers Used for *Vaccinium* spp. Genetic Diversity Studies

Marker Locus	Marker Type	Country, References	No. of Genotypes	*Vaccinium* spp.
UBC823	ISSR	Portugal, Carvalho et al. (2014)	10	*V. corymbosum*
		Portugal, Carvalho et al. (2018)	16	
UBC825	ISSR	Portugal, Carvalho et al. (2014)	10	*V. corymbosum*
		Portugal, Carvalho et al. (2018)	16	
UBC826	ISSR	Canada, An et al. (2015)	5	*V. macrocarpon*
UBC827	ISSR	Chile, Garriga et al. (2013)	13	*V. corymbosum*
		Canada, An et al. (2015)	5	*V. macrocarpon*
UBC835	ISSR	Canada, An et al. (2015)	5	*V. macrocarpon*
UBC844	ISSR	Portugal, Carvalho et al. (2014)	10	*V. corymbosum*
		Portugal, Carvalho et al. (2018)	16	
		Chile, Garriga et al. (2013)	13	
UBC845	ISSR	Chile, Garriga et al. (2013)	13	*V. corymbosum*
UBC848	ISSR	Portugal, Carvalho et al. (2014)	10	*V. corymbosum*
		Portugal, Carvalho et al. (2018)	16	
		Chile, Garriga et al. (2013)	13	
UBC853	ISSR	Portugal, Carvalho et al. (2018)	16	*V. corymbosum*
UBC855	ISSR	Chile, Garriga et al. (2013)	13	*V. corymbosum*
UBC856	ISSR	Portugal, Carvalho et al. (2014)	10	*V. corymbosum*
		Portugal, Carvalho et al. (2018)	16	
UBC867	ISSR	Canada, An et al. (2015)	5	*V. macrocarpon*
UBC873	ISSR	Portugal, Carvalho et al. (2014)	10	*V. corymbosum*
		Portugal, Carvalho et al. (2018)	16	
UBC880	ISSR	Portugal, Carvalho et al. (2014)	10	*V. corymbosum*
		Portugal, Carvalho et al. (2018)	16	
UBC890	ISSR	Canada, An et al. (2015)	5	*V. macrocarpon*
UBC891	ISSR	Canada, An et al. (2015)	5	*V. macrocarpon*
G-ISSR-1[b]	ISSR	Poland, Gawroński et al. (2017)	19	*V. corymbosum*
G-ISSR-2	ISSR	Poland, Gawroński et al. (2017)	19	*V. corymbosum*
G-ISSR-3	ISSR	Poland, Gawroński et al. (2017)	19	*V. corymbosum*
G-ISSR-4	ISSR	Poland, Gawroński et al. (2017)	19	*V. corymbosum*
G-ISSR-5	ISSR	Poland, Gawroński et al. (2017)	19	*V. corymbosum*
G-ISSR-6	ISSR	Poland, Gawroński et al. (2017)	19	*V. corymbosum*
G-ISSR-7	ISSR	Poland, Gawroński et al. (2017)	19	*V. corymbosum*
G-ISSR-8	ISSR	Poland, Gawroński et al. (2017)	19	*V. corymbosum*
G-ISSR-9	ISSR	Poland, Gawroński et al. (2017)	19	*V. corymbosum*
G-ISSR-10	ISSR	Poland, Gawroński et al. (2017)	19	*V. corymbosum*
G-ISSR-11	ISSR	Poland, Gawroński et al. (2017)	19	*V. corymbosum*
BE12_175	SCAR	South Korea, Cho et al. (2017)	45	*V. corymbosum*
B268_262	SCAR	South Korea, Cho et al. (2017)	45	*V. corymbosum*
BA19_288	SCAR	South Korea, Cho et al. (2017)	45	*V. corymbosum*
BE17_354	SCAR	South Korea, Cho et al. (2017)	45	*V. corymbosum*
B504_416	SCAR	South Korea, Cho et al. (2017)	45	*V. corymbosum*
BR09_465	SCAR	South Korea, Cho et al. (2017)	45	*V. corymbosum*
BF07_502	SCAR	South Korea, Cho et al. (2017)	45	*V. corymbosum*
B168_520	SCAR	South Korea, Cho et al. (2017)	45	*V. corymbosum*
B333_561	SCAR	South Korea, Cho et al. (2017)	45	*V. corymbosum*

(*Continued*)

TABLE 6.2 (*Continued*)
Dominant Molecular Markers Used for *Vaccinium* spp. Genetic Diversity Studies

Marker Locus	Marker Type	Country, References	No. of Genotypes	*Vaccinium* spp.
BR09_612	SCAR	South Korea, Cho et al. (2017)	45	*V. corymbosum*
BN03_637	SCAR	South Korea, Cho et al. (2017)	45	*V. corymbosum*
B268_668	SCAR	South Korea, Cho et al. (2017)	45	*V. corymbosum*
BN07_727	SCAR	South Korea, Cho et al. (2017)	45	*V. corymbosum*
B550_750	SCAR	South Korea, Cho et al. (2017)	45	*V. corymbosum*
B297_754	SCAR	South Korea, Cho et al. (2017)	45	*V. corymbosum*
BE17_831	SCAR	South Korea, Cho et al. (2017)	45	*V. corymbosum*
5	SCAR	Polashock and Vorsa (2002)	9	*V. macrocarpon*
6	SCAR	Polashock and Vorsa (2002)	9	*V. macrocarpon*
9	SCAR	Polashock and Vorsa (2002)	9	*V. macrocarpon*
10	SCAR	Polashock and Vorsa (2002)	9	*V. macrocarpon*
21	SCAR	Polashock and Vorsa (2002)	9	*V. macrocarpon*
22	SCAR	Polashock and Vorsa (2002)	9	*V. macrocarpon*
25	SCAR	Polashock and Vorsa (2002)	9	*V. macrocarpon*
26	SCAR	Polashock and Vorsa (2002)	9	*V. macrocarpon*
48	SCAR	Polashock and Vorsa (2002)	9	*V. macrocarpon*
49	SCAR	Polashock and Vorsa (2002)	9	*V. macrocarpon*
50	SCAR	Polashock and Vorsa (2002)	9	*V. macrocarpon*
51	SCAR	Polashock and Vorsa (2002)	9	*V. macrocarpon*
56	SCAR	Polashock and Vorsa (2002)	9	*V. macrocarpon*
57	SCAR	Polashock and Vorsa (2002)	9	*V. macrocarpon*
58	SCAR	Polashock and Vorsa (2002)	9	*V. macrocarpon*
59	SCAR	Polashock and Vorsa (2002)	9	*V. macrocarpon*
64	SCAR	Polashock and Vorsa (2002)	9	*V. macrocarpon*
65	SCAR	Polashock and Vorsa (2002)	9	*V. macrocarpon*

[a,b] Markers are named according to numbering and marker type in reference.

of both blueberry and cranberry genotypes (Novy and Vorsa 1996; Giongo et al. 2006). There are now studies for cranberries by using AFLP markers, and only in one study for each *Vaccinium* spp., the SCAR primers were evaluated (Cho et al. 2017; Polashock and Vorsa 2002).

6.3.2.2 Allelic Diversity of Blueberries and Cranberries by Codominant Molecular Markers

Another class of molecular markers used for fingerprinting of *Vaccinium* spp. is codominant markers, like EST-SSR polymerase chain reaction (EST-PCR) or SSR markers derived from ESTs, namely EST-SSR markers. The SNP belongs to this class of markers also. The development of such markers for species-specific PCR amplification requires the *a priori* information about DNA. Therefore, only a limited number of studies for blueberries and cranberries were conducted at the beginning of this century (Boches et al. 2005a, 2005b; Boches et al. 2006; Rowland et al. 2003a, 2003b; Palmieri et al. 2013), where in some cases SSR markers were first developed for blueberries (Boches et al. 2005b) and then cross-amplified for cranberries (Bassil et al. 2009; An et al. 2015) or other *Vaccinium* species (Bell et al. 2008) or wild populations (Bian et al. 2014; Bassil et al. 2020) (Table 6.3). The increase of studies evaluating genetic diversity, cultivar identification, relatedness of genotypes and linkage mapping started first with the increase of data in EST databases and with the emergence of NGS technologies. Using

TABLE 6.3
SSR Markers Used for *Vaccinium* spp. Genetic Diversity Studies

SSR Locus	Country, References[a]	Size Range, bp	No. of Alleles	Sample Size	*Vaccinium* spp.
CA23F	USA, [1]	150–170	2	11	*V. corymbosum*
	USA, [2]	155–164	nd	69	
	Canada, [5]	nd	nd	32	
	USA, [6]	157–175	7	130	
CA112F	USA, [1]	140–200	5	11	*V. corymbosum*
	USA, [2]	142–184	nd	69	
	Canada, [3]	150–650	26	8	
	Canada, [4]	nd	nd	5	
	Canada, [5]	nd	nd	32	
CA169F	USA, [1]	109–130	5	11	*V. corymbosum*
	USA, [2]	109–136	nd	69	
	Canada, [4]	nd	nd	5	
	Canada, [5]	nd	nd	32	
CA190R	USA, [1]	250–280	3	11	*V. corymbosum*
	USA, [2]	233–251	nd	69	
CA236F	USA, [1]	250–280	6	11	*V. corymbosum*
	USA, [2]	214–241	nd	69	
	Canada, [4]	nd	nd	5	
	Canada, [5]	nd	nd	32	
	Canada, [7]	nd	3	63	
CA344F	USA, [1]	170–190	8	11	*V. corymbosum*
	USA, [2]	134–179	nd	69	
CA421F	USA, [1]	180–250	14	11	*V. corymbosum*
	USA, [2]	153–230	nd	69	*V. corymbosum*
	Canada, [4]	nd	nd	5	*V. marcrocarpon*
	Canada, [5]	nd	nd	32	*V. corymbosum*
	Canada, [7]	nd	10	63	*V. corymbosum*
CA483F	USA, [1]	300–370	9	11	*V. corymbosum*
	USA, [2]	292–347	nd	69	
	Canada, [3]	300–800	8	8	
	Canada, [5]	nd	nd	32	
	Canada, [7]	nd	16	63	
CA642F	USA, [1]	350–390	23	11	*V. corymbosum*
	USA, [2]	302–387	nd	69	
CA741	Canada, [7]	nd	6	63	*V. corymbosum*
CA787F	USA, [1]	270–300	5	11	*V. corymbosum*
	USA, [2]	278–376	Nd	69	
	Canada, [5]	nd	nd	32	
CA794F	USA, [1]	220–290	10	11	*V. corymbosum*
	USA, [2]	141–226	nd	69	*V. marcrocarpon*
	Canada, [4]	nd	3	5	
CA855F	USA, [1]	250–300	10	11	*V. corymbosum*
	USA, [2]	225–258	nd	69	
CA94F	USA, [1]	390–420	7	11	*V. corymbosum*
	USA, [2]	368–420	nd	69	
NA41	USA, [1]	200–275	11	11	*V. corymbosum*
	USA, [2]	186–242	nd	69	

(Continued)

TABLE 6.3 (*Continued*)
SSR Markers Used for *Vaccinium* spp. Genetic Diversity Studies

SSR Locus	Country, References[a]	Size Range, bp	No. of Alleles	Sample Size	*Vaccinium* spp.
NA398	USA, [1]	210–240	5	11	*V. corymbosum*
	USA, [2]	211–232	nd	69	
	USA, [6]	213–238	7	130	
NA172	USA, [6]	295–313	6	130	*V. corymbosum*
NA741	USA, [1]	240–290	14	11	*V. corymbosum*
	USA, [2]	247–300	nd	69	
NA800	USA, [1]	230–290	31	11	*V. corymbosum*
	USA, [2]	180–287	nd	69	*V. corymbosum*
	Canada, [4]	nd	6	5	*V. marcrocarpon*
	Canada, [5]	nd	nd	32	*V. corymbosum*
	Canada, [7]	nd	15	63	*V. corymbosum*
NA824	USA, [1]	150–220	8	11	*V. corymbosum*
	USA, [2]	158–190	nd	69	
NA961	USA, [1]	205–220	6	11	*V. corymbosum*
	USA, [2]	176–201	nd	69	*V. corymbosum*
	Canada, [4]	nd	3	5	*V. marcrocarpon*
	Canada, [5]	nd	nd	32	*V. corymbosum*
	Canada, [7]	nd	nd	63	*V. corymbosum*
NA1040	USA, [1]	180–270	15	11	*V. corymbosum*
	USA, [2]	173–263	nd	69	*V. marcrocarpon*
	Canada, [4]	nd	nd	5	
VCC_B3	USA, [1]	250–275	nd	11	*V. corymbosum*
	Canada, [5]				
VCC_H9	USA, [1]	250–275	13	11	*V. corymbosum*
	USA, [2]	152–279	nd	69	
VCC_I2	USA, [1]	200–275	10	11	*V. corymbosum*
	USA, [2]	205–245	nd	69	
	Canada, [5]	nd	nd	32	
	Canada, [7]	nd	13	63	
VCC_I8	USA, [1]	120–140	4	11	*V. corymbosum*
	USA, [2]	97–149	nd	69	
	Canada, [5]	nd	nd	32	
VCC_J1	USA, [1]	225–275	22	11	*V. corymbosum*
	USA, [2]	213–215	nd	69	
	Canada, [5]				
VCC_J3	USA, [1]	150–200	8	11	*V. corymbosum*
	USA, [2]	115–166	Nd	69	
	Canada, [5]	nd	nd	32	
VCC_J5	USA, [1]	250–300	12	11	*V. corymbosum*
	USA, [2]	250–303	nd	69	
VCC_J9	USA, [1]	100–200	10	11	*V. corymbosum*
	USA, [2]	215–250	nd	69	
	Canada, [5]	nd	nd	32	
VCC_K4	USA, [1]	150–300	14	11	*V. corymbosum*
	USA, [2]	169–300	Nd	69	
	Canada, [5]	nd	nd	32	

(*Continued*)

TABLE 6.3 (*Continued*)
SSR Markers Used for *Vaccinium* spp. Genetic Diversity Studies

SSR Locus	Country, References[a]	Size Range, bp	No. of Alleles	Sample Size	*Vaccinium* spp.
VCC_S10	USA, [1]	200–300	29	11	*V. corymbosum*
	Canada, [5]	nd	nd	32	
	Canada, [7]	nd	40	63	
VCB-C-04624	USA, [6]	94–112	7	130	*V. corymbosum*
GVC-C571	USA, [6]	120–129	4	130	*V. corymbosum*
GVC-V41f03	USA, [6]	177–192	5	130	*V. corymbosum*
GVC-C179	USA, [6]	200–240	15	130	*V. corymbosum*
KAN-262	USA, [6]	237–287	17	130	*V. corymbosum*
GVC-C428	USA, [6]	248–281	9	130	*V. corymbosum*
Pr031818819	USA, [6]	292–333	11	130	*V. corymbosum*
CL2216.Contig1	China, [8]	147–207	10	13	*V. marcrocarpon*
CL2216.Contig4	China, [8]	145–200	9	13	*V. marcrocarpon*
CL2296.Contig2	China, [8]	190–1000	11	13	*V. marcrocarpon*
CL3473.Contig1	China, [8]	112–198	11	13	*V. marcrocarpon*
CL3889.Contig2	China, [8]	155–336	12	13	*V. marcrocarpon*
CL993.Contig1	China, [8]	98–277	7	13	*V. marcrocarpon*
CU15788	China, [8]	127–182	12	13	*V. marcrocarpon*

[a] [1] Boches et al. (2005); [2] Boches et al. (2006); [3] Debnath et al. (2014); [4] An et al. (2015), [5] Bath et al. (2021), [6] Bassil et al. (2020), [7] Tailor et al. (2017), [8] Xu et al. (2021).

17267ESTs from public database NCBI, 54 high-quality EST-SSR primer pairs (with codes Vcxxx) were generated and experimentally tested for blueberries (Liu et al. 2014). A set of 43 polymorphic SSR primers was generated using 454-sequencing data of blueberry transcriptomes from blueberry fruit at different stages of development and flower buds at different stages of cold acclimation (Rowland et al. 2012). Another set of 42 genomic SSR and EST-SSR primers was developed from the whole genome sequencing of a diploid *V. corymbosum* (Bian et al. 2014). A minimal set of five SSR multiplexed primers was developed for effective fingerprinting of blueberry genotypes (Bidani et al. 2017).

For cranberries, the SSR primer sets were created using SOLiD sequencing (Georgi et al. 2012) and Roche 454 pyrosequencing (Zhu et al. 2012) with 21 and 48 polymorphic SSR loci, respectively. Later, the large primer sets were developed from transcriptome data where SSR markers were related to putative defense-related genes, putative conserved orthologous set genes, flavonoid-biosynthetic pathway enzymes and transcription factors (Georgi et al. 2013). Schlautman et al. (2015a) developed in total 697 polymorphic nuclear and EST-SSR markers using publicly available transcriptome data. Recently another seven EST-SSR primers putatively linked to some functions were developed from cranberry transcriptome data (Xu et al. 2021). The SSR primers were used for evaluation of cranberries cultivars (Fajardo et al. 2012) and wild genotypes (Zalapa et al. 2015) genetic diversity. For cranberries, 16 multiplexing panels containing three to four SSR markers (in total 61 SSR markers distributed among the 12 cranberry linkage groups) were developed and validated (Schlautman et al. 2017a). Many of identified SSR markers were used for linkage mapping, in some cases including other molecular markers also (Georgi et al. 2013; Schlautman et al. 2015b, 2017b; Covarrubias-Pazaran et al. 2016; Daverdin et al. 2017; Vorsa and Zalapa 2020).

6.3.2.3 Genomic-Assisted Breeding in Blueberries

In genome-assisted breeding, several strategies are used to identify preferable traits of crops, namely linkage mapping, QTL mapping, and genome-wide association study (GWAS). Linkage mapping is the method to find the molecular markers related to some valuable crop traits. Probably the first reported genetic linkage map for diploid blueberry was created using RAPD markers (Rowland and Levi 1994). The map was constructed using 70 RAPD markers, linked to controlling chilling requirement and cold hardiness genes, comprising 12 linkage groups and covering over 950 cM genetic distance. Another RAPD markers-based linkage map was constructed for tetraploid blueberry (Qu and Hancock 1997). By introduction of SSR markers derived from EST libraries, the preliminary linkage map, covering nine linkage groups and including in total 30 SSRs, was introduced (Brevis et al. 2007) and extended by McCallum et al. (2012). In total, 110 EST-PCR markers which were associated with cold hardiness, chilling requirement, and fruit quality traits (Rowland et al. 2012) were added to previous map of tetraploid blueberry (Rowland and Levi 1994). Later, the linkage map for diploid blueberry population, comprising 265 different type molecular markers (RAPD, EST-PCR, SSR and SNP), was created (Rowland et al. 2014). However, the high-resolution map, including 11,292 SNP markers, comprising 12 linkage groups, and spanning a total genetic distance of 1,953.97 cM with an average markers density of 5.78 cM, was developed recently for autotetraploid blueberry (Cappai et al. 2020). Another high-resolution map, including 17,486 SNP markers and spanning a total genetic distance of 1,539.4 cM, was developed for diploid blueberry (Qi et al. 2021).

For the blueberries breeding, the target traits are chilling requirement, diseases resistance, yield, fruit quality, and adaption to environment traits like cold hardiness, self-fruitfulness, and tolerance to soil pH (Edger et al. 2022). The first QTLs for blueberries were discovered for environment adoption of blueberries, namely cold hardiness and chilling requirement (Rowland et al. 2014) (Table 6.4). Significant variation of many fruit quality traits was evaluated for tetraploid (Hancock et al. 2018) and diploid blueberry populations (Rowland et al. 2020). Ferrão et al. (2018) by GWAS analysis found out 15 significant SNPs associated with five different traits: fruit size (7 SNPs), scar diameter (2), soluble solids content (3), pH (1), and flower bud density (2). Cappai et al. (2020) found five QTLs related to the traits suitable for blueberry machinery harvesting: fruit detachment force, firmness of fruits, and firmness retention (Table 6.4). It was determined that the amount of metabolites varies within and between ploidy groups by assessing variability for bioactive metabolites (anthocyanins, phenolicacids, flavanols, and flavonols) and fruit quality traits (pH, total soluble solid, titratable acidity, and fruit weight) in different ploidy level blueberry accessions. Multivariate analysis of the traits led to the separation of diploid, tetraploid, and hexaploid accessions (Mengist et al. 2020). In total 18 QTLs, associated with chilling requirement, cold hardiness, and fruit quality traits of color, scar size, and firmness, were found in a diploid blueberry mapping population (Qi et al. 2021). Caffeic acid *O*-methyltransferases (COMTs) were found to play an important role for higher fruit firmness and longer shelf life of blueberries by investigating the *COMT* family genes during the development of blueberry fruits (Liu et al. 2021). For the improvement of fruit flavor, GWAS analysis was conducted to identify loci for flavor-related volatile organic compounds (VOS) expression (Ferrão et al. 2020). In total, 11 out of 17 compound associations were found to be significant (Table 6.4, Ferrão et al. 2020).

6.3.2.4 Genomic-Assisted Breeding in Cranberries

For cranberry breeding, the target traits are like blueberry, namely diseases resistance, fruit quality, and adaption to environment traits. Additionally, not only yield amount, but also the stability of yield is an important trait, as many cranberry cultivars bear biennially (Vorsa and Zalapa 2020). The first linkage map for cranberries was comprising 14 linkage groups and covering 879.9 cM genetic distance with an estimated coverage of 82.2% (Georgi et al. 2013). The map was constructed with 129 SSRs, including 33 blueberry-derived and 96 cranberry-derived markers, and nine SCAR

TABLE 6.4
Quantitative Trait Loci Detected in Blueberries

Trait Category	Trait	Ploidy of Blueberry	References	Main QTL Linkage Group	Main QTL Name
Environment adoption	Cold hardiness	2	Rowland et al. (2014)	4	CH2009
		2	Rowland et al. (2014)	4	CH2010
		2	Qi et al. (2021)	5	ColdHard
		2	Qi et al. (2021)	10	ColdHard
	Chilling requirement	2	Rowland et al. (2014)	6	CR2012
		2	Rowland et al. (2014)	6	CR2013a
		2	Rowland et al. (2014)	8	CR2013b
		2	Qi et al. (2021)	5	ChillReq
Fruit quality	Color	2	Qi et al. (2021)	2, 4, 7, 10, 12	Color
	Scar	2	Qi et al. (2021)	12	Scar
	Firmness	2	Qi et al. (2021)	2	Firm
		4	Cappai et al. (2020)	7	Firmness 2018
		4	Cappai et al. (2020)	7	Firmness 2019
	Firmness retention	4	Cappai et al. (2020)	8	FR
	Flavor	2	Qi et al. (2021)	8	Flavor
	Weight	2	Qi et al. (2021)	1	Weight
	Diameter	2	Qi et al. (2021)	2, 5, 8, 12	Dia
	Fruit detachment force (FDF)	4	Cappai et al. (2020)	12	Ripe FDF
		4	Cappai et al. (2020)	8	Delta FDF
	Soluble solids	2	Qi et al. (2021)	nd	SS
Development (growth stage) traits	Shoot elongation	2	Qi et al. (2021)	nd	SE
	Early bloom	2	Qi et al. (2021)	5	EB
	Full bloom	2	Qi et al. (2021)	5, 6	FB
	Petal fall	2	Qi et al. (2021)	5, 6	PF
	Early green	2	Qi et al. (2021)	2, 5, 6	EG
Volatile organic compounds (VOCs)	(*E*)-2-Hexenal	2/4	Ferrão et al. (2020)	6	E2-Hex
	Decanal	2/4	Ferrão et al. (2020)	1, 9	Dec
	Hexanal	2/4	Ferrão et al. (2020)	6, 6, 6	Hexa
	1-Hexanol	2/4	Ferrão et al. (2020)	6	1-Hex
	Geranyl acetone	2/4	Ferrão et al. (2020)	4, 5	G.Ace
	2-Heptanone	2/4	Ferrão et al. (2020)	7, 11, 12	2Hep
	2-Nonanone	2/4	Ferrão et al. (2020)	2, 3	2Non
	2-Undecanone	2/4	Ferrão et al. (2020)	2	2Und
	Linalool	2/4	Ferrão et al. (2020)	3, 4	Lina
	D-limonene	2/4	Ferrão et al. (2020)	3, 7	Dlimo
	Eucalyptol	2/4	Ferrão et al. (2020)	1, 2, 11	Euc

markers. The QTLs on this map were detected for field fruit-rot resistance, fruit weight, titratable acidity, and sound fruit yield (Table 6.5). Another linkage map was also created with SSR markers also, but with a much higher density (Schlautman et al. 2015b). In total, 541 SSR markers were used to cover 12 chromosomes of cranberry with a coverage of 94.5%. The QTLs on this map were identified for fruit yield and quality traits. The use of genotyping by sequencing (GBS) approaches in cranberry allowed to construct high-density linkage maps (Covarrubias-Pazaran et al. 2016; Daverdin et al. 2017; Schlautman et al. 2017b; Diaz-Garcia et al. 2020). These maps additional to SSR markers include thousands of SNPs. Daverdin et al. (2017) identified in total 60 QTLs related to fruit rot (19 QTLs), yield (27), and cranberry weight across four populations. By the search of

TABLE 6.5
Quantitative Trait Loci Detected in Cranberries

Trait Category	Trait	References	Main QTL	
			Linkage Group[a]	**Name: Position**
Disease resistance	Field fruit-rot resistance (FFRR)	Georgi et al. (2013)	V7	FFRR:scf9025
		Georgi et al. (2013)	V11	FFRR:scf6341 and FFRR:scf3072
	Rot rating	Daverdin et al. (2017)	V1, V3, V5, V6, V8	RR-11–13
	Percent rot	Daverdin et al. (2017)	V1, V5, V7, V8, V11	PR-11–13
		Daverdin et al. (2017)	V11	PR14–15
		Daverdin et al. (2017)	V5, V9, V12	PR16
Fruit yield and quality	Fruit weight	Georgi et al. (2013)	V5	FW:2ms2a02
	Mean fruit weight	Georgi et al. (2013)	V8	FW:CA187F
		Schlautman et al. (2015b)	LG6	MFW:scf171621
		Schlautman et al. (2015b)	LG10	MFW:scf35507
		Schlautman et al. (2015b)	LG11	MFW:scf122746
		Schlautman et al. (2015b)	LG11	MFW:scf118608
		Daverdin et al. (2017)	V6, V8, V12	BW14–15
		Daverdin et al. (2017)	V1, V5, V7, V9, V10, V11, V12	BW16
	Sound fruit yield	Georgi et al. (2013)	V8	SFY:scf300f
	Total yield	Schlautman et al. (2015b)	LG4	TY:scf11084
		Schlautman et al. (2015b)	LG6	TY:1trimcontig 344502
		Schlautman et al. (2015b)	LG11	
		Daverdin et al. (2017)	V6	TY: scf204332
		Daverdin et al. (2017)	V1, V2, V3, V6, V8, V9, V10, V11, V12	Y14–15 Y16
	Yield rating	Daverdin et al. (2017)	V2, V3, V7, V8, V10, V11, V12	YR11–13 YR14–15
			V11	
	Biennial bearing index	Schlautman et al. (2015b)	LG11	BBI:scf204332
	Titratable acidity	Georgi et al. (2013)	V3	TA:scf2s
	Proanthocyanidin	Georgi et al. (2013)	V9	Proan:scf262a

[a] The nomenclature of Wisconsin (LG) is analogous to Rutgers (V) as follows: LG1=V9, LG2=V8, LG3=V7, LG4=V2, L5 = V10, LG6=V5, LG7=V3, LG8=V4, LG9=V1, LG10=V12, LG11 = V11, and LG12=V6 (Vorsa and Zalapa 2020).

QTLs related to fruit size and shape, in total 252 QTLs were found in a three-year period (Diaz-Garcia et al. 2018a). The QTLs for total anthocyanin content, titratable acidity, proanthocyanidin content, Brix, and mean fruit weight were found additionally (Diaz-Garcia et al. 2018b). Based on a collection of 323 shared SSR markers, there was a high degree of collinearity between the genetic maps of cranberries and blueberries, suggesting the possibility of QTL and the transferability of markers between species (Schlautman et al. 2018).

6.3.3 Characterization of Allelic Diversity in Currants and Gooseberries

The utilization of the available currant and gooseberry plant material in further research and breeding activities depends on characterization, genetic diversity evaluation, and plant material identification. Germplasm identification is also essential for international cooperation and the successful and safe exchange of plant material. DNA molecular markers are recognized as the most appropriate tool (Nybom and Lácis 2021) for characterizing allelic diversity. Therefore, a diversity of molecular marker systems has been implemented for *Ribes* studies: AFLP, ISSR, RAPD, SNP, SCAR, and SSR or microsatellites.

6.3.3.1 Characterization of Currants and Gooseberries Allelic Diversity by Molecular Markers

Considering the limited genomic information, currant and gooseberry allelic diversity was initiated using molecular markers that do not require preliminary sequence information. It is interesting that in the study of *Ribes* plants, several marker systems are used simultaneously in different combinations to obtain the most accurate understanding of diversity (Table 6.6). All the most important grown *Ribes* species are included in the research. At the same time, a wide variety of markers were used with a relatively small coverage between different studies and a limited number of analyzed samples in an individual study. Only a few of these markers have been applied in at least two studies.

TABLE 6.6
Inter-Simple Sequence Repeat (ISSR), Random Amplified Polymorphic DNA (RAPD), and Sequence-Characterized Amplified Region (SCAR) Application in the Fingerprinting of *Ribes* Species Germplasm

Marker Locus	Marker Type	Country, References	Sample Size	Plant Material
UBC808	ISSR	Italy, Gentini et al. (2015)	88	*Ribes* sp.[a]
UBC810	ISSR	Italy, Gentini et al. (2015)	88	*Ribes* sp.[a]
UBC811	ISSR	Italy, Gentini et al. (2015)	88	*Ribes* sp.[a]
UBC812	ISSR	Poland, Korbin et al. (2002)	24	*R. nigrum*
UBC816	ISSR	UK, Lanham and Brennan (1999)	12	*R. grossularia*
		Poland, Korbin et al. (2002)	24	*R. nigrum*
UBC821	ISSR	Poland, Wójcik et al. (2021)	54	*R. grossularia*
UBC822	ISSR	Poland, Wójcik et al. (2021)	54	*R. grossularia*
UBC823	ISSR	Poland, Wójcik et al. (2021)	54	*R. grossularia*
UBC825	ISSR	Italy, Gentini et al. (2015)	88	*Ribes* sp.[a]
		Poland, Wójcik et al. (2021)	54	*R. grossularia*
UBC826	ISSR	Italy, Gentini et al. (2015)	88	*Ribes* sp.[a]
UBC830	ISSR	Poland, Wójcik et al. (2021)	54	*R. grossularia*
UBC834	ISSR	UK, Lanham and Brennan (1999)	12	*R. grossularia*
		Poland, Wójcik et al. (2021)	54	*R. grossularia*
UBC840	ISSR	Poland, Wójcik et al. (2021)	54	*R. grossularia*
UBC842	ISSR	UK, Lanham and Brennan (1999)	12	*R. grossularia*
UBC845	ISSR	Poland, Korbin et al. (2002)	24	*R. nigrum*
UBC847	ISSR	UK, Lanham and Brennan (1999)	12	*R. grossularia*
UBC848	ISSR	Poland, Wójcik et al. (2021)	54	*R. grossularia*
UBC849	ISSR	Poland, Wójcik et al. (2021)	54	*R. grossularia*
UBC850	ISSR	UK, Lanham and Brennan (1999)	12	*R. grossularia*
		Poland, Korbin et al. (2002)	24	*R. nigrum*
UBC853	ISSR	Poland, Wójcik et al. (2021)	54	*R. grossularia*
UBC861	ISSR	UK, Lanham and Brennan (1999)	12	*R. grossularia*
UBC862	ISSR	UK, Lanham and Brennan (1999)	12	*R. grossularia*
UBC888	ISSR	UK, Lanham and Brennan (1999)	12	*R. grossularia*

(*Continued*)

TABLE 6.6 (*Continued*)
Inter-Simple Sequence Repeat (ISSR), Random Amplified Polymorphic DNA (RAPD), and Sequence-Characterized Amplified Region (SCAR) Application in the Fingerprinting of *Ribes* Species Germplasm

Marker Locus	Marker Type	Country, References	Sample Size	Plant Material
UBC891	ISSR	UK, Lanham and Brennan (1999)	12	*R. grossularia*
UCB816	ISSR	Poland, Korbin et al. (2002)	12	*R. grossularia*
			10	*R. rubrum*
UCB844	ISSR	Poland, Korbin et al. (2002)	12	*R. uva-crispa*
UCB845	ISSR	Poland, Korbin et al. (2002)	12	*R. uva-crispa*
			10	*R. rubrum*
UCB848	ISSR	Poland, Korbin et al. (2002)	12	*R. uva-crispa*
UCB850	ISSR	Poland, Korbin et al. (2002)	10	*R. rubrum*
UCB888	ISSR	Poland, Korbin et al. (2002)	10	*R. rubrum*
UCB891	ISSR	Poland, Korbin et al. (2002)	10	*R. rubrum*
OP marker[b]	RAPD	Russia, Pikunova et al. (2012)	47	*Ribes* sp.[c]
OPA-02	RAPD	Poland, Korbin et al. (2002)	24	*R. nigrum*
			10	*R. rubrum*
OPA-04	RAPD	UK, Lanham and Brennan (1999)	12	*R. grossularia*
OPA-05	RAPD	UK, Lanham and Brennan (1999)	12	*R. grossularia*
OPA-07	RAPD	UK, Lanham and Brennan (1999)	12	*R. grossularia*
OPA-08	RAPD	UK, Lanham and Brennan (1999)	12	*R. grossularia*
OPA-10	RAPD	Poland, Korbin et al. (2002)	24	*R. nigrum*
			10	*R. rubrum*
OPA-18	RAPD	UK, Lanham and Brennan (1999)	12	*R. grossularia*
OPA-19	RAPD	UK, Lanham and Brennan (1999)	12	*R. grossularia*
OPB-01	RAPD	UK, Lanham and Brennan (1999)	12	*R. grossularia*
OPB-02	RAPD	UK, Lanham and Brennan (1999)	12	*R. grossularia*
OPB-03	RAPD	UK, Lanham and Brennan (1999)	12	*R. grossularia*
OPB-04	RAPD	UK, Lanham and Brennan (1999)	12	*R. grossularia*
OPB-08	RAPD	Poland, Korbin et al. (2002)	24	*R. nigrum*
OPB-11	RAPD	Poland, Korbin et al. (2002)	24	*R. nigrum*
			24	*R. nigrum*
OPC-02	RAPD	Poland, Korbin et al. (2002)	12	*R. uva-crispa*
			10	*R. rubrum*
OPC-04	RAPD	UK, Lanham and Brennan (1999)	12	*R. grossularia*
OPC-05	RAPD	UK, Lanham and Brennan (1999)	12	*R. grossularia*
OPC-06	RAPD	UK, Lanham and Brennan (1999)	12	*R. grossularia*
		Poland, Korbin et al. (2002)	10	*R. rubrum*
OPC-07	RAPD	UK, Lanham and Brennan (1999)	12	*R. grossularia*
		Poland, Korbin et al. (2002)	12	*R. uva-crispa*
OPC-08	RAPD	UK, Lanham and Brennan (1999)	12	*R. grossularia*
		Poland, Korbin et al. (2002)	12	*R. uva-crispa*
OPC-09	RAPD	UK, Lanham and Brennan (1999)	12	*R. grossularia*
OPC-10	RAPD	UK, Lanham and Brennan (1999)	12	*R. grossularia*
OPC-11	RAPD	UK, Lanham and Brennan (1999)	12	*R. grossularia*
		Poland, Korbin et al. (2002)	12	*R. uva-crispa*
OPC-12	RAPD	UK, Lanham and Brennan (1999)	12	*R. grossularia*
OPC-13	RAPD	UK, Lanham and Brennan (1999)	12	*R. grossularia*
OPG-16	RAPD	Romania, Butiuc-Keul et al. (2015)	4	*R. nigrum*

(Continued)

TABLE 6.6 (*Continued*)
Inter-Simple Sequence Repeat (ISSR), Random Amplified Polymorphic DNA (RAPD), and Sequence-Characterized Amplified Region (SCAR) Application in the Fingerprinting of *Ribes* Species Germplasm

Marker Locus	Marker Type	Country, References	Sample Size	Plant Material
OPK-06	RAPD	Romania, Butiuc-Keul et al. (2015)	4	*R. nigrum*
OPK-20	RAPD	Romania, Butiuc-Keul et al. (2015)	4	*R. nigrum*
OPP-17	RAPD	Romania, Butiuc-Keul et al. (2015)	4	*R. nigrum*
OPP-19	RAPD	Romania, Butiuc-Keul et al. (2015)	4	*R. nigrum*
Y11	RAPD	China, Huo et al. (2013)	37	*R. nigrum*
Y22	RAPD	China, Huo et al. (2013)	37	*R. nigrum*
Y30	RAPD	China, Huo et al. (2013)	37	*R. nigrum*
Y34	RAPD	China, Huo et al. (2013)	37	*R. nigrum*
Y41	RAPD	China, Huo et al. (2013)	37	*R. nigrum*
Y45	RAPD	China, Huo et al. (2013)	37	*R. nigrum*
Y48	RAPD	China, Huo et al. (2013)	37	*R. nigrum*
Y59	RAPD	China, Huo et al. (2013)	37	*R. nigrum*
A1	SCAR	Romania, Butiuc-Keul et al. (2015)	4	*R. nigrum*
D2	SCAR	Romania, Butiuc-Keul et al. (2015)	4	*R. nigrum*
L	SCAR	Romania, Butiuc-Keul et al. (2015)	4	*R. nigrum*

[a] *R. multiflorum* subsp. *sandalioticum*, *R. sardoum*.
[b] OPA, OPD, OPE, OPH, OPK, and OPN primer series.
[c] *R. grossularia L., R. × nidigrolaria Rud. Bauer & A. Baue, R. nigrum L. var. sibiricum W. Wolf, R. nigrum L. var. europaeum Jancz.. R. nigrum L. var. scandinavicum, R. sanguineum Pursh, R. multiflorum Kit., R. petraeum Wulf. var. atropurpureum (C. A. May.), R. rubrum L., R. sativum Syme, R. sativum Symevar. macrocarpum Jancz., R. warszewiczii Jancz., R. palczewskii (Jancz.) Pojark., R. aureum Pursh.*

In addition to the types of markers already mentioned in Table 6.1, AFLP markers have also been used in *Ribes* species germplasm research using the restrictases EcoRI (Lanham and Brennan 1999; De Mattia et al. 2008; Ipek et al. 2010), MseI (Lanham and Brennan 1999; De Mattia et al. 2008; Ipek et al. 2010; Wójcik et al. 2021), and PstI (Wójcik et al. 2021). All types of markers used were able to ensure the identification of samples of different species of *Ribes*, the analysis of mutual relationships, and the separation of species based on the appropriate level of polymorphism of the markers. In some cases, AFLP markers provided species resolution, but their wild spontaneous hybrid populations could not be resolved (Ipek et al. 2010).

6.3.3.2 Characterization of Currants and Gooseberries Allelic Diversity by Microsatellite or SSR Markers

The group of markers, which, like for other plant species, is the most widely used in the study of currants and gooseberries, is SSRs or microsatellites because they are abundant in most genomes, uniformly distributed, multi-allelic and codominant, as well with a high rate of reproducibility (Nadeem et al. 2018). This choice is also determined by the fact that a small amount of DNA is needed for amplification and that SSRs ensure a high PIC. In the germplasm analysis of different *Ribes* species, mapping studies mainly involve two sets of SSR markers: RJL-1 … RJL-11 – developed in *R. nigrum* (Brennan et al. 2002) and the markers created on the blackcurrant mapping population (Brennan et al. 2008, primer sequences can be found https://fruitbreeding.hutton.ac.uk/RibesGenomicsSSRs.asp). Only in some cases have other microsatellite markers been developed;

for example, the 'MTT' markers group developed within the project "Multinational Approach for Conserving the European Genetic Resources of Currants and Gooseberry" (Ribesco) (Antonius et al. 2012), which were later used in other studies (Droz et al. 2019; Leišová-Svobodová and Kissling 2021; Nordlander et al. 2022). A set of 17 'Ruc' series primers were additionally developed for gooseberry research (Vidyagina et al. 2021). An interesting and atypical case of interspecies transfer of SSR markers was described in a study conducted in Romania when markers developed by *Prunus* were used for the genetic fingerprinting of blackcurrants (Butiuc-Keul et al. 2015). In any case, the interspecies transfer of *Ribes* SSR markers has been very effective, thus overcoming one of the disadvantages of these markers – the unique nucleotide sequences flanking the repeat regions must be determined (Guichoux et al. 2011). It also indicates possible sequence conservatism between species in genera. A detailed summary of SSR marker application in *Ribes* germplasm fingerprinting is shown in Table 6.7, describing the applied markers, the detected allele ranges, and numbers, as well as the number of study samples, which can significantly affect the polymorphism level of the markers and the species in which they are used.

TABLE 6.7
Microsatellite (SSR) Application in the Fingerprinting of *Ribes* Species Germplasm

SSR Locus	Country, References[a]	Size Range, bp	No. of Alleles	Sample Size	Plant Material
816[b]	UK [21]	nd	39	16	*R. rubrum*, *R. petraeum*, *R. multiflorum*, *R. sativum*
888[b]	UK [21]	nd	60	16	*R. rubrum*, *R. petraeum*, *R. multiflorum*, *R. sativum*
891[b]	UK [21]	nd	85	16	*R. rubrum*, *R. petraeum*, *R. multiflorum*, *R. sativum*
BGT 23	Romania [12]	75–200	2	4	*R. nigrum*
ccmp2[c]	Latvia [10]	219–278	3	145	*Ribes* species[e]
ccmp3[c]	Latvia [10]	108–112	5	145	*Ribes* species[e]
ccmp6[c]	Italy [5]	115–116	2	16	*R. petraeum*, *R. uva-crispa*
ccmp7[c]	Latvia [10]	134–143	9	145	*Ribes* species[e]
ccmp10	Latvia [10]	100–109	8	145	*Ribes* species[e]
	Italy [5]	103–106	4	16	*R. petraeum*, *R. uva-crispa*
ccSSR-1[c]	Latvia [10]	160–171	3	145	*Ribes* species[e]
ccSSR-3[c]	Latvia [10]	103–107	5	145	*Ribes* species[e]
ccSSR-6[c]	Latvia [10]	229–301	3	145	*Ribes* species[e]
ccSSR-7[c]	Latvia [10]	320–334	7	145	*Ribes* species[e]
ccSSR-8[c]	Latvia [10]	247–248	2	145	*Ribes* species[e]
ccSSR-9[c]	Latvia [10]	163–177	2	145	*Ribes* species[e]
ccSSR-10[c]	Latvia [10]	152–161	6	145	*Ribes* species[e]
ccSSR-11[c]	Latvia [10]	158–362	9	145	*Ribes* species[e]
ccSSR-12[c]	Latvia [10]	196–232	8	145	*Ribes* species[e]
ccSSR-13[c]	Latvia [10]	313–314	10	145	*Ribes* species[e]
ccSSr-14[c]	Latvia [10]	202–211	2	145	*Ribes* species[e]
ccSSR-15[c]	Latvia [10]	265–270	9	145	*Ribes* species[e]
ccSSR-16[c]	Latvia [10]	189–212	3	145	*Ribes* species[e]
ccSSR-19[c]	Latvia [10]	354–355	8	145	*Ribes* species[e]
Cra-489	Russia [15]	nd	nd	mp	*R. rubrum*
	Russia [13]	nd	2.34[d]	127	*R. rubrum*, *R. nigrum*, *R. uva-crispa*
Cra-531	Russia [15]	nd	nd	mp	*R. rubrum*
	Russia [13]	nd	2.34[d]	127	*R. rubrum*, *R. nigrum*, *R. uva-crispa*

(*Continued*)

TABLE 6.7 (*Continued*)
Microsatellite (SSR) Application in the Fingerprinting of *Ribes* Species Germplasm

SSR Locus	Country, References[a]	Size Range, bp	No. of Alleles	Sample Size	Plant Material
e1–201	USA [23]	nd	2.34[d]	102	*R. echinellum*
e1-O01	Belarus [1]	144–166	8	60	*R. nigrum*
	Russia [14]	138–149	6	27	*R. nigrum*
	Italy [4]	136–153	9	41	*R. nigrum, R. rubrum, R. grossularia, R. missouriense, R. × nidigrolaria*
	Turkey [19]	nd	nd	10	*R. alpinum, R. uva-crispa, R. bieberstenii*
	Russia [15]	nd	nd	mp	*R. rubrum*
	Switzerland [18]	nd	nd	286	*Ribes* sp.
	Italy [7]	nd	nd	17	*R. sardoum*
	USA [23]	nd	2.34[d]	102	*R. echinellum*
	Russia [13]	nd	2.34[d]	127	*R. rubrum, R. nigrum, R. uva-crispa*
	Italy [7]	141–227	12	88	*R. multiflorum* subsp. *sandalioticum, R. sardoum*
	Belarus [2]	142–166	9	65	*R. nigrum, R. uva-crispa, R. alpinum*
	Sweden [17]	131–160	7	242	*R. uva-crispa*
e1-O20	Italy [8]	192–230	4	91	*R. nigrum, R. rubrum*
	Czech Republic [3]	198–230	9	118	*R. rubrum, R. petraeum, R. spicatum and R. multiflorum*
e1-O21	Russia [14]	288–294	3	27	*R. nigrum*
	Italy [4]	286–317	11	41	*R. nigrum, R. rubrum, R. grossularia, R. missouriense, R. × nidigrolaria*
	Russia [15]	nd	nd	mp	*R. rubrum*
	Switzerland [18]	nd	nd	286	*Ribes* sp.
	USA [23]	nd	2.34[d]	102	*R. echinellum*
	Russia [13]	nd	2.34[d]	127	*R. rubrum, R. nigrum, R. uva-crispa*
	Italy [8]	286–306	4	91	*R. nigrum, R. rubrum*
	Sweden [17]	290–318	11	242	*R. uva-crispa*
	Czech Republic [3]	291–313	11	118	*R. rubrum, R. petraeum, R. spicatum, R. multiflorum*
e2-L15	Turkey [19]	nd	nd	10	*R. alpinum, R. uva-crispa, R. bieberstenii*
e3-B02	Russia [14]	188–194	4	27	*R. nigrum*
	Russia [15]	nd	nd	mp	*R. rubrum*
	Switzerland [18]	nd	nd	286	*Ribes* sp.
	USA [23]	nd	2.34[d]	102	*R. echinellum*
	Russia [13]	nd	2.34[d]	127	*R. rubrum, R. nigrum, R. uva-crispa*
	Belarus [1]	161–183	4	60	*R. nigrum*
	Belarus [2]	151–183	6	65	*R. nigrum, R. uva-crispa, R. alpinum*
	Italy [8]	145–176	7	91	*R. nigrum, R. rubrum*
	Czech Republic [3]	160–170	5	118	*R. rubrum, R. petraeum, R. spicatum, R. multiflorum*

(*Continued*)

TABLE 6.7 (*Continued*)
Microsatellite (SSR) Application in the Fingerprinting of *Ribes* Species Germplasm

SSR Locus	Country, References[a]	Size Range, bp	No. of Alleles	Sample Size	Plant Material
e3-M04	Latvia [9]	254–335	21	110	*R. nigrum*
	Italy [4]	293–334	15	41	*R. nigrum, R. rubrum, R. grossularia, R. missouriense, R. × nidigrolaria*
	Turkey [19]	nd	nd	10	*R. alpinum, R. uva-crispa, R. biebersteinii*
	USA [23]	nd	2.34[d]	102	*R. echinellum*
	Italy [8]	294–340	9	91	*R. nigrum, R. rubrum*
	Czech Republic [3]	252–322	6	118	*R. rubrum, R. petraeum, R. spicatum, R. multiflorum*
e3-M04-b	Czech Republic [3]	300–338	16	118	*R. rubrum, R. petraeum, R. spicatum, R. multiflorum*
e4-D03	Russia [14]	197–224	8	27	*R. nigrum*
	Italy [4]	165–211	13	41	*R. nigrum, R. rubrum, R. grossularia, R. missouriense, R. × nidigrolaria*
	Switzerland [18]	nd	nd	286	*Ribes* sp.
	Italy [8]	164–242	11	91	*R. nigrum, R. rubrum*
	Sweden [17]	160–238	9	242	*R. uva-crispa*
	Czech Republic [3]	168–210	16	118	*R. rubrum, R. petraeum, R. spicatum, R. multiflorum*
g1-A01	USA [23]	nd	2.34[d]	102	*R. echinellum*
	Russia [15]	nd	nd	mp	*R. rubrum*
	Italy [4]	204–243	12	41	*R. nigrum, R. rubrum, R. grossularia, R. missouriense, R. × nidigrolaria*
	Switzerland [18]	nd	nd	286	*Ribes* sp.
	Italy [7]	nd	nd	17	*Ribes sardoum*
	Russia [13]	nd	2.34[d]	127	*R. rubrum, R. nigrum, R. uva-crispa.*
	Italy [6]	203–227	5	88	*R. multiflorum* subsp. *sandalioticum, R. sardoum*
	Belarus [1]	209–213	3	60	*R. nigrum*
	Belarus [2]	209–241	5	65	*R. nigrum, R. uva-crispa, R. alpinum*
	Sweden [17]	205–254	14	242	*R. uva-crispa*
g1-B02	Russia [14]	203–207	3	27	*R. nigrum*
	Latvia [9]	175–193	6	110	*R. nigrum*
	Turkey [19]	nd	nd	10	*R. alpinum, R. uva-crispa, R. biebersteinii*
	Switzerland [18]	nd	nd	286	*Ribes* sp.
g1-D11	Italy [4]	236–246	nd	41	*R. nigrum, R. rubrum, R. grossularia, R. missouriense, R. × nidigrolaria*

(*Continued*)

TABLE 6.7 (*Continued*)
Microsatellite (SSR) Application in the Fingerprinting of *Ribes* Species Germplasm

SSR Locus	Country, References[a]	Size Range, bp	No. of Alleles	Sample Size	Plant Material
g1-E03	Russia [14]	232–270	7	27	*R. nigrum*
	Italy [4]	224–251	9	41	*R. nigrum, R. rubrum, R. grossularia, R. missouriense, R. × nidigrolaria*
	Belarus [1]	233–270	8	60	*R. nigrum*
	Belarus [2]	207–270	10	65	*R. nigrum, R. uva-crispa, R. alpinum*
g1-F04	Turkey [19]	nd	nd	10	*R. alpinum, R. uva-crispa, R. bieberstenii*
g1-G11	Latvia [9]	214–244	12	110	*R. nigrum*
g1-H09	Turkey [19]	nd	nd	10	*R. alpinum, R. uva-crispa, R. bieberstenii*
g1-I02	Latvia [9]	183–212	8	110	*R. nigrum*
	Italy [8]	120–130	3	91	*R. nigrum, R. rubrum*
	Czech Republic [3]	123–203	6	118	*R. rubrum, R. petraeum, R. spicatum and R. multiflorum*
g1-K04	Russia [14]	285–298	4	27	*R. nigrum*
	Latvia [9]	277–298	12	110	*R. nigrum*
	Italy [4]	267–299	10	41	*R. nigrum, R. rubrum, R. grossularia, R. missouriense, R. × nidigrolaria*
	Russia [15]	nd	nd	mp	*R. rubrum*
	Switzerland [18]	nd	nd	286	*Ribes* sp.
	Italy [7]	nd	nd	17	*R. sardoum*
	Russia [13]	nd	2.34[d]	127	*R. rubrum, R. nigrum, R. uva-crispa*
	Italy [6]	276–296	8	88	*Ribes multiflorum* subsp. *sandalioticum, R. sardoum*
	Belarus [1]	284–300	6	60	*R. nigrum*
	Belarus [2]	284–300	8	65	*R. nigrum, R. uva-crispa, R. alpinum*
	Italy [8]	278–300	4	91	*R. nigrum, R. rubrum*
	Sweden [17]	277–298	10	242	*R. uva-crispa*
	Czech Republic [3]	282–302	9	118	*R. rubrum, R. petraeum, R. spicatum, R. multiflorum*
g1-L12	Russia [15]	nd	nd	mp	*R. rubrum*
	Russia [13]	nd	2.34[d]	127	*R. rubrum, R. nigrum, R. uva-crispa*
g1-M07	Russia [14]	200–222	7	27	*R. nigrum*
	Italy [4]	201–223	9	41	*R. nigrum, R. rubrum, R. grossularia, R. missouriense, R. × nidigrolaria*
	Russia [15]	nd	nd	mp	*R. rubrum*
	Switzerland [18]	nd	nd	286	*Ribes* sp.
	USA [23]	nd	2.34[d]	102	*R. echinellum*
	USA [23]	nd	2.34[d]	102	*R. echinellum*

(*Continued*)

TABLE 6.7 (*Continued*)
Microsatellite (SSR) Application in the Fingerprinting of *Ribes* Species Germplasm

SSR Locus	Country, References[a]	Size Range, bp	No. of Alleles	Sample Size	Plant Material
	Russia [13]	nd	2.34[d]	127	*R. rubrum, R. nigrum, R. uva-crispa*
	Belarus [1]	200–230	11	60	*R. nigrum*
	Belarus [2]	200–230	11	65	*R. nigrum, R. uva-crispa, R. alpinum*
	Sweden [17]	199–232	15	242	*R. uva-crispa*
g1-P05	Turkey [19]	nd	nd	10	*R. alpinum, R. uva-crispa, R. bieberstenii*
g2-B20	Russia [14]	147–185	5	27	*R. nigrum*
	Latvia [9]	124–169	10	110	*R. nigrum*
	Russia [15]	nd	nd	mp	*R. rubrum*
	Switzerland [18]	nd	nd	286	*Ribes* sp.
	Sweden [17]	139–191	3	242	*R. uva-crispa*
g2-G12	Russia [14]	190–215	7	27	*R. nigrum*
	Latvia [9]	140–192	14	110	*R. nigrum*
	Turkey [19]	nd	nd	10	*R. alpinum, R. uva-crispa, R. bieberstenii*
	Russia [15]	nd	nd	mp	*R. rubrum*
	Switzerland [18]	nd	nd	286	*Ribes* sp.
	Russia [13]	nd	2.34[d]	127	*R. rubrum, R. nigrum, R. uva-crispa*
	Belarus [1]	167–191	11	60	*R. nigrum*
	Belarus [2]	191–197	11	65	*R. nigrum, R. uva-crispa, R. alpinum*
	Italy [8]	168–196	7	91	*R. nigrum, R. rubrum*
	Sweden [17]	164–194	2	242	*R. uva-crispa*
	Czech Republic [3]	174–194	11	118	*R. rubrum, R. petraeum, R. spicatum and R. multiflorum*
g2-H21	Russia [14]	265–273	3	27	*R. nigrum*
	Switzerland [18]	nd	nd	286	*Ribes* sp.
	Russia [13]	nd	2.34[d]	127	*R. rubrum, R. nigrum, R. uva-crispa*
	Sweden [17]	238–272	11	242	*R. uva-crispa*
g2-J08	Russia [14]	158–164	4	27	*R. nigrum*
	Latvia [9]	151–172	9	110	*R. nigrum*
	Italy [4]	142–184	9	41	*R. nigrum, R. rubrum, R. grossularia, R. missouriense, R. × nidigrolaria*
	Russia [15]	nd	nd	mp	*R. rubrum*
	USA [23]	nd	2.34[d]	102	*R. echinellum*
	Russia [13]	nd	2.34[d]	127	*R. rubrum, R. nigrum, R. uva-crispa*
	Italy [8]	140–178	3	91	*R. nigrum, R. rubrum*
	Sweden [17]	140–181	11	242	*R. uva-crispa*
	Czech Republic [3]	142–182	13	118	*R. rubrum, R. petraeum, R. spicatum and R. multiflorum*
g2-J11	Turkey [19]	nd	nd	10	*R. alpinum, R. uva-crispa, R. bieberstenii*

(*Continued*)

TABLE 6.7 (*Continued*)
Microsatellite (SSR) Application in the Fingerprinting of *Ribes* Species Germplasm

SSR Locus	Country, References[a]	Size Range, bp	No. of Alleles	Sample Size	Plant Material
g2-L17	Russia [14]	150–171	6	27	*R. nigrum*
	Italy [4]	122–165	10	41	*R. nigrum, R. rubrum, R. grossularia, R. missouriense, R.* × *nidigrolaria*
	Switzerland [18]	nd	nd	286	*Ribes* sp.
	Italy [7]	nd	nd	17	*R. sardoum*
	Russia [13]	nd	2.34[d]	127	*R. rubrum, R. nigrum, R. uva-crispa*
	Italy [6]	105–163	13	88	*R. multiflorum* subsp. *sandalioticum, R. sardoum*
	Sweden [17]	114–166	19	242	*R. uva-crispa*
g3-A17	Latvia [9]	118–177	15	110	*R. nigrum*
	Turkey [19]	nd	nd	10	*R. alpinum, R. uva-crispa, R. biebersteinii*
gr2-J05	Russia [15]	nd	nd	mp	*R. rubrum*
	Switzerland [18]	nd	nd	286	*Ribes* sp.
	Russia [13]	nd	2.34[d]	127	*R. rubrum, R. nigrum, R. uva-crispa*
	Italy [8]	158–185	4	91	*R. nigrum, R. rubrum*
	Czech Republic [3]	167–177	5	118	*R. rubrum, R. petraeum, R. spicatum, R. multiflorum*
	Sweden [17]	160–186	8	242	*R. uva-crispa*
LfMYB-1	Russia [16]	na	nd	22	*R. uva-crispa*
LfMYB-2	Russia [16]	na	nd	22	*R. uva-crispa*
MTT-32	North Europe [11]	250–320	nd	846[g]	*R. rubrum group*
		280–309	nd	846[g]	*R. uva-crispa*
	Czech Republic [3]	294–320	9	118	*R. rubrum, R. petraeum, R. spicatum, R. multiflorum*
MTT-32b	Czech Republic [3]	304–352	16	118	*R. rubrum, R. petraeum, R. spicatum, R. multiflorum*
MTT-5	Sweden [17]	144–244	30	242	*R. uva-crispa*
	Switzerland [18]	nd	nd	286	*Ribes* sp.
	North Europe [11]	156–194	nd	846[g]	*R. rubrum group*
		145–261	nd	846[g]	*R. uva-crispa*
	Czech Republic [3]	137–177	12	118	*R. rubrum, R. petraeum, R. spicatum, R. multiflorum*
MTT-7	North Europe [11]	151–193	nd	846[g]	*R. uva-crispa*
MTT-9	North Europe [11]	308–322	nd	846[g]	*R. rubrum group*
	Czech Republic [3]	303–319	8	118	*R. rubrum, R. petraeum, R. spicatum, R. multiflorum*
NTCP9[c]	Latvia [10]	237–286	10	145	*Ribes* species[e]
pchgm 12	Romania [12]	40–200	4	4	*R. nigrum*
pchgm 24	Romania [12]	75–250	3	4	*R. nigrum*
RJL-1	UK [20]	167	3	29	*Ribes* species[f]
	USA [22]	nd	1	12	*R. nigrum*
	Czech Republic [3]	109–135	4	118	*R. rubrum, R. petraeum, R. spicatum, R. multiflorum*

(*Continued*)

TABLE 6.7 (*Continued*)
Microsatellite (SSR) Application in the Fingerprinting of *Ribes* Species Germplasm

SSR Locus	Country, References[a]	Size Range, bp	No. of Alleles	Sample Size	Plant Material
RJL-1b	Czech Republic [3]	123–165	5	118	*R. rubrum, R. petraeum, R. spicatum, R. multiflorum*
RJL-1c	Czech Republic [3]	137–197	5	118	*R. rubrum, R. petraeum, R. spicatum, R. multiflorum*
RJL-2	UK [20]	207	16	29	*Ribes* species[f]
	USA [22]	nd	nd	12	*R. nigrum*
	Latvia [9]	192–219	8	110	*R. nigrum*
	Russia [16]	222–254	8	22	*R. uva-crispa*
	North Europe [11]	197–238	nd	846[g]	*R. nigrum*
		192–252	nd	846[g]	*R. rubrum group*
		205–279	nd	846[g]	*R. uva-crispa*
	Czech Republic [3]	205–227	10	118	*R. rubrum, R. petraeum, R. spicatum, R. multiflorum*
RJL-3	UK [20]	136	6	29	*Ribes* species[f]
	USA [22]	nd	1	12	*R. nigrum*
	Latvia [9]	132–172	13	110	*R. nigrum*
	Czech Republic [3]	124–172	6	118	*R. rubrum, R. petraeum, R. spicatum, R. multiflorum*
RJL-4	UK [20]	198	4	29	*Ribes* species[f]
	Latvia [9]	167–220	13	110	*R. nigrum*
	Czech Republic [3]	185–259	5	118	*R. rubrum, R. petraeum, R. spicatum, R. multiflorum*
RJL-5	UK [20]	278	18	29	*Ribes* species[f]
	USA [22]	nd	1	12	*R. nigrum*
	Latvia [9]	256–287	15	110	*R. nigrum*
	North Europe [11]	205–304	nd	846[g]	*R. nigrum*
	Czech Republic [3]	261–283	0	118	*R. rubrum, R. petraeum, R. spicatum, R. multiflorum*
RJL-6	Latvia [9]	270–302	11	110	*R. nigrum*
	UK [20]	291	13	29	*Ribes* species[f]
	USA [22]	nd	1	12	*R. nigrum*
	Russia [16]	283–301	7	22	*R. uva-crispa*
	North Europe [11]	191–309	nd	846[g]	*R. nigrum*
		276–301	nd	846[g]	*R. rubrum group*
		215–308	nd	846[g]	*R. uva-crispa*
	Czech Republic [3]	280–300	9	118	*R. rubrum, R. petraeum, R. spicatum, R. multiflorum*
RJL-7	UK [20]	200	16	29	*Ribes* species[f]
	Latvia [9]	193–229	15	110	*R. nigrum*
	Sweden [17]	200–238	16	242	*R. uva-crispa*
	Switzerland [18]	nd	nd	286	*Ribes* sp.
	North Europe [11]	194–233	nd	846[g]	*R. nigrum*
	Czech Republic [3]	200–236	12	118	*R. rubrum, R. petraeum, R. spicatum, R. multiflorum*

(*Continued*)

TABLE 6.7 (*Continued*)
Microsatellite (SSR) Application in the Fingerprinting of *Ribes* Species Germplasm

SSR Locus	Country, References[a]	Size Range, bp	No. of Alleles	Sample Size	Plant Material
RJL-8	UK [20]	193	2	29	*Ribes* species[f]
	Czech Republic [3]	207–225	5	118	*R. rubrum, R. petraeum, R. spicatum, R. multiflorum*
RJL-8b	Czech Republic [3]	231–233	2	118	*R. rubrum, R. petraeum, R. spicatum, R. multiflorum*
RJL-9	UK [20]	205	3	29	*Ribes* species[f]
	Latvia [9]	203–207	3	110	*R. nigrum*
	Czech Republic [3]	196–234	7	118	*R. rubrum, R. petraeum, R. spicatum, R. multiflorum*
RJL-10	UK [20]	202	9	29	*Ribes* species[f]
	Latvia [9]	180–297	13	110	*R. nigrum*
	Italy [4]	114–165	7	41	*R. nigrum, R. rubrum, R. grossularia, R. missouriense, R. × nidigrolaria*
	North Europe [11]	149–298	nd	846[g]	*R. nigrum*
	Czech Republic [3]	181–291	11	118	*R. rubrum, R. petraeum, R. spicatum, R. multiflorum*
RJL-10b	Czech Republic [3]	195–257	21	118	*R. rubrum, R. petraeum, R. spicatum, R. multiflorum*
RJL11	UK [20]	215	10	29	*Ribes* species[f]
	Latvia [9]	215–232	8	110	*R. nigrum*
	Italy [4]	113	1	41	*R. nigrum, R. rubrum, R. grossularia, R. missouriense, R. × nidigrolaria*
	Sweden [17]	215+	14	242	*R. uva-crispa*
	North Europe [11]	205–229	nd	846[g]	*R. nigrum*
		201–241	Nd	846[g]	*R. rubrum group*
		229–263	nd	846[g]	*R. uva-crispa*
	Czech Republic [3]	213–235	10	118	*R. rubrum, R. petraeum, R. spicatum, R. multiflorum*
RnMYB-1	Russia [16]	nd	nd	22	*R. uva-crispa*
RucANS	Russia [16]	237–241	2	22	*R. uva-crispa*
RucCHS-1	Russia [16]	343	1	22	*R. uva-crispa*
RucCHS-2	Russia [16]	268	1	22	*R. uva-crispa*
RucDFR1-1	Russia [16]	399	1	22	*R. uva-crispa*
RucDFR1-2	Russia [16]	364–397	8	22	*R. uva-crispa*
RucDFR1-3	Russia [16]	318–323	5	22	*R. uva-crispa*
RucDFR1-4	Russia [16]	326	1	22	*R. uva-crispa*
RucDFR2-1	Russia [16]	257–261	4	22	*R. uva-crispa*
RucHLH-1	Russia [16]	198–208	2	22	*R. uva-crispa*
RucHLH-2	Russia [16]	243–248	3	22	*R. uva-crispa*
RucHLH-3	Russia [16]	314–319	3	22	*R. uva-crispa*
RucHLH-4	Russia [16]	246	1	22	*R. uva-crispa*
RucUFGT	Russia [16]	343–355	3	22	*R. uva-crispa*
RucWD-1	Russia [16]	189	1	22	*R. uva-crispa*
RucWD-2	Russia [16]	378	1	22	*R. uva-crispa*

(*Continued*)

TABLE 6.7 (*Continued*)
Microsatellite (SSR) Application in the Fingerprinting of *Ribes* Species Germplasm

SSR Locus	Country, References[a]	Size Range, bp	No. of Alleles	Sample Size	Plant Material
RucWD-3	Russia [16]	204	1	22	*R. uva-crispa*
UDP98–406	Romania [12]	75–250	4	4	*R. nigrum*
UDP98–411	Romania [12]	75–600	4	4	*R. nigrum*

[a] [1] Mezhnina and Urbanovich (2017a); [2] Mezhnina and Urbanovich (2017b); [3] Leišová-Svobodová and Kissling (2021); [4] Cavanna et al. (2009); [5] De Mattia et al. (2008); [6] Gentili et al. (2015); [7] Gentili et al. (2018); [8] Palimieri et al. (2013); [9] Lācis et al. (2021); [10] Lācis et al. (2022); [11] Antonius et al. (2012); [12] Butiuc-Keul et al. (2015); [13] Pavlenko et al. (2021); [14] Pikunova et al. (2015); [15] Pikunova et al. (2022); [16] Vidyagina et al. (2021); [17] Nordlander et al. (2022); [18] Droz et al. (2019); [19] Çelenk and Özcan (2016); [20] Brennan et al. (2002); [21] Lanham and Brennan (1998); [22] Burnes et al. (2008); [23] Oleas et al. (2014).

[b] Anchored SSR.

[c] Chloroplast SSR.

[d] Average value of all tested SSR markers.

[e] *R. nigrum* and interspecies hybrids (interspecific crosses of *R. nigrum* with *R. americanum R. fontaneum*, *R. procumbens*, *R. nigrum*, *R. dikusha*), *R. rubrum*, *R. uva-crispa*, *R. alpinum*, *R. aureum*, *R.×nidigrolaria*, *R. ussuriense*, *R. komarovii*, *R. sanguineum*, *R. latifolium*, *R. fasciculatum* var. *chinense*, *R. fragrans*, *R. laxiflorum*.

[f] *R. nigrum*, *R. nigrum×R. aureum*, *R. dikuscha*, *R. nigrum×R. dikuscha*, *R. nevadense*, *R. sanguineum*, *R. cereum*, *R. petraeum*, *R. grossularia*.

[g] Total number of samples, including 400 black currants (*R. nigrum*), 202 red and white currants (*R. rubrum* group), 242 gooseberries (*R. uva-crispa*) and 2 jostaberries (*R. × nidigrolaria*).

mp, mapping population; nd, no data.

Both sets of SSR markers, which were created in blackcurrants, were successfully applied in a large variety of *Ribes* species (for example, *R. alpinum*, *R. aureum*, *R. bieberstenii*, *R. cereum*, *R. dikuscha*, *R. echinellum*, *R. fasciculatum* var. *chinense*, *R. fragrans*, *R. komarovii*, *R. latifolium*, *R. missouriense*, *R. multiflorum*, *R. nevadense*, *R. nigrum*, *R. petraeum*, *R. rubrum*, *R. sanguineum*, *R. sardoum*, *R. spicatum*, *R. ussuriense*, *R. uva-crispa*, *R.×nidigrolaria*, as well as interspecific hybrids of *R. nigrum* with *R. americanum R. fontaneum*, *R. procumbens*, *R. nigrum*, *R. dikusha*, *R. aureum*) (Table 6.2). Germplasm analysis of *Ribes* shows a great variety of applications in terms of the number of samples and markers and the composition of the studied species. In the collected information, the number of samples varied from four samples of blackcurrant cultivars (Butiuc-Keul et al. 2015) to 846 samples of different *Ribes* species (Antonius et al. 2012), showing an average number of samples analyzed per the study of 128, which is relatively large, and allows to evaluate the suitability of applied markers. The average number of applied markers is also representative – 10.6 markers per study, starting with two (Butiuc-Keul et al. 2015) and up to 30 (Leišová-Svobodová and Kissling 2021), which was determined by the goals of the specific study. Markers g1-K04, e1-O01, g2-G12, g1-A01, and g1-M07 were used in most studies, 13, 11, 11, 10, and 10 cases, respectively. They are followed by markers e1-O21, e3-B02, g2-J08, and g2-L17, each with nine application cases.

In most cases, both the number of identified alleles and the allele range in bp are indicated (Figure 6.1), which allows evaluation of the efficiency of interspecies transfer. In most cases, the sample allele range of different species overlaps, showing only differences in range depending on the number of samples studied and the variety of species. Pronounced range shifts were observed only in some cases – g2-G12 and e3-B02, indicating the specificity of the plant material included in these studies, local adaptation because in these cases, samples of Belarusian and Russian origin prevail (Pikunova et al. 2015; Mezhnina and Urbanovich 2017b). The developed SSR marker

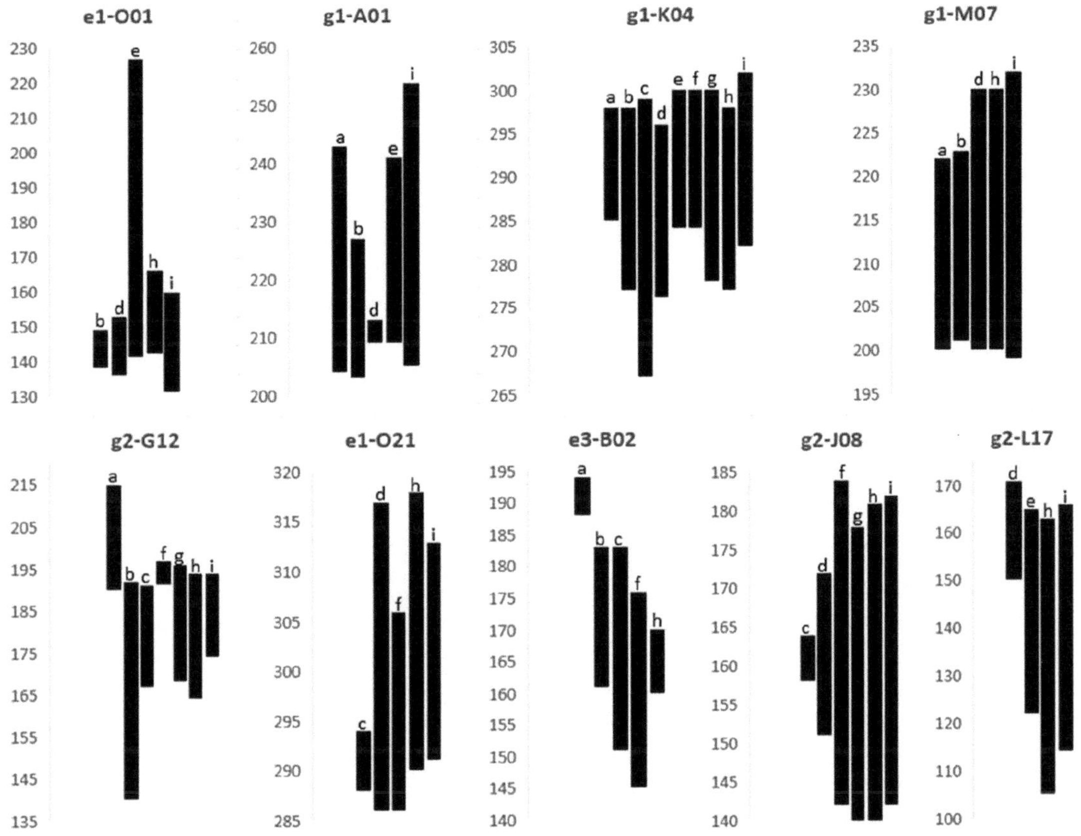

FIGURE 6.1 Base pair range of *Ribes* germplasm for the most widely applied SSR markers. The letters in the figure indicate the base pair range reference and the analyzed *Ribes* species: (a) Mezhnina and Urbanovich (2017a), *R. nigrum* (b) Mezhnina and Urbanovich (2017b), *R. nigrum, R. uva-crispa, R. alpinum* (c) Leišová-Svobodová and Kissling (2021), *R. rubrum, R. petraeum, R. spicatum, R. multiflorum* (d) Cavanna et al. (2009), *R. nigrum, R. rubrum, R. grossularia, R. missouriense, R.* × *nidigrolaria* (e) Gentili et al. (2015), *R. multiflorum subsp. sandalioticum, R. sardoum* (f) Palmieri et al. (2013), *R. nigrum, R. Rubrum* (g) Lācis et al. (2021), *R. nigrum* (h) Pikunova et al. (2015), *R. Nigrum* (i) Nordlander et al. (2022), *R. uva-crispa.*

systems cover all *Ribes* linkage groups and have already proven their effectiveness in QTL mapping (Brennan et al. 2008; Hackett et al. 2010) and are capable and perspective for further research and development of these berry crops.

6.3.3.3 Allele Mining of Currants and Gooseberries Chloroplast DNA

Although SSR markers have successfully ensured the differentiation of *Ribes* species, they cannot always fulfill the raised tasks. The most significant limitations are in work with complex interspecies crosses, in which it is necessary to assess the presence of the genome of a species (Sedlák et al. 2022). One solution is using chloroplast DNA (cpDNA) markers, which do not have these limitations. The cpDNA has a low evolutionary rate and an almost non-existent recombination rate. Besides that, they have maternal inheritance, providing the ability to follow the parentage of plant material. However, even considering these advantages of cpDNA, their application in currants and gooseberries is relatively limited. Grouping of *Ribes* species by genetic relatedness and correspondence to taxonomic groups was analyzed for 22 genera samples (Messinger et al. 1999) using cpDNA restriction with 16 enzymes in combination with rbel and 18S sequencing data.

Similarly, cpDNA restriction and ITS sequence analyses were used to create a *Ribes* molecular phylogeny (Schultheis and Donoghue 2004). An important study of cpDNA was carried out using the NGS approach on *R. odoratum* (Wang et al. 2021), identifying and annotating 56 new SSR markers that can be used to develop the *Ribes* genus phylogeny. The second approach uses already-known cpSSR markers due to their high polymorphism and transferability among a wide range of plant species (Provan et al. 2001; Huang et al. 2018). Thus, when testing nine SSR markers developed in other plant species, two polymorphic ones with four alleles each (ccmp6 and ccmp10) were selected, which ensured the discrimination of wild spontaneous interspecific hybrids (De Mattia et al. 2008). A significantly more extensive set of general cpSSR markers (50 pcs.) was tested on 145 *Ribes* genotypes (*R. nigrum* and interspecies hybrids (interspecific crosses of *R. nigrum* with *R. americanum*, *R. fontaneum*, *R. procumbens*, *R. nigrum*, *R. dikusha*), *R. rubrum*, *R. uva-crispa*, *R. alpinum*, *R. aureum*, *R.×nidigrolaria*, *R. ussuriense*, *R. komarovii*, *R. sanguineum*, *R. latifolium*, *R. fasciculatum* var. *chinense*, *R. fragrans*, *R. laxiflorum*) (Table 6.2, Lācis et al. 2022). Nineteen of these markers showed high polymorphism in samples of *Ribes* species (two to ten alleles) and provided their discrimination and determination of mutual relationship but could not determine the presence of any particular species in interspecies hybrids.

6.3.3.4 Characterization of Currants and Gooseberries Allelic Diversity by Sequencing

With the development of sequencing technologies, they are becoming more and more accessible even for genetically less studied species like *Ribes*, and one of the methods of increasing interest is GBS. This approach for sequencing the species *R. odoratum* chloroplast genome has already been mentioned (Wang et al. 2021), but this method has also been used in genomic DNA sequencing and identification of SNP markers. Redcurrants form a rather complex group of currents, which consists of several species and their mutual hybrids, which often complicates their study and breeding. They were sequenced using 75 redcurrant genotypes from 13 countries, obtaining more than 1.5 million reads for each sample (Pikunova et al. 2022). Thanks to these data, 8.5 K SNPs were identified, of which 7.5 K could be used in genotyping the samples. Sequence data can be directly used for pedigree clarification and trait-specific selection – in this case, white and pink-fruited redcurrants were grouped separately, indicating the inheritable nature of the trait. This approach would also be acceptable for explaining the composition of blackcurrant interspecies crosses and, in the future, would allow solving the study of the heredity of traits related to a specific species. This would also promote the broader use of SNP markers in *Ribes* research, which are currently used in some QTL mapping projects for blackcurrants (Brennan et al. 2008, marker list: https://fruitbreeding.hutton.ac.uk/RibesGenomicsSSRs.asp; Russell et al. 2014; Abreu et al. 2020) and redcurrants (Pikunova et al. 2020) increasing genome coverage and importance in breeding.

6.3.3.5 Allelic Mining and QTL Mapping of Agronomically Important Traits

One of the main interests in crop gene and allele mining is developing trait-specific markers for breeding. Usually, these studies start with QTL mapping. The first QTL map for blackcurrant (*R. nigrum*) was developed in the UK (population SCRI 9328, Brennan et al. 2008), which was later updated (Hackett et al. 2010) and saturated by the introduction of new markers (as SNPs) (Russell et al. 2014). The mapping includes blackcurrant phenological (time of budbreak, full leaf, first flower, full flower) and fruit quality characteristics (average weight of 100 berries (HBW), the content of soluble solids, ascorbic acid and anthocyanins, titratable juice acidity, specific gravity of the extracted juice, pH of the juice). The results of these mapping activities are summarized in Table 6.8. Blackcurrant SCRI 9328 (Brennan et al. 2008) mapping population was also used to study the genetic control of polyphenols and other juice chemical components important for the processing of blackcurrants (Abreu et al. 2020): anthocyanins (19 compounds), flavanols (18 compounds), other metabolites (phenylalanine, epigallocatechin, tryptophan, chlorogenic acid, caffeic acid, nigrumin-coumarate, nigrumin-ferulate), and other traits (content ascorbate, citrate, malate,

TABLE 6.8
The Quantitative Trait Loci (QTLs) Mapped for Blackcurrants (*R. nigrum*)

Trait	References	Main QTL	
		Linkage Group	PVE Max[a]
Time of budbreak	Brennan et al. (2008); Hackett et al. (2010)	3, 8	20.9%, 19.7%
Time of full leaf	Brennan et al. (2008)	1, 3, 7	21.6%, 16.2%, 15.2%
Time of first flower	Brennan et al. (2008)	3, 8	14.6%, 19.3%
Time of full flower	Brennan et al. (2008)	1	21.9%
Average weight of 100 berries (HBW)	Brennan et al. (2008); Abreu et al. (2020)	1, 2, 5	20.5%, 21.7%, nd
The content of soluble solids	Russell et al. (2014)	2, 3	16.7%–43.9%
Concentration of ascorbic acid	Brennan et al. (2008); Abreu et al. (2020)	1, 2, 4, 5	19.7%, 14.1%, nd, nd
The content of anthocyanins in fruits	Hackett et al. (2010); Abreu et al. (2020)	1, 2, 3, 4, 5, 6	nd
Titratable juice acidity	Brennan et al. (2008)	3, 6, 7	26.6%, 24.8%, 15.7%
Specific gravity of the extracted juice	Brennan et al. (2008)	2, 3	19.3%, 15.8%
pH of the juice	Brennan et al. (2008)	3, 7	28.4%, 16.3%
The content of total acids	Abreu et al. (2020)	1, 3	nd
The content of total phenols	Abreu et al. (2020)	1, 3, 4	nd
The content of epigallocatechin	Abreu et al. (2020)	7	nd
The content of chlorogenic acid	Abreu et al. (2020)	2, 3, 4, 6, 8	nd
The content of caffeic acid	Abreu et al. (2020)	2, 5, 6	nd
The content of phenylalanine	Abreu et al. (2020)	2, 6	nd
The content of tryptophan	Abreu et al. (2020)	1, 3, 6	nd
The content of nigrumin-coumarate	Abreu et al. (2020)	1, 3, 4	nd
The content of nigrumin-ferulate	Abreu et al. (2020)	3, 4	nd
The content of citrate	Abreu et al. (2020)	1, 3, 5, 6, 8	nd
The content of succinate	Abreu et al. (2020)	1, 4, 5, 6	nd
The content of malate	Abreu et al. (2020)	3, 4	nd

[a] Maximum percentage of phenotypic variance explained. nd, no data.

succinate, total acids, anthocyanins and phenols, average weight of 100 berries). QTLs of anthocyanin and flavanol metabolites were identified on all eight linkage groups of black currants.

Among other species of the genus *Ribes*, a QTL map has been created only for red currants (*R. rubrum*) (Pikunova et al. 2020). The mapping used previously developed SSR markers gr2-J05, g1-M07, g1-K04, g2-J08, g2-B20, e1-O21, e3-B02, g1-A01, g1-L12, e1-O01, g2-G12 (Brennan et al. 2008), two newly developed SSRs (Cra-531 and Cra-489) as well as 667 SNP markers. As a result of the study, eight linkage groups were identified.

For direct application in breeding, gene-specific DNA molecular markers suitable for selecting hybrid material are the most suitable. Although *Ribes* plants are threatened by many diseases and pests (Frercks et al. 2022), deeper resistance mechanisms are studied, and specific markers have been developed only for the gall mite (*Cecidophyopsis ribis*) – BRV – *Ribes* pathosystem. Gall mite is the most important pest of blackcurrant, causing direct damage to the plants, known as 'big bud', and is also a vector for the transmission of BRV from plant to plant. Current research on the resistance of *Ribes* plants to pests and viruses indicates a complex and not yet fully understood interaction of the involved species (Moročko-Bičevska et al. 2022). The primary providers of resistance to gall mites, also used in breeding, are the *Ce* gene identified in gooseberries (Knight et al. 1974) and *R. nigrum* spp. *sibiricum* identified *P* gene (Anderson 1971). The use of the *Ce* resistance gene in blackcurrant breeding was started in the UK (Brennan et al. 2009), and the obtained plant material was later included in the breeding programs of many countries, while the use of the *P* gene in the creation of new cultivars was used in Lithuania (Mažeikienė et al. 2012).

Ribes QTL mapping using different DNA markers (Brennan et al. 2008) allowed us to identify the genome region linked to the resistance in the male parent linkage group 2. For implementation in breeding, a specific molecular marker to gall-mite resistance gene *Ce* was developed using bulk segregant analysis and the AFLP method, which were later widely used for selecting breeding material (Brennan et al. 2009). As a valuable source of the *Ce* resistance gene, other *Ribes* species that could be successfully used in blackcurrant breeding are also mentioned, for example, *R. sanguineum* var. *glutinosum* (Knyazev and Ogoltsova 2004). Comparing *Ce* gene-associated amplification fragments from *R. uva-crispa* with species such as *R. sanguineum* and *R. aureum* showed the sequence homology (Mazeikiene et al. 2017), indicating potential applicability in resistance breeding. Similarly, QTL mapping using AFLP and SSR markers was carried out for the *P* resistance gene, and by applying the AFLP method, a gene-specific marker was developed for use in breeding (Mažeikienė et al. 2012). The *P* gene molecular marker was later validated also in *R. nigrum* spp. *sibiricum*, *R. americanum*, and *R. aureum* (Mazeikiene et al. 2017), proving broad applicability in currants MAS (Mazeikiene et al. 2019).

6.4 FUTURE SCOPE

The use of molecular markers closely related to desirable features or QTLs may facilitate the development of superior berry cultivars with economically significant traits (taste, nutritional quality). Thus, more or better markers must be generated. Now that several *Fragaria* genomes have been analyzed using the latest molecular methods, there is a unique opportunity to analyze the evolution of gene content and function at the genus level. With the improvement of research methods, it will be possible to quickly analyze many genotypes of a species and reveal the entire diversity of gene alleles, including homoeologous alleles, and their expression depending on various factors and interactions with other genes. This will make it possible to explore the function of individual genes and the mechanisms of adaptability in detail and to create synthetic (hybrid or biotechnologically modified) plant lines adapted to the ever-increasing nutritional and health needs of humans, and the challenges of climate change. Genomic tools have been developed to support cranberry and blueberry improvement. New transcriptome sequences and markers are being used to identify QTLs associated with cold hardiness, chilling requirements and fruit yield and quality traits for marker-assisted breeding in cranberry and blueberry, as well as for genetic structure and diversity studies and marker transferability between *Vaccinium* species. Transcriptome sequences will also be useful for developing microarrays to study the expression of thousands of genes in response to biotic and abiotic stresses and during development. Although high-quality chromosome-scale genome assemblies are available for cranberry and blueberry, the development of new genotyping tools is needed, which, combined with new emerging technologies for phenotyping, will help accelerate genetic discoveries and advance breeding. Current gene mining studies of currants and gooseberries have developed sets of molecular markers capable of ensuring successful germplasm research, population analyses, international data exchange and cooperation in genetic resources. The developed sets of markers have shown an outstanding possibility of interspecies transfer and allow for the involvement of more and more new species in research, which would be necessary for future breeding. However, in the case of currants and gooseberries, more genome information is needed, which means more development of sequencing platforms that would help develop new markers. Breeding faces new challenges, and the broader introduction of trait-gene-specific markers, which is still relatively limited in the case of *Ribes* plants, is important in solving them. NGS technology significantly improves marker discovery and makes breeding line resequencing possible. Therefore, genomics-assisted breeding might replace marker-assisted breeding in the near future.

REFERENCES

Abreu IN, Brennan RM, Kanichukattu EN, Stewart D, Hancock RD, et al. (2020) Quantitative trait loci mapping of polyphenol metabolites in blackcurrant (*Ribes nigrum* L.). *Metabolomics* 16:25. https://doi.org/10.1007/s11306-020-1647-6.

Albani MC, Battey NH, Wilkinson MJ (2004) The development of ISSR-derived SCAR markers around the seasonal flowering locus (SFL) in *Fragaria vesca. Theor Appl Genet* 109:571–579. https://doi.org/10.1007/s00122-004-1654-4.

Albert T, Raspé O, Jacquemart AL (2003) Clonal structure in *Vaccinium myrtillus* L. revealed by RAPD and AFLP markers. *Int J Plant Sci* 164:649–655. https://doi.org/10.1086/375373.

Alger EI, Platts AE, Deb SK, Luo X, Ou S, et al. (2021) Chromosome-scale genome for a red-fruited, perpetual flowering and runnerless woodland strawberry (*Fragaria vesca*). *Front Genet* 12:671371. https://doi.org/10.3389/fgene.2021.671371.

An D, Bykova NV, Debnath SC (2015) EST-PCR, EST-SSR and ISSR markers to identify a set of wild cranberries and evaluate their relationships. *Can J Plant Sci* 95:1155–1165. https://doi.org/10.4141/CJPS-2015-158.

Anderson MM (1971) Resistance to gall mite (*Phytoptusribis* Nal.) in the *Eucoreosma* section of *Ribes. Euphytica* 20:422–426.https://doi.org/10.1007/BF00035668.

Antonius K, Karhu S, Kaldmäe H, Lacis G, Rugenius R, et al. (2012) Development of the Northern European *Ribes* core collection based on a microsatellite (SSR) marker diversity analysis. *Plant Genet Resour* 10:70–73. https://doi.org/10.1017/S1479262111000980.

Aristya GR, Kasiamdari RS, Setyoningrum R, Larasati B (2019) Genetic variations of strawberry cultivars of *Fragaria × ananassa* and *Fragaria vesca* based on RAPD. *Biodivers J* 20:770–775. https://doi.org/10.13057/biodiv/d200322.

Bai C, Alverson WS, Follansbee A, Waller DM (2012) New reports of nuclear DNA content for 407 U.S. plant species. *Ann Bot* 110:1623–1629. https://doi.org/10.1093/aob/mcs222.

Barbey CR, Hogshead M, Schwartz AE, Mourad N, Verma S, et al. (2020) The genetics of differential gene expression related to fruit traits in strawberry (*Fragaria × ananassa*). *Front Genet* 10:1317. https://doi.org/10.3389/fgene.2019.01317.

Barbey CR, Lee S, Verma S, Bird KA, Yocca AE, et al. (2019) Disease resistance genetics and genomics in octoploid strawberry. *G3 (Bethesda)* 9, 3315–3332. https://doi.org/10.1534/g3.119.400597.

Bassil N, Bidani A, Nyberg A, Hummer K, Rowland LJ (2020) Microsatellite markers confirm identity of blueberry (*Vaccinium* spp.) plants in the USDA-ARS National Clonal Germplasm Repository collection. *Genet Resour Crop E* 67:393–409. https://doi.org/10.1007/s10722-019-00873-8.

Bassil N, Oda A, Hummer KE (2009) Blueberry microsatellite markers identify cranberry cultivars. *Acta Hortic* 810:181–186. https://doi.org/10.17660/actahortic.2009.810.23.

Bassil NV, Davis TM, Zhang H, Ficklin S, Mittmann M, et al. (2015) Development and preliminary evaluation of a 90 K Axiom(r) SNP array for the allo-octoploid cultivated strawberry *Fragaria × ananassa. BMC Genomics* 16:155. https://doi.org/10.1186/s12864-015-1310-1.

Bell DJ, Rowland LJ, Polashock JJ, Drummond FA (2008) Suitability of EST-PCR markers developed in highbush blueberry for genetic fingerprinting and relationship studies in lowbush blueberry and related species. *J Am Soc Hortic* 133:701–707. https://doi.org/10.21273/jashs.133.5.701.

Bhatt DS, Debnath SC (2021) Genetic diversity of blueberry genotypes estimated by antioxidant properties and molecular markers. *Antioxidants* 10:1–29. https://doi.org/10.3390/antiox10030458.

Bian Y, Ballington J, Raja A, Brouwer C, Reid R, et al. (2014) Patterns of simple sequence repeats in cultivated blueberries (*Vaccinium* section *Cyanococcus* spp.) and their use in revealing genetic diversity and population structure. *Mol Breed* 34:675–689. https://doi.org/10.1007/s11032-014-0066-7.

Bidani A, Hummer KE, Rowland LJ, Bassil NV (2017) Development of an efficient DNA test for genetic identity confirmation in blueberry. *Acta Hortic* 1180:363–367. https://doi.org/10.17660/ActaHortic.2017.1180.49.

Biswas A, Melmaiee K, Elavarthi S, Jones J, Reddy U (2019) Characterization of strawberry (*Fragaria* spp.) accessions by genotyping with SSR markers and phenotyping by leaf antioxidant and trichome analysis. *Sci Hortic* 256:108561. https://doi.org/10.1016/j.scienta.2019.108561.

Boches P, Bassil NV, Rowland L (2006) Genetic diversity in the highbush blueberry evaluated with microsatellite markers. *J Am Soc Hortic* 131:674–686. https://doi.org/10.21273/jashs.131.5.674.

Boches P, Rowland LJ, Kim H, Bassil N (2005a) Microsatellite markers developed from 'Bluecrop' reveal polymorphisms in the genus *Vaccinium* and are suitable for cultivar fingerprinting. *Hort Science* 40:1093–1147. https://doi.org/10.21273/hortsci.40.4.1143a.

Boches PS, Bassil NV, Rowland LJ (2005b) Microsatellite markers for *Vaccinium* from EST and genomic libraries. *Mol Ecol Notes* 5:657–660. https://doi.org/10.1111/j.1471-8286.2005.01025.x.

Bors RH, Sullivan JA (2005) Interspecific hybridization of *Fragaria vesca* subspecies with *F. nilgerrensis, F. nubicola, F. pentaphylla*, and *F. viridis*. *J Am Soc Hortic* 130:418–423. https://doi.org/10.21273/JASHS.130.3.418.

Brennan R, Jorgensen L, Gordon S, Loades K, Hackett C, Russ J (2009) The development of a PCR-based marker linked to resistance to the blackcurrant gall mite (*Cecidophyopsisribis*Acari: *Eriophyidae*). *Theor Appl Genet* 118:205–211. https://doi.org/10.1007/s00122-008-0889-x.

Brennan R, Jorgensen L, Hackett C, Woodhead M, Gordon S, Russell J (2008) The development of a genetic linkage map of blackcurrant (*Ribes nigrum* L.) and the identification of regions associated with key fruit quality and agronomic traits. *Euphytica* 61:19–34. https://doi.org/10.1007/s10681-007-9412-8.

Brennan R, Jorgensen L, Woodhead M, Russell J (2002) Development and characterisation of SSR markers in *Ribes* species. *Mol Ecol Notes* 2:327–330. https://doi.org/10.1046/j.1471-8286.2002.00233.x.

Brennan RM (2008) Currants and gooseberries. In: Hancock JF (ed) *Temperate Fruit Crop Breeding*. Springer, Dordrecht, pp. 177–196. https://doi.org/10.1007/978-1-4020-6907-96.

Brevis P, Hancock J, Rowland LJ (2007) Development of a genetic linkage map for tetraploid highbush blueberry using SSR and EST-PCR markers. *HortScience* 42:963.

Bruederle LP, Vorsa N (1994) Genetic differentiation of diploid blueberry, *Vaccinium* sect. *Cyanococcus* (*Ericaceae*). *Syst Bot* 19:337–349.

Brunings AM, Moyer C, Peres N, Folta KM (2010) Implementation of simple sequence repeat markers to genotype Florida strawberry varieties. *Euphytica* 173:63–75. https://doi.org/10.1007/s10681-009-0112-4.

Burnes TA, Blanchette RA, Smith JA, Luby JJ (2008) Black currant clonal identity and white pine blister rust resistance. *HortScience* 43(1):200–202. https://doi.org/10.21273/Hortsci.43.1.200.

Butiuc-Keul AL, Crăciunaş C, Dobrotă C, Clapa D (2015) Molecular characterisation of new black currant cultivars bred in Romania exposed to changing climate conditions. *Acta Hortic* 1100:155–159. https://doi.org/10.17660/ActaHortic 2015.1100.24.

Cappai F, Amadeu RR, Benevenuto J, Cullen R, Garcia A, et al. (2020) High-resolution linkage map and qtl analyses of fruit firmness in autotetraploid blueberry. *Front Plant Sci* 11:1–11. https://doi.org/10.3389/fpls.2020.562171.

Carvalho M, Matos M, Carnide V (2014) Fingerprinting of *Vaccinium corymbosum* cultivars using DNA of fruits. *HortScience* 41:175–184. https://doi.org/10.17221/21/2014-hortsci.

Carvalho M, Matos M, Carnide V (2018) Identification of cultivated and wild *Vaccinium* species grown in Portugal. *Span J Agric Res* 16:1–6. https://doi.org/10.5424/sjar/2018163-12502.

Cavanna M, Marinoni Torello D, Beccaro GL, Bounous G (2009) Microsatellite-based evaluation of *Ribes* spp. germplasm. *Genome* 52:839–848. https://doi.org/10.1139/g09-057.

Çelenk AG, Özcan T (2016) Characterisation of some *Ribes* L. accessions from Turkey based on SSRs patterns. *Int Jenviron Agricres* 2:12–19.

Čereković N, Jarret D, Pagter M, Cullen DW, Morris JM, et al. (2015) The effects of drought stress on leaf gene expression during flowering in blackcurrant (*Ribes nigrum* L.). *Eur J Hort Sci* 80(1):39–46.

Chavez DJ, Lyrene PM (2009) Interspecific crosses and backcrosses between diploid *Vaccinium darrowii* and tetraploid southern highbush blueberry. *J Amer Soc Hort Sci* 134(2):273–280.

Chen P, Liu Q, (2019) Genome-wide characterization of the *WRKY* gene family in cultivated strawberry (*Fragaria* × *ananassa* Duch.) and the importance of several group III members in continuous cropping. *Sci Rep* 9:8423. https://doi.org/10.1038/s41598-019-44479-7.

Chiche J, Brown SC, Leclerc J-C, Siljak-Yakovlev S (2003) Genome size, heterochromatin organisation, and ribosomal gene mapping in four species of *Ribes*. *Can J Bot* 81:1049–1057. https://doi.org/10.1139/B03-088.

Cho KH, Park SJ, Kim SJ, Kim SH, Lee HC, et al. (2017) Blueberry cultivar identification using random amplified polymorphic DNA and sequence-characterized amplified region markers. *HortScience* 52:1483–1489. https://doi.org/10.21273/HORTSCI12105-17.

Cockerton HM, Vickerstaff RJ, Karlström A, Wilson F, Sobczyk M, et al. (2018) Identification of powdery mildew resistance QTL in strawberry (*Fragaria* × *ananassa*). *Theor Appl Genet* 131:1995–2007. https://doi.org/10.1007/s00122-018-3128-0.

Colle M, Leisner CP, Wai CM, Ou S, Bird KA, et al. (2019) Haplotype-phased genome and evolution of phytonutrient pathways of tetraploid blueberry. *Giga Sci* 8(3):1–15. https://doi.org/10.1093/gigascience/giz012.

Corrêa JVW, Weber GG, Zeist AR, de Resende JTV, Da-Silva PR (2021) ISSR analysis reveals high genetic variation in strawberry three-way hybrids developed for tropical regions. *Plant Mol Biol Rep* 39:566–576. https://doi.org/10.1007/s11105-020-01270-7.

Cortez RE, Gonzalez de Mejia E. (2019) Blackcurrants (*Ribes nigrum*): areview on chemistry, processing, and health benefits. *J Food Sci* 84(9):2387–401. https://doi.org/10.1111/1750-3841.14781.

Covarrubias-Pazaran G, Diaz-Garcia L, Schlautman B, Deutsch J, Salazar W, et al. (2016) Exploiting genotyping by sequencing to characterize the genomic structure of the American cranberry through high-density linkage mapping. *BMC Genomics* 17:1–16. https://doi.org/10.1186/s12864-016-2802-3.

Covarrubias-Pazaran G, Schlautman B, Diaz-Garcia L, Grygleski E, Polashock J, et al. (2018) Multivariate GBLUP improves accuracy of genomic selection for yield and fruit weight in biparental populations of *Vaccinium macrocarpon* ait. *Front Plant Sci* 9:1–13. https://doi.org/10.3389/fpls.2018.01310.

Crizel RL, Perin EC, Vighi IL, Woloski R., Seixas A, et al. (2020) Genome-wide identification, and characterization of the *CDPK* gene family reveal their involvement in abiotic stress response in *Fragaria × ananassa*. *Sci Rep* 10:11040. https://doi.org/10.1038/s41598-020-67957-9.

Cui F, Ye X, Li X, Yang Y, Hu Z, et al. (2022). Chromosome-level genome assembly of the diploid blueberry Vaccinium darrowii provides insights into its subtropical adaptation and cuticle synthesis. *Plant Comm* 3:100307.

Darwish O, Shahan R, Liu Z, Slovin JP, Alkharouf NW (2015) Re-annotation of the woodland strawberry (*Fragaria vesca*) genome. *BMC Genomics* 16:29. https://doi.org/10.1186/s12864-015-1221-1.

Daverdin G, Johnson-Cicalese J, Zalapa J, Vorsa N, Polashock J (2017) Identification and mapping of fruit rot resistance QTL in American cranberry using GBS. *Mol Breed* 37:1–15. https://doi.org/10.1007/s11032-017-0639-3.

De Mattia F, Grassi F, Imazio S, Labra M (2008) Chloroplast and nuclear DNA markers to characterise cultivated and spontaneous *Ribes*. *Plant Biosyst* 142(2):204–212. https://doi.org/10.1080/11263500802150290.

Debnath SC (2005) Differentiation of *Vaccinium* cultivars and wild clones using RAPD markers. *J Plant Biochem Biotechnol* 14:173–177. https://doi.org/10.1007/BF03355954.

Debnath SC (2007) An assessment of the genetic diversity within a collection of wild cranberry (*Vaccinium macrocarpon* Ait.) clones with RAPD-PCR. *Genet Resour Crop E* 54:509–517. https://doi.org/10.1007/s10722-006-0007-3.

Debnath SC (2014) Structured diversity using EST-PCR and EST-SSR markers in a set of wild blueberry clones and cultivars. *Biochem Syst Ecol* 54:337–347. https://doi.org/10.1016/j.bse.2014.03.018.

Debnath SC, Khanizadeh S, Jamieson AR, Kempler C (2008) Inter simple sequence repeat (ISSR) markers to assess genetic diversity and relatedness within strawberry genotypes. *Can J Plant Sci* 88:313–322. https://doi.org/10.4141/CJPS07088.

Diaz-Garcia L, Covarrubias-Pazaran G, Johnson-Cicalese J, Vorsa N, Zalapa J (2020) Genotyping-by-sequencing identifies historical breeding stages of the recently domesticated American cranberry. *Front Plant Sci* 11:1–13. https://doi.org/10.3389/fpls.2020.607770.

Diaz-Garcia L, Covarrubias-Pazaran G, Schlautman B, Grygleski E, Zalapa J (2018a) Image-based phenotyping for identification of QTL determining fruit shape and size in American cranberry (*Vaccinium macrocarpon* L.). *Peer J* 6:1–19. https://doi.org/10.7717/peerj.5461.

Diaz-Garcia L, Schlautman B, Covarrubias-Pazaran G, Johnson-Cicalese J, Grygleski E, et al. (2018b) Massive phenotyping of multiple cranberry populations reveals novel QTLs for fruit anthocyanin content and other important chemical traits. *Mol Genet Genom* 293:1379–1392. https://doi.org/10.1007/s00438-018-1464-z.Massive.

Die JV, Jones RW, Ogden EL, Ehlenfeldt MK, Rowland LJ (2020). Characterization and analysis of anthocyanin-related genes in wild-type blueberry and the pink-fruited mutant cultivar 'Pink Lemonade': new insights into anthocyanin biosynthesis. *Agronomy* 10:1296.

Droz E, Frei M, Julmi-Moreillon C, Niggli C (2019) First attempt to characterise *Ribes* genetic resources with SSR markers in Switzerland. ISHS XII *Rubus&Ribes* Symposium, June 25–28, Zurich, Switzerland, Book of Abstracts, pp. 94.

Edger PP, Iorizzo M, Bassil N, Benevenuto J, Ferrão LF, et al. (2022) There and back again; historical perspective and future directions for *Vaccinium* breeding and research studies. *Hortic Res* 9. https://doi.org/10.1093/hr/uhac083.

Edger PP, McKain MR, Yocca AE, Knapp SJ, Qiao Q, et al. (2020) Reply to: revisiting the origin of octoploid strawberry. *Nat Genet* 52:5–7. https://doi.org/10.1038/s41588-019-0544-2.

Edger PP, Poorten TJ, VanBuren R, Hardigan MA, Colle M, et al. (2019) Origin and evolution of the octoploid strawberry genome. *Nat Genet* 51, 541–547. https://doi.org/10.1038/s41588-019-0356-4.

Edger PP, VanBuren R, Colle M, Poorten TJ, Wai CM, et al. (2018) Single-molecule sequencing and optical mapping yields an improved genome of woodland strawberry (*Fragaria vesca*) with chromosome-scale contiguity. *Gigascience* 7:1–7. https://doi.org/10.1093/gigascience/gix124.

Ehlenfeldt MK, Luteyn JL (2021) Fertile intersectional F1 hybrids of 4× *Vaccinium meridionale* (section *Pyxothamnus*) and highbush blueberry, *V. corymbosum* (section *Cyanococcus*) *HortScience* 56(3):318–323.

Ehlenfeldt MK, Polashock JJ, Ballington JR (2018) *Vaccinium corymbodendron* Dunal as a bridge between taxonomic sections and ploidies in *Vaccinium*: a work in progress. Proceeding papers of North American Blueberry Research and Extension Workers Conference 15:1–6.

Fajardo D, Morales J, Zhu H, Steffan S, Harbut R, et al. (2012) Discrimination of American cranberry cultivars and assessment of clonal heterogeneity using microsatellite markers. *Plant Mol Biol Report* 31:264–271. https://doi.org/10.1007/s11105-012-0497-4.

FAOSTAT (2022) Food and Agriculture Organization of the United Nations. FAOSTAT Statistical Database. https://www.fao.org/faostat/en/#home.

Feng C, Wang J, Harris AJ, Folta KM, Zhao M, et al. (2021) Tracing the diploid ancestry of the cultivated octoploid strawberry. *Mol Biol E* 38: 478–485. https://doi.org/10.1093/molbev/msaa238.

Ferrão LF, Benevenuto J, Oliveira I,Cellon C, Olmstead J, et al. (2018) Insights into the genetic basis of blueberry fruit-related traits using diploid and polyploid models in a GWAS context. *Front Ecol E* 6:1–16. https://doi.org/10.3389/fevo.2018.00107.

Ferrão LFV, Johnson TS, Benevenuto J, Edger PP, Colquhoun TA, et al. (2020) Genome-wide association of volatiles reveals candidate loci for blueberry flavor. *New Phytol* 226:1725–1737. https://doi.org/10.1111/nph.16459.

Fierascu RC, Temocico G, Fierascu I, Ortan A, Babeanu NE (2020) Fragaria genus: chemical composition and biological activities. *Molecules* 25:498. https://doi.org/10.3390/molecules25030498.

Fong SK, Kawash J, Wang Y, Johnson-Cicalese J, Polashock J, Vorsa N (2021) A low malic acid trait in cranberry fruit: genetics, molecular mapping, and interaction with a citric acid locus. *Tree Genet Genomes* 17: 4. https://doi.org/10.1007/s11295-020-01482-8

Frercks B, Gelvonauskienė D, Juškytė AD, Sikorskaitė-Gudžiūnienė S, Mažeikienė I, et al. (2022) Development of biotic stress tolerant berries. In: Kole C (ed) *Genomic Designing for Biotic Stress Resistant Fruit Crops*. Springer, Cham, pp. 331–384. https://doi.org/10.1007/978-3-030-91802-6_9.

García MG, Ontivero M, Diaz Ricci JC, Castagnaro A (2002) Morphological traits and high resolution RAPD markers for the identification of the main strawberry varieties cultivated in Argentina. *Plant Breed* 121:76–80. https://doi.org/10.1046/j.1439-0523.2002.00671.x.

Garriga M, Parra PA, Caligari PDS, Retamales JB, Carrasco BA, et al. (2013) Application of inter-simple sequence repeats relative to simple sequence repeats as a molecular marker system for indexing blueberry cultivars. *Can J Plant Sci* 93:913–921. https://doi.org/10.4141/CJPS2013-057.

Gaston A, Perrotte J, Lerceteau-Köhler E, Rousseau-Gueutin M, Petit A, et al. (2013) PFRU, a single dominant locus regulates the balance between sexual and asexual plant reproduction in cultivated strawberry. *J Exp Bot* 64:1837–1848. https://doi.org/10.1093/jxb/ert047.

Gawroński J, Kaczmarska E, Dyduch-Siemińska M (2017) Assessment of genetic diversity between *Vaccinium corymbosum* L. cultivars using RAPD and ISSR markers. *Acta Sci Pol* 16:129–140. https://doi.org/10.24326/asphc.2017.3.13.

Genome Database for Rosaceae (n.d.) *Fragaria* all species. [WWW Document], URL https://www.rosaceae.org/species/fragaria/all (accessed 12.15.22).Gentili R, Fenu G, Mattana E, Citterio S, De Mattia F, et al. (2015) Conservation genetics of two island endemic *Ribes* spp. (*Grossulariaceae*) of Sardinia: survival or extinction? *Plant Biol* 17(5):1085–1094. https://doi.org/10.1111/plb.12330.

Gentili R, Fenu G, Porceddu M, Bruni I, Citterio S, et al. (2018) Genetic variability of the first-generation of *Ribes sardoum*, a threatened relic plant requiring translocation measures. *Plant Biosyst* 153(1):1–4. https://doi.org/10.1080/11263504.2018.1435574.

Georgi L, Herai RH, Vidal R, Carazzolle MF, Pereira GG, et al. (2012) Cranberry microsatellite marker development from assembled next-generation genomic sequence. *Mol Breed* 30:227–237. https://doi.org/10.1007/s11032-011-9613-7.

Georgi L, Johnson-Cicalese J, Honig J, Das SP, Rajah VD, et al. (2013) The first genetic map of the American cranberry: exploration of synteny conservation and quantitative trait loci. *Theor Appl Genet* 126:673–692. https://doi.org/10.1007/s00122-012-2010-8.

Giongo L, Vendramin E, Quarta R, Dettori MT, Verde I, et al. (2006) Use of RAPD analysis to characterize *Vaccinium* accessions. *Acta Hortic* 715:153–156. https://doi.org/10.17660/ActaHortic.2006.715.22.

Gor MC, Candappa C, de Silva T, Mantri N, Pang E, (2017) Identification and validation of FaP1D7, a putative marker associated with the biosynthesis of methyl butanoate in cultivated strawberry (*Fragaria × ananassa*). *Sci Rep* 7:17454. https://doi.org/10.1038/s41598-017-17448-1.

Gor MC, Mantri N, Pang E (2016) Application of subtracted gDNA microarray-assisted bulked segregant analysis for rapid discovery of molecular markers associated with day-neutrality in strawberry (*Fragaria × ananassa*). *Sci Rep* 6:32551. https://doi.org/10.1038/srep32551.

Govan CL, Simpson DW, Johnson AW, Tobutt KR, Sargent DJ (2008) A reliable multiplexed microsatellite set for genotyping *Fragaria* and its use in a survey of 60 *F. × ananassa* cultivars. *Mol Breed* 22:649–661. https://doi.org/10.1007/s11032-008-9206-2.

Guichoux E, Lagache L, Wagner S, Chaumeil P, Leger P, et al. (2011) Current trends in microsatellite genotyping. *Mol Ecol Resour* 11(4):591–611. https://doi.org/10.1111/j.1755-0998.2011.03014.x.

Hackett CA, Russell J, Jorgensen L, Gordon SL, Brennan RM (2010) Multienvironment QTL mapping in blackcurrant (*Ribes nigrum* L.) using mixed models. *Theor Appl Genet* 121(8):1483–1488. https://doi.org/10.1007/s00122-010-1404-8.

Hadonou AM, Sargent DJ, Wilson F, James CM, Simpson DW (2004) Development of microsatellite markers in *Fragaria*, their use in genetic diversity analysis, and their potential for genetic linkage mapping. *Genome* 47:429–438. https://doi.org/10.1139/g03-142.

Haider MS, Khan N, Pervaiz T, Zhongjie L, Nasim M, et al. (2019) Genome-wide identification, evolution, and molecular characterization of the *PP2C* gene family in woodland strawberry. *Gene* 702, 27–35. https://doi.org/10.1016/j.gene.2019.03.025.

Hancock JF (1999) *Strawberries*. CABI Pub University Press, Cambridge, MA, 237 p.

Hancock JF, Lyrene P, Finn CE, Vorsa N, Lobos GA (2008) Blueberries and cranberries. In: Hancock JF, Lyrene P, Finn CE, Vorsa N, Lobos GA (eds) *Temperate Fruit Crop Breeding*. Springer Science+Business Media B.V, pp. 115–149. ISBN : 978-1-4020-6906-2

Hancock JF, Olmstead JW, Itle RA, Callow PW, Neils-Kraft S, et al. (2018) Performance of an elite, hybrid family of a northern × southern highbush cross ('Draper' × 'Jewel'). *Euphytica* 214:1–16. https://doi.org/10.1007/s10681-018-2173-8.

Hanson L, McMahon KA, Johnson MAT, Bennett MD (2001) First nuclear DNA C-values for another 25 angiosperm families. *Ann Bot* 88:851–858. https://doi.org/10.1006/anbo.2000.1325.

Harrison RE, Luby JJ, Furnier GR, Hancock JF (2000) Differences in the apportionment of molecular and morphological variation in North American strawberry and the consequences for genetic resource management. *Genet Resour Crop E* 47:647–657. https://doi.org/10.1023/A:1026530125493.

Hawkins C, Caruana J, Schiksnis E, Liu Z (2016) Genome-scale DNA variant analysis and functional validation of a SNP underlying yellow fruit color in wild strawberry. *Sci Rep* 6:29017. https://doi.org/10.1038/srep29017.

Hedley PE, Russell JR, Jorgensen L, Gordon S, Morris JA, et al. (2010) Candidate genes associated with bud dormancy release in blackcurrant (*Ribes nigrum* L.). *BMC Plant Biol* 10:202. https://www.biomedcentral.com/1471-2229/10/202.

Hilmarsson HS, Hytönen T, Isobe S, Göransson M, Toivainen T, et al. (2017) Population genetic analysis of a global collection of *Fragaria vesca* using microsatellite markers. *PLoS One* 12, e0183384. https://doi.org/10.1371/journal.pone.0183384.

Hirakawa H, Shirasawa K, Kosugi S, Tashiro K, Nakayama S, et al. (2014) Dissection of the octoploid strawberry genome by deep sequencing of the genomes of *Fragaria* species. *DNA Res* 21:169–181. https://doi.org/10.1093/dnares/dst049.

Honjo M, Nunome T, Kataoka S, Yano T, Yamazaki H, et al. (2011) Strawberry cultivar identification based on hypervariable SSR markers. *Breed Sci* 61:420–425. https://doi.org/10.1270/jsbbs.61.420.

Horvath A, Sánchez-Sevilla JF, Punelli F, Richard L, Sesmero-Carrasco R, et al. (2011) Structured diversity in octoploid strawberry cultivars: importance of the old European germplasm. *Ann Appl Biol* 159:358–371. https://doi.org/10.1111/j.1744-7348.2011.00503.x.

Hosseini SA, Karami E, Talebi R (2013) Molecular variation of Iranian local and exotic strawberry (*Fragaria × ananassa* Duch.) varieties using SSR markers. *Biodivers J* 4: 243–252.

Huang LS, Sun YQ, Jin Y, Gao Q, Hu XG, et al. (2018) Development of high transferability cpSSR markers for individual identification and genetic investigation in *Cupressaceae* species. *Ecol E* 8:4967–4977. https://doi.org/10.1002/ece3.4053.

Huo J, Yang G, Zhang Y, Li F (2013) A new strategy for identification of currant (*Ribes nigrum* L.) cultivars using RAPD markers. *Genet Mol Res* 12(2):2056–2067. https://doi.org/10.4238/2013.June.21.1.

Inostroza-Blancheteau C, Aquea F, Reyes-Díaz M, Alberdi M, Arce-Johnson P (2011) Identification of aluminum-regulated genes by cDNA-AFLP analysis of roots in two contrasting genotypes of highbush blueberry (*Vaccinium corymbosum* L.). *Mol Biotechnol* 49:32–41. https://doi.org/10.1007/s12033-010-9373-3.

Ipek A, Barut E, Gulen H, Ipek M (2010) Genetic diversity among some currants (*Ribes* spp.) cultivars as assessed by AFLP markers. *Pak J Bot* 42(2):1009–1012.

Isobe SN, Hirakawa H, Sato S, Maeda F, Ishikawa M, et al. (2013) Construction of an integrated high density simple sequence repeat linkage map in cultivated strawberry (*Fragaria × ananassa*) and its applicability. *DNA Res* 20:79–92. https://doi.org/10.1093/dnares/dss035.

Jarret D, Rowbottom S, Morris J, Jorgensen L, Verrall SR, et al. (2013) Comparative transcriptome and metabolome analysis of developing blackcurrant seed provides an insight into pathway-specific gene-metabolite co-expression. *Asp Appl Biol* 119:157–162.

Jarret DA, Morris J, Cullen DW, Gordon SL, Verrall SR, et al. (2018) A transcript and metabolite atlas of blackcurrant fruit development highlights hormonal regulation and reveals the role of key transcription factors. *Front Plant Sci* 9:1235. https://doi.org/10.3389/fpls.2018.01235.

Joldersma D, Sadowski N, Timp W, Liu Z, Joldersma D, et al. (2022) Assembly and annotation of *Fragaria vesca* "Yellow Wonder" genome, a model diploid strawberry for molecular genetic research. *Fruit Res* 2:13. https://doi.org/10.48130/FruRes-2022-0013.

Jung S, Lee T, Cheng C-H, Buble K, Zheng P, et al. (2019) 15 years of GDR: new data and functionality in the Genome Database for Rosaceae. *Nucleic Acids Res* 47:D1137–D1145. https://doi.org/10.1093/nar/gky1000.

Juškytė AD, Mažeikienė I, Stanys V (2022) Putative genes of pathogenesis-related proteins and coronatine-insensitive *Protein 1* in *Ribes* spp. *Plants* 11:355. https://doi.org/10.3390/plants11030355.

Karlsons A, Osvalde A, Čekstere G, Pormale J (2018) Research on the mineral composition of cultivated and wild blueberries and cranberries. *Agron Res* 16:454–463.

Kawash J, Colt K, Hartwick NT, Abramson BW, Vorsa N, Polashock JJ, et al. (2022) Contrasting a reference cranberry genome to a crop wild relative provides insights into adaptation, domestication, and breeding. *PLoS ONE* 17(3): e0264966. https://doi.org/10.1371/journal. pone.0264966

Kim HJ, Lee JN, Cho KS, Won HS, Suh JT (2019) Genetic diversity and population structure analysis of everbearing and June-bearing strawberry cultivars using SSR markers. *Hortic Sci Biotechnol* 37:108–118. https://doi.org/10.12972/kjhst.20190010.

Knight RL, Keep E, Briggs JB, Parker JH (1974) Transference of resistance to blackcurrant gall mite, *Cecidophyopsisribis*, from gooseberry to black currant. *Ann Appl Biol* 76:123–130. https://doi.org/10.1111/j.1744-7348.1974.tb01362.x.

Knyazev SD, Ogoltsova TP (2004) *Breeding of blackcurrant nowadays*. Orel: Orel GAU, p. 238 (in Russian).

Korbin M, Kuras A, Żurawicz E (2002) Fruit plant germplasm characterisation using molecular markers generated in RAPD and ISSR-PCR. *CMBL* 7(2B):785–794.

Kunihisa M (2011) Studies using DNA markers in *Fragaria × ananassa*: genetic analysis, genome structure, and cultivar identification. *J Jpn Soc Hort Sci* 80:231–243. https://doi.org/10.2503/jjshs1.80.231.

Kunihisa M, Fukino N, Matsumoto S (2003) Development of cleavage amplified polymorphic sequence (CAPS) markers for identification of strawberry cultivars. *Euphytica* 134:209–215. https://doi.org/10.1023/B:Euph.0000003884.19248.33.

Labadie M, Vallin G, Petit A, Ring L, Hoffmann T, et al. (2020a) The genetic architecture of fruit colour in strawberry (*Fragaria × ananassa*) uncovers the predominant contribution of the *F. vesca*subgenome to anthocyanins and reveals underlying genetic variations. *bioRxiv* https://doi.org/10.1101/2020.06.25.171496.

Labadie M, Vallin G, Petit A, Ring L, Hoffmann T, et al. (2020b) Metabolite quantitative trait loci for flavonoids provide new insights into the genetic architecture of strawberry (*Fragaria × ananassa*) fruit quality. *J Agric Food Chem* 68:6927–6939. https://doi.org/10.1021/acs.jafc.0c01855.

Lācis G, Kārkliņa K, Bartulsons T, Stalažs A, Jundzis M, et al. (2022) Genetic structure of a *Ribes* genetic resource collection: inter- and intraspecific diversity revealed by chloroplast DNA simple sequence repeats (cpSSRs). *Sci Hortic* 304:111285. https://doi.org/10.1016/j.scienta.2022.111285.

Lācis G, Kārkliņa K, Kota-Dombrovska I, Strautiņa S (2021) Evaluation of blackcurrant (*Ribes nigrum*) germplasm structure by microsatellite-based fingerprinting for the diversification of the breeding material. *J Berry Res* 11:497–510. https://doi.org/10.3233/JBR-210743.

Langdon QK, Peris D, Kyle B, Hittinger CT (2018) sppIDer: a species identification tool to investigate hybrid genomes with high-throughput sequencing. *Mol Biol E* 35:2835–2849. https://doi.org/10.1093/molbev/msy166.

Lanham PG, Brennan RM (1998) Characterisation of the genetic resources of redcurrant (*Ribes rubrum*: subg. Ribesia) using anchored microsatellite markers. *Theor Appl Genet* 96:917-921. https://doi.org/10.1007/s001220050820.

Lanham PG, Brennan RM (1999) Genetic characterisation of gooseberry (*Ribes grossularia* subgenus Grossularia) germplasm using RAPD, ISSR and AFLP markers. *J Hortic Sci Biotechnol* 74(3):361–366. https://doi.org/10.1080/14620316.1999.11511122.

Lei D, Lin Y, Luo M, Zhao B, Tang H, et al. (2022) Genome-wide investigation of *G6PDH* gene in strawberry: evolution and expression analysis during development and stress. *Int J Mol Sci* 23:4728. https://doi.org/10.3390/ijms23094728.

Leišová-Svobodová L, Kissling P (2021) The European species of *Ribes* subg. *Ribes*: population genetic testing of classical systematics. *Plant Syst E* 307:41. https://doi.org/10.1007/s00606-021-01763-2.

Lerceteau-Köhler E, Roudeillac P, Markocic M, Guérin G, Praud K, Denoyes B (2002) The use of molecular markers for durable resistance breeding in the cultivated strawberry (*Fragaria* × *ananassa*). *Acta Hortic* 615–618. https://doi.org/10.17660/ActaHortic.2002.567.132.

Li H, Yue M, Jiang L, Liu Y, Zhang N, et al. (2022) Genome-wide identification of strawberry *C2H2-ZFP C1-2i* subclass and the potential function of *FaZAT10* in abiotic stress. *Int J Mol Sci* 23:13079. https://doi.org/10.3390/ijms232113079.

Li M, Li B, Yang M, Wang L, Hou G, et al. (2022) Genome-wide identification and expression of *MAPK*gene family in cultivated strawberry and their involvement in fruit developing and ripening. *Int J Mol Sci* 23:5201. https://doi.org/10.3390/ijms23095201.

Li M, Xiao Y, Mount S, Liu Z (2021) An atlas of genomic resources for studying *Rosaceae* fruits and ornamentals. *Front Plant Sci* 12:644881.

Li XG, Yang TT, Yu ZY, Huo JW, Dong Q, et al. (2019a) Cloning of flavonoid 3′,5′-hydroxylase and 3′-hydroxylase homologs from black currant (*Ribes nigrum*) and their differential expression at various fruit maturation stages. *J For Res* 30(2):463–470. https://doi.org/10.1007/s11676-017-0560-y.

Li Y, Pi M, Gao Q, Liu Z, Kang C (2019b) Updated annotation of the wild strawberry *Fragaria vesca* V4 genome. *Hortic Res* 6:1–9. https://doi.org/10.1038/s41438-019-0142-6.

Li Y, Wei W, Feng J, Luo H, Pi M, et al. (2018) Genome re-annotation of the wild strawberry *Fragaria vesca* using extensive Illumina- and SMRT-based RNA-seq datasets. *DNA Res* 25:61–70. https://doi.org/10.1093/dnares/dsx038.

Lin Y, Zhang J, Wu L, Zhang Y, Chen Q, et al. (2021) Genome-wide identification of GMP genes in *Rosaceae* and functional characterization of FaGMP4 in strawberry (*Fragaria* × *ananassa*). *Genes Genom* 43:587–599. https://doi.org/10.1007/s13258-021-01062-7.

Lin Y, Zhang L, Zhang J, Zhang Y, Wang Y, et al. (2020) Identification of anthocyanins-related glutathione *S-transferase (GST)*genes in the genome of cultivated strawberry (*Fragaria* × *ananassa*). *Int J Mol Sci* 21:8708. https://doi.org/10.3390/ijms21228708.

Liston A, Wei N, Tennessen JA, Li J, Dong M, et al. (2020) Revisiting the origin of octoploid strawberry. *Nat Genet* 52:2–4. https://doi.org/10.1038/s41588-019-0543-3.

Liu H, Wei L, Ni Y, Chang L, Dong J, et al. (2022) Genome-wide analysis of ascorbic acid metabolism related genes in *Fragaria* × *ananassa* and its expression pattern analysis in strawberry fruits. *Front Plant Sci* 13:954505.

Liu HT, Ji Y, Liu Y, Tian SH, Gao QH, et al. (2020) The sugar transporter system of strawberry: genome-wide identification and expression correlation with fruit soluble sugar-related traits in a *Fragaria* × *ananassa* germplasm collection. *Hortic Res* 7:1–19. https://doi.org/10.1038/s41438-020-00359-0.

Liu J, Wang J, Wang M, Zhao J, Zheng Y, et al. (2021a) Genome-wide analysis of the *R2R3-MYB*gene family in *Fragaria* × *ananassa* and its function identification during anthocyanins biosynthesis in pink-flowered strawberry. *Front Plant Sci* 12:702160.

Liu T, Li M, Liu Z, Ai X, Li Y (2021b) Reannotation of the cultivated strawberry genome and establishment of a strawberry genome database. *Hortic Res* 8:1–9. https://doi.org/10.1038/s41438-021-00476-4.

Liu Y, Liu S, Liu D, Wei Y, Liu C, et al. (2014) Exploiting EST databases for the development and characterization of EST-SSR markers in blueberry (*Vaccinium*) and their cross-species transferability in Vaccinium spp. *Sci Hortic* 176:319–329. https://doi.org/10.1016/j.scienta.2014.07.026.

Liu Y, Wang Y, Pei J, Li Y, Sun H (2021c) Genome-wide identification and characterization of COMT gene family during the development of blueberry fruit. *BMC Plant Biol* 21:1–16. https://doi.org/10.1186/s12870-020-02767-9.

Lobos GA, Hancock JF (2015) Breeding blueberries for a changing global environment: a review. *Front Plant Sci* 6:782. https://doi.org/10.3389/fpls.2015.00782

Luk'yanchuk IV, Lyzhin AS, Kozlova II (2018) Analysis of strawberry genetic collection (*Fragaria* L.) for *Rca2* and *Rpfl*genes with molecular markers. *Vavilov J Genet Breed* 22:795–799. https://doi.org/10.18699/VJ18.423.

Luo G, Xue L, Guo R, Ding Y, Xu W, et al. (2018) Creating interspecific hybrids with improved cold resistance in *Fragaria. Sci Hortic* 234:1–9. https://doi.org/10.1016/j.scienta.2018.02.023.

Lyzhin A, Luk'yanchuk I (2021) Marker-assisted screening of promising forms in the strawberry breeding. *E3S Web Conf* 254:03002. https://doi.org/10.1051/e3sconf/202125403002.

Marangelli F, Pavese V, Vaia G, Lupo M, Bashir MA, et al. (2022) *In vitro* polyploid induction of highbush blueberry through *de novo* shoot organogenesis. *Plants* 11:2349. https://doi.org/10.3390/plants11182349.

Marwal A, Gaur RK (2020) Molecular markers: tool for genetic analysis. In: Verma AS, Singh A (eds) *Animal Biotechnology* (2nd Edition). Academic Press, Boston, pp. 353–372.

Mazeikiene I, Bendokas V, Baniulis D, Staniene G, Juskyte DA, et al. (2017) Genetic background of resistance to gall mite in *Ribes* species. *Agric Food Sci* 26(2):111–117. https://doi.org/10.23986/afsci.59410.

Mažeikienė I, Bendokas V, Stanys V, Siksnianas T (2012) Molecular markers linked to resistance to the gall mite in blackcurrant. *Plant Breed* 131:762-766. https://doi.org/10.1111/j.1439-0523.2012.01995.x.

Mažeikienė I, Juškytė AD, Bendokas V, Stanys V (2022) *De novo*transcriptome analysis of *R. nigrum* cv. Aldoniai in response to blackcurrant reversion virus infection. *Int J Mol Sci* 23:9560. https://doi.org/10.3390/ijms23179560.

Mazeikiene I, Juskyte DA, Stanys V (2019). Application of marker-assisted selection for resistance to gall mite and Blackcurrant reversion virus in *Ribes* genus. *Zemdirbyste Agric* 106(4):359–366. https://doi.org/10.13080/z-a.2019.106.046.

McCallum S, Woodhead M, Jorgensen L, Gordon S, Brennan R, et al. (2012) Developing tools for long-term breeding of blueberry germplasm for UK production. *Int J Fruit Sci* 12:294–303. https://doi.org/10.1080/15538362.2011.619447.

Mengist MF, Grace MH, Xiong J, Kay CD, Bassil N, et al. (2020) Diversity in metabolites and fruit quality traits in blueberry enables ploidy and species differentiation and establishes a strategy for future genetic studies. *Front Plant Sci* 11:1–17. https://doi.org/10.3389/fpls.2020.00370.

Messinger W, Hummer K, Liston A (1999) *Ribes (Grossulariaceae)* phylogeny as indicated by restriction-site polymorphisms of PCR-amplified chloroplast DNA. *Plant Syst E* 217:185–195. https://doi.org/10.1007/BF00984364.

Mezhnina OA, Urbanovich OY (2017a) Genetic variability of blackcurrant varieties (*Ribes nigrum*) in Belarus. *Proc Nat Acad Sci Belarus Biol Ser* 1:62–69. (in Russian).

Mezhnina OA, Urbanovich OY (2017b) Study of genetic variability of representatives of *Ribes* L. grown in Belarus. *Cytol Genet* 51(6):441–447.

Michalska A, Łysiak G (2015) Bioactive compounds of blueberries: Post-harvest factors influencing the nutritional value of products. *Int J Mol Sci* 16:18642–18663. https://doi.org/10.3390/ijms160818642

Mohamed A-RMA-W, Jęcz T, Korbin M (2014) The history of genome mapping in *Fragaria* spp. *Hortic Res* 22:93–103. https://doi.org/10.2478/johr-2014-0026.

Morgil H, Gercek YC, Tulum I (2020) Single nucleotide polymorphisms (SNPs) in plant genetics and breeding. In: Caliskan M (ed) *The Recent Topics in Genetic Polymorphisms*. Intech Open, Turkey, p. 148.

Moročko-Bičevska I, Stalažs A, Lācis G, Laugale V, Baļķe I, et al. (2022) *Cecidophyopsis* mites and blackcurrant reversion virus on *Ribes* hosts: current scientific progress and knowledge gaps. *Ann Appl Biol* 180(1):26–43. https://doi.org/10.1111/aab.12720.

Moyano E, Martínez-Rivas FJ, Blanco-Portales R, Molina-Hidalgo FJ, Ric-Varas P, et al. (2018) Genome-wide analysis of the NAC transcription factor family and their expression during the development and ripening of the *Fragaria × ananassa* fruits. *PLoS One* 13. https://doi.org/10.1371/journal.pone.0196953.

Nadeem MA, Nawaz MA, Shahid MQ, Doğan Y, Comertpay G, et al. (2018) DNA molecular markers in plant breeding: current status and recent advancements in genomic selection and genome editing. *Biotechnol Biotechnol Equip* 32:261–285. https://doi.org/10.1080/13102818.2017.1400401.

Nellist CF, Vickerstaff RJ, Sobczyk MK, Marina-Montes C, Wilson FM, et al. (2019) Quantitative trait loci controlling *Phytophthora cactorum* resistance in the cultivated octoploid strawberry (*Fragaria × ananassa*). *Hortic Res* 6, 1–14. https://doi.org/10.1038/s41438-019-0136-4.

Njuguna W, Liston A, Cronn R, Ashman TL, Bassil N (2013) Insights into phylogeny, sex function and age of *Fragaria* based on whole chloroplast genome sequencing. *Mol Phylogenet E* 66: 17–29. https://doi.org/10.1016/j.ympev.2012.08.026.

Noguchi Y, Mochizuki T, Sone K (2002) Breeding of a new aromatic strawberry by interspecific hybridization *Fragaria × ananassa × F. nilgerrensis*. *J Jpn Soc Hortic Sci* 71:208–213. https://doi.org/10.2503/jjshs.71.208.

Norden EH, Lyrene PM, Chaparro JX (2020) Ploidy, fertility, and phenotypes of F1 hybrids between tetraploid highbush blueberry cultivars and diploid *Vaccinium elliottii*. *Hort Sci* 55(3):281–286. https://doi.org/10.21273/HORTSCI14597-19.

Nordlander J, Skytte af Satra J, Mattisson H, Udovychenko K, Lushpigan O, et al. (2022) Genetic diversity in gooseberry (*Ribes uva-crispa*), as estimated with SSR markers. *Sci Hortic* 306:111438. https://doi.org/10.1016/j.scienta.2022.111438.

Novy RG, Vorsa N (1996) Evidence for RAPD heteroduplex formation in cranberry: implications for pedigree and genetic-relatedness studies and a source of co-dominant RAPD markers. *Theor Appl Genet* 92:840–849. https://doi.org/10.1007/BF00221896.

Novy RG, Vorsa N, Kobak C, Goffreda J (1994) RAPDs identify varietal misclassification and regional divergence in cranberry [*Vaccinium macrocarpon* (Ait.) Pursh]. *Theor Appl Genet* 88:1004–1010. https://doi.org/10.1007/BF00220808.

Nunes CF, Ferreira JL, Generoso AL, Dias MSC, Pasqual M, et al. (2014) The genetic diversity of strawberry (*Fragaria ananassa* Duch.) hybrids based on ISSR markers. *Acta Sci Agron* 35:443–452.

Nybom H, Lācis G (2021) Recent large-scale genotyping and phenotyping of plant genetic resources of vegetatively propagated crops. *Plants* 10:415. https://doi.org/10.3390/plants10020415.

Oleas NH, von Wettberg EJB, Negrón-Ortiz V (2014) Population genetics of the federally threatened miccosukee gooseberry (*Ribes echinellum*), an endemic North American species. *Conserv Genet* 15:749–755. https://doi.org/10.1007/s10592-014-0575-2.

Palmieri L, Grando MS, Sordo M, Grisenti M, Martens S, et al. (2013) Establishment of molecular markers for germplasm management in a worldwide provenance *Ribes* spp. collection. *Plant Omics* 6:165–174.

Pavlenko A, Dolzhikova M, Pikunova A, Bakhotskaya A (2021) Use of SSR markers to study genetic polymorphism in members of the *Ribes* L. genus from the VNIISPK collection. E3S Web of Conferences 254:02009. https://doi.org/10.1051/e3sconf/202125402009.

Pellicer J, Leitch IJ (2020) The plant DNA C-values database (release 7.1): an updated online repository of plant genome size data for comparative studies. *New Phytol* 226:301–305.

Peng M, Zong X, Wang C, Meng F (2015) Genetic diversity of strawberry (*Fragaria ananassa* Duch.) from the Motuo County of the Tibet Plateau determined by AFLP markers. *Biotechnol Biotechnol Equip* 29:876–881. https://doi.org/10.1080/13102818.2015.1050968.

Perrotte J, Gaston A, Potier A, Petit A, Rothan C, Denoyes B (2016) Narrowing down the single homoeologous *FaPFRU* locus controlling flowering in cultivated octoploid strawberry using a selective mapping strategy. *Plant Biotechnol J* 14:2176–2189. https://doi.org/10.1111/pbi.12574.

Pikunova AV, Goryunova SV, Goryunov DV, Dolzhikova MA, Golyaeva OD (2020) The development of a genetic linkage map of redcurrant (*Ribes rubrum* L.) by means of SSR and SNP DNA markers. *Russ J Genet* 56(11):1387–1390.

Pikunova AV, Goryunova SV, Goryunov DV, Golyaeva OD, Dolzhikova MA, Pavlenko A (2022) Genetic diversity and pedigree analysis of red currant germplasm. *Plants* 11:1623. https://doi.org/10.3390/plants11131623

Pikunova AV, Knyazev SD, Golyaeva OD, Bahotskaya AU, Kalinina OV (2019) Genome studies by means of DNA markers of the blackcurrant. *Russ J Genet* 55(9):1061–1071.

Pikunova AV, Knyazev SD, Golyaeva OD, Bahotskaya AY, Kochumova AA (2015) Microsatellite loci polymorphism in Russian black currant (*Ribes nigrum* L.) varieties from collection of all-Russian research institute of breeding fruit crops. *Agric Biol* 50:46–54.

Pikunova AV, Martirosian EV, Kniazev SD, Ryzhova NN (2012) Application of the RAPD analysis for the study of genetic polymorphism and phylogenetic relationships in the *Ribes* L. genus. *Russ J Genet* 2(2):141–151. https://doi.org/10.1134/S2079059712020098.

Pincot DDA, Feldmann MJ, Hardigan MA, Vachev MV, Henry PM, et al. (2022) Novel *Fusarium* wilt resistance genes uncovered in natural and cultivated strawberry populations are found on three non-homoeologous chromosomes. *Theor Appl Genet* 135, 2121–2145. https://doi.org/10.1007/s00122-022-04102-2.

Polashock JJ, Vorsa N (2002) Development of SCAR markers for DNA fingerprinting and germplasm analysis of American cranberry. *J Am Soc Hortic* 127:677–684. https://doi.org/10.21273/jashs.127.4.677.

Provan J, Powell W, Hollingsworth PM (2001) Chloroplast microsatellites: new tools for studies in plant ecology and evolution. *Trends Ecol E* 16:142–147. https://doi.org/10.1016/S0169-5347(00)02097-8.

Qarni A, Muhammad K, Wahab A, Ali A, Khizar C, et al. (2022) Molecular characterization of wild and cultivated strawberry (*Fragaria × ananassa*) through DNA barcode markers. *Genet Res (Camb)* 2022:9249561. https://doi.org/10.1155/2022/9249561.

Qi X, Ogden EL, Bostan H, Sargent DJ, Ward J, et al. (2021) High-density linkage map construction and QTL identification in a diploid blueberry mapping population. *Front Plant Sci* 12:1–13. https://doi.org/10.3389/fpls.2021.692628.

Qiao Q, Edger PP, Xue L, Qiong L, Lu J, et al. (2021) Evolutionary history and pan-genome dynamics of strawberry (*Fragaria* spp.). *Proc Natl Acad Sci USA* 118:e2105431118. https://doi.org/10.1073/pnas.2105431118.

Qu L, Hancock JF (1997) Randomly amplified polymorphic DNA- (RAPD-) based genetic linkage map of blueberry derived from an interspecific cross between diploid *Vaccinium darrowi* and tetraploid *V. corymbosum*. *J Am Soc Hortic* 122:69–73.

Quililongo CM, Wheeler PV, Cumplido LV, Ramírez MR, Fica JE (2013) Efficiency of RAPD in assessing genetic variation in commercially viable local varieties of blueberry (*Vaccinium*). *Agric Sci* 04:238–243. https://doi.org/10.4236/as.2013.45034.

Reddy MP, Sarla N, Siddiq EA (2002) Inter simple sequence repeat (ISSR) polymorphism and its application in plant breeding. *Euphytica* 128:9–17.

Redpath LE, Aryal R, Lynch N, Spencer JA, Hulse-Kemp AM, et al. (2022) Nuclear DNA contents and ploidy levels of North American Vaccinium species and interspecific hybrids. *Sci Hortic* 297:110955.

Rigotti S (2002) Characterization of molecular markers for specific and sensitive detection of *Botrytis cinerea* Pers.: Fr. in strawberry (*Fragaria × ananassa* Duch.) using PCR. *FEMS Microbiol Lett* 209:169–174. https://doi.org/10.1016/S0378-1097(02)00491-3.

Rousseau-Gueutin M, Gaston A, Aïnouche A, Aïnouche ML, Olbricht K, et al. (2009) Tracking the evolutionary history of polyploidy in *Fragaria* L. (strawberry): new insights from phylogenetic analyses of low-copy nuclear genes. *Mol Phylogenet E* 51: 515–530. https://doi.org/10.1016/j.ympev.2008.12.024.

Rowland LJ, Alkharouf N, Darwish O, Ogden EL, Polashock JJ, et al. (2012) Generation and analysis of blueberry transcriptome sequences from leaves, developing fruit, and flower buds from cold acclimation through deacclimation. *BMC Plant Biol* 12:46.

Rowland LJ, Dhanaraj AL, Polashock JJ, Arora R (2003b) Utility of blueberry-derived EST-PCR primers in related *Ericaceae*species. *HortScience* 38:1428–1432. https://doi.org/10.21273/hortsci.38.7.1428.

Rowland LJ, Levi A (1994) RAPD-based genetic linkage map of blueberry derived from a cross between diploid species (*Vaccinium darrowi* and *V. elliottii*). *Theor Appl Genet Int J Plant Breed Res* 87:863–868. https://doi.org/10.1007/BF00221139.

Rowland LJ, Mehra S, Dhanaraj AL, Ogden EL, Slovin JP, Ehlenfeldt MK (2003a) Development of EST-PCR markers for DNA fingerprinting and genetic relationship studies in blueberry (*Vaccinium*, section *Cyanococcus*). *J Am Soc Hortic* 128:682–690. https://doi.org/10.21273/jashs.128.5.0682.

Rowland LJ, Ogden EL, Bassil N, Buck EJ, McCallum S, et al. (2014) Construction of a genetic linkage map of an interspecific diploid blueberry population and identification of QTL for chilling requirement and cold hardiness. *Mol Breed* 34:2033–2048. https://doi.org/10.1007/s11032-014-0161-9.

Rowland LJ, Ogden EL, Vinyard BT (2020) Phenotypic evaluation of a hybrid diploid blueberry population for plant development and fruit quality traits. *Agronomy* 10:1–11. https://doi.org/10.3390/agronomy10081067.

Rugienius R, Šikšnianienė JB, Frercks B, Stanienė G, Stepulaitienė I, et al. (2015) Characterization of strawberry (*Fragaria × ananassa* Duch.) cultivars and hybrid clones using SSR and AFLP markers. *Zemdirbyste Agric* 102:177–184. https://doi.org/10.13080/z-a.2015.102.023.

Rui S, GuiXia W, LinLin C, Jian S, Jing D, et al. (2018) Development and application of molecular marker technique in strawberry (*Fragaria × ananassa*). *J Agric Biotechnol* 26:1588–1600.

Ruiz-Rojas JJ, Sargent DJ, Shulaev V, Dickerman AW, Pattison J, et al. (2010) SNP discovery and genetic mapping of T-DNA insertional mutants in *Fragaria vesca* L. *Theor Appl Genet* 121:449–463. https://doi.org/10.1007/s00122-010-1322-9.

Russell J, Hackett C, Hedley P, Liu H, Milne L, et al. (2014) The use of genotyping by sequencing in blackcurrant (*Ribes nigrum*): developing high-resolution linkage maps in species without reference genome sequences. *Mol Breed* 33(4):835–849. https://doi.org/10.1007/s11032-013-9996-8.

Russell JR, Bayer M, Booth C, Cardle L, Hackett CA, et al. (2011) Identification, utilisation and mapping of novel transcriptome-based markers from blackcurrant (*Ribes nigrum*). *BMC Plant Biol* 11:147. https://doi.org/10.1186/1471-2229-11-147.

Salinas N, Verma S, Peres N, Whitaker VM (2019) FaRCa1: a major subgenome-specific locus conferring resistance to *Colletotrichum acutatum* in strawberry. *Theor Appl Genet* 132:1109–1120. https://doi.org/10.1007/s00122-018-3263-7.

Salinas NR, Zurn JD, Mathey M, Mookerjee S, Denoyes B, et al. (2017) Validation of molecular markers associated with perpetual flowering in Octoploid *Fragaria* germplasm. *Mol Breed* 37:70. https://doi.org/10.1007/s11032-017-0672-2.

Sangiacomo MA, Sullivan JA (1994) Introgression of wild species into the cultivated strawberry using synthetic octoploids. *Theor Appl Genet* 88, 349–354. https://doi.org/10.1007/BF00223644.

Sargent DJ, Fernandéz-Fernandéz F, Ruiz-Roja JJ, Sutherland BG, Passey A, et al. (2009) A genetic linkage map of the cultivated strawberry (*Fragaria × ananassa*) and its comparison to the diploid *Fragaria* reference map. *Mol Breed* 24:293–303. https://doi.org/10.1007/s11032-009-9292-9.

Sargent DJ, Kuchta P, Girona EL, Zhang H, Davis TM, et al. (2011) Simple sequence repeat marker development and mapping targeted to previously unmapped regions of the strawberry genome sequence. *Plant Genome* 4:165–177. https://doi.org/10.3835/plantgenome2011.05.0014.

Sargent DJ, Passey T, Surbanovski N, Lopez Girona E, Kuchta P, et al. (2012) A microsatellite linkage map for the cultivated strawberry (*Fragaria × ananassa*) suggests extensive regions of homozygosity in the genome that may have resulted from breeding and selection. *Theor Appl Genet* 124:1229–1240. https://doi.org/10.1007/s00122-011-1782-6.

Schlautman B, Bolivar-Medina J, Hodapp S, Zalapa J (2017a) Cranberry SSR multiplexing panels for DNA horticultural fingerprinting and genetic studies. *Sci Hortic* 219:280–286. https://doi.org/10.1016/j.scienta.2017.03.005.

Schlautman B, Covarrubias-Pazaran G, Diaz-Garcia L, Iorizzo M, Polashock J, et al. (2017b) Construction of a high-density American Cranberry (*Vaccinium macrocarpon* Ait.) composite map using genotyping-by-sequencing for multi-pedigree linkage mapping. *G3* 7:1177–1189. https://doi.org/10.1534/g3.116.037556.

Schlautman B, Covarrubias-Pazaran G, Diaz-Garcia LA, Johnson-Cicalese J, Iorrizo M, et al. (2015b) Development of a high-density cranberry SSR linkage map for comparative genetic analysis and trait detection. *Mol Breed* 35:177. https://doi.org/10.1007/s11032-015-0367-5.

Schlautman B, Diaz-Garcia L, Covarrubias-Pazaran G, Schlautman N, Vorsa N, et al. (2018) Comparative genetic mapping reveals synteny and collinearity between the American cranberry and diploid blueberry genomes. *Mol Breed* 38:9. https://doi.org/10.1007/s11032-017-0765-y.

Schlautman B, Fajardo D, Bougie T, Wiesman E, Polashock J, et al. (2015a) Development and validation of 697 novel polymorphic genomic and EST-SSR markers in the American cranberry (*Vaccinium macrocarpon* Ait.). *Molecules* 20:2001–2013. https://doi.org/10.3390/molecules20022001.

Schultheis LM, Donoghue MJ (2004) Molecular phylogeny and biogeography of *Ribes* (Grossulariaceae), with an emphasis on gooseberries (subg. *Grossularia*). *Syst Bot* 29(1):77–96.

Sedlák P, Sedlaková V, Vašek J, Zeka D, Čílová D, et al. (2022) Phenotypic, molecular and biochemicalevaluation of somatic hybrids between *Solanum tuberosum* and *S. bulbocastanum*. *Sci Rep* 12:4484. https://doi.org/10.1038/s41598-022-08424-5.

Seppey M, Manni M, Zdobnov EM (2019) BUSCO: assessing genome assembly and annotation completeness. *Methods Mol Biol* 1962:227–245. https://doi.org/10.1007/978-1-4939-9173-0_14.

Sharma S, Kaur R, Kumar K, Prasad H (2021) Genetic variability in strawberry (*Fragaria × ananassa* Duch.) cultivars assessed by morphological traits and EST-SSR markers of *Rubus ellipticus*. *Indian J Biotechnol* 20:81–90.

Sheng L, Ma C, Chen Y, Gao H, Wang J. (2021) Genome-wide screening of AP2 transcription factors involving in fruit color and aroma regulation of cultivated strawberry. *Genes (Basel)* 12(4):530. https://doi.org/10.3390/genes12040530.

Shimomura K, Hirashima K (2006) Development and characterization of simple sequence repeats (SSR) as markers to identify strawberry cultivars (*Fragaria × ananassa* Duch.). *J Japan Soc Hort Sci* 75:399–402. https://doi.org/10.2503/jjshs.75.399.

Shulaev V, Sargent DJ, Crowhurst RN, Mockler TC, Folkerts O, et al. (2011) The genome of woodland strawberry (*Fragaria vesca*). *Nat Genet* 43, 109–116. https://doi.org/10.1038/ng.740.

Skrovankova S, Sumczynski D, Mlcek J, Jurikova T, Sochor J (2015) Bioactive compounds and antioxidant activity in different types of berries. *Int J Mol Sci* 16(10):24673–706. https://doi.org/10.3390/ijms161024673

Starkevič P, Ražanskienė A, Starkevič U, Kazanavičiūtė V, Denkovskienė E, et al. (2020) Isolation and analysis of anthocyanin pathway genes from *Ribes* genus reveals *MYB* gene with potent anthocyanin-inducing capabilities. *Plants* 9, 1078. https://doi.org/10.3390/plants9091078.

Stewart CN, Excoffier L (1996) Assessing population genetic structure and variability with RAPD data: application to *Vaccinium macrocarpon* (American Cranberry). *J Evol Biol* 9:153–171. https://doi.org/10.1046/j.1420-9101.1996.9020153.x.

Sun J, Sun R, Liu H, Chang L, Li S, et al. (2021) Complete chloroplast genome sequencing of ten wild *Fragaria* species in China provides evidence for phylogenetic evolution of *Fragaria*. *Genomics* 113:1170–1179. https://doi.org/10.1016/j.ygeno.2021.01.027.

Tailor S, Bykova NV, Igamberdiev AU, Debnath SC (2017) Structural pattern and genetic diversity in blueberry (*Vaccinium*) clones and cultivars using EST-PCR and microsatellite markers. *Genet Resour Crop E* 64:2071–2082. https://doi.org/10.1007/s10722-017-0497-1.

Tennessen JA, Govindarajulu R, Ashman TL, Liston A (2014) Evolutionary origins and dynamics of octoploid strawberry subgenomes revealed by dense targeted capture linkage maps. *Genome Biol E* 6:3295–3313. https://doi.org/10.1093/gbe/evu261.

Tlatelpa MA, Zavala GC, Espinosa MAG, et al. (2019) DNA fingerprint of strawberry varieties developed at Colegio de Postgraduados, Mexico. *Nova Scientia* 11:186–206. https://doi.org/10.21640/ns.v11i22.1794.

Tyrka M, Dziadczyk P, Hortynski JA (2002) Simplified AFLP procedure as a tool for identification of strawberry cultivars and advanced breeding lines. *Euphytica* 125:273–280.

van Dijk T, Pagliarani G, Pikunova A, Noordijk Y, Yilmaz-Temel H, et al. (2014) Genomic rearrangements and signatures of breeding in the allo-octoploid strawberry as revealed through an allele dose based SSR linkage map. *BMC Plant Biol* 14:55. https://doi.org/10.1186/1471-2229-14-55.

Verma S, Zurn JD, Salinas N, Mathey MM, Denoyes B, et al. (2017) Clarifying sub-genomic positions of QTLs for flowering habit and fruit quality in U.S. strawberry (*Fragaria × ananassa*) breeding populations using pedigree-based QTL analysis. *Hortic Res* 4:1–9. https://doi.org/10.1038/hortres.2017.62.

Vidyagina EO, Lebedev VG, Subbotina NM, Treschevskaya EI, Lebedeva TN, et al. (2021) The development of the genic SSR markers for analysis of genetic diversity in gooseberry cultivars. *Agronomy* 11:1050. https://doi.org/10.3390/agronomy11061050.

Vorsa N, Zalapa J (2020) Domestication, genetics, and genomics of the American cranberry. *Plant Breed Rev* 43:279–315.

Wambui N (2010) Development and use of molecular tools in *Fragaria*. Dissertation Oregon State University, USA, pp. 370.

Wang L, Liang J, Sa W, Wang L (2021) Sequencing and comparative analysis of the chloroplast genome of *Ribes odoratum* provide insights for marker development and phylogenetics in *Ribes*. *Physiol Mol Biol Plants* 27(1):81–92. https://doi.org/10.1007/s12298-021-00932-4.

Wei N, Cronn R, Liston A, Ashman TL (2019) Functional trait divergence and trait plasticity confer polyploid advantage in heterogeneous environments. *New Phytol* 221, 2286–2297. https://doi.org/10.1111/nph.15508.

Whitaker VM (2011) Applications of molecular markers in strawberry. *J Berry Res* 1:115–127. https://doi.org/10.3233/BR-2011-013.

Wójcik D, Trzewik A, Kucharska D (2021) Field performance and genetic stability of micropropagatedgooseberry plants (*Ribes grossularia* L.). *Agronomy* 11:45. https://doi.org/10.3390/agronomy11010045.

Xu J, Huo Y, Dong K, Geng J, Dong M, et al. (2021) Development of novel polymorphic EST-SSR markers from the cranberry fruit transcriptome. *Iran J Biotechnol* 19:48–55. https://doi.org/10.30498/IJB.2021.2499.

Yang Y, Davis TM (2017) A new perspective on polyploid *Fragaria* (strawberry) genome composition based on large-scale, multi-locus phylogenetic analysis. *Genome Biol E* 9:3433–3448. https://doi.org/10.1093/gbe/evx214.

Yoon M-Y, Moe KT, Kim D-Y, Rho IR, Kim S, et al. (2012) Genetic diversity and population structure analysis of strawberry (*Fragaria × ananassa* Duch.) using SSR markers. *Elect J Biotechnol* 15:6–6. https://doi.org/10.2225/vol15-issue2-fulltext-5.

Yuan H, Yu H, Huang T, Shen X, Xia J, et al. (2019) The complexity of the *Fragaria × ananassa* (octoploid) transcriptome by single-molecule long-read sequencing. *Hortic Res* 6:1–14. https://doi.org/10.1038/s41438-019-0126-6.

Zalapa JE, Bougie TC, Bougie TA, Schlautman BJ, Wiesman E, et al. (2015) Clonal diversity and genetic differentiation revealed by SSR markers in wild *Vaccinium macrocarpon* and *Vaccinium oxycoccos*. *Ann Appl Biol* 166:196–207. https://doi.org/10.1111/aab.12173.

Zdepski A, Debnath SC, Howell A, Polashock J, Oudemans P, Vorsa N, et al (2016) *Cranberry. Genetics, genomics and Breeding of Berries*. CRC Press, pp. 41–63.

Zhang H, Jing X, Chen Y, Liu Z, Xin Y, et al. (2020a) The genome-wide analysis of *RALF*-like genes in strawberry (wild and cultivated) and five other plant species (*Rosaceae*). *Genes* 11:174. https://doi.org/10.3390/genes11020174.

Zhang J, Lei Y, Wang B, Li S, Yu S, et al. (2020b) The high-quality genome of diploid strawberry (*Fragaria nilgerrensis*) provides new insights into anthocyanin accumulation. *Plant Biotechnol J* 18:1908–1924. https://doi.org/10.1111/pbi.13351.

Zhang L, Cao B, Bai C (2013) New reports of nuclear DNA content for 66 traditional Chinese medicinal plant taxa in China. *Caryologia* 66:375–383.https://doi.org/10.1080/00087114.2013.859443.

Zhang Y, Miao L, Yang X, Jiang G (2022) Genome-wide characterization and expression of the TLP gene family associated with *Colletotrichum gloeosporioides* inoculation in *Fragaria × ananassa*. *Peer J* 10:e12979. https://doi.org/10.7717/peerj.12979.

Zhou B, Wang J, Lou H, Wang H, Xu Q (2019) Comparative transcriptome analysis of dioecious, unisexual floral development in *Ribes diacanthum* Pall. *Gene* 699:43–53. https://doi.org/10.1016/j.gene.2019.03.009.

Zhu H, Senalik D, McCown BH, Zeldin EL, Speers J, et al. (2012) Mining and validation of pyrosequenced simple sequence repeats (SSRs) from American cranberry (*Vaccinium macrocarpon* Ait.). *Theor Appl Genet* 124:87–96. https://doi.org/10.1007/s00122-011-1689-2.

7 Allele Mining and Candidate Gene Identification in Cherries

Basis for Crop Improvement

Birutė Frercks, Raminta Antanynienė, Edvinas Misiukevičius, Jurgita Vinskienė, Augustina Kolytaitė, and Gunars Lacis

7.1 INTRODUCTION

7.1.1 Taxonomical Relationships of Cherry Species

Sweet cherry (*Prunus avium*) and sour cherry (*Prunus cerasus*), both with edible fruit trees, belong to the *Prunus* subg. *Cerasus* Pers. according to Rehder's classification (Dondini et al. 2018). Sweet cherry is diploid ($2n=2x=16$, AA genome), and the genome size is believed to be around 350 Mb (Soundararajan et al. 2019), but the biggest genome of sweet cherry cv. Tieton sequenced up to date at chromosome scale is 344.29 Mb (Wang et al. 2000). Sour cherry is tetraploid ($2n=4x=32$, AAFF genome), and the genome size is believed to be 599 Mb (Dirlewanger et al. 2007), but the sequencing is still lacking, probably due to the challenges of sequencing and analyzing genomic data for polyploid genomes (Bird et al. 2021). The ground cherry (*Prunus fruticosa*) is a wild scrub that belongs to this subgenus and is closely related to both cherries (Dirlewanger et al. 2007). Ground cherry is a self-incompatible tetraploid ($2n=4x=32$, FFFF genome), and the recently sequenced genome size is 366 Mb (Wöhner et al. 2021). The ground cherry and sweet cherry are progenitors of sour cherry (Figure 7.1), which resulted from the natural hybridization between ground cherry and unreduced gametes of sweet cherry (Tavaud et al. 2004).

Although the close relatedness between the three *Prunus* species was suggested after observation of the artificial hybridization of sweet and ground cherry more than 50 years ago (Oldén and Nybom 1968), many studies with different techniques have been initiated to date to approve it. Based on karyotype analysis and with genomic *in situ* hybridization (GISH) of *P. avium* and *P. fruticosa* DNA, *P. avium* chromosomes were detected in *P. cerasus* genome (Schuster and Schreiber 2000). In studies based on chloroplast DNA, three species of subgenus *Cerasus* separated into one group from the reminding *Prunus* species (Badenes and Parfitt 1995; Bortiri et al. 2001). Initially, it was discovered that the ground cherry should be a maternal progenitor (Badenes and Parfitt 1995; Brettin et al. 2000), but there are contrary results (Shi et al. 2013). The relationships between the three cherries were also confirmed by flow cytometry (Macková et al. 2018). The presence of the sweet cherry genome in the sour cherry genome was confirmed with AFLP molecular markers (Tavaud et al. 2004) and *S*-alleles (Lisek et al. 2017). The latest research by sequencing plastid genomes and nuclear genes on the transcriptome level confirmed the parental origins of sour cherry and the maternal inheritance of chloroplast in sour cherry from ground cherry (Bird et al. 2022).

DOI: 10.1201/9781003386490-7

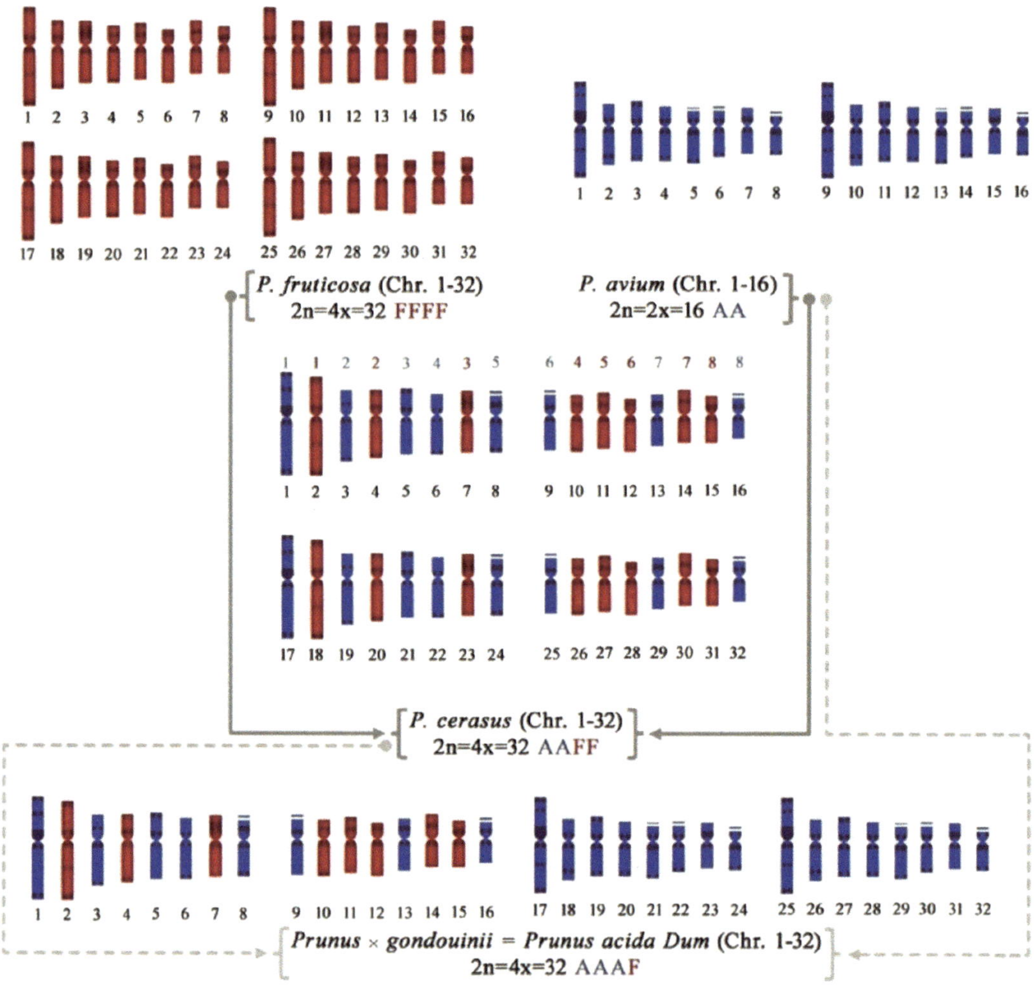

FIGURE 7.1 Taxonomical relationships of cherry species. The ground cherry (*Prunus fruticosa*) and sweet cherry (*P. avium*) are progenitors of sour cherry (*P. cerasus*), which resulted from the natural hybridization between ground cherry and unreduced gametes of sweet cherry. *P.* × *gondouinii* resulted from the pollination of sour cherry by unreduced gametes of sweet cherry (Tavaud et al. 2004). The idiotypes of chromosomes are adapted from Schuster and Schreiber (2000). The numbers in different colors above the idiotype of sour cherry indicate the progenitor of inherited chromosomes.

7.1.2 Distribution of Cherries

It is believed that the center of origin of the wild cherry is in Anatolia around the Caspian and Black Seas, and through the migration of people and birds, cherries slowly spread towards the west (Webster 1996; Dirlewanger et al. 2007; Iezzoni et al. 2017). However, there are some findings of cherry seeds in many European sites dating back to 3,000–4,000 years B.C. Therefore, the cherry is considered native to the whole Europe (Faust and Surányi 2011). However, the cherry's natural cherry range in Europe covers the temperate region from North Spain to Turkey and south Russia (Dirlewanger et al. 2007). There is no consensus either when the cherry arrived overseas in America: some authors indicate the 1900s (Dondini et al. 2018), and some others say the mid-1500s (Jajasankar and Kappel 2011). In Europe, the native ranges of the three closely related cherries overlap. The sweet cherry is the least cold-hardy species compared to the ground and sour cherry, and grows throughout mainland Europe's temperate regions and southern Russia (Brettin et al. 2000). Sweet cherry prefers regions where summers

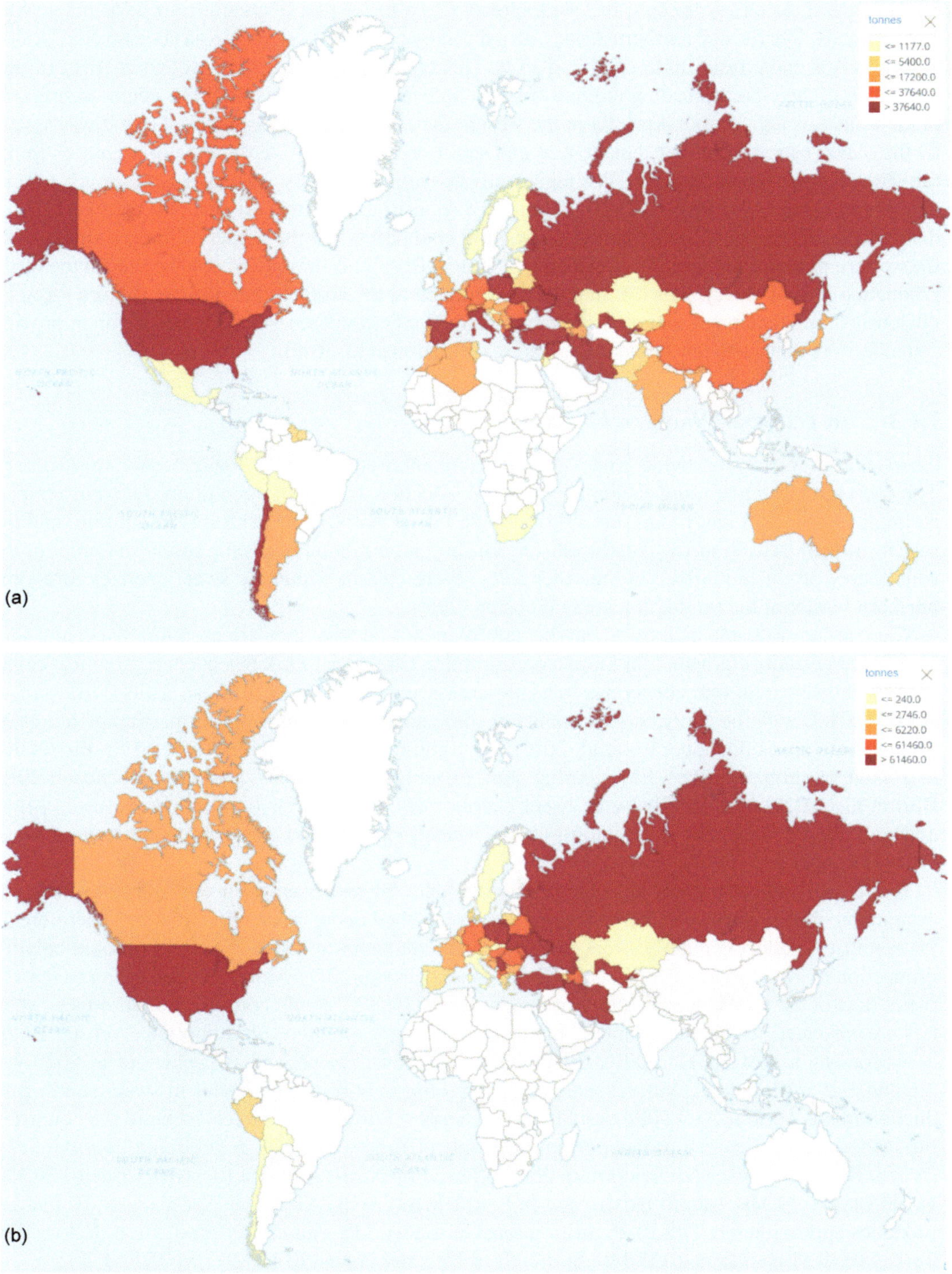

FIGURE 7.2 World cherry production by country according to FAOSTAT (2020) database. (a) sweet cherry, (b) sour cherry.

are warm and dry, but it still needs adequate rainfall during the growing season (Dirlewanger et al. 2007). Therefore, the production of sweet cherries is distributed through all continents in temperate climate zones in the northern and southern hemispheres (Figure 7.2a). The ground cherry is the coldest hardy of the three cherry species. It is widespread over a major part of central Europe as far west as northern Italy, Siberia, and northern Asia, with a diversity center in western Russia (Brettin et al. 2000). Sour cherry is more resistant to cold than sweet cherry but slightly less than ground cherry (Tavaud et al. 2004). Therefore, spreading regions of sweet and ground cherry overlap. The center of origin is in the Caucasus region from Anatolia in the west to the north of Iran. It is believed that it was spread by the Slavic people from west Asia to east and southeast Europe during the time of migration in the 6th–8th centuries (Schuster 2019). The highest diversity of sour cherry was observed in south-eastern and central Europe (Brettin et al. 2000). Due to its cold resistance, sour cherry extends the sweet cherry distribution range up into higher latitudes (Dondini et al. 2018), but the border of the growing area in the western part of Europe ends in Germany (Schuster 2019). This limitation could be seen in the world production of sour cherry, which is mainly concentrated in the northern hemisphere (Figure 7.2b). So obviously the cultivation of sour cherry established in Nordic countries is mainly focused on improving sour cherry resistance to low winter temperatures (Dondini et al. 2018).

7.1.3 The Economic Value of Cherries

Cherries are grown worldwide where the environmental conditions are suitable for this species. The most important limiting factor for cherry production is climatic condition. Only in temperate regions, tree dormancy is ensured through the fall of average temperatures in autumn, and the regrow of fruit trees in spring is induced. The chilling requirement limits the southern border of the sour cherry production area. On the other side, severe cold in winter for sweet cherries limits the northern border of the production area (Webster 1996).

Cherries are valuable fruit trees, but the value depends on what they are used for. Sweet and sour cherries are usually cultivated for human consumption (Webster 1996). Sweet cherry, in most cases, is more valuable than sour cherry due to its sweetness and consumption of fresh fruits (Dirlewanger et al. 2007). Due to the many health benefits of sour cherry, their commercial importance increases in temperate zones (Ganopoulos et al. 2016), but traditionally, they are mostly used for processing in a range of culinary dishes, like canning, jam, juice, bakery products, and spirits (Schuster 2019; Bird et al. 2022). Additionally, wild sweet cherry trees are used for their wood (Webster 1996), and sour cherry rootstocks were helpful for the development of new commercial cultivars of sweet cherry (Kappel et al. 2012).

It is a challenge to present accurate production data for sweet and sour cherry fruits. First, the production of sour and sweet cherry is sometimes published under one name, "cherry." Therefore, in the countries producing mainly sweet cherries, the production is given for both cherries, but only the production of sweet cherry is presented (Bujdosó and Hrotkó 2017). Moreover, according to the statistics data of the food and agriculture organization (FAOSTAT), only commercially produced cherries are presented, as the trees planted in gardens are assigned to home consumption (Webster 1996).

According to FAOSTAT (2020), the global production of sweet and sour cherries in 2020 was 2.6 and 1.48 million tons (Mt), respectively. According to regions, the leader in sweet cherry production in the world is Asia (48%), and for sour cherry – Europe (66%). According to the countries, the leader of sweet cherry production in the world is Turkey (0.7 Mt) by realizing approximately 28% of the world's sweet cherry production, followed by North America (0.3 Mt), Chile (0.26 Mt), Uzbekistan (0.19 Mt), Iran (0.16 Mt), and Italy (0.1 Mt) (FAOSTAT 2020). The Russian Federation produces approximately 17% (0.26 million tons) of the world's sour cherry production, followed by Turkey (0.19 Mt), Ukraine (0.17 Mt), Serbia (0.16 Mt), and Poland (0.15 Mt) (FAOSTAT 2020).

Every year, global cherry production rises slightly. For example, from 3.3 million tons in 2019 (Quero-García et al. 2019) to slightly more than 4 million tons in 2020 (FAOSTAT 2020). But there are shifts between regions. In this millennium, the market for cherries joined Chile, Argentina, and China (Dirlewanger et al. 2007; Bujdosó and Hrotkó 2017; Quero-García et al. 2019). On the other hand, production in France and Germany has largely declined (Bujdosó and Hrotkó 2017).

7.1.4 Molecular Breeding of Cherries

Controlled breeding programs for the genetic improvement of cherries began in the early 20th century. As the most ancient areas of cherry cultivation are believed to be Europe countries and Anatolia, the highest number of governmental-supported cherry breeding programs are in Europe (Table 7.1). The breeding programs of sweet cherry were investigated by conducted and reviewed by Quero-García et al. (2017) and Dondini et al. (2018). Kappel (2008) described the situation regarding cherry breeding in the countries of the "New world". Some of the sour cherry breeding programs in the world are listed by Iezzoni (2008). There are some inconsistencies in the number of sour cherry breeding programs (Iezzoni 2008; Schuster 2019). This could be due to the fact that some of the breeding programs have been completely terminated, and some of them are keeping the existing selections for evaluation without making new crosses. Additionally, private breeding programs exist, at least in the USA.

The aims of breeding programs vary a little depending on the geographical location and what kind of cherries are used (sour or sweet).

One important task for all breeding programs is to evaluate the genetic diversity of plant material. In the beginning, the local cultivars (landraces) of cherries were selected as parents for the breeding programs in Europe due to the broad landrace diversity. The sour cherry cultivars cultivated today, in most cases are old landraces selections or next-generation of landraces selections (Iezzoni 2008).

The main goals of sour and sweet cherry breeding programs are excellent fruit quality and high productivity (Schuster 2019; Quero-García et al. 2019). Excellent fruit quality is characterized by fruit size, firmness, and flavor (Dondini et al. 2018). The suitability for mechanical harvesting, late flowering, tolerance or resistance to biotic and abiotic stresses, and the extension of harvesting season are major priorities in the breeding programs investigating sour cherries (Schuster 2019). The cultivars of sour and sweet cherry with easy peduncle removal are suitable for mechanical harvesting (Sansavini and Lugli 2008). However, evaluating the genetic diversity of plant material is an important part of all breeding programs. At first, the local cultivars (also called landraces) of cherries were selected as parents for the breeding programs in Europe due to the broad landrace diversity. Most of the sour cherry cultivars grown today are either old landrace selections or landrace selections of the next generation (Iezzoni 2008). The genetic diversity of cherries, characterized by simple sequence repeats (SSRs) markers, will be reviewed in Section 7.2.1.

Late flowering is also important for sweet cherry breeding, but mostly in northern regions, where the risk of spring frost is high. Several studies were induced on the evaluation of cherries' resistance to spring frost and/or winter hardiness in the Czech Republic (Vávra et al. 2018), Germany (Chmielewski et al. 2018), Lithuania (Gelvonauskienė et al. 2013; Stepulaitienė et al. 2013), and Russia (Ozherelieva 2020; Upadysheva et al. 2021; Gulyaeva et al. 2022). Molecular markers were found for spring frost resistant and susceptible sour cherry (Gelvonauskienė et al. 2013).

Another environmental challenger, especially for sweet cherry breeding, is rain-induced fruit cracking (reviewed in Correia et al. 2017; Hakeem 2015). It was studied that skin cracks only occur in specific areas of the fruit and that this is due to direct water entry through the fruit skin (Rodríguez et al. 2018). Winkler et al. (2016) suggested the zipper model, which explains the local impact of the fruit skin to liquid-phase water. For the evaluation of sweet cherry resistance to fruit cracking, the cracking index was introduced (Christensen 1972). Later, a more reliable method for determining skin cracking was developed, and it was named the waterfall method (Michailidis et al. 2020). Already in this decade, several sweet cherry cultivars were evaluated for cracking resistance (Pereira et al. 2020; Quero-García et al. 2021; Demirsoy et al. 2022), and the first stable quantitative trait loci (QTLs) were found for fruit cracking (Quero-García et al. 2021). More about fruit cracking will be discussed in Section 7.2.3.

The other objectiveness common to most cherry breeding programs is to increase the self-fertility of cherries, especially sweet ones. Compared to other traits, the basis of self-fertility or infertility is well explored on a molecular level. The gametophytic system of incompatibility (GSI) is controlled by a single gametophytic expressed multiallelic *S*-locus (Kivistik et al. 2021), and

TABLE 7.1
Public Cherry Breeding Programs

Region	Country	Location	Sweet Cherry	Sour Cherry	Continuing
Europe	1. Belarus	Research Institute for Fruit Growing, Minsk	✓	✓	Yes
	2. Bulgaria	Fruit Growing Institute, Plovdiv; Institute of Agriculture, Kyustendil	✓	✓	Yes
	3. Czech Republic	Research and Breeding Institute of Pomology, Holovosy	✓		Yes
	4. Estonia	Polli HRC-Polli Research Center of Horticulture, Karksi	✓	✓	Yes
	5. France	P. Argot and Delbard Pepinières, Commentry	✓	✓	No
	6. France	INRA, Station de Recherches Fruitières, Bordeaux UK – John Innes Institute, Norwich	✓	✓	Yes
	7. Germany	BAZ, Institute for Fruit Breeding, Dresden	✓	✓	Yes
	8. Germany	Fruit Research Station, York	✓		Yes
	9. Hungary	Research Institute for Fruitgrowing and Ornamentals, Budapest	✓	✓	Yes
	10. Italy	DipartimentoColtureArboree, University of Bologna	✓		Yes
	11. Italy	Italy – CRA – IstitutoSperimentale Frutticoltura, Ministry of Agriculture, Roma	✓		No
	12. Italy	Italy – IstitutoSperimentaleFrutticoltura, Verona Province	✓		No
	13. Latvia	Institute of Horticulture, Dobele	✓	✓	No
	14. Lithuania	LAMMC SDI, Lithuanian Research Centre for Agriculture and Forestry, Institute of Horticulture, Babtai	✓	✓	Yes
	15. Poland	Research Institute of Horticulture, Skierniewice	✓	✓	Yes
	16. Romania	Research Institute for Fruitgrowing, Pitesti, Iasi, Bistrita	✓	✓	Yes
	17. Russia	All-Russian Scientific Research Institute of Horticultural Breeding, Orel	✓		Yes
	18. Russia	Russian Research Institute of Genetics and Breeding of Fruit Plants, Michurinsk	✓		Yes
	19. Russia	Horticultural Experiment Station, Krymsk	✓	✓	Yes
	20. Serbia	Fruit and Grape Research Centre, Čačak	✓	✓	Yes
	21. Sweden	Department of Horticultural Plant Breeding, Balsquard	✓	✓	No
	22. Switzerland	Swiss Federal Research Station for Fruit Growing, Wadensville	✓		No
	23. Ukraine	Institute of Horticulture of National Academy of Agrarian Sciences, Kiev	✓	✓	Yes

(Continued)

TABLE 7.1 (*Continued*)
Public Cherry Breeding Programs

Region	Country	Location	Sweet Cherry	Sour Cherry	Continuing
Asia	24. Turkey	Horticultural Central Research Institute, Yalova Atatürk	✓	✓	Yes
	25. China	Beijing Academy of Agricultural and Forest Sciences, Institute of Pomology Institute of Pomology an Forestry, Beijing	✓		Yes
	26. China	Dalian Academy of Agricultural Sciences, Institute of Pomology, Dalian	✓		Yes
	27. Japan	Horticultural Experiment Station, Yamagata Integrated Agricultural Research Center, Yamagata	✓		Yes
	28. Japan	NARO – Hokkaido Agricultural Research Center, Hokkaido	✓		Yes
	29. Japan	Apple Research Institute, Aomori Prefectural Industrial Technology Research Center, Aomori	✓		Yes
	30. Korea	RDA – National Horticultural Research Institute, Suwon	✓		Yes
North America	31. USA	Washington State University, Prosser	✓		Yes
	32. USA	Cornell University, Geneva	✓		Yes
	33. USA	Michigan State University, Michigan		✓	Yes
	34. Canada	Agriculture and Agri-Food Canada, Summerland	✓		Yes
	35. Canada	University of Guelph, Ontario	✓		Yes
	36. Canada	University of Saskatchewan, Saskatchewan		✓	Yes
South America	37. Chile	Politecnica Universidad Catolica, Valparaiso	✓		Yes
	38. Chile	Instituto Nacional de Investigaciones, Agropecuarias & Biofrutales, Santiago	✓		Yes

this locus is mapped on linkage group 6 (Quero-García et al. 2022). *S*-locus genotyping in cherry collections enables the identification of self-(in)compatible genotypes and is a useful technique for DNA-informed breeding (Iezzoni et al. 2017). The *S*-locus is very important for sweet cherry growers and breeders, as knowledge about the cross-compatibility of different cultivars is needed. The last update for sweet cherry accessions was done by Schuster (2020), including 63 incompatibility groups (IGs) in 1,483 sweet cherry accessions. The occurrence of self-incompatible cultivars in sweet cherry is much higher compared to sour cherry. Sour cherry has been found to have 12 functional and nine non-functional *S*-haplotypes (Lisek et al. 2017). Some of the sour cherry *S*-haplotypes are found in sweet cherry, indicating the genetic relationship between both species. More about self-(in)compatibility and *S*-allele typing in cherries will be described in Section 7.2.2.

Regarding resistance to biotic factors, there is a range of diseases and pests limiting cherry growing (Kappel et al. 2012; Papadopoulos et al. 2017; Dondini et al. 2018). According to Iezzoni (2008), fungal diseases cause the biggest yield losses for growers. The most damaging fungal disease for cherries is brown rot blossom blight caused by *Monilinia* spp., as it causes damage through all stages of cherry vegetation and even at postharvest around the world. According to Sansavini and Lugli (2008), only in several cherry breeding programs in Eastern Europe (Bulgaria, the Czech Republic, Hungary, and Lithuania), resistance and tolerance for *Monilinia* spp. diseases were included in breeding. However, the situation is changing as the need for environmentally friendly fruit-growing practices is growing (Quero-García et al. 2022). Another very important cherry disease is bacterial cancer caused by *Pseudomonas* spp. Despite the fact that this disease is included in breeding programs in Germany (Sansavini and Lugli 2008), the damage is widespread in all the Nordic hemisphere, from Canada to Russia (Dondini et al. 2018). The efforts in gene mining for resistance research will be discussed in Section 7.2.3.

DNA-informed breeding is indispensable to improving the quality of cultivars. After a millennium, advances in molecular biology, especially genome sequencing, increased knowledge of cherry genetics. Molecular markers are the most used tool for genetic diversity studies. In just a few years, starting with RAPD and AFLP markers, probably due to their dominant origin and unreliable repeatability, they were replaced with more precise SSR markers (Dondini et al. 2018). SSR markers are polymorphic, span multiple alleles, and are codominant (Conner and Hartl 2004). They are used for cultivar identification, genetic mapping, and phylogenetic studies (Iezzoni et al. 2017). They were used to investigate cherry germplasm genetic diversity (see Section 7.2.1) and for *S*-allele genotyping (see Section 7.2.2).

The introduction of relatively low-cost and high-throughput next-generation sequencing technology made it possible to develop a huge number of single nucleotide polymorphism (SNP) markers in cherry species and to construct genomic maps of genomes (Dondini et al. 2018; Quero-García et al. 2022). The first powerful tool for cherries was made within the RosBREED project in the USA (Iezzoni et al. 2020), where the SNP chip of 6K SNP markers for the Illumina's Infinium platform was designed (Peace et al. 2012). This chip covers the sweet cherry genome with 4212 and the sour cherry genome with 1482 SNPs. Later, the chip was improved and named the 6K+9K SNP chip by increasing marker density with additional 2.132 and 1.161 SNPs in sweet and sour cherry, respectively (Vanderzande et al. 2020). Additionally, the entire genomes of cherries have been sequenced and re-sequenced already (Shirasawa et al. 2017; Pinosio et al. 2020; Xanthopoulou et al. 2020; Wöhner et al. 2021; Le Dantec et al. 2020; Wang et al. 2020) which identify QTLs and candidate genes controlling important horticultural traits (see Section 7.2.3) like fruit size (Rosyara et al. 2013), fruit firmness and weight (Campoy et al. 2014), or fruit cracking (Quero-García et al. 2021).

7.2 ALLELE MINING

The allele mining technique involves the identification of polymorphism by comparative analysis of fragments or sequences of various genotypes. Such polymorphism allows characterizing the germplasm diversity of wild or cultivated pools. SSRs, also called microsatellite markers, were recommended as the most suitable marker system for genetic diversity studies because they have high polymorphism and the capability to reflect the genetic relationships between genotypes of a given germplasm pool with just a relatively small number of informative SSR loci (Clarke and Tobutt 2009; Quero-García et al. 2022). One of the reasons for the slower breeding of cherry than other Rosaceae family members is the GSI controlled by a single gametophytic expressed multiallelic *S*-locus, a specially for sweet cherry cultivars as most of them are self-incompatible (Kivistik et al. 2021). The *S*-locus encodes two genes, one in the pistil and one in the pollen, and self-incompatibility (SI) occurs if both carry the same *S*-allele (Matsumoto and Tao 2019). Therefore, the characterization of sweet cherry diversity with *S*-allele genotyping is another useful technique of allele mining (Iezzoni et al. 2017). Besides the SI, several challenges are limiting successful cherry breeding/growing: climate changes, the quality of the yield, and the public's demand for sustainable and safe food (Aranzana et al. 2019). Gene mining technique includes the identification of sequences of novel or known genes and their variation within germplasm. This should help find gene variants that control important agronomic traits and could be used in programs to improve crops (Ashkani et al. 2015).

7.2.1 Characterization of Cherries' Allelic Diversity

Characterizing germplasm diversity of sweet and sour cherry and wild cherry cultivars using SSR genotyping is well documented (Xuan et al. 2009; Iezzoni et al. 2017; Quero-García et al. 2022). In most of these studies, samples from one region or country are analyzed, with common standard cultivars from different countries added. A variety of common SSR markers were used, including a standard set of 16 SSR loci and 8 reference cherry accessions (*P. avium* F12/1, 'Goodnestone Black', 'Napoleon', 'Noble', and 'Noir de Meched', *P. incisa*, *P. mahaleb*, and *P. nipponica*) for cherry genetic diversity analysis (Clarke and Tobutt 2009).

The *Prunus* Group of the European Collaborative Programme for Genetic Resources (ECPGR) facilitates the comparison of *Prunus* genetic resources in European collections by analyzing alleles using microsatellite markers worldwide. Sixteen microsatellite markers have been generated by selecting the diploid species *Prunus avium*, *P. incisa*, *P. mahaleb*, and *P. nipponica*, but studies on the adaptation of these primers in species with higher levels of ploidy, such as sour cherries, are still lacking (Clarke and Tobutt 2009).

However, the most important is to unify the complex use of microsatellite markers in different laboratories and datasets to genotype cherry collections and later use them on a large scale for cherry germplasm. When performing the same assay in different laboratories, small differences (one base pair) in allele size estimates can occur due to various causes (Quero-García et al. 2022). Different microsatellite detection methods have a major impact on fragment sizes and can result in differences of one to two base pairs. Another reason for differences in cultivar genotypes can be explained by clonal differences within the analyzed cultivars from different collections, which is typical for widely distributed local cultivars (Ordidge et al. 2021).

Microsatellite markers or simple repetitive sequences (SSR) are recommended for allele mining studies because they are polymorphic, span multiple alleles, are codominant, and are PCR-based (Conner and Hartl 2004). The degree of polymorphism of SSRs is high due to the high variation rate in the number of repeat units. Simple repeat sequences are used for cultivar identification and genetic mapping and as an important genetic resource for phylogenetic studies (Struss et al. 2003). Germplasm studies using such markers accurately represent accessions and genetic relationships (Clarke and Tobutt 2009; Quero-García et al. 2022) and can maintain core collections with high genetic diversity (Patzak et al. 2019).

Within *Prunus* species, molecular markers have been developed and are most actively used in peach (*P. persica*) as a model plant, having a relatively short juvenile period (Cipriani et al. 1999). SSR primer pairs have been developed for peach, sweet cherry (Cipriani et al. 1999; Sosinski et al. 2000; Struss et al. 2003), apricot (Messina et al. 2004), plum (Mnejja et al. 2004), and sour cherry

(Downey and Iezzoni 2000) and adapted to another *Prunus* species. In total, 108 microsatellite loci were used for sweet, sour, and wild cherry genetic diversity analysis in 24 countries. In 2005, the ECPGR group hosted the European Prunus Database (EPDB), containing 5,585 cherry accessions from 17 different countries. Researchers worldwide analyzed their own sweet, sour, and wild cherry germplasm collection using standardized SSR loci set (Clarke and Tobutt 2009).

The genetic fingerprint of a species is essential for genotype identification and classification and provides a sufficient amount of genetic variability and unique alleles worldwide (Cantini et al. 2001). Analysis of molecular markers helps to assess inter- and intraspecific genetic variability (Gril et al. 2008).

7.2.1.1 Sweet Cherry Cultivars

SSR markers are highly polymorphic and are recommended for genetic diversity studies. With dozens of studies analyzing sweet cherries germplasm worldwide, most were based on local cultivars and germplasm collections just in that area, along with commercial, domestic cultivars and wild populations (Table 7.2).

Microsatellites were first designed for peach (*Prunus persica*) and then adapted to sweet cherries. 26 (UDP-) microsatellites were developed from peach 'Redhaven' by Testolin et al. (2000), and 15 were successfully used for sweet cherry genetic fingerprinting (Table 7.2). Microsatellite markers named BPPCT- were designed by Dirlewanger et al. (2002) for peach (*Prunus persica*) and their intraspecific hybrids. The same markers were evaluated for sweet cherry. Eleven out of 41 designed loci were monomorphic, while in peach, all showed heterozygous features (Dirlewanger et al. 2002). Sweet cherry could be distinguished with sets of only three microsatellites: BPPCT034, BPPCT039, and BPPCT040, or by BPPCT034, BPPCT038, and BPPCT040 due to higher observed heterozygosity. Aranzana et al. (2002) developed CPP CT-primers for peach (*Prunus persica*), and out of 36 primers that were suitable for peaches, only four were informative for sweet cherries.

TABLE 7.2
SSR Fingerprinting of Sweet Cherry

SSR Locus	Country [Author][a]	Size Range	No. of Alleles
BPPCT002	France [14], Greece [16], Iran [15,22]	168–231	3–23
BPPCT004	France [14], Iran [15,22]	172–201	2–17
BPPCT005	Czech Republic [1], France [14], Greece [16]	154–204	2–7
BPPCT006	France [14]	99	1
BPPCT007	France [14], Iran [22]	110–234	2–25
BPPCT010	France [14], Iran [22]	94–146	2–15
BPPCT012	France [14], Greece [16]	154–156	2–4
BPPCT013	France [14], China [24]	140–242	15
BPPCT014	Canada [2], France [14], Iran [22]	175–235	3–22
BPPCT017	France [14]	197	1
BPPCT018	France [14]	193	1
BPPCT022	France [14]	99	1
BPPCT024	France [14]	91	1
BPPCT026	Czech Republic [1], Chile [3], France [14], Greece [16], Iran [22], China [24]	130–222	3–20
BPPCT027	France [14]	246	1
BPPCT028	France [14], Greece [16]	151–185	2–3
BPPCT030	France [14]	140	1

(Continued)

TABLE 7.2 (*Continued*)
SSR Fingerprinting of Sweet Cherry

SSR Locus	Country [Author][a]	Size Range	No. of Alleles
BPPCT032	France [14]	158	1
BPPCT034	France, Germany, Italy, Sweden, Switzerland [6,7], Czech Republic [1], China [4], France [14], Iran [15], Greece [16], Turkey [30], Germany [33]	204–300	3–32
BPPCT036	France [14]	246	1
BPPCT037	France, Germany, Italy, Sweden, Switzerland [6,7], Austria [5], Chile [3], UK [8], France [14]	120–172	5–30
BPPCT038	France [14], Greece [16]	100–133	4–5
BPPCT039	Chile [3], France [14], Iran [15], Greece [16], Ukraine [34]	134–160	4–8
BPPCT040	Chile [3], France [14], Iran [15], Greece [16], Turkey [30], Germany [33]	116–202	3–9
BPPCT041	France [14]	201	1
BPPCT042	France [14]	215	1
CN89269	Czech Republic [1]	294–300	3
CN911135	Czech Republic [1]	178–180	2
CPPCT006	France, Germany, Italy, Sweden, Switzerland [6,7], UK [8], Austria [5], Czech Republic [9], Iran [22]	173–212	6–27
CPPCT022	France, Germany, Italy, Sweden, Switzerland [6,7], UK [8], Austria [5], Czech Republic [9]	228–259	4–29
CPSCT010	Turkey [26]	n/d	5
CPSCT038	France, Germany, Italy, Sweden, Switzerland [6,7], Austria [5], China [24]	182–209	3–18
EMPA001	Italy [10,18,28], UK [19]	143–164	4–9
EMPA002	France, Germany, Italy, Sweden, Switzerland [6,7], UK [8,19], Austria [5], Czech Republic [9], Lithuania [11], Italy [18,28], Iran [22]	93–219	1–30
EMPA003	UK [8,19], Lithuania [11], Czech Republic [9], Italy [18,28]	157–179	2–4
EMPA004	France, Germany, Italy, Sweden, Switzerland [6,7], Turkey [12], Czech Republic [1], China [4], Italy [10,18,28], Sicily (Italy) [25], Iran [15], Greece [17], UK [19]	160–212	3–15
EMPA005	Turkey [12], Czech Republic [1], Italy [10,18,28], Iran [15], Greece [17], UK [19], Sicily (Italy) [25]	230–264	2–15
EMPA006	Italy [18,28], UK [19]	94–101	1–2
EMPA007	Italy [18,28], UK [19]	172–179	1–6
EMPA008	Italy [18,28], UK [19]	160–172	2–3
EMPA009	Italy [18,28], UK [19]	201–244	2–7
EMPA010	Italy [18,28], UK [19]	127–136	2–5
EMPA011	Italy [10,18,28], UK [19]	234–252	2–9
EMPA012	Italy [10,18,28], UK [19]	225–233	2–6
EMPA013	Italy [18,28], UK [19]	187–200	2–4
EMPA014	UK [19]	215–227	6
EMPA015	Turkey [12], Italy [10,18,28], Iran [15], Greece [16,17], UK [19], Sicily (Italy) [25], Ukraine [34]	203–253	1–21
EMPA016	Italy [18,28], UK [19]	177–191	1–8
EMPA017	France, Germany, Italy, Sweden, Switzerland [6,7], UK [8,19], Austria [5], Czech Republic [9], Lithuania [11], Italy [18,28]	221–246	1–17

(*Continued*)

TABLE 7.2 (*Continued*)
SSR Fingerprinting of Sweet Cherry

SSR Locus	Country [Author][a]	Size Range	No. of Alleles
EMPA018	France, Germany, Italy, Sweden, Switzerland [6,7], Czech Republic [9], Lithuania [11], China [4], Italy [10,18,28], Iran [15], Greece [16,17], Sicily (Italy) [25]	76–163	4–14
EMPA019	Italy [18,28], UK [19]	101–121	2–3
EMPA020	UK [19]	194	1
EMPA021	Italy [18,28], UK [19]	120–123	1–2
EMPA026	UK [8], Czech Republic [9]	189–234	4
EMPAS01	Czech Republic [1,9], UK [8], Turkey [12], Austria [5], Lithuania [11], Italy [10,18,28], Iran [15], Greece [17], China [24], Sicily (Italy) [25]	210–259	4–18
EMPAS02	France, Germany, Italy, Sweden, Switzerland [6,7], UK [8], Turkey [12], Austria [5], Lithuania [11], Italy [10,18,28], Czech Republic [9], Canada [2], Iran [15,22], Greece [17], China [24], Sicily (Italy) [25], Ukraine [34]	109–192	2–27
EMPAS04	Italy [18,28]	223–238	4–6
EMPAS06	France, Germany, Italy, Sweden, Switzerland [6,7], Czech Republic [1,9], Turkey [12], Lithuania [11], China [4], Italy [10,18,28], Greece [17], Sicily (Italy) [25], Ukraine [34]	116–241	1–22
EMPAS10	UK [8], Turkey [12], Austria [5], Lithuania [11], Czech Republic [9], Italy [10,18,28], Greece [16,17], Sicily (Italy) [25]	134–195	2–21
EMPAS11	Turkey [12], Lithuania [11], Greece [17], Italy [18,28], Sicily (Italy) [25]	62–114	4–18
EMPAS12	France, Germany, Italy, Sweden, Switzerland [6,7], Czech Republic [1,9], UK [8], Turkey [12], Lithuania [11], Italy [10,18,28], Greece [16,17], Sicily (Italy) [25]	112–155	2–28
EMPAS13	Italy [10,18,28]	203–223	1–7
EMPAS14	France, Germany, Italy, Sweden, Switzerland [6,7], Czech Republic [9], UK [8], Turkey [12], Italy [10,18,28], Greece [17], Sicily (Italy) [25]	164–213	3–27
EMPAS16	Italy [10]	n/d	5
EPDCU5060	China [24]	98–148	10
EPDCU5183	China [24]	95–175	20
EPPCU4092	China [24]	124–262	13
EPPCU9168	Ukraine [34]	162–192	5
PAV-Rf-SSR	France, Germany, Italy, Sweden and Switzerland [6,7]	343–363	7–9
PceGA25	China [4], Latvia [23], Sweden [23], Denmark [27], Turkey [30]	144–364	4–9
PceGA34	UK [8], Lithuania [11], Greece [17], Sicily (Italy) [25], Ukraine [34]	100–237	7–11
PceGA59	Denmark [27], Turkey [30]	176–208	2–6
PceGA77	Turkey [21]	165–180	4
Pchcms2	Spain [29]	160–180	2
Pchcms4	Spain [29]	260–290	2
Pchgms1	Turkey [20,26], Iran [22], Spain [29]	94–291	2–22
Pchgms5	Iran [22]	212–275	30
Pchpgms3	Turkey [13], Denmark [27]	164–216	2–7
PMS02	China [4], Turkey [13,21,30], Denmark [27], Ukraine [34]	118–180	2–11

(*Continued*)

TABLE 7.2 (*Continued*)
SSR Fingerprinting of Sweet Cherry

SSR Locus	Country [Author][a]	Size Range	No. of Alleles
PMS03	Canada [2], Chile [3], Turkey [13,21,30], Latvia [23], Sweden [23], Denmark [27]	114–232	3–10
PMS30	China [4], Canada [2], Chile [3], Turkey [21], Denmark [27]	131–200	4–8
PMS40	China [4], Denmark [27], Turkey [30]	88–108	4–5
PMS49	Turkey [21], Latvia [23], Sweden [23], Denmark [27]	127–142	1–5
PMS67	China [4], Canada [2], Chile [3], Turkey [13,21], Denmark [27]	144–189	2–7
Pru-T2/S132	China [4]	n/d	9
PS05C03	UK [8], Lithuania [11]	101–163	8
PS08E08	Iran [22], Denmark [27]	159–211	7–12
ps12A02	Turkey [20,21,26]30], Iran [22], Ukraine [34]	122–450	3–17
SC2	China [24]	136–177	14
SC3	China [24]	140–172	18
UCD-CH10	United States [31]	135–148	2
UCD-CH11	Iran [15], United States [31]	134–175	4–5
UCD-CH12	Czech Republic [1], China [4], Canada [2], Iran [15], United States [31]	165–200	3–6
UCD-CH13	Turkey [26], United States [31]	133–139	2–5
UCD-CH14	Czech Republic [1], Lithuania [11], United States [31]	135–162	3–4
UCD-CH15	United States [31]	94–112	3
UCD-CH16	United States [31]	123	1
UCD-CH17	Czech Republic [1], Turkey [26], United States [31]	186–190	3–10
UCD-CH18	Chile [3], United States [31]	178–218	4–6
UCD-CH19	United States [31]	122	1
UCD-CH21	Czech Republic [1], Turkey [26], United States [31]	114–122	4–5
UCD-CH26	United States [31]	110–124	4
UCD-CH31	Czech Republic [1], Iran [15], Turkey [26], United States [31]	111–180	4–10
UCD-CH36	Iran [15], United States [31]	110–128	4
UCD-CH39	Iran [15], United States [31]	122–160	4–12
UCD-CH62	Turkey [20]	n/d	4
UCD-CH70	Turkey [20]	n/d	3
UCD-CH79	Turkey [20]	n/d	3
UCD-CH83	Turkey [20]	n/d	3
UDP96-001	Czech Republic [1], Chile [3], Greece [15], Turkey [26], Germany [32]	108–140	2–7
UDP96-005	Greece [17], Iran [22], Spain [29], Germany [32]	84–156	3–22
UDP97–403	Germany [32], Turkey [30]	126–170	5–8
UDP98-021	Czech Republic [1], Greece [16], Germany [32]	100–118	2–6
UDP98-022	Czech Republic [1], Canada [2], Greece [17]	91–114	4–7
UDP98–409	Iran [22], Spain [29]	117–160	2–22
UDP98–410	Czech Republic [9], Greece [16], Iran [22], Germany [32,33]	116–160	2–16
UDP98–411	Czech Republic [9], Germany [32]	146–180	3–9
UDP98–412	Czech Republic [1,9], UK [8], Austria [5], Lithuania [11], Greece [17], Sicily (Italy) [25], Germany [32,33], Ukraine [34]	87–157	5–9
UPD96-005	Turkey [26]	n/d	6
UPD97–407	Spain [29]	120–150	7

(*Continued*)

TABLE 7.2 (*Continued*)
SSR Fingerprinting of Sweet Cherry

SSR Locus	Country [Author][a]	Size Range	No. of Alleles
UPD98-021	Spain [29]	100–110	2
UPD98-022	Spain [29]	90–110	4
UPD98–412	Spain [29]	100–140	4
UPPCU961	Canada [2]	145–149	3

[a] [1] Patzak et al. (2019); [2] Wiersma et al. (2018); [3] Guajardo et al. (2021); [4] Liu et al. (2018); [5] Schüller et al. (2021); [6] Barreneche et al. (2021); [7] Ordidge et al. (2021); [8] Clarke and Tobutt (2009); [9] Sharma et al. (2015); [10] Muccillo et al. (2019); [11] Stanys et al. (2012); [12] Uzan Eken et al.(2022); [13] Demir et al. (2011); [14] Dirlewanger et al. (2002); [15] Farsad and Esna-Ashari (2016); [16]; Ganopoulos et al. (2010); [17] Ganopoulos et al. (2011); [18] Guarino et al. (2010); [19] Clarke and Tobutt (2003); [20] Gulen et al. (2010); [21] Aka Kaçar et al. (2005); [22] Khadivi-Khub (2014); [23] Lacis et al. (2009); [24] Liang et al. (2018); [25] Marchese et al. (2007); [26] Öz et al. (2013); [27] Pedersen (2006); [28] Guarino et al. (2009); [29] Wünsch and Hormaza (2002); [30] Turet-Sayar et al. (2012); [31] Struss et al. (2003); [32] Schueler et al. (2003); [33] Höltken and Gregorius (2006); [34] Ivanovych and Volkov (2018).

PMS primers were explicitly developed for sweet cherry (*Prunus avium*) by Cantini et al. (2001) and Struss et al. (2002). Clarke and Tobutt (2003) created 21 microsatellite primers EMPA- from *Prunus avium* cv. 'Napoleon'. All cultivars evaluated could be separated with primer pair set combination of EMPA014, EMPA015, and EMPA018. A total of 16 EMPAS primers were developed from *Prunus avium* cv. 'Charger' by Vaughan and Russell (2004). Fifteen microsatellite UCDCH primers were developed for the sweet cherry (*Prunus avium*) cultivar 'Valerij Tschkalov' by Struss et al. (2003).

Primers PceGA were obtained from a *Prunus cerasus* 'Erdi Botermo' genetic library and developed for sour cherry by Iezzoni (unpublished). Four of these primers were suitable for sweet cherry as well (Table 7.2). For black cherry (*Prunus serotina*) fingerprinting the primers of sweet cherry (PS12A) and peach (pchpgms) were used by Downey and Iezzoni (2000).

Not all primer pairs produce stable and repeatable polymorphic amplification fragments. For the experimental proving of primers' usefulness, usually, the first electrophoresis on agarose gel is conducted as it is cost-efficient and less time-consuming than fluorescence-based capillary electrophoresis. However, the automated sequencer capillary electrophoresis method obtains more reliable data separating fragments by one base pair. Even though it is considered expensive and applied to mid- to high-throughput laboratories with more costly and advanced equipment, the precision of detection and unification protocols among laboratories makes it very useful in identifying and characterizing sweet cherry germplasm.

The main areas of sweet cherry natural spread are western Asia and southern Europe. Iran is the center of the most diverse sweet cherry genotypes, cultivated and wild. The allelic range of Iranian sweet cherries is from 4 to 30 with various microsatellite markers (Table 7.2), followed by Turkey with a range of 1–21 alleles (Figure 7.3). Four to twenty alleles were observed in China's germplasm collections. Chinese scientists analyzed a set of 50 previously developed primers from almond, plum, peach, and sweet cherry with ten representative sweet cherry cultivars. Only 10 SSR markers covered all eight chromosomes of sweet cherry with a high degree of polymorphism and good amplification. For further studies, used in combination with primer Pru-T2/S132 for detecting *S*-alleles, eleven selected markers were highly polymorphic, resulting in 67 alleles with an average of 6.09 per locus (range from 4 to 9). The highest PIC value was 0.80 (PMS30 and Pru-T2/S132) (Liu et al. 2018). Four to twelve alleles were observed in Ukraine. In Ukraine, the most informative loci were EMPA015, EMPAS02, EMPAS06, PCEGA34, and PS12A12. Locus BPPCT30 resulted in only a single monomorphic allele (128 bp) for tested genotypes and is unsuitable for discrimination of sweet cherry cultivars. Ukrainian sweet cherry cultivars contain western European, local,

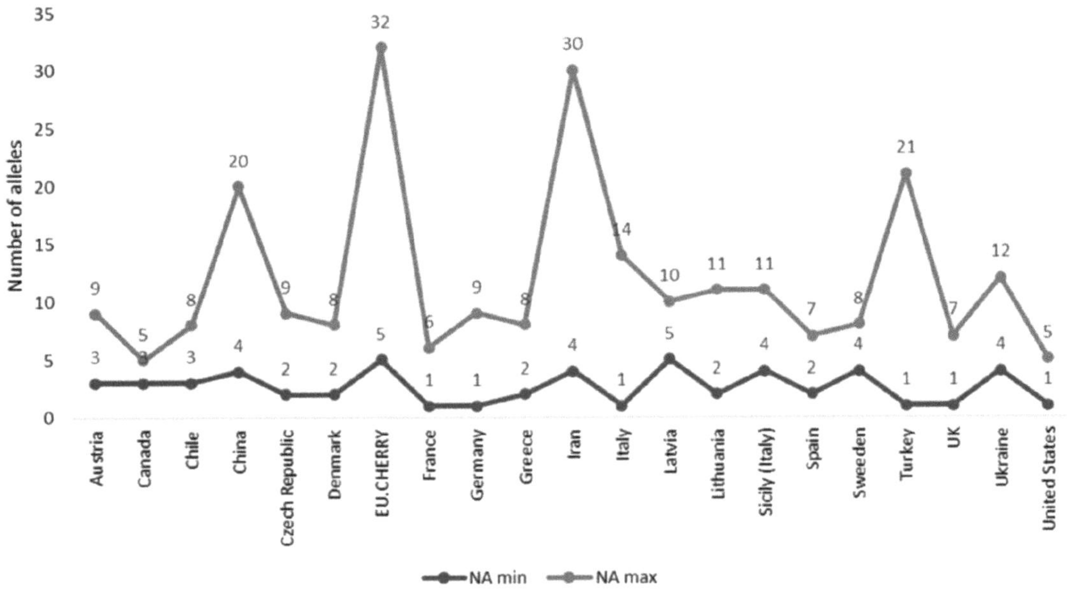

FIGURE 7.3 Number of alleles in sweet cherry germplasm collections worldwide.

and probably Caucasian origin. With the estimation that cultivars are related, results could not be interpreted well. Hence, it is essential to add the probability of identity (*PI*) or unrelated cultivars and the likelihood of siblings' identity (*PIsibs*) by computer programs (Ivanovych and Volkov 2018). Italy germplasm collections and wild accessions represent 1–14 alleles with base pairs ranging from 66 to 258 (Figure 7.4). Campania's (Italy) sweet cherries exhibit a remarkably different pattern of alleles compared with other international breeding program cultivars (Muccillo et al. 2019), which means eastern and western Asia, along with the southern Europe region, have very diverse collections. However, the range of base pairs in France was remarkably high, from 91 to 500 (Figure 7.4).

In central Europe, the allelic range varies from 1 to 9 with different sets of primers. CPPCT022 and EMPAS10 detect unique alleles in different genotypes analyzed in Austria, while BPPCT034 allele lengths seemed ambiguous, and this locus was not valuable in the study (Schüller et al. 2021). However, BPPCT034 was used in the EU.CHERRY project resulted in a high allele range (4–32). Selected SSR fingerprinting offers ways to confirm or reject morphological identification of germplasm accessions. It pointed out the identical genotypes under different names that could be homonyms and synonyms, as well as mark errors in labeling, grafting, and even phenotyping, where visually same phenotype groups gave different genotypes. Sharma et al. (2015) described that primer pairs EmpaS12, EmpaS01, Empa2, UDP-98–412, and Empa17 were monomorphic. EMPA2 and EMPA3 locus had the lowest alleles, while UDP-98–412 obtained the highest alleles. Over 90% of primers produced clear polymorphic bands for all loci, and they agreed with the expected size ranges. Study shows that 'Kordia' was extensively used in Czech breeding programs for nearly half a century. The analysis also offers a clear picture of the origin of germplasm collection from all over the world with distinct groups of Canadian, US, and UK cultivars with a high effect of European germplasm.

Northern Europe germplasm collections are unique for specific traits for climate conditions. The allelic range is greater than in central Europe (Austria, Czech Republic, Germany, and the UK), ranging from 2 to 11 (Figure 7.4). Twenty unique alleles were observed for Lithuanian cultivars with a selected set of primers, implying the genetic variation needed to adapt to northern ecological conditions. These alleles spread in the northern border populations and can be observed in the offspring of these populations (Stanys et al. 2012). Lithuanian cultivars' loci varied from 1 to 7, while foreign

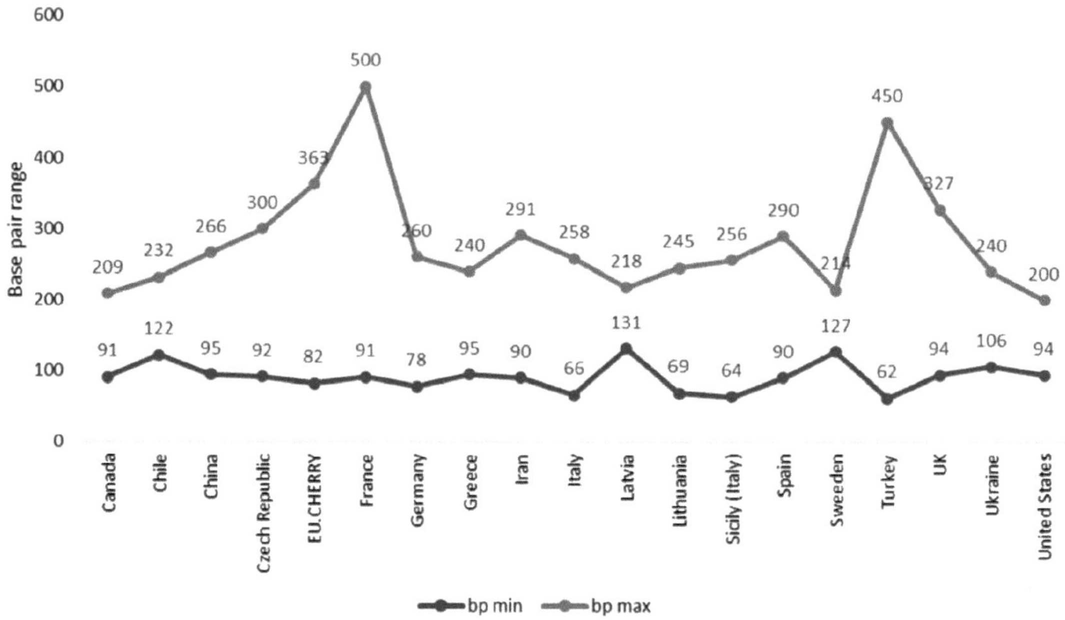

FIGURE 7.4 Base pair range of sweet cherry germplasm collections worldwide.

cultivars varied from 3 to 8. Loci EMPAS01, EMPAS06, and EMPAS12 show equal allelic size ranges for both foreign and Lithuanian cultivars. In contrast, in other cases, Lithuanian cultivars showed a mostly more extensive allelic size range than the group of foreign cultivars.

Sweet cherry genotypes were tested in South and North America. In Chile, a total of 56 alleles were detected using 13 microsatellites on a group of 37 sweet cherry cultivars. An average of six alleles per locus were observed, ranging from 3 in UDP96-001 to 8 in PMS-3, PMS-30, and BPPCT-040. In total twenty-three rare alleles (frquency $p_i \leq 0.1$) were detected in Hungarian cultivars 'Solymari Gombolyu' and 'Katalin'. Among studied markers locus UDP96-001 had one over-represented allele with the highest frequency value ($p_i = 0.545$), while locus BPPCT-039 allele distribution was more uniform with no overrepresented alleles. These rare alleles did not affect heterozygosity overall. BPPCT-040 was the most informative, with a polymorphic information content (PIC) of 0.75, while UDP96-001 was the least informative marker, with a PIC value of 0.55. However, on average, this value was 0.68 (Guajardo et al. 2021). Canadian and US germplasm collections had up to five alleles with tested primer sets, meaning that their genetic diversity could be more diverse.

EMPA004, EMPA005, EMPAS001, EMPAS06, EMPAS10, EMPAS11, and EMPAS12 showed more alleles in Turkey than in northern countries (Table 7.2). Allele ranges of primers BPPCT06 EMPAS002, EMPAS14 were similar in all geographic regions. Primer EMPA018 was tested in the Czech Republic, China, Lithuania, and Italy, and the highest allele number was observed in Italy. PCEGA34 and UDP97–402 locus have high differentiation power of genotype, and usage of these markers enables the identification of autochthonous genotypes of sweet cherry (Krmpot et al. 2020). However, closely related cultivars could not be described using a set of primers. They should be selected broader set of primers and specific to determine the same origin cultivars (Wiersma et al. 2018).

Sweet cherry morpho-physiological features are linked to SSR markers (Table 7.3). Harvest time is linked to primer pairs EMPAS02, PceGA34, EMPA015, PceGA34, EMPAS06, and EMPAS12, which correspond to alleles 147, 160, 241, 143, 207, and 142, respectively. Fruit polar diameter associated with EMPA005, EMPAS14, EMPAS11, and EMPA004 with alleles respectfully 248, 199, 110, and 185. EMPAS02 with allele 147 is associated with fruit weight (Ganopoulos et al. 2011), as well as UDP96-005 and BPPCT002 with alleles 122 and 168. Stone weight is linked to alleles 184

TABLE 7.3
Primers and Alleles Associated with Traits of Sweet Cherry

Primer	Allele	Trait	References
BPPCT002	168	Fruit weight	Khadivi-Khub (2014)
BPPCT004	181,199	Fruit width	Khadivi-Khub (2014)
BPPCT004	197	Fruit cracking	Khadivi-Khub (2014)
BPPCT004	197	Fruit shape	Khadivi-Khub (2014)
BPPCT004	199	Titratable acidity	Khadivi-Khub (2014)
BPPCT004	197	Fruit firmness	Khadivi-Khub (2014)
BPPCT004	197,199	Total soluble solid	Khadivi-Khub (2014)
BPPCT007	234	Doubled fruit	Khadivi-Khub (2014)
BPPCT007	234	Fruit flesh color	Khadivi-Khub (2014)
BPPCT014	175	Titratable acidity	Khadivi-Khub (2014)
BPPCT014	193	Total soluble solid	Khadivi-Khub (2014)
BPPCT014	175,193	Anthocyanin	Khadivi-Khub (2014)
BPPCT014	191	Fruit flesh color	Khadivi-Khub (2014)
BPPCT026	149	Anthocyanin	Khadivi-Khub (2014)
CPPCT6	184,189	Stone weight	Khadivi-Khub (2014)
CPPCT6	196	Anthocyanin	Khadivi-Khub (2014)
EMPAS02	136	Titratable acidity	Khadivi-Khub (2014)
Pchcms5	231	Fruit shape	Khadivi-Khub (2014)
Pchcms5	231	Doubled fruit	Khadivi-Khub (2014)
Pchcms5	231	Total soluble solid	Khadivi-Khub (2014)
Pchcms5	231	Fruit skin color	Khadivi-Khub (2014)
Pchcms5	231,271	Fruit flesh color	Khadivi-Khub (2014)
PS08E08	171	Anthocyanin	Khadivi-Khub (2014)
PS08E08	171	Fruit skin color	Khadivi-Khub (2014)
PS12A02	175	Anthocyanin	Khadivi-Khub (2014)
UDP96-005	122	Fruit weight	Khadivi-Khub (2014)
UDP96-005	106	Stone weight	Khadivi-Khub (2014)
UDP96-005	122	Fruit length	Khadivi-Khub (2014)
UDP96-005	106	Fruit length	Khadivi-Khub (2014)
UDP96-005	94	Titratable acidity	Khadivi-Khub (2014)
UDP96-005	110	Anthocyanin	Khadivi-Khub (2014)
UDP96-005	122	Fruit diameter	Khadivi-Khub (2014)
UDP98–409	104	Fruit length	Khadivi-Khub (2014)
EMPAS02	147	Harvest time	Ganopoulos et al. (2011)
PceGA34	160	Harvest time	Ganopoulos et al. (2011)
EMPA015	241	Harvest time	Ganopoulos et al. (2011)
PceGA34	143	Harvest time	Ganopoulos et al. (2011)
EMPAS06	207	Harvest time	Ganopoulos et al. (2011)
EMPAS12	142	Harvest time	Ganopoulos et al. (2011)
EMPA005	248	Fruit polar diameter	Ganopoulos et al. (2011)
EMPAS14	199	Fruit polar diameter	Ganopoulos et al. (2011)
EMPAS11	110	Fruit polar diameter	Ganopoulos et al. (2011)
EMPA004	185	Fruit polar diameter	Ganopoulos et al. (2011)
EMPAS02	147	Fruit weight	Ganopoulos et al. (2011)

and 189 in CPPCT6 and 106 in UDP96-005. Alleles 106 and 122 with primer UDP96-005 and 104 with UDP98–009 are associated with fruit length. Alleles 181 and 199 in BPPCT004 associated with fruit width. In total, 122 with primer UDP96-005 associated with fruit diameter (Khadivi-Khub 2014). Assay Pav-Rf-SSR described by Sandefur et al. (2016) lies in close vicinity to the PavMYB10.1 gene with five unique SSR alleles observed by Cmejla et al. (2021) and was used to predict cherry skin color. BPPCT034 is fruit size SSR marker and heterozygous in parental cultivars 'Ambrunés' 222/229 and 'Sweetheart' 222/332 (Calle et al. 2020a).

Different average allele number ranges can be explained by the number of cultivars, their genetic background, and the set of microsatellites used in each study (Guajardo et al. 2021). Few highly informative markers could be sufficient in most collections. Still, closely related cultivars can show no difference (Wiersma et al. 2018), making it clear to use more loci to determine the polymorphism of these genotypes. SSR analysis has the potential to reliably identify duplicates and genetically heterogeneous groups that exhibit similar phenotypes in breeding and germplasm collecting. (Schüller et al. 2021). EU.CHERRY project aimed to align datasets using standard marker sets (Ordidge et al. 2021) and helps conserve Europe's germplasm. Then the number of errors in prior datasets was resolved (Barreneche et al. 2021). However, there are still unresolved issues with synonyms, mislabeling, sports, and clones that have value in themselves (Ordidge et al. 2021). Alleles in sweet cherry range from 1 to 6. Sweet cherry genetic diversity is highly dependable on geolocation. Allele number in northern countries is lower than that in more southern countries; however, EU.CHERRY project offers a broader range of alleles from 17 European germplasms (Belgium, Czech Republic, Germany, Spain, France, the UK, Greece, Hungary, Israel, Italy, Latvia, Poland, Portugal, Romania, Slovakia, Turkey, and Ukraine). It reveals the collection's large genetic diversity, justifying the efforts made over decades for their conservation (Barreneche et al. 2021).

7.2.1.2 Sour Cherry Cultivars

Identification and characterization of sour cherry cultivars are important for the efficient selection of genotypes for breeding (Xuan et al. 2009). Characterization of allelic diversity in sour cherry (*Prunus cerasus*) with SSR markers is used worldwide. Sour cherry is native to Europe and distributed from Turkey to northern Russia (Bird et al. 2022).

The major limitation in developing SSR for *Prunus cerasus* is the time and cost required to isolate and characterize individual loci, while the complete genome is not yet available. Cloning and screening of the plasmid are required to develop and identify new markers for *P. cerasus* (Struss et al. 2003).

The transferability of the developed microsatellites to other *Prunus* species was very high, with 90% of the developed SSRs from peach DNA libraries amplifying fragments in sour cherry (Dirlewanger et al. 2002). The SSRs applicable to *P. cerasus* are polymorphic and have 1–4 alleles, and heterozygosity ranges from 0.26 to 0.82 (Cipriani et al. 1999; Sosinski et al. 2000; Dirlewanger et al. 2002; Struss et al. 2003; Messina et al. 2004; Mnejja et al. 2005).

A total of 73 SSR markers were used for genetic analysis of sour cherry cultivars from 12 different countries – Belarus (Urbanovich et al. 2014), Denmark (Pedersen 2006), Finland (Antonius et al. 2012), Germany (Höfer and Peil 2015; Xuan et al. 2009), Hungary (Baris et al. 2017; György et al. 2010), Iran (Najafzadeh et al. 2016), Latvia (Lacis et al. 2011; Lacis and Kota 2013), Lithuania (Rugienius et al. 2015), Russia (Kulikov et al. 2018), Sweden (Lacis et al. 2011), the USA (Cantini et al. 2001), and Turkey (Turkoglu et al. 2010). These are provided in Table 7.4 together with the allele number and allele size ranges. Microsatellite (SSR) loci generate between 1 and 36 alleles. In total, 71 SSR loci are polymorphic, and two loci (EMPA006 and EMPA026) generate monomorphic patterns (Table 7.4).

The highest number of alleles (36) was generated in Danish cultivars (Pedersen 2006). In Lithuanian and Latvian cultivars, the highest number of SSR alleles generated is 20 (Lacis and Kota 2013; Rugienius et al. 2015), followed by 16 alleles in the US and Finnish cultivars (Cantini et al. 2001; Antonius et al. 2012). The highest number of alleles in Iranian cultivars is 15, followed by 14

TABLE 7.4
Sour Cherry SSR Loci Analysis Worldwide

	SSR Locus[a]	Country (Author)	References	Size Range	No. of Alleles
1.	BPPCT002	Finland, Hungary	Antonius et al. (2012), Baris et al. (2017)	167–183	3
2.	BPPCT004	Belarus	Urbanovich et al. (2014)	182–200	3
3.	BPPCT005	Finland, Belarus	Antonius et al. (2012), Войтенкова et al. (2014)	129–201	6–11
4.	BPPCT007	Finland	Antonius et al. (2012)	114–185	7
5.	BPPCT016	Belarus	Urbanovich et al. (2014)	94–96	2
6.	BPPCT017	Belarus	Urbanovich et al. (2014)	196–204	3
7.	BPPCT021[A]	Iran	Najafzadeh et al. (2016)	185–292	10
8.	BPPCT025	Belarus	Urbanovich et al. (2014)	152–194	14
9.	BPPCT026	Finland, Belarus	Antonius et al. (2012), Urbanovich et al. (2014)	132–189	6–9
10.	BPPCT027	Iran, Belarus	Najafzadeh et al. (2016), Urbanovich et al. (2014)	248–285	2
11.	BPPCT030	Hungary	Baris et al. (2017)	140–162	3
12.	BPPCT032	Belarus	Urbanovich et al. (2014)	162–199	8
13.	BPPCT034	Finland	Antonius et al. (2012)	207–241	7
14.	BPPCT037[a]	Finland, Latvia	Antonius et al. (2012), Lacis and Kota (2013)	129–165	16–18
15.	BPPCT038	Finland	Antonius et al. (2012)	98–132	7
16.	BPPCT039	Finland, Belarus	Antonius et al. (2012), Urbanovich et al. (2014)	128–150	7–10
17.	BPPCT040	Finland, Belarus	Antonius et al. (2012), Urbanovich et al. (2014)	121–152	8-7
18.	BPPCT041	Hungary	Baris et al. (2017)	201–229	2
19.	CPPCT006[a]	Latvia	Lacis and Kota (2013)	N/A	13
20.	CPPCT022[a]	Latvia	Lacis and Kota (2013)	N/A	10
21.	CPSCT010	Turkey	Turkoglu et al. (2010)	170–176	2
22.	EMPA001	Belarus	Urbanovich et al. (2014)	133–165	13
23.	EMPA002[a]	Latvia	Lacis and Kota (2013)	N/A	11
24.	EMPA003[a]	Latvia	Lacis and Kota (2013)	N/A	5
25.	EMPA004	Iran, Belarus	Najafzadeh et al. (2016), Urbanovich et al. (2014)	145–192	4–9
26.	EMPA005	Belarus	Urbanovich et al. (2014)	244–282	7
27.	EMPA006	Belarus	Urbanovich et al. (2014)	106	1
28.	EMPA007	Belarus	Urbanovich et al. (2014)	174–178	3
29.	EMPA010	Belarus	Urbanovich et al. (2014)	134–142	4
30.	EMPA015	Iran, Belarus	Najafzadeh et al. (2016), Urbanovich et al. (2014)	85–228	5–7
31.	EMPA017[a]	Latvia	Lacis and Kota (2013)	N/A	10
32.	EMPA018	Iran, Belarus	Najafzadeh et al. (2016), Urbanovich et al. (2014)	75–135	8–10
33.	EMPA019	Belarus	Urbanovich et al. (2014)	132–137	4
35.	EMPA026	Latvia	Lacis and Kota (2013)		1
34.	EMPA026[a]	Latvia, Belarus	Lacis and Kota (2013), Urbanovich et al. (2014)	184–238	10
36.	EMPaJ04	Germany	Xuan et al. (2009)	164–194	11
37.	EMPaJ05	Germany	Xuan et al. (2009)	242–276	8
38.	EMPaJ15	Germany	Xuan et al. (2009)	191–255	11
39.	EMPaS01[a]	Latvia Germany	Höfer and Peil (2015), Lacis and Kota (2013), Xuan et al. (2009)	168–240	7–10
40.	EMPaS02[a]	Germany, Latvia, Iran, Lithuania	Höfer and Peil (2015), Lacis and Kota (2013), Najafzadeh et al. (2016), Rugienius et al. (2015), Xuan et al. (2009)	119–270	7–14
41.	EMPaS06[a]	Germany, Latvia, Iran, Lithuania	Höfer and Peil (2015), Lacis and Kota (2013), Najafzadeh et al. (2016), Rugienius et al. (2015), Xuan et al. (2009)	112–285	7–9

(*Continued*)

TABLE 7.4 (*Continued*)
Sour Cherry SSR Loci Analysis Worldwide

	SSR Locus[a]	Country (Author)	References	Size Range	No. of Alleles
42.	EMPaS10[a]	Lithuania, Latvia Germany	Höfer and Peil (2015), Lacis and Kota (2013), Rugienius et al. (2015), Xuan et al. (2009)	145–188	8–10
43.	EMPaS11	Lithuania, Germany	Rugienius et al. (2015), Xuan et al. (2009)	59–121	9–11
44.	EMPaS12[a]	Lithuania, Latvia Germany	Höfer and Peil (2015), Lacis and Kota (2013), Rugienius et al. (2015), Xuan et al. (2009)	99–188	8–9
45.	EMPaS14[a]	Lithuania, Latvia Germany	Höfer and Peil (2015), Lacis and Kota (2013), Rugienius et al. (2015), Xuan et al. (2009)	153–216	6–10
46.	PceGA25	Latvia, Sweden, USA, Iran, Denmark, Hungarian	Cantini et al. (2001), György et al. (2010), Lacis et al. (2011), Najafzadeh et al. (2016), Pedersen (2006)	108–222	5–21
47.	PceGA34[a]	Lithuania, Latvia, Iran	Lacis and Kota (2013), Rugienius et al. (2015), Najafzadeh et al. (2016)	75–360	12–20
48.	PceGA59	Latvia, USA, Iran, Denmark	Cantini et al. (2001), Lacis and Kota (2013), Najafzadeh et al. (2016), Pedersen (2006)	181–226	5–36
49.	Pchgms1	Turkey, Iran	Turkoglu et al. (2010), Najafzadeh et al. (2016)	138–184	3–8
50.	Pchpgms3	USA, Iran, Denmark, Hungarian	Cantini et al. (2001), Pedersen (2006), György et al. (2010), Najafzadeh et al. (2016)	152–191	4–23
51.	PMS2	Latvia, USA, Denmark	Cantini et al. (2001), Pedersen (2006), Lacis and Kota (2013)	132–152	8–20
52.	PMS3	Latvia, Sweden, USA, Iran, Denmark, Hungarian	Cantini et al. (2001), Pedersen (2006), György et al. (2010), Lacis et al. (2011), Najafzadeh et al. (2016)	152–212	5–31
53.	PMS30	USA, Denmark	Cantini et al. (2001), Pedersen (2006)	119–175	11–26
54.	PMS40	Latvia, USA, Denmark	Cantini et al. (2001), Pedersen (2006), Lacis and Kota (2013)	88–118	8–25
55.	PMS49	Latvia, Sweden, USA, Iran, Denmark, Hungarian	Cantini et al. (2001), Pedersen (2006), György et al. (2010), Lacis et al. (2011), Najafzadeh et al. (2016)	70–185	4–20
56.	PMS67	Latvia, USA, Iran Denmark	Cantini et al. (2001), Pedersen (2006), Lacis and Kota (2013), Najafzadeh et al. (2016)	85–191	4–22
57.	PS05C03[a]	—	—	—	—
58.	PS08E08	Latvia, USA, Denmark, Hungarian	Cantini et al. (2001), Pedersen (2006), György et al. (2010), Lacis and Kota (2013)	172–185	3–16
59.	PS12A02	Turkey, Latvia, Iran	Turkoglu et al. (2010), Lacis and Kota (2013), Najafzadeh et al. (2016)	112–182	2–13
60.	UCDCH 17	Hungary, Turkey, Germany	Xuan et al. (2009), Turkoglu et al. (2010), Baris et al. (2017)	155–205	4–15
61.	UCDCH21	Turkey	Turkoglu et al. (2010)	103–113	3
62.	UCDCH31	Turkey, Iran	Turkoglu et al. (2010), Najafzadeh et al. (2016)	124–184	4
63.	UDAP401	Turkey	Turkoglu et al. (2010)	262–272	3
64.	UDAP404	Turkey	Turkoglu et al. (2010)	154–168	2
65.	UDP 96 001	Hungary, Turkey, Germany, Iran, Russia	Xuan et al. (2009), Turkoglu et al. (2010), Najafzadeh et al. (2016), Baris et al. (2017), Kulikov et al. (2018)	95–132	2–9
66.	UDP 96 005	Iran, Russia, Hungary, Turkey, Latvia	Turkoglu et al. (2010), Lacis and Kota (2013), Najafzadeh et al. (2016), Baris et al. (2017), Kulikov et al. (2018)	75–180	3–9
67.	UDP96-003	Russia	Kulikov et al. (2018)	88–93	4

(*Continued*)

TABLE 7.4 (*Continued*)
Sour Cherry SSR Loci Analysis Worldwide

	SSR Locus[a]	Country (Author)	References	Size Range	No. of Alleles
68.	UDP97–402	Latvia, Russia	Lacis and Kota (2013), Kulikov et al. (2018)	121–144	6–9
69.	UDP97–403	Russia	Kulikov et al. (2018)	102–149	4
70.	UDP97–406	Russia	Kulikov et al. (2018)	87–106	5
71.	UDP98–410	Germany	Xuan et al. (2009)	127–139	5
72.	UDP98–411	Germany	Xuan et al. (2009)	129-8	8
73.	UDP98–412[a]	Lithuania, Latvia Germany	Xuan et al. (2009), Lacis and Kota (2013), Rugienius et al. (2015)	85–127	8–12

[a] SSR locus includes to ECPGR primer set.

alleles in Belarusian and Swedish cultivars (Lacis et al. 2011; Urbanovich et al. 2014; Najafzadeh et al. 2016). The highest number of alleles in Turkey is 10 (Turkoglu et al. 2010). The lowest number of alleles is produced in Russia (6) and Hungary (5).

According to the number of alleles, the most informative SSR loci for Danish and Lithuanian cultivars are PceGA59 and PceGA34 (Pedersen 2006; Rugienius et al. 2015). For Finnish cultivars, the most informative loci are BPPTC-037, and for Belarusian cultivars, BPPCT025 locus (Antonius et al. 2012; Urbanovich et al. 2014). For Iranian, US, Hungarian, and Latvian sour cherry cultivars, the PMS3 locus is the most informative (Cantini et al. 2001; György et al. 2010; Lacis et al. 2011; Lacis and Kota 2013; Najafzadeh et al. 2016). Locus PS12A02 is the most informative in Turkish cultivars (Turkoglu et al. 2010). In sour cherry cultivars in Russia, the SSR locus UDP97–402 is the most informative (Kulikov et al. 2018).

Markers with a very high level of polymorphism were selected for the general genetic characterization of sour cherry (PceGA25, PMS3, and PMS49) (Table 7.4) (Cantini et al. 2001; Lacis et al. 2011). The set of three SSR markers was used to screen cultivars from six different countries – Latvia, Sweden (Lacis et al. 2011), the USA (Cantini et al. 2001), Iran (Najafzadeh et al. 2016), Denmark (Pedersen 2006), and Hungary (György et al. 2010). The highest number of alleles was observed in Danish cultivars and the lowest in Hungarian varieties. The countries in Europe (Denmark, Sweden, Latvia, and Hungary) have similar allelic diversity according to the PceGA25 SSR loci, and the Iranian cultivars have a larger allele size range (Figure 7.5). This SSR-loci combination is suitable for rapid and effective genotyping and genetic relatedness screening (Lacis et al. 2011).

Since the *Prunus* Group of the ECPGR generated 16 microsatellite markers for the diploid species of *Prunus*, studies on the adaptation of these primers in species with higher levels of ploidy, such as sour cherries, are still lacking (Clarke and Tobutt 2009). However, loci from standard microsatellites set were performed on sour cherry cultivars from six countries – Germany ('Staatliche Lehr- und Versuchsanstalt für Wein- und Obstbau Weinsberg' and Fruit Genebank of the Institute for Breeding Research on Fruit Crops) (Höfer and Peil 2015; Xuan et al. 2009), Finland (Antonius et al. 2012), Latvia (Institute of Horticulture (LatHort)) (Lacis and Kota 2013), Lithuania (The Institute of Horticulture of Lithuanian Research Center for Agriculture and Forestry (LRCAF IH)) (Rugienius et al. 2015), Belarus (Urbanovich et al. 2014), and Iran (Najafzadeh et al. 2016).

The complete set of 16 SSR markers was used only for the genetic analysis of Latvian cultivars. Only 14 primers were polymorphic, generating between 5 and 18 alleles. The PS05C03 locus did not amplify a PCR product and was excluded from the study. The EMPA026 locus showed a monomorphic pattern in Latvian varieties and was not used for further studies. However, in Belarusian varieties, this marker produced ten alleles in the size range of 184–238 bp (Lacis and Kota 2013; Urbanovich et al. 2014).

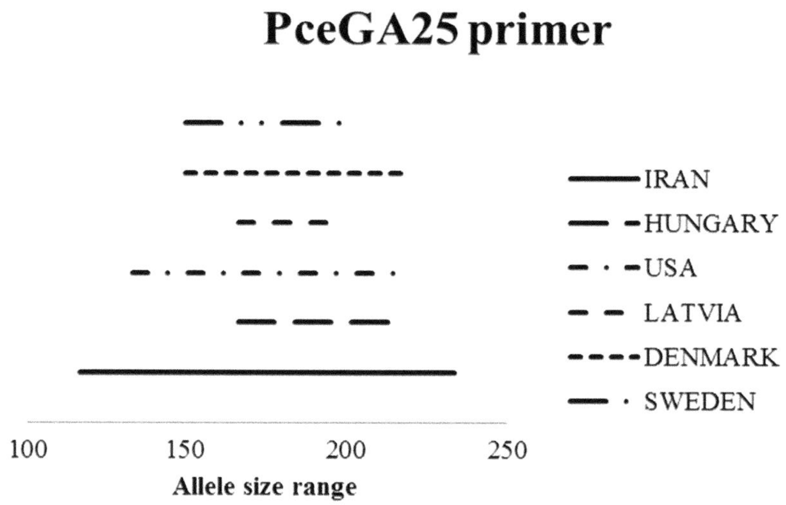

FIGURE 7.5 Allelic diversity according to PceGA25 loci.

Two SSR loci (EMPaS02 and EMPaS06) were performed for German, Latvian, Lithuanian, and Iranian cultivars. The range of allele size of European cultivars according to EMPaS02 loci is 119–147 bp, while Iranian cultivars have a different range of 185–270 bp. EMPaS06 loci in European cultivars are in the range of 176–236 bp, while Iranian cultivars have a wider range of 112–285 bp (Lacis and Kota 2013; Najafzadeh et al. 2016). According to the alleles generated, SSR loci pceGA34 and BPPCT037 have the highest polymorphism and are suitable for the genetic characterization of sour cherries.

7.2.1.3 Wild Cherries

Wild cherries are important to the timber industry because they have straight, easy-to-machine, and fine-grained wood (Vaughan and Russell 2004). Nowadays, it is important to maintain and increase the productivity of wild cherry plantations for the timber industry (Licea-Moreno et al. 2019). In addition, wild cherries are an important and potential gene pool for sweet cherry cultivars and breeding purposes (de Rogatis et al. 2013). Molecular markers such as microsatellites are used to identify and classify genetic diversity and resources and to conserve the germplasm of the species to counteract future abiotic, biotic, and climatic changes (de Rogatis et al. 2013; Licea-Moreno et al. 2019).

In 2004, 14 SSR markers were designed from wild cherry for the first time (Vaughan and Russell 2004). Seven primers were polymorphic (EMPaS01, EMPaS02, EMPaS06, EMPaS10, EMPaS11, EMPaS12, and EMPaS14) and were used in combination of four polymorphic loci (EMPa004, 005, 015, and 018) identified in sweet cherry (Clarke and Tobutt 2009) for further study (Table 7.5). In a *Prunus* mapping progeny (*P. avium* 'Napoleon'×*P. nipponica*), the genetic linkage of eleven loci was evaluated and found to be unfavorable (Vaughan and Russell 2004). Scientists used these primer pairs to analyze wild cherries' genetic diversity worldwide, and there are some markers (EMPaS01, EMPaS02, EMPaS06, EMPaS10, and EMPaS14) included in ECPGR standardized primer set for *Prunus* genetic fingerprinting (Clarke and Tobutt 2009).

In central and northern Italy, ten primer pairs were used for 278 cultivars' genetic diversity analysis (Marchese et al. 2007; de Rogatis et al. 2013). The number of alleles was in the range of 4–25 per locus (Table 7.4), which is higher than that in other wild cherry populations (Vaughan and Russell 2004; Stoeckel et al. 2006; Guarino et al. 2009; Licea-Moreno et al. 2019). Also, the number of alleles and the size ranges of alleles are higher than those in sweet cherries (de Rogatis et al. 2013). The alleles' size ranges overlap with other scientists' results (Clarke and Tobutt 2003;

TABLE 7.5
Wild Cherry SSR Loci Analysis Worldwide

SSR Locus	Country (Author)	Author	Size Range	No. of Alleles
EMPaS01	Italy, UK, Turkey	de Rogatis et al. (2013), Vaughan and Russell (2004), Uzan Eken et al. (2022)	222–259	4–1
EMPaS02	Italy, UK, Turkey	de Rogatis et al. (2013), Guarino et al. (2009), Vaughan and Russell (2004), Uzan Eken et al. (2022)	133–148	5–16
EMPaS04	Italy, UK	Guarino et al. (2009), Vaughan and Russell (2004)	223–248	2–4
EMPaS05	UK	Vaughan and Russell (2004)	161–171	3
EMPaS06	Italy, UK, Turkey	de Rogatis et al. (2013), Guarino et al. (2009), Vaughan and Russell (2004), Uzan Eken et al. (2022)	200–230	1–24
EMPaS07	UK	Vaughan and Russell (2004)	192–194	2
EMPaS08	UK	Vaughan and Russell (2004)	240	1
EMPaS10	Italy, UK, Turkey	Rogatis et al. (2009), Guarino et al. (2009), Vaughan and Russell (2004), Uzan Eken et al. (2022)	148–192	5–25
EMPaS11	Italy, UK, Turkey	Guarino et al. (2009), Vaughan and Russell (2004), Uzan Eken et al. (2022)	62–114	4–18
EMPaS12	Italy, UK, Turkey	de Rogatis et al. (2013), Guarino et al. (2009), Vaughan and Russell (2004), Uzan Eken et al. (2022)	121–155	5–23
EMPaS13	Italy, UK	Guarino et al. (2009), Vaughan and Russell (2004)	203–219	1–3
EMPaS14	Italy, UK, Turkey	de Rogatis et al. (2013), Guarino et al. (2009), Vaughan and Russell (2004), Uzan Eken et al. (2022)	188–213	3–6
EMPaS15	UK	Vaughan and Russell (2004)	185–236	2
EMPaS18	UK	Vaughan and Russell (2004)	222	1
EMPA001	Italy	Guarino et al. (2009)	143–158	4
EMPA002	Italy	Guarino et al. (2009)	106	1
EMPA003	Italy	Guarino et al. (2009)	175–179	2
EMPaO04	Italy	de Rogatis et al. (2013), Guarino et al. (2009), Uzan Eken et al. (2022)	171–198	3–13
EMPaO05	Italy, Turkey	de Rogatis et al. (2013), Guarino et al. (2009), Uzan Eken et al. (2022)	224–264	2–15
EMPaO06	Italy	Guarino et al. (2009)	101	1
EMPaO07	Italy	Guarino et al. (2009)	175	1
EMPaO08	Italy	Guarino et al.2009	170–172a	2
EMPaO09	Italy	Guarino et al. (2009)	202–244	2
EMPaO10	Italy	Guarino et al. (2009)	127–130	3
EMPaO11	Italy	Guarino et al. (2009)	244–252	2
EMPaO12	Italy	Guarino et al. (2009)	230–232	2
EMPaO13	Italy	Guarino et al. (2009)	189–200	4
EMPa015	Italy, Turkey	de Rogatis et al. (2013), Guarino et al. (2009), Uzan Eken et al. (2022)	202–300	1–24
EMPa016	Italy	Guarino et al. (2009)	178–191	1–2
EMPa017	Italy	Guarino et al. (2009)	240	1
EMPa018	Italy	Guarino et al. (2009)	95–107	4
EMPa019	Italy	Guarino et al. (2009)	119–121	2
EMPa021	Italy	Guarino et al. (2009)	120–121	2
UDP96-001	Spain	Licea-Moreno et al. (2019)	107–135	7
UDP97–402	Spain	Licea-Moreno et al. (2019)	113–143	9
PCHGMS 1	Spain	Licea-Moreno et al. (2019)	121–151	6

(Continued)

TABLE 7.5 (*Continued*)
Wild Cherry SSR Loci Analysis Worldwide

SSR Locus	Country (Author)	Author	Size Range	No. of Alleles
UDP98-021	Spain	Licea-Moreno et al. (2019)	101–117	5
UDP98–410	Spain	Licea-Moreno et al. (2019)	123–135	6
BPPCT 002	Spain	Licea-Moreno et al. (2019)	167–187	6
BPPCT 005	Spain	Licea-Moreno et al. (2019)	140–198	10
BPPCT 026	Spain	Licea-Moreno et al. (2019)	163–185	5
BPPCT 037	Spain	Licea-Moreno et al. (2019)	133–170	10
BPPCT 038	Spain	Licea-Moreno et al. (2019)	98–133	10
BPPCT 039	Spain	Licea-Moreno et al. (2019)	130–147	4
UCD-CH12	Spain	Licea-Moreno et al. (2019)	175–197	9
UCD-CH18	Spain	Licea-Moreno et al. (2019)	182–211	10

Vaughan and Russell 2004; Guarino et al. 2009). *Prunus avium* populations in Germany (Höltken and Gregorius 2006) showed more alleles (range from 4 to 9 alleles per locus) than in Italy populations (1–5 per locus, respectively) (Guarino et al. 2009).

In Spain, 36 wild cherry cultivars were analyzed using 20 microsatellite markers. All loci showed high polymorphism, and up to 97 alleles were detected, averaging 7.7 alleles per locus. In the wild cherry genotype, some new alleles were observed that were not present in the original species. At loci UDP96-001 and UDP97-402, six and seven more alleles, respectively, were detected than in Cipriani et al. (1999). Loci UCD-CH18 (Struss et al. 2003) recorded seven new alleles (Licea-Moreno et al. 2019).

Cerasus subgenus of *Prunus* species studied by Iranian scientists showed high polymorphism rates with chosen microsatellite markers, indicating high inter- and intraspecific genetic variations (Khadivi-Khub et al. 2014a). The gene pool of cherries may benefit greatly from the various wild *Cerasus* germplasm for breeding purposes, as provides few benefits in terms of creating dwarfing rootstocks resistant to unfavorable climatic conditions and drought in soil. *Prunus yazdiana* has shown low inter- and intraspecific genetic intraspecies variation, while other species resulted in high variation in its individuals.

7.2.2 Self-(In)Compatibility/*S*-Allele Typing in Cherries

7.2.2.1 Self-Incompatibility of Sweet Cherry

The ability of plants to prevent self-pollination forms the basis of incompatibility genetics. In many studies, complex genetic locus structure and haplotypes were analyzed, and their interaction models were hypothesized at the biological, genetical, and biochemical levels (Crane and Brown 1938; Crane and Lewis 1942; Silva and Goring 2001; Wheeler et al. 2001; Yamane and Tao 2009). Our understanding of the molecular gametophytic SI system in *Prunus* has advanced significantly.

There are two types of incompatibility: heteromorphic (diallelic) and homomorphic (multiallelic) (McCubbin 2008). Heteromorphic SI is determined by morphological (e.g., different lengths of styles and stamens within the same flower) and physiological (e. g., different maturation times of pistils and anthers) features of flowers. Meanwhile, homomorphic SI is determined by the specific interaction of *S*-locus proteins, and each plant has the same flowers in the population (Charlesworth 2010; Miljuš-Đukić et al. 2010). The homomorphic SI will be discussed in this section.

7.2.2.1.1 Genetics of Incompatibility, S-Locus, and Haplotypes

Most land flowering plants are hermaphrodites. This means that either male or female flowers are produced on the same plant, or more often, the male and female reproductive organs are within the same flower. The main task of SI is to prevent the hermaphrodite plant from fertilizing its pollen. Hence, the germination of self-pollen has to be inhibited by a specific mechanism. The genotype of a sporophyte or gametophyte determines whether or not it is self-incompatible based on how the pollen behaves during SI interaction. Plants of the Solanaceae, Plantaginaceae, and Rosaceae families are characterized by gametophytic self-incompatibility (GSI) (McCubbin and Kao 2000; Matsumoto and Tao 2016a). The majority of *Prunus* species have a homomorphic GSI system. SI is controlled by a single multi-allelic *S*-locus and is determined by gene combinations termed haplotypes and denoted as S_x, where x can take values from 1 to n. In this system, the specificity of self-/nonself-recognition is regulated by products encoded in the *S* locus (Tao and Iezzoni 2010). Two polymorphic genes, *S-RNase* and *SFB*, were identified (Sorkheh 2015). Physically, these two genes are located close to each other (30–380 kb in *Prunus*) and may be inherited as a single unit (Ushijima et al. 2003; Ikeda et al. 2004; Hegedűs et al. 2012; Matsumoto and Tao 2016a). During pollination, a SI reaction is induced when the same *S* allele specificity is expressed in both the pollen and pistil (Figure 7.6) (Tao and Iezzoni 2010).

T2-type *S*-ribonuclease's gene (*S-RNase*) codes for an enzyme called *S*-ribonuclease, which is responsible for self-pollen recognition and rejection and is defined as a determinant of a pistil. When the pollen tube and stigma/style haplotypes match, the enzyme (glycoprotein) digests the pollen tube's RNA, and the pollen is directed for degradation. *Prunus S-RNase* genes contain two introns (Igic and Kohn 2001) (Figure 7.7) that present allele-specific length variations. In Rosaceae, the diversity of protein-coding alleles depends on five conserved (C1–C3, RC4, and C5) and a single hypervariable region (RHV) (Sorkheh 2015).

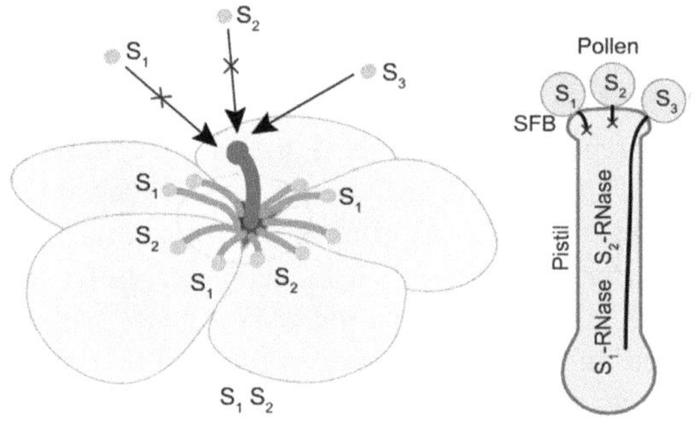

FIGURE 7.6 Gametophytic self-incompatibility in *Prunus avium* (Kivistik et al. 2021), during pollination, the pollen is rejected (S_1 or S_2) when its haplotype coincides with one of the two diploid stigma's haplotypes (S_1S_2). The pollen tube starts growing and reaches the ovary when the pollen's haplotype (S_3) does not match the stigma's haplotype (S_1S_2). *S-RNase*, *S*-ribonuclease; SFB, *S* F-box.

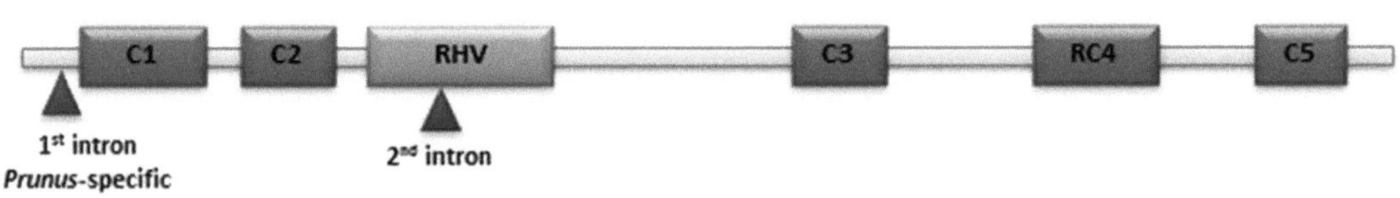

FIGURE 7.7 The structure of Rosaceae – Prunoideae (*Prunus*) *S*-RNase protein according to Sorkheh (2015) and Yamane and Tao (2009). C1, C2, C3, C5, conserved regions; RC4, Rosaceae conserved region; RHV, Rosaceae hyper variable region.

The evolutionary relationships between *Prunus*, *Pyrus*, and *Malus* were evaluated and polymorphism analysis of *S-RNase* gene and protein (RC4 and C5 domain) sequences from sweet cherry and almond was performed (Ma and Oliveira 2002), Yamane et al. (2003c) identified a putative protein, PA1, as a possible ancestor of S-RNases in the genus *Prunus*. Ishizaka et al. (2003) characterized S_3-, S_4-, and S_6-*RNase* promoters of sweet cherry varieties 'Benisyuhou' and 'Satounishiki' and their activity in the pistil. Each *S* allele (*S-RNase* gene) can be described by distinct intron size. Determination is done by developing and applying consensus and *S* allele-specific molecular primers (Schuster 2014).

Tao et al. (1999) first used two-dimensional polyacrylamide gel electrophoresis (2D PAGE) to identify basic stylar glycoproteins and developed a PCR-based *S*-allele typing method. Consensus primers according to the *S-RNase* gene's conserved regions were designed. Using different combinations of primers (e.g., PruT2-PruC4R; PruC2-PruC4R) helped to distinguish the S_1 allele from S_4 and S_6 and S_2 from S_5. Later, another set of consensus primers (e.g., SI19-SI20 and SI31-SI32) and degenerative primers were designed, and they were very valuable for describing S_1–S_6 and S_9 alleles in sweet cherry cultivars (Wiersma et al. 2001; Sonneveld et al. 2001). The scientists announced that S-RNases are likely involved in GSI in sweet cherries. PaConsIF/R and PaConsIIF/R consensus primers were useful for distinguishing S_1–S_{16} alleles (Sonneveld et al. 2003). Allele-specific primers were designed after the *S-RNase* gene's alleles were cloned and sequenced (Sonneveld et al. 2001, 2003; Schuster 2012; Szikriszt et al. 2013). These findings were valuable for further *S*-allele composition studies of sweet cherry in many European, Asian, and North American regions.

Six new *S-RNases* (S_{17}–S_{22}) were identified in wild cherry in Belgium and three (S_{23}–S_{25}) in local Spanish and Italian sweet cherry cultivars (Wünsch and Hormaza 2004; De Cuyper et al. 2005). Vaughan et al. (2008) characterized six more *S-RNase* alleles (from 27 to 32) in British wild cherry. In 2009, 28 *S-RNase* alleles of sweet cherry were announced (Yamane and Tao 2009). More than 30 *S*-alleles are currently characterized by wild and cultivated sweet cherries (Schuster 2020).

The *S* F-box gene codes for a protein, *S*-ribonuclease inhibitor, and is defined as a pollen determinant. If pollen is compatible, a protein inhibits *S*-ribonuclease activity in the pollen tube, and pollen germinates. Gene composition and allelic sequence polymorphism differ between species. Based on this, different definitions are given to it. For example, in *Prunus* (sweet cherry, sour cherry, almond, peach), the *S* F-box genes are termed the *S haplotype-specific F-box* (*SFB*), and the *S locus F-box* (*SLF*) or the *S locus F-box-like* (*SLFL*) genes in Solanaceae (*Petunia*) and Plantaginaceae (*Antirrhinum*) species (Lai et al. 2002; Ushijima et al. 2004; Matsumoto et al. 2008; Matsumoto and Tao 2016a). Besides, in *Pyrus* (pear) and *Malus* (apple), F-box genes are named *S-locus F-box brothers* (*SFBBs*) (Sassa et al. 2007). The evolution of *SFB* genes in the *Prunus* genome was studied by Akagi et al. (2016). Based on their results, it was proposed that the *Prunus SFB* gene originated from a "recent *Prunus*-specific gene duplication event."

SFB in *P. avium* and *P. cerasus* contains only one intron, and its size is minimal. The differences in intron size are not detectable on agarose gels. For detection, a fluorescent-labeled primer and an automatic sequencer are necessary (Schuster 2014).

Two hypervariable HVa and HVb and two variable V1 and V2 regions were detected in the *Prunus* SFB (Figure 7.8) (Ushijima et al. 2003, 2004; Ikeda et al. 2004). Regions located at the C terminus are supposed to influence allele-specific recognition of the substrate (Matsumoto and Tao 2016a). However, the precise functions of each motif in the C-terminal sequences are poorly understood (Akagi et al. 2016).

FIGURE 7.8 Structure of *Prunus S* F-box protein (Matsumoto and Tao 2016a). V1, V2, Vn, variable regions; HVa, HVb, hypervariable regions.

Yamane et al. (2003b) used genomic Southern blotting and confirmed that SFB^1-SFB^6 are linked to appropriate *S*-haplotypes of sweet cherry. These alleles were cloned and sequenced by Ikeda et al. (2004). Putative amino acid sequences were aligned with almond and Japanese apricot sequences. Later, Vaughan et al. (2006) optimized PCR analysis, and 12 *SFB* alleles were sequenced by a semi-automated sequencer. The *SFB* and *S-RNase* intron analysis revealed the sequence length polymorphism of known sweet cherry cultivars and wild cherry populations.

In *Prunus*, outside the *S*-locus, three SLFLs are not involved in determining pollen specificity (Ushijima et al. 2003, 2004). Phylogenetic analysis of *SFB* and *SLFL* genes, biochemical features of SLFL1, SLFL2, and SLFL3, and their roles in the GSI mechanism were investigated (Matsumoto et al. 2008; Matsumoto and Tao 2016b). Possible models of S-RNase 'degradation' or 'compartmentalization' were reviewed by McClure et al. (2011). According to scientists, *S-RNase*-based SI depends on the interaction between S-RNase and modifier gene products, *SLFs*.

Different mechanisms of SI in Brassicaceae, Solanaceae (e. g., plants of *Solanum*, *Nicotiana*, and *Petunia* genus), Rosaceae (e.g., plants of *Malus*, *Prunus*, and *Pyrus* genus), and Papaveraceae (*Papaver rhoes*) were studied at the molecular level (McClure et al. 2011; Matsumoto and Tao 2016a, b). It was proposed a "general inhibitor" degradation model by SFB, which in *Prunus* is different compared to other taxa (Tao and Matsumoto 2012). SLFL genes were proposed to code for putative proteins called F-box proteins (Matsumoto and Tao 2016b; Chen et al. 2018).

From earlier studies, it is known that SFB is a component of a conserved ubiquitin-protein ligase complex (Skpl, Cullin1, F-box protein, and Rbx1), which is essential for regulatory and signaling protein destruction (Deshaies 1999; Zheng et al. 2002). This led to the assumption that the ubiquitin/26S proteasome proteolytic system may be essential for RNase-based GSI (Ushijima et al. 2003). *In vitro* findings showed that SLFLs could interact with *Prunus avium* SSK1 (SLF-interacting Skpl-like 1) and Cul1A and form SCF. According to scientists, SLFL2 mediates polyubiquitination of S-RNases and SFB, its substrate, during the SI process, thus maintaining a 'clean' environment in the pistil (Matsumoto and Tao 2016b, 2019).

7.2.2.1.2 Self-(In)Compatibility Groups and S-Allele Identification Methods

Information about the *S* genotype of sweet cherries is necessary for growers and breeders (Schuster 2017). In planting commercial orchards, appropriate cross-cultivars from different pollen-incompatibility groups that bloom simultaneously must be interplanted to ensure a sufficient fruit set (Sorkheh 2015).

Cultivars that share the same *S*-genotype are cross-incompatible and assigned to the same IG. Meanwhile, cultivars sharing different *S*-genotypes are defined as cross-compatible and are assigned to a different group. Cultivars having distinct *S*-genotypes but are compatible with all other known *S*-genotype cultivars are categorized into group O and are defined as universal donors (Herrero et al. 2017). Studies of sweet cherries haplotypes, which assign varieties to a certain (in)compatibility group, are of great importance in developing breeding material (Wünsch and Hormaza 2004; Sorkheh 2015).

Researchers have long been fascinated by SI and cross-incompatibility in plants. Correns (1913), East (1925, 1926), Lehmann (1926), and Sirks (1926) were the first scientists who used genetic analysis methods in order to find out the answer to this question. They demonstrated that incompatibility is determined by genetic and morphological factors (Crane and Lawrence 1928). In the beginning, the SI of varieties was investigated by artificial pollination of flowers or by observing the growth of the pollen tube (Crane and Brown 1938; Mahmoudi and Arzani 2006; Mahmoodi et al. 2008; Radičević et al. 2021). For a long time, only six alleles (S_1–S_6) have been known in sweet cherry (Tao et al. 1999). It is partly because these methods are time-consuming, require several years of study, and depend on environmental conditions (Ganopoulos et al. 2010; Herrero et al. 2017).

Later, more rapid molecular methods were applied for *S*-alleles' identification. Bošković and Tobbut al. (1996) used isoelectric focusing (IEF) and ribonuclease activity staining to identify new *S-RNase* alleles. The *S-RNase* gene's cloning and sequence characterization allowed the development of PCR-based methods for the *S*-genotyping of sweet cherries. Each *S-RNase* allele has two introns which vary in size and several conserved domains. These intron differences can be detected by PCR using consensus primers or allele-specific primers (Tao et al. 1999; Wiersma et al. 2001; Sonneveld et al. 2001, 2003, 2006). Restriction fragment length polymorphism (RFLP) was applied for the *S*-genotyping of two sweet cherry cultivars in Japan and sour cherry (Yamane et al. 2000a; Hauck et al. 2002). A genotyping method based on pollen *S*-allele by RFLP has been used by Yamane et al. (2003b) and Ikeda et al. (2004). Marchese et al. (2007) used microsatellite fingerprinting or SSR analysis to distinguish Sicilian sweet cherry accessions.

Kitashiba et al. (2008) developed a dot-blot analysis method for determining *S* genotypes and screening Japanese sweet cherry and plum. It was based on SFB-HVa region polymorphism. Researchers succeeded in discriminating ten haplotypes (S_1–S_4, $S_{4'}$, S_5–S_7, S_9, and S_{16}) in sweet cherry. As it had some limitations, Wang et al. (2013) developed a similar method where PCR products were immobilized on streptavidin-coated magnetic beads. It allowed identifying of sweet cherry cultivars in Japan by utilizing additional *S*-RNase allele-specific oligonucleotide probes. This method was more reliable for discriminating between S_4 and S_4' alleles.

Molecular markers were immutable for determining the *S*-genotype of a very large number of sweet cherry cultivars, including wild and local sweet cherry cultivars and accessions in various countries of Europe, Asia, and North America (Schuster 2012, 2020).

7.2.2.1.3 S-Genotyping, (In)Compatibility Groups, and S-Allele/S-Haplotype Distribution

The *S*-allele composition and *S*-genotypes of sweet cherry have been studied in many European regions, Asia, Canada, and the USA (Ipek et al. 2011). Sweet cherry breeding programs in Europe and Asia provided an overview of the cultivar situation and forecast the future direction of the cherry industry (Sansavini and Lugli 2008). The worldwide distribution of *S*-genotypes is shown in Figures 7.9–7.11.

A single system replaced the numbering employed by various writers for newer alleles. *S*-ribonuclease and PCR findings were combined, and a table of XXVI SI groups was announced (Wünsch and Hormaza 2002; Sonneveld et al. 2003; Tobutt et al. 2004b).

Near the center of sweet cherry origin in Turkey, *S*-allele combinations of local sweet cherry genotypes were first determined by Ipek et al. (2011). Studies revealed that the S_3 (30%) allele was very frequent among genotypes, whereas the S_1 (2.5%) allele was not so common. Some cultivars have been assigned to three new groups XLII-XLIV (Ipek et al. 2011). Old Turkish landrace cultivars and wild cherries in the Black Sea region were *S*-genotyped by Szikriszt et al. (2013). S_{17}–S_{19}, $S_{21/25}$, and S_{31} alleles were detected among genotypes. Previously, these alleles were found in wild cherries (De Cuyper et al. 2005; Vaughan et al. 2008). The sour cherry-specific allele S_{34} was found in one selection. Meanwhile, the allele S_{16} was not detected.

Sweet cherry is quite popular in Greece; probably here, it was cultivated first (Marshall 1954). The largest area under sweet cherry plantations in 2019 was 16.1 thousand hectares (Gorina et al. 2021). Ganopoulos et al. (2010) detected eight different alleles (S_1–S_5, S_9, S_{13}, and S_{30}) in Greek sweet cherry cultivars. S_4 (39%) and S_3 (29%) occurred with a high frequency, and the S_{30} allele was found in one cultivar. Two fully compatible cultivars carrying rare (S_2S_5, S_3S_{13}) genotypes were detected. The authors mentioned that many cultivar combinations were semi-compatible and were associated with low yield.

In Croatia, sweet cherries are mainly popular in home gardens and are less often grown on more extensive commercial plantations (Biško et al. 2017). *S*-genotyping was first studied by Ercisli et al. (2012). A study of local sweet cherry cultivars revealed S_3 and S_{12} being the most frequent alleles and S_3S_{12} as the most frequent genotype. The findings suggest that Turkish and Croatian sweet cherry varieties may share genetic ancestry. The S_{12} allele was common among sweet cherry cultivars in Turkey (Ipek et al. 2011). Self-compatible sweet cherry cultivars were not found in Croatia (Radunic 2010).

Békefi et al. (2003) noticed that the S_2 allele was rare in Hungarian sweet cherry cultivars and accessions. The S_x allele was detected only in one cultivar, 'Rita', and was assigned as S_{22}. S_3 and S_4 were found in many of the studied sweet cherries. Later research showed that S_6 was a frequent allele in the local Hungarian sweet cherry collection (Békefi et al. 2010).

Earlier studies of SI and cross-incompatibility of Iranian sweet cherry cultivars were determined by monitoring fruit set percent under field conditions or recording pollen tube growth rate by fluorescent microscopy (Mahmoudi and Arzani 2006; Mahmoodi et al. 2008). It is a time-consuming process and could lead to false results. Molecular studies based on allele-specific PCR and RAPD markers were much more accurate for identifying compatible/incompatible groups and analyzing of genetic variations. It revealed three alleles (S_3, S_4, and S_{14}) in local sweet cherry cultivars, and the most frequently observed genotype was S_3S_4 (Ershadi and Moghadam 2009). According to Gharaghani et al. (2017), all Iranian sweet cherry germplasm is self-incompatible, except 'Sefid-90', which has better traits than self-incompatible genotypes.

Marić et al. (2020) overviewed sweet cherry *S*-allele composition in the Balkan region. The authors announced that the most frequent allele was S_3 (~39%). According to Ershadi and Moghadam (2009), S_3 (59%) and S_4 (63%) alleles were very frequent in Iranian sweet cherry cultivars. Khadivi-Khub et al. (2014b) also detected the S_3 allele in a few sweet cherry cultivars. In China, the most frequent alleles were S_3 (27%) and S_9 (20%), but the S_4 (16%) allele was not so common (Wang et al. 2010). The S_7 allele was not detected, while in local Turkish cultivars, its frequency was 3.7% (Ipek et al. 2011). The S_3 allele was also frequent among Russian sweet cherry cultivars (Bezlepkina et al. 2020). *S*-haplotypes' distribution in Asian countries is presented in Figure 7.10.

XXXVII-IL groups were proposed by Marchese et al. (2007). It was reported that S_3 and S_6 were the most frequent alleles in Sicilian sweet cherry germplasm, and many accessions shared S_{16} and S_{22} alleles. The number of groups reached XLI (41) ($S_{12}S_{22}$) after the *S*-allele characterization of sweet cherry cultivars in southeast Spain (Alicante Mountain area) (Gisbert et al. 2008). *S*-genotypes' distribution in European countries is presented in Figure 7.9.

In Italy, sweet cherry plantations cover a large part of a region (29,200 hectares), and a significant amount of fruits is harvested (98,600 tons) every year (Gorina et al. 2021). However, the genetic diversity of Italian genetic resources was not investigated broadly. Marchese et al. (2017b, 2017a) characterized local sweet cherry accessions and cultivars from different Italian regions. Among accessions, the most common alleles were S_3, S_6, and S_{13}. S_7, S_{17}, S_{19}, S_{21}, and S_{22} (0.3%–0.6%) alleles were the rarest in the Italian sweet cherry germplasm. Meanwhile, a few accessions and cv. Maiolina were proposed to have a mutant allele S_5', the same as in 'Kronio'. S_1 and S_2 were rare alleles among local cultivars, and S_4 was found just in one genotype. After these studies, the SI groups of *S*-alleles were expanded to 56. According to Schuster (2017), *S*-alleles of cultivated sweet cherries were arranged into 60 SI groups, and more than 22 different *S*-allele of cultivated *Prunus avium* were identified. An S_2S_9 genotype was assigned to group O, and two cultivars ('Endiabla' and 'Runar') with determined S_3S_y genotypes were assigned to the self-compatibility (SC) group as S_3S_3.

Schuster (2012) genotyped cultivars from a sweet cherry collection in Germany. Forty-seven IGs were announced, and 15 genotypes were assigned to universal donor group O. According to the relative occurrence of *S*-alleles, three groups were separated: Europe (I), West, Middle Asia (II), and North America, East Asia and Australia (III). The most frequent among countries was allele S_3. The lowest number of different *S*-alleles was detected in the third group. The largest number of IGs occurred among sweet cherry cultivars from Europe.

Sharma et al. (2014) designed new allele-specific primers which helped them to identify the S_3, S_4, and S_9 alleles of sweet cherry cultivars in the Czech Republic more clearly. Patzak et al. (2019) used universal and allele-specific primers set and announced that S_3 (63%) and S_1 (~37%) alleles were very frequent and just one variety of each had S_7, S_{13}, S_{14}, S_{16}, and S_{19} alleles in their *S*-genotypes. Some new *S*-haplotypes (S_4S_x, $S_{19}S_x$, S_2S_{13}, S_1S_6) were found. The III and II IGs were the most common among analyzed cherry genotypes.

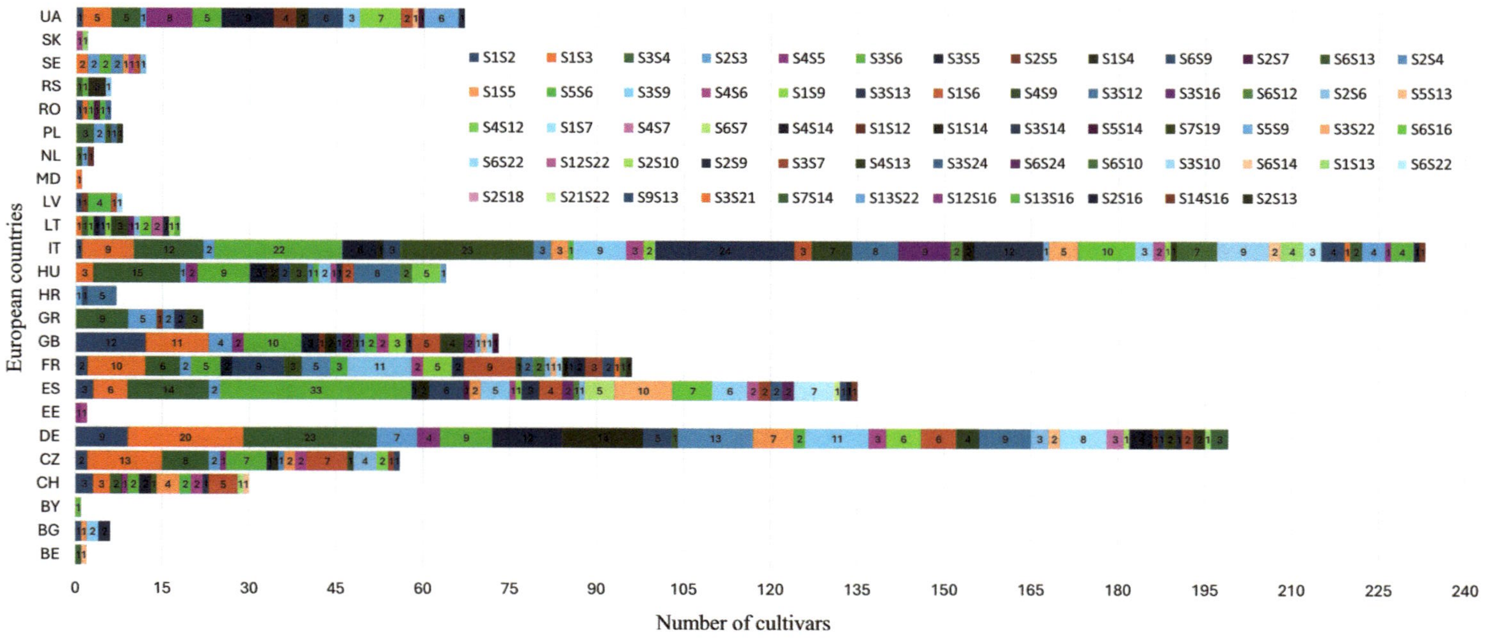

FIGURE 7.9 Distribution of self-incompatible sweet cherry cultivars in European countries (according to Schuster (2020) data). BE, Belgium; BG, Bulgaria; CH, Switzerland; CZ, The Czech Republic; DE, Germany; EE, Estonia; ES, Spain; FR, France; GB, Great Britain; GR, Greece; HR, Croatia; HU, Hungary; IT, Italy; LT, Lithuania; LV, Latvia; MD, Moldova; NL, Holland; PL, Poland; RO, Romania; RS, Serbia; RU, Russia; SE, Sweden; SK, Slovakia; UA, Ukraine.

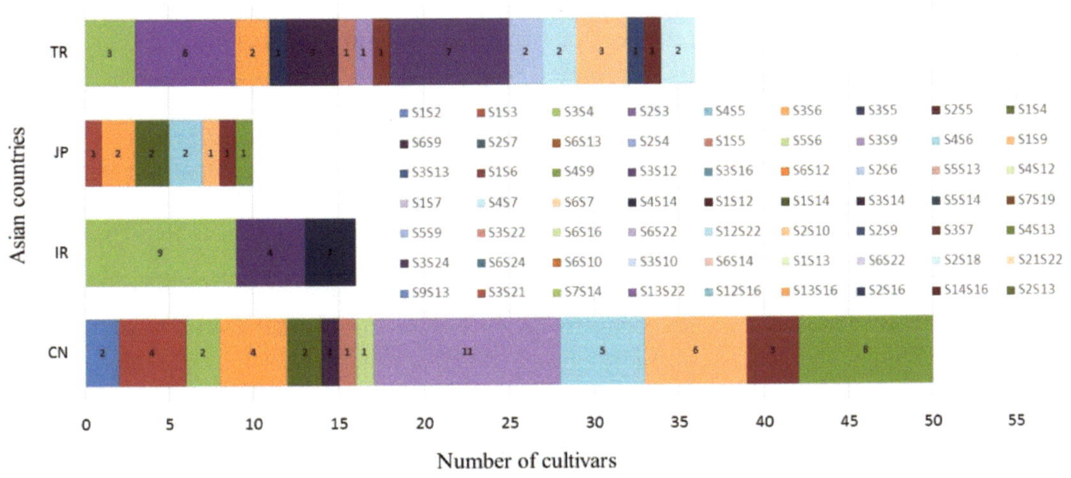

FIGURE 7.10 Distribution of self-incompatible sweet cherry cultivars in Asian countries (according to Schuster (2020) data). CN, China; IR, The Islamic Republic of Iran; JP, Japan; TR, Turkey.

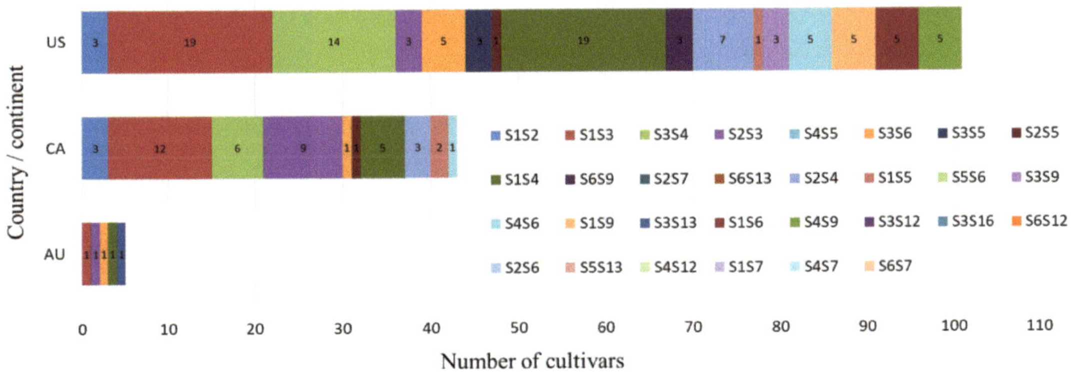

FIGURE 7.11 Distribution of self-incompatible sweet cherry cultivars in North America and Australia (according to Schuster (2020) data). AU, Australia; CA, Canada; US, The United States of America.

Stanys et al. (2008) published the first data on the *S*-allele composition of Lithuanian sweet cherry cultivars. Two new *S*-allele combinations (S_1S_{13} and S_6S_{16}) were announced. The most frequent alleles were S_3, S_6, and S_9 (~18%), and the less common were S_4 and S_{16} (12%) alleles. S_{13} and S_{16} were identified in a few landraces. One of them (Hrebnickio Nr. 1) is very resistant to cold (Stanys et al. 2008). In a later study, Kolytaitė et al. (2017) identified eight different *S*-alleles in sweet cherry cultivars and selections. In addition to the S_6 and S_9 alleles, the most frequent (~15%) was the S_4 allele.

In Latvia, cherries are an important fruit crop widely distributed in home gardens and commercial areas. Lacis et al. (2008) used degenerative *S-RNase* primers for S_1–S_6 allele composition identification from the Latvian and Swedish sweet cherry germplasm. It was shown that between these two collections, the most frequent was the S_6 allele and the S_3S_6 genotype. Meanwhile, S_1S_2 and S_2S_5 genotypes were only specific to Latvian germplasm, while S_1S_5, S_2S_3, and S_3S_4 were only found in Swedish germplasm. In Latvia, the S_3 allele was the most frequent, while in Sweden, the S_6.

Numerous sweet cherry cultivars, both local and foreign, are grown in Estonia, and they are popular in home gardens and private farms. PCR-based analysis and sequencing helped Kivistik et al. (2021) detect S_6 and S_{17} alleles with a high frequency (~27%) in Estonian cultivars. S_1, S_2, and S_9 alleles were not found. Allele frequencies were compared to those of foreign and European sweet cherry cultivars. The results showed that in other European countries (Belgium, Turkey, Italy, and Ukraine), the S_{17} allele was rare (De Cuyper et al. 2005; Szikriszt et al. 2013; Ivanovych 2016; Marchese et al. 2017a). In the Russian variety 'Orlovskaja', S_{17} has to be proven (Bezlepkina et al. 2020). In Ukraine, S_3, S_5, S_6, and S_9 (14.9%–19.1%) alleles were widespread. However, the occurrence of the S_{17} allele was only 1.6%. Eremina et al.'s (2020) study revealed that S_5 and S_9 alleles were associated with large fruit size (9 g and larger). Among two Belgian wild populations, common alleles were detected with a similar frequency: S_2, S_7, and S_{10}. S_{17} and S_{18} were mentioned as new and rare.

Breeding takes an important place in the agricultural sciences in Poland. Polish farmers are very experienced in sweet cherry growing. Sweet cherry trees accounted for 4% of all fruit trees grown in Poland in 2013 (https://www.kowr.gov.pl/uploads/rynek-owocow2014-en.pdf). Lisek et al. (2016) used PCR analysis to amplify *S-RNase* I and II intron and S_1–S_{16} alleles. This was the first time *S*-alleles (S_1–S_6, S_9, and S_{13}) and their compositions were identified in sweet cherry cultivars and selections from Central and Eastern Europe. It was shown that S_5 and S_9 alleles were very common between sweet cherry from Ukraine and S_3, S_1, and S_4 from the Czech Republic. The S_{13} allele was the rarest among Polish and Ukrainian sweet cherry cultivars and very rare in other European countries reported by Schuster (2012) but quite frequent (20%) in Italian accessions (Marchese et al. 2017a). According to Marchese et al. (2010), the S_{13} allele is found in some sour cherries.

A new update in 2020 announced 63 IGs (Schuster 2020). Ninety-one self-compatible sweet cherry cultivars and 26 universal donors were reported. The last four IGs were proposed by Kivistik et al. (2021), ending in 67 (S_6S_{17}).

7.2.2.1.4 Self-Compatibility of Sweet Cherry

SC is an interesting trait of some sweet cherry varieties and is one of the major goals of many breeders (Cachi and Wünsch 2014; Gharaghani et al. 2017). The use of self-compatible cultivars is important from the horticultural point of view. Such cultivars can be planted in orchards without requiring a pollinator cultivar and usually produce more fruit than SI cultivars (Zhu et al. 2004; Herrero et al. 2017). In addition, global warming can affect the pollination process of self-incompatible sweet cherries and reduce the yield.

Many sweet cherry genotypes are self-compatible, and most have inherited mutations in the *S*-locus (Schuster 2020). The molecular basis of SC was first studied with two pollen-part mutants (PPM) that originated in the John Innes Institute using X-ray irradiation (Lewis and Crowe 1954). The loss of pollen *S*-function was caused by the mutation of different alleles in the *SFB* gene. *SFB4'* gene of *JI2420* selection had a 4 bp deletion (TTTA and TTAT), which leads to a defective *SFB* (lacks HVa and HVb regions). Meanwhile, *JI2434* had a complete deletion of S_3 in the *SFB3'* gene (Ushijima et al. 2004; Zhu et al. 2004; Sonneveld et al. 2005). Zhu et al. (2004) provided a molecular marker for simply detecting the S_4' allele in self-compatible sweet cherry genotypes. Kolytaitė et al. (2017) noticed that among the studied Lithuanian varieties, S_4' allele was not detected. Later, the one-step method was used to develop a new molecular marker suitable for S_4 and S_4' allele discrimination (Muñoz-Espinoza et al. 2017). Sequencing analysis of proposed partly self-compatible Estonian sweet cherry cultivars carrying S_3 or S_4 alleles did not confirm S_3' and S_4' haplotypes, but mostly all of them had the S_{17} allele (Kivistik et al. 2021). Interestingly, it was mentioned that S_3, S_6, and S_{17} alleles were present in the cultivars that might be pollinated by or can pollinate another cultivar from the same IG.

The other two self-compatible genotypes occurred naturally in Italy ('Kronio') and Spain ('Cristobalina'). Marchese et al. (2007) confirmed the *S*-allele haplotype of cv. Kronio (S_5S_6). Sequencing analysis revealed that the mutated allele (S_5') contained a premature stop codon, resulting in an 80 amino acid shorter protein. Cachi and Wünsch (2014) attempted to discover the SC process in other local Spain varieties by crossing experiments.

Some stylar-part mutations (SPM) have been reported in *Prunus avium* and other fruit trees belonging to the Rosaceae family (Yamane and Tao 2009; Martí et al. 2012; Hegedűs et al. 2012; Socias et al. 2015). Though it is obvious that the dysfunction of the pistil or pollen *S* determinant leads to SC, the fundamental mechanism of GSI is currently not fully explored in *Prunus* species. Molecular studies showed that other genes expressed in pistil and pollen are required for SI (McClure et al. 2011; Sorkheh 2015). Such genes are termed modifiers and are expressed at several phases of SI/SC (Ono et al. 2018). Studies with sweet cherry cv. Cristobalina (S_3S_6) showed that modifiers may be related to SI breakdown (Wünsch et al. 2010). Cachi and Wünsch (2011) identified a molecular marker, EMPaS02 locus, linked to SC. Large-scale DNA sequencing was used to identify a putative pollen-part modifier gene M locus-encoded GST (Ono et al. 2018). This study was in agreement with Wünsch et al. (2010), who announced that disruption of this gene leads to SC in sweet cherries.

The deep study of SC mechanism in sweet cherry cv. Lapins was made by Li et al. (2020). It was demonstrated that SLFL2 interacts with ubiquitin and still, by an unknown mechanism, recognizes *S-RNase* and directs it for degradation.

7.2.2.2 Self-(In)Compatibility of Sour Cherry

One of the options to achieve high fertility is to increase the fertility of the tree and its flowers. In earlier times, the fertility of sour cherry was determined by evaluating pollen germination and pollen tube growth in the pistil (Lansari and Iezzoni 1990). Three types of fertility in sour cherry were characterized:

- Self-compatible sour cherry (SC). Genotypes with a good fruit set (more than 15%) after open pollination and self-pollination (e.g., 'Fanal', 'Röhrigs Weichsel', 'Schattenmorelle') (Schuster 2014);
- Partial self-compatible sour cherry (pSC). Some pollen gets through and fertilizes the ovule, but many get inhibited along the way (Lansari and iezzoni 1990). These are genotypes with a low to average fruit set after open pollination and self-pollination when the fruit set is less than 15% (e.g., 'Karneol', 'Kelleriis 16', 'Morina') (Schuster 2014);
- Self-incompatible sour cherry (SI). The pollen would germinate, but it is suppressed while growing down the pistil style. Self-incompatible cultivars mainly consist of old landrace cultivars that are being substituted in commercial production by self-compatible types (Lansari and Iezzoni 1990). These are genotypes with an average fruit set after open pollination and no fruit set after self-pollination when the fruit set is less than 1% (e.g., 'SchwäbischeWeinweichsel', 'Tschernokorka') (Schuster 2014).

Most of the sour cherry cultivars are self-compatible, but self-incompatible cultivars have been described recently. The allotetraploidy of sour cherry makes the SI separation more complex in the progeny (Joulani et al. 2020). The reasons for the pSC might be two: pollen (tube) and pistil interaction or inbreeding (Schuster 2014).

7.2.2.2.1 Fertility of Sour Cherry

Fertility studies of sour cherry cultivars were started at the beginning of the 20th century. Nine self-fertile sour cherry cultivars were characterized by Middlebrook in 1916 and other nine by Kostina in 1926 (Davarynejad et al. 2010). Later, cultivars of duke cherry and sour cherry that were both fully self-fertile and self-sterile were found, according to Crane and Lawrence (1928). According to Davarynejad et al. (2014), Husz and Maliga were the first ones who mentioned sour cherry self-fertility in the middle of last century, in years 1943 and 1953, respectively. The highest percentage of fruit set because of self-fertility could reach 50%. Nyéki and colleagues (2008a) achieved a higher result (64%).

The rain and cold weather during blooming time negatively affect the yield of self-sterile sour cherry cultivars. According to Nyéki et al. (2008a), yearly fluctuations affect the yield of self-sterile sour cherries more than that in self-fertile ones. This means that the yield of self-fertile cultivars is more stable. The highest fruit set rate is reached when flowers are pollinated right after opening. Hence, pollination should happen after emasculation in the balloon stage (up to 1–2 days later) (Ansari and Davarynejad 2008).

It was shown that when flowers were open-pollinated, self-sterile and highly self-sterile varieties produced 12.1% and 6.3% of the fruits, respectively. Meanwhile, self-fertile and highly self-fertile sour cherry varieties produced 29% and 29.4% of fruits, respectively (Nyéki et al. 2002).

According to Nyéki and other researchers (2008b), fruit set from open and self-pollination is usually associated with climate. For this reason, it fluctuates during seasons, and the changes in fruit set fluctuation by self-pollination are greater than open pollination. The open-pollination results of the self-sterile cultivar 'Pandy 7' varied more than self-fertile cultivars. Variation among self- and open pollination ranged from 9.8% to 25.2%. This explains the role of honeybees even for self-fertile varieties.

The most important environmental factor is air temperature. It affects ovule longevity and pollen tube growth. Fluctuating environmental conditions (such as warmer winters and the risk of frost in the early spring) lead to various problems related to flowering and fruit set. Radičević et al. (2021) tested pollen performance in the pistils of the autochthonous sour cherry cultivar 'Feketićka' in vivo. Different pollination modes were applied. Self- and cross-pollination were done. Four different sour cherry cultivars were tested: 'Cigančica', 'Čačanski Rubin', 'Érdi Bötermö', and 'Majurka'. The influence of various factors on pollen tube growth efficacy showed an overlapping effect of pollinizer genotype, its dependency on flowering temperatures, the impact of female genotype on pollen performance, and the response of male–female relations to air temperature. The authors classified 'Feketićka' as partially self-compatible because it had less than 6% of fruit set after self-pollination. The best pollinator for this cultivar was 'Cigančica'. 'Čačanski Rubin' and 'Érdi Bötermö' demonstrated the lowest parameters of pollen tubes growth in vivo, so they were not suggested as pollinizers. Also, 'Feketićka' showed opposite results, preferring higher or lower temperatures as pollinizer or pollinated cultivar, respectively (Radičević et al. 2021).

Open pollination of highly self-fertile cultivars could greatly impact higher fruit sets. Brozik and Nyéki (1980) demonstrated cultivars' ability to set fruit after open pollination, as explained further:

1. low: less than 10%;
2. medium: 10.1%–20%;
3. high: 20.1%–30%;
4. extremely high: more than 30%.

Based on the fruit set after natural self-pollination, Nyéki et al. (2008b) pointed out five fertility groups:

I. entirely self-sterile (0%);
II. self-sterile (0.1%–1%);
III. partially self-fertile (1.1%–10%);
IV. self-fertile (10.1%–20%);
V. highly self-fertile (more than 20%).

Sour cherry blossoms should be pollinated within 24 h after anther dehiscence. Later, pollination could affect the rate of fruit set (Ansari and Davarynejad 2008). Nyéki et al.'s (2008b) study demonstrated that artificial self-pollination improved fruit sets even in self-fertile varieties when compared to natural self-pollination. Ansari and Davarynejad (2008) compared and evaluated the results of controlled cross-pollination with open pollination. This allowed them to describe fine and weak pollinators according to their fertilization potential.

Even in self-compatible cultivars, mutual and unilateral cross-incompatibility has been detected. The heredity of SI in sour cherry is not fully understood. According to Bekefi (2006), the SI system in polyploid species is not so well defined. For instance, self-incompatible seedlings can be attained when crossing two self-compatible individuals.

7.2.2.2.2 Gametophytic Self-(In)Compatibility of Sour Cherry

As mentioned before, sour cherry fertility is similar to other species of *Prunus* and is regulated by a GSI system (Joulani et al. 2020). It arrests self-fertilization and enables fertilization with pollen of different genotypes within the same species. Successful pollination happens when the *S* alleles in pollen and pistil (or style) do not match (Figure 7.12) (Schuster 2014).

Sour cherry (*Prunus cerasus*) is a tetraploid (allopolyploid). In this species, 12 *S* alleles have been detected (S_1, S_3, S_4, S_6, S_9, S_{12}, S_{13}, S_{26}, S_{33}, S_{34}, S_{35}, and S_{36}). SC is a consequence of non-functional or mutated *S* alleles. The pistil *S* or part of pollen *S* could be mutated (Figure 7.12) (Schuster 2014).

GSI is, in most cases, disturbed because of polyploidy in many Solanaceous species. This results in self-compatible tetraploid plants (Hauck et al. 2002). Nevertheless, sour cherry (*Prunus cerasus* L.) is an exception. Sour cherry is a tetraploid resulting from spontaneous hybridization of the diploid sweet cherry (*P. avium* L.) and the tetraploid dwarf cherry (*P. fruticosa* Pall.) (Joulani et al. 2020), containing both SI and SC individuals. Thirteen *S*-ribonucleases (*S-RNases*) have been identified as the stylar *S*-locus products among sweet cherry. The authors compared sweet and sour cherry *S*-allele function, *S-RNase* sequences, and linkage map location to understand better the genetic basis of SI and SC in sour cherry. Interspecific crosses among sweet and sour cherries revealed that GSI exists in sour cherry and that the recognition of common *S*-alleles has been preserved despite polyploidization. It is hypothesized that SC in sour cherry resulted from the presence of non-functional *S-RNases* and pollen *S*-genes that may have originated from natural mutations (Hauck et al. 2002).

Inter-specific crossing studies have shown that sour cherry shares eight sweet cherry *S* haplotypes: S_1, S_4, S_6, S_9, S_{12}, S_{13}, S_{14}, and S_{16}. As mentioned before, natural sour cherry selections contain both SI and SC types, suggesting that polyploidy per se does not cause the breakdown of SI. This is opposite to Solanaceae and Plantaginaceae. In those plants, polyploidy ends in SC related to competitive interaction. Several genetic and molecular studies revealed that the genotype-controlled loss of SI in *P. cerasus* is induced by the accumulation of non-functional *S* haplotypes. Two

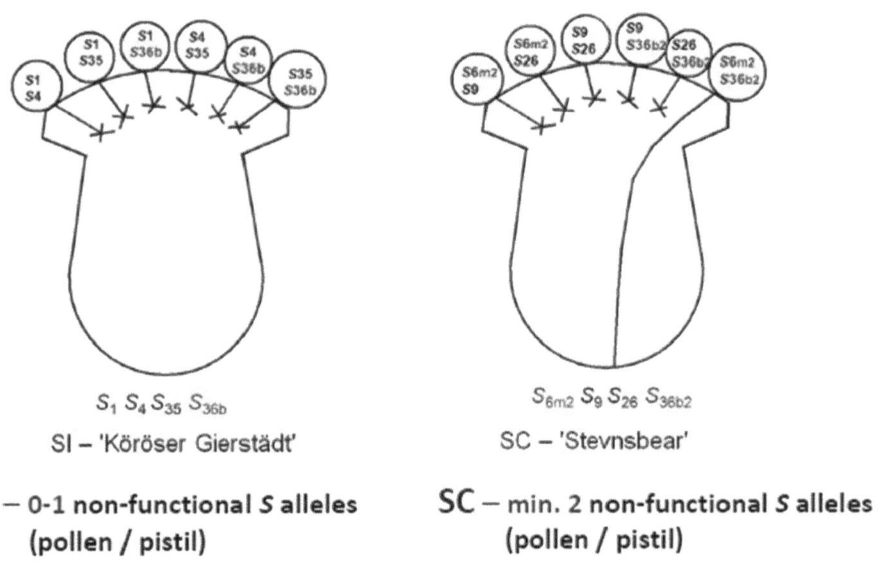

FIGURE 7.12 Self-incompatibility and self-compatibility system in *Prunus cerasus* (Schuster 2014).

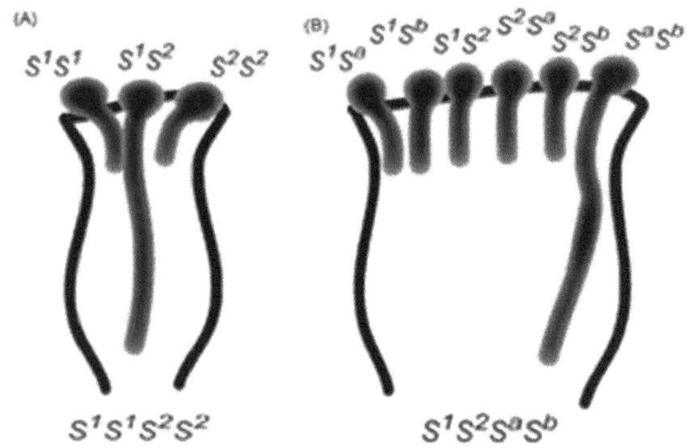

FIGURE 7.13 (a) Competitive interaction, (b) one allele match model. *Solanaceae* and *Plantaginaceae* tetraploid plants are self-compatible because of the compatibility of heteroallelic pollen. In *P. cerasus*, polyploidy outcome is a non-direct breakdown of self-incompatibility. It is required that for successful self-compatibility would be at least two copies of non-functional *S*-haplotypes (S_a and S_b). Take notice of the difference in S_1S_2 pollen tube growth in competitive interaction (a) and one allele match model (b) (Tao and Iezzoni 2010).

non-functional *S* haplotypes in sour cherry 2× pollen cause that pollen to be SC. Heteroallelic sour cherry pollen is shown to be SI, which is opposite to the competitive interaction in *Solanaceae* and *Plantaginaceae* (Figure 7.3) (Tao and Iezzoni 2010).

According to the findings, the one allele match model has been suggested to explain the SI recognition in tetraploid sour cherry. According to the model, an incompatible reaction would occur when a functional pollen *S* gene product in the pollen and its corresponding functional *S-RNase* in the style match, this would result in an incompatible reaction. Hence, this means a fundamental difference in the effect of polyploidy on SI between *Prunus* and the *Solanaceae*, and the *Plantaginaceae* (Figure 7.13). In the *Solanaceae* and the *Plantaginaceae* plants, polyploidy is a direct source of SC because of competitive interaction. In *Prunus*, polyploidy does not directly cause SC, as heteroallelic pollen maintains its SI phenotype. Instead, in *P. cerasus*, mutations of the pistil- and pollen-specificity elements have eventuated and then accumulated as an outcome of SC. The fact that sour cherry is SI though it has one non-functional *S* haplotype, is also important for evolution. This is because non-functional *S* haplotypes could be preserved in the population without causing an accidental shift to SC (Tao and Iezzoni 2010).

Competitive interaction of the alleles in diploid pollen grains could explain the SC of some sour cherry cultivars (Békefi 2006). Lewis and Modlibowska (1942) submitted that in pear, heteroallelic pollen (e.g., S_1S_2) of a tetraploid individual would be effective on a style even if it has both alleles (e.g., $S_1S_1S_2S_2$) together, whereas homoallelic pollen (e.g., S_1S_1 or S_2S_2) degrades.

The first research experiments with sour cherry *S-RNases* were done by Yamane et al. (2000b). In the 13 cultivars that were genotyped, RFLP analysis showed the existence of S_1, S_4, S_6, S_9, and S_{12} sweet cherry alleles. The other five alleles were identified; they differed from S_1 to S_{12} sweet cherry alleles and were labeled S_a to S_e. The similarity between SC and certain allele combinations could not be determined. In sour cherry, unlike in some *Solanaceous* species, heteroallelic pollen does not induce SC.

Hauck et al. (2002) crossed between sour and sweet cherry cultivars and analyzed pollen tube growth. The cross between the sour cherry 'Crisana' ($S_1S_4S_d$) and the sweet cherry 'Rainier' (S_1S_4) was incompatible, as was the reciprocal cross. This suggested that the stylar and pollen components in both species regarding these cultivars are functionally identical. The sequences of the S_4- and S_6-*RNases* in sour cherry were homogeneous with the corresponding sweet cherry alleles.

Progeny experiments from sour and sweet cherry crosses indicated disomic inheritance of the *P. fructicosa* and *P. avium* genomes in sour cherry (Békefi 2006).

Segregation analysis in a progeny from the cross 'Rheinische Shattenmorelle' ($S_aS_bS_cS_6$)×'Erdi Botermo' ($S_aS_4S_{6m}$) allowed learning the SI system in sour cherry (Yamane et al. 2000b). According to segregation data, it was concluded that in 'Erdi Botermo', there are two copies of the allele S_4. Hence, the genotype was $S_aS_4S_4S_{6m}$. What is important, the S_{6m}-RNase allele has not been found in the progeny. It was considered that the S_{6m}-RNase was not functionally active compared with its pollen component, which was functional. Yamane et al. (2003a) supported the statement by conducting expression analyses of the S_6-*RNase* and the pollen-part gene SFB_6. SFB_6 was transcribed in both haplotypes S_6 and S_{6m} of pollen. S_{6m} transcription of S_6-*RNase* failed because "Erdi Botermo" separated from its pollen parent "Nagy Angol" and it was thought to be the result of a naturally mutation. Still, heteroallelic pollen was not universally compatible. Since none of the alleles was separated by SC, SC could not be ascribed to a mutation in the *S-RNase* gene (Békefi 2006).

The work with sour cherry incompatibility was continued by Tobutt et al. (2004a). The *S-RNase* pattern of 36 accessions was investigated by IEF and NEPHGE (non-equilibrium pH gel electrophoresis). Four non-sweet cherry alleles were discovered and marked *SA*, *SB*, *SC*, and *SD*. The allele *SB* was also found in some *P. fructicosa* accessions.. These alleles and attained genotypes could not be matched with the results and labeling of Yamane et al. (2000b). Pollen tube growth and field test crosses between sour and sweet cherry demonstrated that the studied *S-RNases* are functional (Békefi 2006).

Six *S*-haplotypes (S_1, S_4, S_6, S_9, S_{12}, and S_{13}) are present in sweet cherries. The identical haplotypes are also found in sour cherry. S_1, S_6, and S_{13} are non-functional variants in sour cherry, in which pollen and/or stylar function was lost. Loss of function happened because of structural changes in the *S-RNase*, SFB, or *S-RNase* upstream sequences (Tsukamoto et al. 2008).

Genetic segregation data showed that the S_{16}, S_{33}, S_{34}, and S_{35} sour cherry haplotypes are completely functional. Phylogenetic tests of the SFB and *S-RNase* sequences from existing *Prunus* species uncover that there were no similarities to known organismal relationships at any taxonomic level within *Prunus*. This means that polymorphisms at the *S*-locus have been preserved during the evolution of the genus. Moreover, the phylogenetic relationships among SFB sequences are generally incompatible with those among *S-RNase* sequences for the same *S*-haplotypes (Tsukamoto et al. 2008).

Tsukamoto et al. (2008) showed that S_{14} and S_{16} haplotypes are also shared between sweet and sour cherries. It was suggested that the *S*-haplotypes were given by the ground cherry (*P. fruticosa*) parent. Because of this statement, an investigation of the *P. fruticosa* present in the Michigan State University cherry germplasm collection followed the identification of multiple accessions (S_{33} and S_{35}) haplotypes. Even so, any individual with the S_{34} haplotype has been discovered. There is a probability that the S_{33}-*RNase* identified in 'Meteor' is equally specific as in *P. mume* S_{10}-*RNase*. If these genes code for identical proteins, this would be another example of ancestral specificities shared between *Prunus* subgenera. *Prunus* species from the same subgenus share more ancestral specificities on a percentage than *Prunus* species from other subgenera. *S-RNase* amino acid sequences from *P. cerasus* S_{33} and *P. mume* S_{10} were compared, and identical sequences were found between *P. avium* S_1-*RNase* and *P. tenella* S_8-RNase, also *P. salicina Sf-RNase* and *P. tenella* S_3-*RNase*. Differences between the SFB of *P. tenella* S_8 and *P. avium* S_1-haplotypes were by 12 amino acids distributed along the gene.

The inferred rate of synonymous and non-synonymous mutation for related haplotypes was not higher at the SFB comparing the *S-RNase* gene. Tsukamoto et al. (2008) did not succeed in determining where the putative recombination event happened.

P. cerasus S_{35}, *P. dulcis Sa-RNases*, and SFBs were classified as a sister group according to phylogenetic analyses.

Khadivi-Khub (2014b) investigated the *S*-allele diversity of eight wild and two commercial species of the *Cerasus* subgenus in Iran. A high level of *S*-allele polymorphism was discovered among and within these species. Furthermore, the majority of wild species had 2–4 alleles based on *S*-allele primers and might be tetraploid. Wild cherry (mazzard) had a wide range of alleles such as S_1, S_2, S_7, S_{14}, S_{20}, and unknown alleles. Sour cherries appeared with S_6, S_9, S_{13}, and S_{27} alleles. Daneshkadeh's genotype had identical alleles to the Montmorency genotype.

P. avium, *P. cerasus*, and *P. fruticosa* share several *S-RNase* alleles, including S_{13}. Non-functional form, $S_{13°}$, is found in some sour cherries. Two microsatellites are related to allele S_{13}-*RNase*, one in the first intron and the other in the second. The lengths of the two studied microsatellites were correlated. The sweet, wild, and sour cherries had different combinations, and the four examples of $S_{13°}$-RNase were linked with three combinations. Particular sequences related to the microsatellites correspond to footprints of transposons. It might be possible that in the diploid sweet and wild cherries, S_{13} variants are a consequence of slippage during replication. In the tetraploid sour cherries, which can have more copies of S_{13} or $S_{13°}$, intra-allelic crossing over may have produced novel variants. The feasible implication of transposable elements in the origin of these microsatellites is considered (Marchese et al. 2010).

7.2.2.2.3 S-Allele/S-Haplotype Distribution

S-alleles in sour cherries are investigated by applying PCR analysis with *S* allele markers. Consensus and *S* allele-specific primers are used. Consensus primers are designed for the first *S-RNase* intron: PaSPcons-F1/R1 (Sonneveld et al. 2003) and the second *S-RNase* intron PaC2cons-F2/R2 (Sonneveld et al. 2003), and PruC2(F)/PCE-R (Yamane et al. 2000b). Specific primers are designed for 16 alleles (Table 7.6) (Schuster 2014).

7.2.3 Gene Mining of Agronomically Important Traits in Cherries

The most direct demand for gene mining approaches comes from cherry breeding, which requires new knowledge to solve current challenges and create suitable cultivars. Particular interest is agronomically important traits, which need to be introduced in new advanced cultivars. These characteristics have the ones that the end consumers and cherry growers are most directly confronted with. Unfortunately, they are also the most difficult to introduce in breeding material because quantitative heredity, the involvement of many genes, and their mutual interaction or modifying action are characteristic. Contrary, this process is facilitated by the high synteny level of the genus *Prunus* species genomes, which allows the knowledge of genes identified in one species to be transferred to others (Dirlewanger et al. 2004; Arús et al. 2006), as well as the *Prunus* genome synteny with model plant species is supportive. For example, trait-encoding genes identified in peach, the best-genetically studied *Prunus* species (Aranzana et al. 2019), are also transferred to other species of the genus (Arús et al. 2012). The second thing that facilitates gene mining of agronomically important traits is the ever-increasing information about the genome, the sequencing of the entire genome, including the cherries (Shirasawa et al. 2017; Pinosio et al. 2020; Xanthopoulou et al. 2020; Wöhner et al. 2021).

The current primary cherry growing challenges are (1) climate changes, (2) the quality of the yield – thus the competitiveness and economy of production, and (3) the public's demand for

TABLE 7.6
Consensus, Specific, Non-functional, or Mutated Allele Primers Designed for Sour Cherries

Primer	Primer Sequence or Allele	References
Consensus	PaSPcons-F1/R1	Sonneveld et al. (2003)
	PaC2cons-F2/R2	Sonneveld et al. (2003)
	PruC2(F)/PCE-R	Yamane et al. (2000b)
Specific	S_1, S_2, S_3, S_4, S_5, S_6, S_7, S_9, S_{10}, S_{12}, S_{14}, S_{16}	Sonneveld et al. (2003)
	S_{26}	Hauck et al. (2006)
	S_{33}, S_{34}, S_{35}	Tsukamoto et al. (2008)
For non-functional	$S_{1'}$, $S_{6/6m}$, S_{6m}, $S_{13/13'}$, S_{13m}, $S_{13'}$	Tsukamoto et al. (2008b)
or mutated alleles	S_{36a}, $S_{36b/b2/b3}$, S_{36b}, S_{36b2}, S_{36b3}	Tsukamoto et al. (2010)

sustainable and safe food (Aranzana et al. 2019). Some of these challenges can be mitigated by improving growing technologies. However, the most sustainable and efficient way is to create new genotypes through breeding. On the other hand, breeding today is unthinkable without using DNA-based selection methods. They include knowledge about the traits of interest, their encoding genes, and related molecular markers – technologies called differently but with similar meanings: marker-assisted breeding (MAB) or DNA-informed breeding, including marker-assisted parent selection (MAPS), marker-assisted seedling selection (MASS), marker-assisted introgression (MAI), and genomic selection (GS) (Peace 2017; Quero-García et al. 2022).

Therefore, considering the challenges mentioned above, breeding priorities, and available DNA technologies, cherry gene mining for agronomically important traits could be as follows:

1. Reduce the impact of climate changes – phenology, flowering time, fruit maturity,
2. Increase harvest quality – fruit size and weight, firmness, color, chemical composition,
3. Improve cherry growing technologies for more efficient cultivation – tree crown architecture,
4. Implementing requirements protecting the environment and consumer health – resistance to diseases.

Various DNA-based methods can study the heredity of agronomically important traits. However, considering these traits' polygenic and quantitative nature, the QTL methodology is widely applied. The heredity of most agronomically important traits has been studied by this approach, and it has served as an entering point for the development of specific molecular markers already applied in breeding. Another group of methods applied to gene mining of agronomically important traits is transcriptome analysis, which allows studying functional implications of the genetic variability (Wang et al. 2019). The application of these technologies is constantly increasing, and they will play a vital role in the study of heredity and gene mining of agronomically important traits of cherries.

7.2.3.1 Phenology (Flowering Time, Fruit Maturity)

Successful blooming and pollination are essential processes ensuring growers with a high and quality cherries yield. These processes can be significantly affected by climate changes – as the temperature rises, the course of cherry development phases can change with a further impact on the yield. Cherry, typical temperate climate fruit trees, need to go through a dormancy period in which meristem growth is inactive to avoid frost damage in winter (Lang et al. 1987). Dormancy in cherries can be divided into three stages: paradormancy and endodormancy (bud growth is inhibited during autumn and winter) and ecodormancy (bud growth resumed under favorable climatic conditions in late winter and early spring) (Lang et al. 1987). This can be directly affected by climate changes, especially in increased temperatures, since the length of the dormant period depends on the environmental temperatures, specifically from amounts of chill and heat (called chilling and heat requirements) (Cooke et al. 2012). In addition, these requirements are genotype-specific, vary according to the environmental conditions, and directly affect the flowering time. This can cause the shift in phenological phases: the change of flowering time, which in turn affects the pattern of blooming, and pollination; as a result, the flowers suffer more and more from late frosts, affecting the size and quality of the harvest (Wenden and Mariadassou 2017; Drogoudi et al. 2020; Haokip et al. 2020; Paltineanu and Chitu 2020). Such changes have already been observed, and sweet cherry phenology has even been used as a model to demonstrate climate change, based on 60 years of weather records and 30 years of phenology observations (Blanke and Kunz 2017). The analysis of the climate change impact on phenology showed that one of the possibilities for their mitigating is to choose or select tolerant varieties (Chawla et al. 2021). For this purpose, effective selection methods are needed, while their development requires knowledge about the heredity of involved traits and the mining of controlling genes. A series of physiology and genetics studies have been conducted on these traits in cherry and other fruit trees (Calle et al. 2021a), finding that bloom time is a quantitative trait with very high heritability (Dirlewanger et al. 2012; Castède et al. 2014; Calle et al. 2020b). This has enabled the development of a series of major QTLs associated with bloom time and its components.

Carrying out a genetic determinism study on five populations of peaches, apricots, and sweet cherries (Dirlewanger et al. 2012), several QTLs linked to flowering time were identified: major QTLs on LG4 for apricots and sweet cherry and LG6 for peach. Meanwhile, minor QTLs for all species were identified on LG2, LG3, LG4, and LG7. In this study, major QTLs for all species were identified on LG4 for the maturity date. Using the peach genome sequence information, an alignment of QTLs on LG4 and LG6 was performed, which allowed the identification of candidate genes suitable for application in MAS.

Stable major QTLs on LG4 linked to flowering date were also identified in mapping populations 'Regina' × 'Lapins' and 'Regina' × 'Garnét' (Castède et al. 2014; Branchereau et al. 2022). In addition, these studies have proven the relationship between chilling requirements and the flowering date and mapped these traits in one LG4 region. Studies on these mapping populations also included identifying and investigating candidate genes, finding differential expression in gibberellin biosynthesis genes, and negative regulation of flower development pathways (Castède et al. 2014). Differential expression was also found for the genes *PavB01-E3*, *PavSR5a*, and *PavSAUR71*, and two KASP markers were developed that can be used for marker-assisted selection (MAS). In these same mapping populations, 79 candidate genes related to bud dormancy and flowering time were tested and derived from orthologs of peach and sweet cherry sequences (Castède et al. 2015). Of the tested genes, 35 were mapped and found a co-localization with temperature requirements and flowering date QTLs, mapped on LG4. Considering the variability of these candidate genes forms the basis for developing specific molecular markers.

Using six full-sib populations related to the cultivar 'Cristobalina', two major QTLs on LG1 and LG2 (Calle et al. 2020b) were identified in the heritability study of flowering time, while in the case of maturity time, a highly significant and stable QTL was identified on LG4 (Calle and Wünsch 2020). LG1 mapped QTL is associated with Dormancy-associated MADS-box (DAM) genes, consistent with previously published peach results. A later study (Calle et al. 2021a) found the association of DAM mutations with sweet cherry early blooming. Analyzing *PavDAMs* genes, the relationship of their variability with bloom time regulation was determined, which allowed the development and validation in the F_2 population of PCR-based molecular markers. Further research of the DAM genes is a promising direction in understanding the regulation of cherry blooming and maturity time, as well as for the development of markers for breeding. The activity of these genes is affected not only by internal variability but also by epigenetic modifications, like DNA methylations and siRNA, which are involved in *PavMADS1* silencing (Rothkegel et al. 2017). This was also confirmed by another study with histone methylation dynamics of dormancy-related genes (Vimont et al. 2019), in which it was found that the expression of DAM-related genes is affected by environmental temperature and that they are up-regulated in dormant buds, but down-regulated in non-dormant buds.

Less research has been done on sour cherries and their blooming time regulation. A study of five F_1 populations whose parents included sour cherry cultivars such as 'EnglaiseTimpurii', 'Montmorency', 'Rheinische Schattenmorelle', 'Surefire', 'Újfehértói Fürtös', and hybrids of various origins demonstrated blooming date polygenic inheritance with additive effects of different QTLs (Cai et al. 2018). This study identified blooming time-related QTLs on LG1, LG2, LG4, and LG5, creating a basis for further studies of *P. cerasus* in this direction.

Summarizing the research information currently available on cherry phenology genetic regulation and its possible application in climate change mitigation, it should be concluded that the majority of studies were conducted on a relatively narrow genetic background – the studies mainly used two mapping population groups: 'Lapins' × 'Garnet', 'Regina' × 'Lapins', 'Regina' × 'Garnét' (e.g. Castède et al. 2014, 2015; Branchereau et al. 2022) and full-sib populations linked through the cultivar 'Cristobalina' (e.g. Calle et al. 2020b; Calle and Wünsch 2020; Calle et al. 2021a). Since the QTLs mapping results from these population groups differ, to obtain more general results, it is necessary to involve a broader range of plant materials with different origins in order to address the current challenges in cherry cultivation more successfully.

7.2.3.2 Fruit Quality

The final consumer of cherry production is most interested in the fruit and various aspects of its quality (Correia et al. 2017). The appearance of the fruit is the first sign by which the consumer makes their choice. It is especially important for sweet cherries, which are mostly consumed fresh. Therefore, fruit size and color characteristics are still relevant and essential for breeders. On the other hand, growers are interested in cherry cultivars with firm fruits suitable for transportation and storage, which are at the same time cracking resistant to ensure the profitability and efficiency of production (Correia et al. 2018). Besides, public interest in food's functional value is increasing, promoting the introduction of particular chemical compositions with higher functional activity (Antognoni et al. 2020). All these traits are complex and usually polygenically inherited, which significantly complicates their application in breeding to introduce them into new cultivars. Considering the perennial nature of cherries, their long juvenile phase, which can last up to 5 years, requires a long time to evaluate fruit traits. Therefore, understanding heredity mechanisms and introducing molecular markers for the early selection of desired genotypes for fruit traits is particularly important.

7.2.3.2.1 Fruit Size and Weight

Fruit size is one of the first traits the consumer notices. Therefore, it has always been the goal of the breeding programs. As a result, modern sweet cherry fruits reach 14 g, although for wild forms (mazzard), it is ~2 g, and for landraces 6 g. Fruit size in cherry has been considered a classical quantitative trait controlled by multiple loci that are highly influenced by the environment (Aranzana et al. 2019). Cherry fruit anatomy studies have found that the number of mesocarp cells significantly influences fruit size, which is genetically determined and relatively less influenced by the environment. Cell length, on the other hand, was not observed to be related to fruit size, and this trait had a large variation under the influence of the environment (Olmstead et al. 2007). Considering this knowledge, a series of studies on the genetic factors determining the size of cherry fruits have been conducted.

The number of mesocarp cells was also used as a phenotyping character in the first QTL study on sweet cherry fruit size and weight (Zhang et al. 2010). Using the mapping population obtained from the cross 'Emperor Francis' × New York 54 (wild cherry or mazzard), two major QTLs on LG2 and LG6 were identified. In addition, it was found that QTL on LG2 is associated with the number of cells and on LG6 – with pit size. Using the available sweet cherry QTL maps, two SSR molecular markers (CSPSCT038 and BPPCT034) flanking the QTL on LG2 were identified. This discovery made it possible to perform MAS in the sweet cherry breeding process by selecting large fruit sizes. This QTL detected on LG2 was later confirmed in a larger study by Rosyara et al. (2013). In this case, the researchers used four bi-parental full-sib families (New York 54 × 'Emperor Francis', 'Regina' × 'Lapins', 'Namati' × 'Summit', 'Namati' × 'Krupnoplodnaya') and 23 sweet cherry cultivars, and QTL association with fruit size was found on LG2 (three QTLs), LG1, LG3, and LG6. The QTLs identified on LG2 and LG6 corresponded to those previously mapped (Zhang et al. 2010). The QTL described the largest part of phenotypic variability mapped on LG2 in both studies, naming it FW_G2a. Also, in a study with a mapping population obtained from the cross 'Regina' × 'Lapins', the QTLs associated with fruit size identified on LG2 and LG6 were confirmed (de Franceschi et al. 2013). In this case, the linkage of LG2 and LG6 genomic regions was also proven with two of the 'cell number regulator' or CNR genes identified in other plant species. In a study conducted using two mapping populations of 'Regina' × 'Lapins' and 'Regina' × 'Garnét' seedlings, all QTLs identified by Rosyara et al. (2013) were verified, and in addition to them, a fruit weight associated with QTL on LG5 was mapped close to the marker EMPAS14 (Campoy et al. 2014). Considering the potential of the locus FW_G2a for use in the selection of large-fruited sweet cherries, its validation was performed using the KASP (Competitive Allele-Specific PCR) method (Dirlewanger et al. 2020). As a result, the ability of the markers of this locus to discriminate genotypic groups significantly associated with fruit size was proven. The applicability of the SSR markers CSPSCT038 and BPPCT034 linked to the FW_G2a locus was also demonstrated in a study with 15 sweet cherry cultivars from the Hungarian collection (Szilágyi et al. 2022).

QTL mapping studies have also been performed using other mapping populations with different genetic backgrounds of cherries. For example, the population obtained from the cross 'Ambrunés' × 'Sweetheart' allowed the identification of fruit size-related QTLs on LG1 and LG3 (Calle et al. 2020a). Whereas, in crosses of the self-compatible cultivar 'Cristobalina' with such cultivars as 'Brooks', 'Lambert', 'Vic', and 'Cristobalina' self-pollinated seedlings, QTL related to fruit size were identified on LG1, LG2, and LG5 (Calle and Wünsch 2020). These studies identified QTLs as different from those previously found. In general, there is a tendency that when the study includes more plant material (for example, by using the FlexQTL approach, combining data from different mapping populations), the number of discovered QTLs also increases.

Although QTLs suitable for breeding and molecular markers developed on their basis for fruit size have been developed relatively successfully, research to understand fruit development regulatory mechanisms is also ongoing in other directions. One approach is to relate the expression of specific genes to the characteristics of a given phenotype, in this case, fruit development. The *CYP78A* group gene *PaCYP78A9* determines various plant species' seed development (size). A study on sweet cherries found that RNA silencing of this gene causes the formation of small fruits (Qi et al. 2017). The research concluded that this gene is a critical upstream regulator of cell cycle processes and is essential for fruit size. Considering the high syntheny of the *Prunus* genomes and the homology of genes regulating fruit development in different plant species, such studies would be useful for further gene mapping research.

7.2.3.2.2 Fruit Color

One of the quality traits of cherries, especially sweet cherries, is the skin and flesh color. Consumer preferences may differ, but obtaining a predictable result in a targeted breeding process is important. Although the trait is important, there are still many gaps in understanding the mechanisms of its inheritance. Researching segregating populations discovered that red skin color is dominant over yellow one, and there was proposed the presence of one major (A/a) and one minor gene (B/b) that exhibit epistasis (Fogle 1958; Schmidt 1998). It is also concluded that A/a controls a red flesh color, but A- and aa correspond to mahogany and yellow flesh, respectively. Other cross-breeding trials later confirmed these results. However, the color of the fruit's skin and flesh can have different transitions between these two colors, which indicates the involvement of several other minor genes. Considering this trait's characteristics, applying the QTL method was a reasonable choice for further research.

Working with the mapping population of 'Emperor Francis' × NY wild mazzard, Sooriyapathirana et al. (2010) discovered a major QTL on LG3 controlling skin and flesh color that was inherited according to Mendelian pattern. This study also led to the identification of a candidate gene, *PavMYB10*, a homolog of the *MdMYB10* gene identified in apples and responsible for the fruits' red flesh. LG3 linkage with fruit color was also confirmed by recent studies on the mapping population 'Vic' × 'Cristobalina' (Calle et al. 2021b). In this study, LG3 was involved in the regulation of anthocyanin synthesis through the transcription factor *PavMYB10*, thereby affecting fruit color. QTL identified by Sooriyapathirana et al. (2010) on LG3 was also confirmed by Sandefur et al. (2016). In this study, the *Rf* locus was defined by transforming the information on 14 SNPs and the Pav-Rf-SSR molecular marker related to fruit color, creating a single PCR-based assay. Genotyping of the plant material using the developed marker revealed incomplete dominance of the *Rf* locus, which was manifested in three genotype classes: *rfrf* (reddish skin color and yellow-white flesh), *Rfrf* (red to mahogany skin and flesh color), and *RfRf* (typically dark mahogany skin and flesh color).

7.2.3.2.3 *Fruit Firmness*

Apart from the external appearance, the sensory characteristics of the fruit, such as the consistency of the flesh, are a critical component of cherry quality. Sweet cherries are mainly consumed fresh, so picking is done manually. However, further processing, including transportation, mechanical sorting, warehouse storage, and a sufficiently long shelf-life in the store should be ensured. All these actions can cause damage, and only enough firm fruits can reach the final consumer in high quality. Sour cherries are mostly a raw material for processing, where fruit firmness is not so critical. However, mechanical harvesting and pit removal can be a limiting factors. Therefore, gene mining approaches such as QTL analysis are applied to understand the heredity of cherry fruit firmness and develop selection methods suitable for breeding.

It was working with the populations of the crosses 'Regina' × 'Lapins' and 'Regina' × 'Garnét' already widely used in mapping experiments, and fruit firmness-related QTLs were identified on LG2, LG5, and LG6 (Campoy et al. 2014). In addition, a significant relationship between fruit weight and firmness was observed, but with a negative correlation. For the QTL found on LG5, the predictiveness of the developed KASP marker was tested, which was able to distinguish between the firm and soft fruits well in the mapping populations, but it was not effective in the germplasm collection (Dirlewanger et al. 2020). A hitherto unknown QTL on LG4 was identified by applying the analysis of three sweet cherry populations: (1) the INRA bi-parental F_1 population, (2) the INRA sweet cherry germplasm collection, and (3) RosBREED pedigreed population (Cai et al. 2019). Later, a major QTL on the same LG4 was also found by analyzing six genetically different background mapping populations (Calle and Wünsch 2020). Similar to studies with fruit weight (Zhang et al. 2010), the dominance of the 'soft' phenotype over the 'firm' phenotype was observed, while soft-type homozygous genotypes were observed only in wild samples. By analyzing the mapping population obtained from the cross 'Ambrunés' × 'Sweetheart', a so far unknown QTL related to fruit firmness on LG1 was determined (Balas et al. 2019; Calle et al. 2020a), for which the previously mentioned correlation between fruit firmness and weight was true, only positive. The QTL identified on LG1 was also confirmed by a later study (Crump et al. 2022), which also found a linkage with firmness on LG3, but it was not stable between years. Although a series of QTL mapping studies on fruit firmness have been carried out, markers stable in germplasm of different origins are still not available, and gene mining studies in this direction should be continued.

7.2.3.2.4 *Fruit Cracking Tolerance*

Fruit cracking of sweet cherries is widespread damage that causes significant losses to growers and could only increase given the predicted climate changes and its associated increase in excessive rainfall (Khadivi-Khub 2015). In order to reduce the impact of these damages, various cherry growing strategies are being worked on, for example, installing rain protection covers, mechanical removal of water from fruit surface, applying mineral, anti-transparent, and growth regulator sprays (Correia et al. 2018). Most of these expensive approaches do not always provide the desired effect. A possible reason could be the complex and multi-factorial nature of fruit cracking (Correia et al. 2018), which has not yet allowed us to understand it fully. According to the research, there is no completely resistant cultivar to fruit cracking, and only tolerance variability has been observed. The cultivar 'Regina' is considered the most tolerant, followed by 'Kordia', 'Lapins', and 'Hedelfinger'. Initial studies on cultivar tolerance showed a relationship between fruit cracking and fruit firmness, which was not confirmed later. Later it was found that cracking can be caused by excessive water intake through the roots or directly from the surface of the fruit. In addition, a correlation between cherry fruit cracking and the osmotic potential of the fruit was found. Different physiological processes are involved in fruit cracking, depending on its location on the fruit surface – three types of cracking are distinguished: stem, apical, and side cracks (Balbontín et al. 2013; Correia et al. 2018; Knoche and Winkler 2019). The complexity of the phenomenon has influenced the associated gene mining.

Considering the quantitative nature of the trait, the QTL method was involved in gene mining research and using mapping population of the cross 'Regina' × 'Lapins', and the first QTLs linked to fruit cracking were established on LG5, although they were relatively weak and variable across years (Quero-Garcia et al. 2021). The first stable cracking-tolerance linked QTLs were later identified by applying several phenotyping methods and collecting three mapping population data (Quero-García et al. 2021). The established QTLs were characterized by a relationship with a particular fruit cracking location on fruit: on LG2 – fruit side, on the LG4 – stem end, and on the LG5 – pistillar end of the fruit. By performing a Pedigree-Based Analysis, supplemented by a Genome-Wide Association Study, significant QTLs linked to fruit cracking were identified on LG1 and LG5, respectively *qCrack-LG1.1m* and *qCrack-LG5*.1m. For each detected QTL, at least two significant effect SNPs were defined, which can also be considered as putative functional SNP haplotypes (Crump et al. 2022). These QTLs are sufficiently strong and stable to become specific markers for use in breeding (Khadivi-Khub 2015).

The knowledge obtained from other plant species on fruit cracking transcriptional analysis of cell wall and cuticle-related genes was applied in studying heredity mechanisms of cherry fruit cracking, analyzing the dynamics during fruit development (Alkio et al. 2012; Balbontín et al. 2014). As a result, the association with fruit cracking was found for the genes controlling PaWINA, PaWINB, PaLipase, PaLTG1, PaATT1, PaLCR, PaGPAT4/8, PaLACS2, PaLACS1, and PaCER1, which involved in the formation of the cuticular membrane at early stages of fruit development (Alkio et al. 2012). The relationship with tolerance to cherry fruit cracking was also found in the expression of the genes controlling wax synthase, β-ketoacyl-CoA synthase (PaKCS6), and β-Gal, but the expression of the expansin (*PaEXP1*) gene was higher in more tolerant cultivars (Balbontín et al. 2014). In genotypes with different fruit cracking tolerance, different expression and SNP variability were observed in genes that control such compounds involved in fruit development as pectin, aquaporin, expansin, ABA, and ethylene synthesis (Michailidis et al. 2021).

7.2.3.2.5 Fruit Taste and Chemical Composition

Cherry fruit taste is one of the most important characteristics determining the variety's success and can also affect its further use. Taste is a complex characteristic formed by the combination of many chemical compounds. This fact makes it difficult to find out the regularities of heredity. The modifying effect of the environment also should be mentioned, which is significant for the chemical composition. Soluble solids (SS, representing sugars) and titratable acidity (TA, representing acids) are most often analyzed in the context of taste. These two criteria are also used in gene mining experiments. According to Zhao et al. (2014b), major QTL with sugar content was mapped on LG2 and LG4, while minor QTL was on LG7.

On the other hand, titratable acids QTLs were found on LG2, LG4, and LG6. This study is based on 601 pedigree-linked individuals using a 6K SNP array and SSR markers. Although the study allowed the identification of QTLs for these two traits, they were not stable between years and needed further validation. Later, Querro-Garcia et al. (2019) described several more QTLs for titratable acids, which were more stable and significant, especially those on LG6, but Calle and Wünsch (2020) found a significant QTL for sugar content on LG4.

In addition to the previously described quality characteristics of cherry fruits, there is increasing interest in the chemical composition by identifying specific compounds affecting dietary habits and evaluating their content (Antognoni et al. 2020). This increased an interest in studying biochemical pathways and mining the genes that control them. Currently, the number of such studies is limited but tends to increase. One of the first chemical composition QTL studies was in 2000 (Wang et al. 2000). When using sour cherry cross 'Rheinische Schattenmorelle' × 'Erdi Botermo' mapping population and RFLP molecular marker analysis were identified QTLs for soluble solids on EB7 and concentration °Brix on RS6. Gene mining studies have been conducted on the composition of anthocyanin and other phenolics, both in connection with fruit color and even more - with the growing interest in functionally rich foods. Transcriptome analysis identified enzymes and transcription factors involved in anthocyanin synthesis (Wei et al. 2015). The study found that 46% of the identified unigenes corresponded to those identified in other species, revealing the universal nature of biochemical pathway regulation in plants. Genotyping by 6+9 SNP array was used for QTL mapping for anthocyanins and hydroxycinnamic acid composition (Calle et al. 2021b). Analysis of mapping population 'Vic' × 'Cristobalina' consisting of 161 seedlings showed significant linkage on LG1 (hydroxycinnamic acid) and LG3 (anthocyanins). This study also discovered the relationship of transcription factor *PavMYB10* with cherry fruit color, laying the foundations for MAS for this trait.

7.2.3.3 Tree Architecture

Due to the convenience and economic benefit of cultivation, features that determine the tree's architecture are also crucial for cherry growers. In modern fruit growing, a human determines the tree canopy characteristics. However, this requires a large amount of work and, accordingly, costs. Therefore, the creation of new commercially effective cherry cultivars can also be advanced by introducing traits of tree architecture. Tree canopy architecture is a complex phenomenon based on many traits, respectively controlled by many genes. Currently, studies examine only some of them, for example, trunk diameter (Wei et al. 2015). By conducting a study with a mapping population of 860 F_1 seedlings and using SLAF markers, it was found that four loci control trunk diameter in three different LGs, respectively, LG6, LG7, and LG8.

7.2.3.4 Disease Resistance

Sustainable and environmentally friendly fruit-growing is associated with minimal possible pesticide application, which is becoming increasingly complicated as outbreaks of existing pathogens and the arrival of new ones are observed as a result of climate changes. Due to the economic needs of growers and society's demands, creating cultivars resistant to biotic stress factors would be the most appropriate solution, especially in fruit trees where it is time-consuming to breed new characteristics and especially for resistance. Modern stone fruit, including cherries, are propagated only vegetatively – this ensures the stable preservation of cultivars and the quality of planting material but completely excludes evolutionary processes. Consequently, crop adaptation to new pests can only happen with the participation of human-aided breeding (Quero-García et al. 2022). Knowledge of resistance mechanisms and their gene pathways is required to do this. Cherries are threatened by many pathogens and pests (Kappel et al. 2012; James et al. 2017; Papadopoulos et al. 2017; Puławska et al. 2017). However, resistance research, especially in gene mining, is quite limited.

Research is being carried out on identifying and selecting forms resistant to various diseases to find sources of resistance for the study of heredity of resistance and directly for breeding. For example, the gene *Pmr1* has been identified in sweet cherries, which confers resistance to powdery mildew (*Podosphaeraclandestina*) (Olmstead and Lang 2002). Using artificial inoculation, sweet cherry and wild *Prunus* genotypes were analyzed for resistance to bacterial canker (*Pseudomonas syringae*) (Hulin et al. 2021). The acquired knowledge can be transferred to the QTL mapping studies, which have been carried out for powdery mildew (*Podosphaera clandestina*) and bacterial cancer (*Pseudomonas syringae* pv. Syringae). Using 600 seedlings of the breeding population and 1100 SNP and 4 SSR markers, a stable linkage was found on LG5 in three years of the study, while on LG1, LG3, and LG6- only in one of the years of the study (Zhao et al. 2014a; Mgbechi-Ezeri 2016). Major QTL on LG4 for resistance to cherry leaf spot was found in sweet, sour cherries, and its progenitor species *P. canescens*, discovering it as a resistance source in both cultivated species (Stegmeir et al. 2014).

Resistance gene analog (RGA) and transcript analysis are also used to study resistance against cherry pathogens (Zambounis et al. 2016; Liang et al. 2020). The identified RGAs help to understand the mechanisms of resistance and could be used in the future as a gene-specific marker source for selecting resistant plant material. On the other hand, artificial inoculation experiments and transcriptome analysis discovered the involvement of lignin-related genes in the resistance mechanisms of cherry rootstock (*Prunus mahaleb*) to crown gall disease (Liang et al. 2020).

7.3 FUTURE SCOPE

Given the ancient history of cherry cultivation and its wide distribution, they have left an indelible mark on the culture of many nations and are an essential part of their national economy. Targeted breeding also started at the beginning of the 1900s, thus requiring an understanding of the mechanisms of traits' heredity and crop genetic research. Regardless of the region of breeding activities, its goals are adaptability to local growing conditions, self-fertility, good fruit quality, and disease resistance. These actualities have also determined the main directions of gene mining – development and application of molecular markers, important trait coding gene identification, genetic mapping, and transcriptome analysis. In order to mitigate environmental challenges and meet the demands of society, the requirement for new cherry cultivars will remain, and the circulation of the cultivar assortment will have to be as fast as possible. This can be ensured by modern breeding, which is unthinkable without trait-specific molecular markers. Their development is still challenging because agronomically important traits have polygenic inheritance and extensive mutual linkage. Current information on agronomically important trait gene mining reveals extensive knowledge from various research methods and diverse plant materials. In particular, this process has been facilitated by the availability of whole genome sequences. Taking into account the synteny of the genus *Prunus* genome, various molecular markers have been implemented in the study of cherry germplasm, which has allowed for obtaining a wide range of diverse information. It has served as an opportunity, at least in Europe, to establish good collaboration in germplasm research by sharing data and joint research activities. Thereby, there is a potential to develop further research by including other regions of the world, especially from the crop's centers of origin, allowing the improvement of new germplasm introduction and the maintenance of diversity. Special attention should be paid to gene mining of agronomically significant traits, as it is of the greatest interest from the breeding side. Considering the highly complex inheritance of this group of traits, they mainly were QTL mapping studies, as well as transcriptome studies of the genes involved, using high *Prunus* genus and other plant species DNA synteny. However, these studies should be continued involving more diverse cherry germplasm because, as the experience of existing studies shows, better gene mining can be obtained by involving, for example, in QTL mapping as many populations as possible, preferably of different origins. Also, the applicability of the developed markers needs to be verified in the broadest possible germplasm. Such an approach has limitations such as the perennial nature of cherries, long juvenility, and high plant material maintenance costs. The solution is to include collections or mapping populations located in different places in the study, international cooperation, and data exchange for a common goal. Examples of such cooperation already exist and should be continued and extended to regions with available genetically diverse germplasm. Current studies on gene mining of agronomically significant traits, implementation, and application of molecular markers show that the research was carried out mainly for sweet cherries, only in a quite limited number of cases - for sour cherries. Although the cultivation of sour cherries is not so widespread, in certain regions, as well as for the processing industry, they are significant, and gene mining can significantly contribute to crop development.

REFERENCES

Aka Kaçar Y, Lezzoni AF, Çetiner S (2005) Sweet cherry cultivar identification by using SSR markers. *J Biol Sci* 5:616–619.

Akagi T, Henry IM, Morimoto T, Tao R (2016) Insights into the *Prunus* -specific S-RNase-based self-incompatibility system from a genome-wide analysis of the evolutionary radiation of S locus-related F-box genes. *Plant Cell Physiol* 57:1281–1294. https://doi.org/10.1093/pcp/pcw077.

Alkio M, Jonas U, Sprink T, van Nocker S, Knoche M (2012) Identification of putative candidate genes involved in cuticle formation in *Prunus avium* (sweet cherry) fruit. *Ann Bot* 110:101–112. https://doi.org/10.1093/aob/mcs087.

Ansari M, Davarynejad GH (2008) Marked improvement of Hungarian sour cherries grown in Iran by cross-pollination 1: Fruit set. *Asian J Plant Sci* 7:687–691. https://doi.org/10.3923/ajps.2008.687.691.

Antognoni F, Potente G, Mandrioli R, Angeloni C, Freschi M et al. (2020) Fruit quality characterization of new sweet cherry cultivars as a good source of bioactive phenolic compounds with antioxidant and neuroprotective potential. *Antioxidants* 9:1–22. https://doi.org/10.3390/antiox9080677.

Antonius K, Aaltonen M, Uosukainen M, Hurme T (2012) Genotypic and phenotypic diversity in Finnish cultivated sour cherry (*Prunus cerasus* L.). *Genet Resour Crop E* 59:375–388. https://doi.org/10.1007/s10722-011-9688-3.

Aranzana MJ, Decroocq V, Dirlewanger E, Eduardo I, Gao ZS et al. (2019) *Prunus* genetics and applications after *de novo* genome sequencing: Achievements and prospects. *Hortic Res* 6. https://doi.org/10.1038/s41438-019-0140-8.

Aranzana MJ, Garcia-Mas J, Carbó J, Arús P (2002) Development and variability analysis of microsatellite markers in peach. *Plant Breed* 121:87–92. https://doi.org/10.1046/j.1439-0523.2002.00656.x.

Arús P, Verde I, Sosinski B, Zhebentyayeva T, Abbott AG (2012) The peach genome. *Tree Genet Genomes* 8:531–547. https://doi.org/10.1007/s11295-012-0493-8.

Arús P, Yamamoto T, Dirlewanger E, Abbott AG (2006) Synteny in the Rosaceae. In: *Plant Breeding Reviews*, ed. Janick J. John Wiley and Sons Inc., London, pp. 175–211. https://doi.org/10.1002/9780470650349.ch4.

Ashkani S, Yusop MR, Shabanimofrad M, Azadi A, Ghasemzadeh A et al. (2015) Allele mining strategies: Principles and utilisation for blast resistance genes in rice (*Oryza sativa* L.). *Curr Iss Mol Biol* 17:57–74. https://doi.org/10.21775/cimb.017.057.

Badenes ML, Parfitt DE (1995) Phylogenetic relationships of cultivated *Prunus* species from an analysis of chloroplast DNA variation. *Theor Appl Genet* 90:1035–1041. https://doi.org/10.1007/BF00222918.

Balas F, López-Corrales M, Serradilla MJ, Cai L, Iezzoni A et al. (2019) Firmness QTL mapping using an "Ambrunés" × "Sweetheart" sweet cherry population. *Acta Hortic* 1235:41–46. https://doi.org/10.17660/Acta Hortic.2019.1235.5.

Balbontín C, Ayala H, Bastías RM, Tapia G, Ellena M et al. (2013) Cracking in sweet cherries: A comprehensive review from a physiological, molecular, and genomic perspective. *Chil J Agric Res* 73:66–72. https://doi.org/10.4067/S0718-58392013000100010.

Balbontín C, Ayala H, Rubilar J, Cote J, Figueroa CR (2014) Transcriptional analysis of cell wall and cuticle related genes during fruit development of two sweet cherry cultivars with contrasting levels of cracking tolerance. *Chil J Agric Res* 74:162–169. https://doi.org/10.4067/S0718-58392014000200006.

Baris E, Bedő J, Edosa OS, Kiss E, Veres A (2017) Preliminary results of SSR based characterization of sour (*Prunus cerasus* L.) and sweet cherry (*Prunus avium* L.) genotypes cultivated in Hungary. *Columella J Agric Environ Sci* 4:31–45. https://doi.org/10.18380/szie.colum.2017.4.1.31.

Barreneche T, de la Concepción MC, Blouin-Delmas M, Ordidge M, Nybom H et al. (2021) SSR-based analysis of genetic diversity and structure of sweet cherry (*Prunus avium* L.) from 19 countries in Europe. *Plants* 10:1–17. https://doi.org/10.3390/plants10101983.

Békefi Z (2006) Review of sweet and sour cherry incompatibility. *Int J Hortic Sci* 12:111–116. https://doi.org/10.31421/IJHS/12/2/642.

Békefi Z, Tobutt KR, Sonneveld T (2003) Determination of (in)compatibility genotypes of Hungarian sweet cherry (*Prunus avium* L.) accessions by PCR based methods. *Int J Hortic Sci* 9:37–42. https://doi.org/10.31421/IJHS/9/3-4/403.

Bekefi Z, Vaughan S, Tobutt K (2010) Determination of incompatibility (S) genotypes of sweet cherries in the Hungarian gene-bank by a PCR-based method. *Acta Agron Hung* 58:377–384. https://doi.org/10.1556/AAgr.58.2010.4.6.

Bezlepkina EV, Gulyaeva AA, Pikunova AV (2020) Allelic polymorphism of the self-incompatibility gene in cherry varieties of Vniispk Russian research institute selection (in Russian). *Bull Russian Agric Sci* 4:26–28. https://doi.org/10.30850/vrsn/2020/4/26-28.

Bird KA, Hardigan MA, Ragsdale AP, Knapp SJ, VanBuren R et al. (2021) Diversification, spread, and admixture of octoploid strawberry in the Western Hemisphere. *Am J Bot* 108:2269–2281. https://doi.org/10.1002/ajb2.1776.

Bird KA, Jacobs M, Sebolt A, Rhoades K, Alger EI et al. (2022) Parental origins of the cultivated tetraploid sour cherry (*Prunus cerasus* L.). *Plants People Planet* 4:444–450. https://doi.org/10.1002/ppp3.10267.

Biško A, Savić Z, Jukić L, Leko M, Radunić M et al. (2017) Sweet cherry cultivation in the Republic of Croatia. *Acta Hortic* 1161:185–192. https://doi.org/10.17660/ActaHortic.2017.1161.31.

Blanke MM, Kunz A (2017) Cherry phenology as bioindicator for climate change. *Acta Hortic* 1162:1–7. https://doi.org/10.17660/ActaHortic.2017.1162.1.

Bortiri E, Sang-Hun OH, Jiang UO, Baggett S, Granger A et al. (2001) Phylogeny and systematics of *Prunus* (Rosaceae) as determined by sequence analysis of ITS and the chloroplast trnL-trnF spacer DNA. *Syst Bot* 26:797–807.

Bošković R, Tobutt KR (1996) Correlation of stylar ribonuclease zymograms with incompatibility alleles in sweet cherry. *Euphytica* 90:245–250. https://doi.org/10.1007/BF00023865.

Branchereau C, Quero-García J, Zaracho-Echagüe NH, Lambelin L, Fouche M et al. (2022) New insights into flowering date in *Prunus*: Fine mapping of a major QTL in sweet cherry. *Hortic Res* 9. https://doi.org/10.1093/hr/uhac042.

Brettin TS, Karle R, Crowe EL, Iezzoni AF (2000) Chloroplast inheritance and DNA variation in sweet, sour, and ground cherry. *J Hered* 91:75–79. https://doi.org/10.1093/jhered/91.1.75.

Brozik S, Nyéki J (1980) Sour Cherry. In: *Bloom Biology and Fertility of Fruit Cultivars*, ed. Nyéki, J. Agricultural Publisher, Budapest, pp. 229–233.

Bujdosó G, Hrotkó K (2017) Cherry Production. In: *Cherries: Botany, Production and Uses*, eds. Quero-García J, Iezzoni A, Pulawska J, Lang G. CABI, Boston, pp. 1–13.

Cachi AM, Wünsch A (2011) Characterization and mapping of non-S gametophytic self-compatibility in sweet cherry (*Prunus avium* L.). *J Exp Bot* 62:1847–1856. https://doi.org/10.1093/jxb/erq374.

Cachi AM, Wünsch A (2014) Characterization of self-compatibility in sweet cherry varieties by crossing experiments and molecular genetic analysis. *Tree Genet Genomes* 10:1205–1212. https://doi.org/10.1007/s11295-014-0754-9.

Cai L, Quero-García J, Barreneche T, Dirlewanger E, Saski C et al. (2019) A fruit firmness QTL identified on linkage group 4 in sweet cherry (*Prunus avium* L.) is associated with domesticated and bred germplasm. *Sci Rep* 9:1–14. https://doi.org/10.1038/s41598-019-41484-8.

Cai L, Stegmeir T, Sebolt A, Zheng C, Bink MCAM et al. (2018) Identification of bloom date QTLs and haplotype analysis in tetraploid sour cherry (*Prunus cerasus*). *Tree Genet Genomes* 14:22. https://doi.org/10.1007/s11295-018-1236-2.

Calle A, Balas F, Cai L, Iezzoni A, López-Corrales M et al. (2020a) Fruit size and firmness QTL alleles of breeding interest identified in a sweet cherry 'Ambrunés' × 'Sweetheart' population. *Mol Breed* 40:1–17. https://doi.org/10.1007/s11032-020-01165-1.

Calle A, Cai L, Iezzoni A, Wünsch A (2020b) Genetic dissection of bloom time in low chilling sweet cherry (*Prunus avium* L.) using a multi-family QTL approach. *Front Plant Sci* 10:1–13. https://doi.org/10.3389/fpls.2019.01647.

Calle A, Grimplet J, le Dantec L, Wünsch A (2021a) Identification and characterization of DAMs mutations associated with early blooming in sweet cherry, and validation of DNA-based markers for selection. *Front Plant Sci* 12:621491. https://doi.org/10.3389/fpls.2021.621491.

Calle A, Serradilla MJ, Wünsch A (2021b) QTL mapping of phenolic compounds and fruit colour in sweet cherry using a 6+9K SNP array genetic map. *Sci Hortic* 280:109900. https://doi.org/10.1016/j.scienta.2021.109900.

Calle A, Wünsch A (2020) Multiple-population QTL mapping of maturity and fruit-quality traits reveals LG4 region as a breeding target in sweet cherry (*Prunus avium* L.). *Hortic Res* 7:127. https://doi.org/10.1038/s41438-020-00349-2.

Campoy JA, le Dantec L, Barreneche T, Dirlewanger E, Quero-García J (2014) New insights into fruit firmness and weight control in sweet cherry. *Plant Mol Biol Rep* 33:783–796. https://doi.org/10.1007/s11105-014-0773-6.

Cantini C, Iezzoni AF, Lamboy WF, Boritzki M, Struss D (2001) DNA fingerprinting of tetraploid cherry germplasm using simple sequence repeats. *J Am Soc Hortic Sci* 126:205–209. https://doi.org/10.21273/jashs.126.2.205.

Castède S, Campoy JA, García JQ, le Dantec L, Lafargue M et al. (2014) Genetic determinism of phenological traits highly affected by climate change in *Prunus avium*: Flowering date dissected into chilling and heat requirements. *New Phytol* 202:703–715. https://doi.org/10.1111/nph.12658.

Castède S, Campoy JA, le Dantec L, Quero-García J, Barreneche T et al. (2015) Mapping of candidate genes involved in bud dormancy and flowering time in sweet cherry (*Prunus avium*). *PLoS One* 10:1–18. https://doi.org/10.1371/journal.pone.0143250.

Charlesworth D (2010) Self-incompatibility. *F1000 Biol Rep* 2:68. https://doi.org/10.3410/B2-68.

Chawla R, Sheokand A, Rai MR, Kumar R (2021) Impact of climate change on fruit production and various approaches to mitigate these impacts. *Pharma Innovation* 10:564–571.

Chen Q, Meng D, Gu Z, Li W, Yuan H et al. (2018) SLFL genes participate in the ubiquitination and degradation reaction of S-RNase in self-compatible peach. *Front Plant Sci* 9:227. https://doi.org/10.3389/fpls.2018.00227.

Chmielewski F-M, Götz K-P, Weber KC, Moryson S (2018) Climate change and spring frost damages for sweet cherries in Germany. *Int J Biometeorol* 62:217–228. https://doi.org/10.1007/s00484-017-1443-9.

Christensen VJ (1972) Cracking in cherries. III. Determination of cracking susceptibility. *Acta Agric Scand* 22:128–36.

Cipriani G, Lot G, Huang WG, Marrazzo MT, Peterlunger E et al. (1999) AC/GT and AG/CT microsatellite repeats in peach [*Prunus persica* (L) Batsch]: Isolation, characterisation and cross-species amplification in *Prunus*. *Theor Appl Genet* 99:65–72. https://doi.org/10.1007/s001220051209.

Clarke JB, Tobutt KR (2003) Development and characterization of polymorphic microsatellites from *Prunus avium* "Napoleon." *Mol Ecol Notes* 3:578–580. https://doi.org/10.1046/j.1471-8286.2003.00517.x.

Clarke JB, Tobutt KR (2009) A standard set of accessions, microsatellites and genotypes for harmonising the fingerprinting of cherry collections for the ECPGR. *Acta Hortic* 814:615–618. https://doi.org/10.17660/ActaHortic.2009.814.104.

Cmejla R, Zdarska I, Suran P, Paprstein F, Cmejlova J (2021) A new allelic discrimination real-time PCR assay for PavMYB10.1 genotyping to predict sweet cherry fruit colour: A comparison with the Pav-Rf-SSR assay. *J Hortic Sci Biotechnol* 96:260–269. https://doi.org/10.1080/14620316.2020.1837021.

Conner JK, Hartl DL (2004) *A Primer of Ecological Genetics (in Lithuanian)*. Translator Morkunas V. Vilniaus universiteto leidykla, Vilnius, Lithuania.

Cooke JEK, Eriksson ME, Junttila O (2012) The dynamic nature of bud dormancy in trees: Environmental control and molecular mechanisms. *Plant Cell Environ* 35:1707–1728. https://doi.org/10.1111/j.1365-3040.2012.02552.x.

Correia S, Schouten R, Silva AP, Gonçalves B (2017) Factors affecting quality and health promoting compounds during growth and postharvest life of sweet cherry (*Prunus avium* L.). *Front Plant Sci* 8:1–15. https://doi.org/10.3389/fpls.2017.02166.

Correia S, Schouten R, Silva AP, Gonçalves B (2018) Sweet cherry fruit cracking mechanisms and prevention strategies: A review. *Sci Hortic* 240:369–377. https://doi.org/10.1016/j.scienta.2018.06.042.

Crane MB, Brown AG (1938) Incompatibility and sterility in the sweet cherry, *Prunus avium* L. *J Pomol Hortic Sci* 15:2–116. https://doi.org/10.1080/03683621.1938.11513493.

Crane MB, Lawrence WJC (1928) Genetical and cytological aspects of incompatibility and sterility in cultivated fruits. *J Pomol Hortic Sci* 7:276–301. https://doi.org/10.1080/03683621.1928.11513345.

Crane MB, Lewis D (1942) Genetical studies in pears. *J Genet* 43:31–43. https://doi.org/10.1007/BF02982745.

Crump WW, Peace C, Zhang Z, McCord P (2022) Detection of breeding-relevant fruit cracking and fruit firmness quantitative trait loci in sweet cherry via pedigree-based and genome-wide association approaches. *Front Plant Sci* 13:823250. https://doi.org/10.3389/fpls.2022.823250.

Davarynejad GH, Nyéki J, Szabó T, Tornyai J, Szabó Z et al. (2010). Comparison of varying pollen source on productivity of sour cherry (*Prunus cerasus* L.) cultivars. *Int J Hortic Sci* 16(1):23–26. https://doi.org/10.31421/IJHS/16/1/857.

Davarynejad GH, Szabo Z, Persely SZ, Szabo T, Nyéki J (2014) The fruit set capability of some sour cherry cultivars (*Prunus cerasus* L.). *Acta Hortic* 1020:181–184. https://doi.org/10.17660/ActaHortic.2014.1020.25.

De Cuyper B, Sonneveld T, Tobutt KR (2005) Determining self-incompatibility genotypes in Belgian wild cherries. *Mol Ecol* 14:945–955. https://doi.org/10.1111/j.1365-294X.2005.02460.x.

de Franceschi P, Stegmeir T, Cabrera A, van der Knaap E, Rosyara UR et al. (2013) Cell number regulator genes in *Prunus* provide candidate genes for the control of fruit size in sweet and sour cherry. *Mol Breed* 32:311–326. https://doi.org/10.1007/s11032-013-9872-6.

de Rogatis A, Ferrazzini D, Ducci F, Guerri S, Carnevale S et al. (2013) Genetic variation in Italian wild cherry (*Prunus avium* L.) as characterized by nSSR markers. *Forestry* 86:391–400. https://doi.org/10.1093/forestry/cpt009.

Demir T, Demirsoy L, Demirsoy H, Kaçar YA, Yälmaz M et al. (2011) Molecular characterization of sweet cherry genetic resources in Giresun, Turkey. *Fruits* 66:53–62. https://doi.org/10.1051/fruits/2010041.

Demirsoy H, Demirsoy L, Lang GA (2022) Research on spring frost damage in cherries. *Hortic Sci* 49:89–94. https://doi.org/10.17221/91/2021-HORTSCI.

Deshaies RJ (1999) SCF and Cullin/Ring H2-based ubiquitin ligases. *Annu Rev Cell Dev Biol* 15:435–467. https://doi.org/10.1146/annurev.cellbio.15.1.435.

Dirlewanger E, Claverie J, Iezzoni AF (2007) Cherry. In: *Fruits and Nuts. Genome Mapping and Molecular Breeding in Plants 4*, ed. Kole C. Springer, Berlin, Heidelberg, pp. 103–118. https://doi.org/10.1007/978-3-540-34533-6_3.

Dirlewanger E, Cosson P, Tavaud M, Aranzana MJ, Poizat C et al. (2002) Development of microsatellite markers in peach [*Prunus persica* (L.) Batsch] and their use in genetic diversity analysis in peach and sweet cherry (*Prunus avium* L.). *Theor Appl Genet* 105:127–138. https://doi.org/10.1007/s00122-002-0867-7.

Dirlewanger E, Graziano E, Joobeur T, Garriga-Caldere F, Cosson P et al. (2004) Comparative mapping and marker-assisted selection in Rosaceae fruit crops. *Pnas* 101(26):9891–9896. https://doi.org/10.1073/pnas.0307937101.

Dirlewanger E, Le-Dantec L, Fouché M, Zaracho Echagüe NH, Campoy JA et al. (2020) New sweet cherry genomic tools and their use in marker-assisted breeding. *Acta Hortic* 533–542. https://doi.org/10.17660/ActaHortic.2020.1297.70.

Dirlewanger E, Quero-García J, le Dantec L, Lambert P, Ruiz D et al. (2012) Comparison of the genetic determinism of two key phenological traits, flowering and maturity dates, in three *Prunus* species: Peach, apricot and sweet cherry. *Heredity (Edinb)* 109:280–292. https://doi.org/10.1038/hdy.2012.38.

Dondini L, Lugli S, Sansavini S (2018) Cherry Breeding: Sweet Cherry (*Prunus avium* L.) and Sour Cherry (*Prunus cerasus* L.). In: *Advances in Plant Breeding Strategies: Fruits 3*, eds. Al-Khayri JM, Jain SM, Johnson DV. Springer, Cham, pp. 31–88. https://doi.org/10.1007/978-3-319-91944-7.

Downey SL, Iezzoni AF (2000) Polymorphic DNA markers in black cherry (*Prunus serotina*) are identified using sequences from sweet cherry, peach, and sour cherry. *J Am Soc Hortic Sci* 125:76–80. https://doi.org/10.21273/jashs.125.1.76.

Drogoudi P, Kazantzis K, Kunz A, Blanke MM (2020) Effects of climate change on cherry production in Naoussa, Greece and Bonn, Germany: Adaptation strategies. *Euro Mediterr J Environ Integr* 5. https://doi.org/10.1007/s41207-020-0146-5.

Ercisli S, Radunic M, Gadze J, Ipek A, Skaljac M et al. (2012) S-RNase based S-genotyping of Croatian sweet cherry (*Prunus avium* L.) genotypes. *Sci Hortic* 139:21–24. https://doi.org/10.1016/j.scienta.2012.02.041.

Eremina O, Eremin V, Smirnov R (2020) Genealogical analysis of large-fruited sweet cherry varieties in accordance with the S-locus of parental forms, and the pattern of inheritance of large-fruited in the presence of S5 and S9 alleles in the genome. *BIO Web Conf* 25:03005. https://doi.org/10.1051/bioconf/20202503005.

Ershadi A, Moghadam EH (2009) Identification of s-alleles in 25 sweet cherry (*Prunus avium L.*) cultivars by allele-specific PCR amplification. *Acta Hortic* 814:619–622. https://doi.org/10.17660/ActaHortic.2009.814.105.

FAOSTAT (2020) Food and Agriculture Organization of the United Nations. FAOSTAT Statistical Database. https://www.fao.org/faostat/en/#home.

Farsad A, Esna-Ashari M (2016) Genetic diversity of some Iranian sweet cherry (*Prunus avium*) cultivars using microsatellite markers and morphological traits. *Cytol Genet* 50:8–19. https://doi.org/10.3103/S0095452716010059.

Faust M, Surányi D (2011) Origin and dissemination of cherry. In: *Horticultural Reviews*, ed. Janick J. ISHS, Leuven, Belgium. pp. 55–102.

Fogle H (1958) Inheritance of fruit color in sweet cherries (*Prunus avium*). *J Hered* 49:294–298.

Ganopoulos IV, Argiriou A, Tsaftaris AS (2010) Determination of self-incompatible genotypes in 21 cultivated sweet cherry cultivars in Greece and implications for orchard cultivation. *J Hortic Sci Biotechnol* 85:444–448. https://doi.org/10.1080/14620316.2010.11512695.

Ganopoulos IV, Kazantzis K, Chatzicharisis I, Karayiannis I, Tsaftaris AS (2011) Genetic diversity, structure and fruit trait associations in Greek sweet cherry cultivars using microsatellite based (SSR/ISSR) and morpho-physiological markers. *Euphytica* 181:237–251. https://doi.org/10.1007/s10681-011-0416-z.

Ganopoulos I, Moysiadis T, Xanthopoulou A, Osathanunkul M, Madesis P et al. (2016) Morpho-physiological diversity in the collection of sour cherry (*Prunus cerasus*) cultivars of the Fruit Genebank in Naoussa, Greece using multivariate analysis. *Sci Hortic* 207:225–232. https://doi.org/10.1016/j.scienta.2016.05.018.

Gelvonauskienė D, Frercks B, Stepulaitiene I, Stanys V (2013) Diversity of resistance to *Monilinia laxa* and spring frost of sour cherry germplasm. *Acta Hortic (ISHS)* 976:537–542.

Gharaghani A, Solhjoo S, Oraguzie N (2017) A review of genetic resources of almonds and stone fruits (*Prunus* spp.) in Iran. *Genet Resour Crop E* 64:611–640. https://doi.org/10.1007/s10722-016-0485-x.

Gisbert AD, Badenes ML, Tobutt KR, Llacer G, Romero C et al. (2008) Determination of the S-allele composition of sweet cherry (*Prunus avium* L.) cultivars grown in the southeast of Spain by PCR analysis. *J Hortic Sci Biotechnol* 83:246–252. https://doi.org/10.1080/14620316.2008.11512376.

Gorina V, Lukicheva L, Grigoriev A, Chernen'ky L, Sokolovskaya J et al. (2021) Genetic resources are the base for improving the assortment of sweet cherry and large fruit cherry plum. *E3S Web Conf* 254:01023. https://doi.org/10.1051/e3sconf/202125401023.

Gril T, Celar F, Munda A, Javornik B, Jakse J (2008) AFLP analysis of intraspecific variation between *Monilinia laxa* isolates from different hosts. *Plant Dis* 92:1616–1624. https://doi.org/10.1094/PDIS-92-12-1616.

Guajardo V, Muñoz C, Hinrichsen P (2021) Molecular profiling of sweet cherry cultivars present in Chile using polymorphic microsatellite markers. *Chil J Agric Res* 81:326–337. https://doi.org/10.4067/S0718-58392021000300326.

Guarino C, Santoro S, de Simone L, Cipriani G (2009) *Prunus avium*: Nuclear DNA study in wild populations and sweet cherry cultivars. *Genome* 52:320–337. https://doi.org/10.1139/G09-007.

Guarino C, Santoro S, de Simone L, Cipriani G (2010) Molecular characterisation of ancient *Prunus avium* L. germplasm using sweet cherry SSR markers. *J Hortic Sci Biotechnol* 85:295–304. https://doi.org/10.1080/14620316.2010.11512671.

Gulen H, Ipek A, Ergin S, Emin Akcay M, Eris A (2010) Assessment of genetic relationships among 29 introduced and 49 local sweet cherry accessions in Turkey using AFLP and SSR markers. *J Hortic Sci Biotechnol* 85:427–431. https://doi.org/10.1080/14620316.2010.11512692.

Gulyaeva AA, Berlova TN, Galkova AA, Bezlepkina EV, Efremov IN (2022) Winter resistance of sour cherry and sweet cherry flower buds in extreme winter 2009–2010 and 2020–2021. *BIO Web Conf* 47:02011. https://doi.org/10.1051/bioconf/20224702011.

György Z, Szabó T, Nyéki J, Pedryc A (2010) Analyses of Hungarian sour cherry germplasm with simple sequence repeat markers. *Int J Hortic Sci* 16:27–30. https://doi.org/10.31421/ijhs/16/1/858.

Hakeem KR, ed. (2015) *Crop Production and Global Environmental Issues*. Springer, Cham. https://doi.org/10.1007/978-3-319-23162-4.

Haokip SW, Shankar K, Lalrinngheta J (2020) Climate change and its impact on fruit crops. *J Pharmacogn Phytochem* 9:435–438.

Hauck NR, Yamane H, Tao R, Iezzoni AF (2002) Self-compatibility and incompatibility in tetraploid sour cherry (*Prunus cerasus* L.). *Sex Plant Reprod* 15:39–46. https://doi.org/10.1007/s00497-002-0136-6.

Hegedűs A, Lénárt J, Halász J (2012) Sexual incompatibility in Rosaceae fruit tree species: Molecular interactions and evolutionary dynamics. *Biol Planta* 56:201–209. https://doi.org/10.1007/s10535-012-0077-3.

Herrero M, Rodrigo J, Wünsch A (2017) Flowering, fruit set and development. In: *Cherries: Botany, Production and Uses*, eds. Quero-García J, Lezzoni A, Pławska J, Lang G. CABI, Wallingford, pp. 14–35.

Höfer M, Peil A (2015) Phenotypic and genotypic characterization in the collection of sour and duke cherries (*Prunus cerasus* and *P. gondouini*) of the Fruit Genebank in Dresden-Pillnitz, Germany. *Genet Resour Crop E* 62:551–566. https://doi.org/10.1007/s10722-014-0180-8.

Höltken AM, Gregorius HR (2006) Detecting local establishment strategies of wild cherry (*Prunus avium* L.). *BMC Ecol* 6:1–13. https://doi.org/10.1186/1472-6785-6-13.

Hulin MT, Vadillo Dieguez A, Cossu F, Lynn S, Russell K et al. (2021) Identifying resistance in wild and ornamental cherry towards bacterial canker caused by *Pseudomonas syringae*. *Plant Pathol* 71:949–965. https://doi.org/10.1111/ppa.13513.

Iezzoni A, Wünsch A, Höfer M, Giovannini D, Jensen M et al. (2017) Biodiversity, Germplasm Resources and Breeding Methods. In: *Cherries: Botany, Production and Uses*, eds. Quero-García J, Iezzoni A, Pulawska J, Lang G. CABI, Wallingford, pp. 36–59.

Iezzoni AF (2008) Cherries. In: *Temperate Fruit Crop Breeding*, ed. Hancock JF. Springer Science+Business Media B.V., Berlin, pp. 151–175.

Iezzoni AF, McFerson J, Luby J, Gasic K, Whitaker V et al. (2020) RosBREED: Bridging the chasm between discovery and application to enable DNA-informed breeding in rosaceous crops. *Hortic Res* 7. https://doi.org/10.1038/s41438-020-00398-7.

Igic B, Kohn JR (2001) Evolutionary relationships among self-incompatibility RNases. *Proc Natl Acad Sci USA* 98:13167–13171. https://doi.org/10.1073/pnas.231386798.

Ikeda K, Igic B, Ushijima K, Yamane H, Hauck NR et al. (2004) Primary structural features of the S haplotype-specific F-box protein, SFB, in *Prunus*. *Sex Plant Reprod* 16:235–243. https://doi.org/10.1007/s00497-003-0200-x.

Ipek A, Gulen H, Ipek M, Ipek M, Ergin S et al. (2011) Determination of self-incompatibility groups of sweet cherry genotypes from Turkey. *Genet Mol Res* 10:253–260. https://doi.org/10.4238/vol10-1gmr1024.

Ishizaka T, Nakano H, Suzuki T, Kitashiba H (2003) Characterization of the S-RNase promoters from sweet cherry (*Prunus avium* L.). *Genes Genet Syst* 78:191–194. https://doi.org/10.1266/ggs.78.191.

Ivanovych Y (2016) Sweet cherry genetic fingerprinting: Methods and techniques. Marker-assisted selection (MAS) approaches for selection of sweet cherry varieties. Short term scientific mission report of COST

action FA1104: "Sustainable production of high-quality cherries for the Europeanmarket" https://doi.org/10.13140/RG.2.2.11739.98080.
Ivanovych Y, Volkov R (2018) Genetic relatedness of sweet cherry (*Prunus avium* L.) cultivars from Ukraine determined by microsatellite markers. *J Hortic Sci Biotech* 93:64–72. https://doi.org/10.1080/14620316.2017.1342568.
Jajasankar S, Kappel F (2011) Recent Advances in Cherry Breeding. *Fruit, Veg Cereal Sci Biotech* 5 (1):63–67.
James D, Cieślińska M, Pallás V, Flores R, Candresse T et al. (2017) Viruses, Viroids, Phytoplasmas and Genetic Disorders of Cherry. In: *Cherries: Botany, Production and Uses*, eds. Quero-Garcia J, Iezzoni A, Pulawska J, Lang G. CABI, Boston, pp. 386–419.
Joulani A, Sharafi Y, Sabbaghpour Heris J (2020) Compatibility relationships of some sour cherry cultivars and genotypes using PCR and florescence microscopy. *J Hortic Sci* 34:427–438. https://doi.org/10.22067/jhorts4.v34i3.78796.
Kappel F (2008) Breeding cherries in the "new world." *Acta Hortic* 795(PART 1):59–69. https://doi.org/10.17660/actahortic.2008.795.2.
Kappel F, Granger A, Hrotkó K, Schuster M (2012) Cherry. In: *Fruit Breeding*, eds. Badenes ML, Byrne DH. Springer, New York, pp. 459–504.
Khadivi-Khub A (2014) Regression association analysis of fruit traits with molecular markers in cherries. *Plant Syst E* 300:1163–1173. https://doi.org/10.1007/s00606-013-0953-0.
Khadivi-Khub A (2015) Physiological and genetic factors influencing fruit cracking. *Acta Physiol Plant* 37. https://doi.org/10.1007/s11738-014-1718-2.
Khadivi-Khub A, Zamani Z, Fattahi R, Wünsch A (2014a) Genetic variation in wild *Prunus* L. subgen. *Cerasus* germplasm from Iran characterized by nuclear and chloroplast SSR markers. *Trees Struct Funct* 28:471–485. https://doi.org/10.1007/s00468-013-0964-z.
Khadivi-Khub A, Zamani Z, Fattahi R, Wünsch A (2014b) S-allele diversity in *Prunus* L. *Cerasus* subgenus from Iran. *Biochem Syst Ecol* 53:1–7. https://doi.org/10.1016/j.bse.2013.12.002.
Kitashiba H, Zhang SL, Wu J, Shirasawa K, Nishio T et al. (2008) *S* genotyping and *S* screening utilizing SFB gene polymorphism in Japanese plum and sweet cherry by dot-blot analysis. *Mol Breed* 21:339–349. https://doi.org/10.1007/s11032-007-9134-6.
Kivistik A, Jakobson L, Kahu K, Laanemets K (2021) Wild and rare self-incompatibility allele *S17* found in 24 sweet cherry (*Prunus avium* L.) cultivars. *Plant Mol Biol Report* 40:376–388. https://doi.org/10.1007/s11105-021-01327-1.
Knoche M, Winkler A (2019) The mechanism of rain cracking of sweet cherry fruit. *Italus Hortus* 26:59–65. https://doi.org/10.26353/j.itahort/2019.1.5965.
Kolytaitė A, Vinskienė J, Stanys V (2017) *S*-allele identification by PCR analysis in sweet cherry cultivars, clones and new Lithuanian breed numbers. *Hortic Sci Articles* 36:16–25.
Krmpot T, Rados L, Vokurka A (2020) Genetic characterisation of autochthonous sweet cherry genotypes (*Prunus avium* L.) using SSR markers. *Genetika* 52:43–53. https://doi.org/10.2298/GENSR2001043K.
Kulikov IM, Kudryavtsev AM, Marchenko LA, Morozova NG, Boris KV et al. (2018) Microsatellite loci polymorphism in contemporary Russian sour cherry (*Prunus cerasus* L.) varieties bred in All-Russian Horticultural Institute for Breeding, Agrotechnology and Nursery. *Hortic Vitic* 5:5–9. https://doi.org/10.31676/0235-2591-2018-5-5-9.
Lacis G, Kaufmane E, Rashal I, Trajkovski V, Iezzoni AF et al. (2008) Identification of self-incompatibility (S) alleles in Latvian and Swedish sweet cherry genetic resources collections by PCR based typing. *Euphytica* 160:155–163. https://doi.org/10.1007/s10681-007-9496-1.
Lacis G, Kota I (2013) SSR marker-based fingerprinting for sour cherry (*Prunus cerasus*) genetic resources identification and management. *Acta Hortic* 976:251–256. https://doi.org/10.17660/ActaHortic.2013.976.33.
Lacis G, Rashal I, Ruisa S, Trajkovski V, Iezzoni AF (2009) Assessment of genetic diversity of Latvian and Swedish sweet cherry (*Prunus avium* L.) genetic resources collections by using SSR (microsatellite) markers. *Sci Hortic* 121:451–457. https://doi.org/10.1016/j.scienta.2009.03.016.
Lacis G, Rashal I, Trajkovski V (2011) Implementation of a limited set of SSR markers for screening of genetic variability in Latvian and Swedish sour cherry (*Prunus cerasus* L.) genetic resources collections. *Proc Latvian Acad Sci Section B Nat Exact Appl Sci* 65:21–28. https://doi.org/10.2478/v10046-011-0014-4.
Lai Z, Ma W, Han B, Liang L, Zhang Y et al. (2002) An F-box gene linked to the self-incompatibility (S) locus of *Antirrhinum* is expressed specifically in pollen and tapetum. *Plant Mol Biol* 50:29–42. https://doi.org/10.1023/a:1016050018779.
Lang GA, Early JD, Martin AJ, Darnell RL, Martin G (1987) Endodormancy, paradormancy, and ecodormancy: Physiological terminology and classification for dormancy research. *Hortscience* 22:37–377.

Lansari A, Iezzoni A (1990) A preliminary analysis of self-incompatibility in sour cherries. *Hortscience* 25:1636–1638.
Lewis D, Crowe LK (1954) Structure of the incompatibility gene. *Heredity* 8:357–363. https://doi.org/10.1038/hdy.1954.38.
Lewis D, Modlibowska I (1942) Genetical studies in pears. *J Genet* 43:211–222. https://doi.org/10.1007/BF02982754.
Li Y, Duan X, Wu C, Yu J, Liu C et al. (2020) Ubiquitination of *S4-RNase* by *S*-locus F-box *LIKE2* contributes to self-compatibility of sweet cherry 'Lapins'. *Plant Physiol* 184:1702–1716. https://doi.org/10.1104/pp.20.01171.
Liang C, Wan T, Wu R, Zhao M, Zhao Y et al. (2020) Resistance analysis of cherry rootstock 'CDR-1' (*Prunus mahaleb*) to crown gall disease. *BMC Plant Biol* 20:1–14. https://doi.org/10.1186/s12870-020-02673-0.
Liang C, Wan T, Xu S, Li B, Li X et al. (2018) Molecular identification and genetic analysis of cherry cultivars using capillary electrophoresis with fluorescence-labeled SSR markers. *3 Biotech* 8. https://doi.org/10.1007/s13205-017-1036-7.
Licea-Moreno RJ, Quintana J, Contreras A, Gomez L (2019) Using SSR markers from *Prunus* genus for wild cherry genotyping. *Gabj* 3:1–10. https://doi.org/10.46325/gabj.v3i1.43.
Lisek A, Kucharska D, Głowacka A, Rozpara E (2017) Identification of S-haplotypes of European cultivars of sour cherry. *J Hortic Sci Biotechnol* 92:484–492. https://doi.org/10.1080/14620316.2017.1289071.
Lisek A, Rozpara E, Głowacka A, Kucharska D, Zawadzka M (2016) Identification of S-genotypes of sweet cherry cultivars from Central and Eastern Europe. *Hortic Sci (Prague)* 42:13–21. https://doi.org/10.17221/103/2014-HORTSCI.
Liu C, Qi X, Song L, Li Y, Li M (2018) Species identification, genetic diversity and population structure of sweet cherry commercial cultivars assessed by SSRs and the gametophytic self-incompatibility locus. *Sci Hortic* 237:28–35. https://doi.org/10.1016/j.scienta.2018.03.063.
Ma R-C, Oliveira MM (2002) Evolutionary analysis of S-RNase genes from Rosaceae species. *Mol Genet Genom* 267:71–78. https://doi.org/10.1007/s00438-002-0637-x.
Macková L, Vít P, Urfus T (2018) Crop-to-wild hybridization in cherries-empirical evidence from *Prunus fruticosa*. *Evol Appl* 11:1748–1759. https://doi.org/10.1111/eva.12677.
Mahmoudi M, Arzani K (2006) Study of compatibility of six sweet cherry (*Prunus avium* L.) cultivars with. *Agriculture (Not Published)* 8:47–57.
Mahmoodi M, Arzani K, Bouzari N (2008) Pollination, pollen tube growth and determination of the best pollinizer for sweet cherry (*Prunus avium* L.) cv. Red Rezaeieh. *Acta Hortic* 769:207–210. https://doi.org/10.17660/ActaHortic.2008.769.28.
Marchese A, Bošković R, Caruso T, Tobutt K (2010) Intra-allelic variation in introns of the *S13-RNase* allele distinguishes sweet, wild and sour cherries. *Tree Genet Genom* 6:963–972. https://doi.org/10.1007/s11295-010-0305-y.
Marchese A, Giovannini D, Leone A, Mafrica R, Palasciano M et al. (2017a) *S*-genotype identification, genetic diversity and structure analysis of Italian sweet cherry germplasm. *Tree Genet Genomes* 13:93. https://doi.org/10.1007/s11295-017-1176-2.
Marchese A, Marra FP, Priolo D, Caruso T, Giovannini D et al. (2017b) Identification of (in)compatible S-genotypes and molecular characterisation of Italian sweet cherry cultivars. *Acta Hortic* 1161:41–46. https://doi.org/10.17660/ActaHortic.2017.1161.6.
Marchese A, Tobutt KR, Raimondo A, Motisi A, Bošković RI et al. (2007) Morphological characteristics, microsatellite fingerprinting and determination of incompatibility genotypes of Sicilian sweet cherry cultivars. *J Hortic Sci Biotechnol* 82:41–48. https://doi.org/10.1080/14620316.2007.11512197.
Marić S, Radičević S, Milošević N, Popovska M, Malchev S et al. (2020) An overview of self-incompatibility (*S*) genotypes of autochthonous sweet cherries grown in Balkan region. *J Mount Agric Balkans* 23:168–181.
Marshall RE (1954). Cherries and Cherry Products. In: *Economic Crops* (Fifth Edition), ed. Marshall RE. Inter-Science, New York, 283 pp.
Martí AFI, Kodad O, Segura JMA (2012) Self-compatibility in *Prunus* species: Diversity of mutations. 19th Eucarpia General Congress, Budapest, Hungary.
Matsumoto D, Tao R (2016a) Distinct self-recognition in the *Prunus* S-RNase-based gametophytic self-incompatibility system. *Hortic J* 85:289–305. https://doi.org/10.2503/hortj.MI-IR06.
Matsumoto D, Tao R (2016b) Recognition of a wide-range of S-RNases by S locus F-box like 2, a general-inhibitor candidate in the *Prunus*-specific S-RNase-based self-incompatibility system. *Plant Mol Biol* 91:459–469. https://doi.org/10.1007/s11103-016-0479-2.

Matsumoto D, Tao R (2019) Recognition of *S-RNases* by an *S* locus F-box like protein and an *S* haplotype-specific F-box like protein in the *Prunus*-specific self-incompatibility system. *Plant Mol Biol* 100:367–378. https://doi.org/10.1007/s11103-019-00860-8.

Matsumoto D, Yamane H, Tao R (2008) Characterization of *SLFL1*, a pollen-expressed F-box gene located in the *Prunus S* locus. *Sexual Plant Reprod* 21:113–121.

McClure B, Cruz-García F, Romero C (2011) Compatibility and incompatibility in S-RNase-based systems. *Ann Bot* 108:647–658. https://doi.org/10.1093/aob/mcr179.

McCubbin A (2008) Heteromorphic Self-Incompatibility in *Primula*: Twenty-First Century Tools Promise to Unravel a Classic Nineteenth Century Model System. In: *Self-Incompatibility in Flowering Plants: Evolution, Diversity, and Mechanisms*, ed. Franklin-Tong VE. Springer, Berlin, Heidelberg, pp. 289–308. https://doi.org/10.1007/978-3-540-68486-2_14.

McCubbin AG, Kao T (2000) Molecular recognition and response in pollen and pistil interactions. *Annu Rev Cell Dev Biol* 16:333–364. https://doi.org/10.1146/annurev.cellbio.16.1.333.

Messina R, Lain O, Marrazzo MT, Cipriani G, Testolin R (2004) New set of microsatellite loci isolated in apricot. *Mol Ecol Notes* 4:432–434. https://doi.org/10.1111/j.1471-8286.2004.00674.x.

Mgbechi-Ezeri JU (2016) Phenotypic screening of sweet cherry (*Prunus avium* L.) germplasm and identification of quantitative trait loci (QTL) underlying bacterial canker disease. Washington State University.

Michailidis M, Karagiannis E, Bazakos C, Tanou G, Ganopoulos I et al. (2021) Genotype- and tissue-specific metabolic networks and hub genes involved in water-induced distinct sweet cherry fruit cracking phenotypes. *Comput Struct Biotechnol J* 19:5406–5420. https://doi.org/10.1016/j.csbj.2021.09.030.

Michailidis M, Karagiannis E, Tanou G, Sarrou E, Karamanoli K et al. (2020) Sweet cherry fruit cracking: Follow-up testing methods and cultivar-metabolic screening. *Plant Methods* 16:1–14. https://doi.org/10.1186/s13007-020-00593-6.

Miljuš-Đukić J, Banović B, Ninković S, Radović S (2010) The heteromorphic sporophytic self-incompatibility system of buckwheat (*Fagopyrum esculentum* Moench). *Eur J Plant Sci Biotechnol* 4:51–56.

Mnejja M, Garcia-Mas J, Howad W, Arús P (2005) Development and transportability across *Prunus* species of 42 polymorphic almond microsatellites. *Mol Ecol Notes* 5:531–535. https://doi.org/10.1111/j.1471-8286.2005.00977.x.

Mnejja M, Garcia-Mas J, Howad W, Badenes ML, Arús P (2004) Simple-sequence repeat (SSR) markers of Japanese plum (*Prunus salicina* Lindl.) are highly polymorphic and transferable to peach and almond. *Mol Ecol Notes* 4:163–166. https://doi.org/10.1111/j.1471-8286.2004.00603.x.

Muccillo L, Colantuoni V, Sciarrillo R, Baiamonte G, Salerno G et al. (2019) Molecular and environmental analysis of Campania (Italy) sweet cherry (*Prunus avium* L.) cultivars for biocultural refugia identification and conservation. *Sci Rep* 9:1–11. https://doi.org/10.1038/s41598-019-43226-2.

Muñoz-Espinoza C, Espinosa E, Bascuñán R, Tapia S, Meneses C et al. (2017) Development of a molecular marker for self-compatible S4′ haplotype in sweet cherry (*Prunus avium* L.) using high-resolution melting. *Plant Breed* 136:987–993. https://doi.org/10.1111/pbr.12546.

Najafzadeh R, Arzani K, Bouzari N, Saei A (2016) Genetic variation and identification of promising sour cherries inferred from microsatellite markers. *Russ J Genet* 52:64–73. https://doi.org/10.1134/S1022795415110113.

Nyéki J, Soltesz M, Szabo T (2008a) *Establishment and Production Techniques of Sour Cherry Orchards* (First Edition). University of Debrecen Hungary, Debrecen.

Nyéki J, Soltész M, Szabó Z (2008b) *Morphology, Biology and Fertility of Flowers in Temperate Zone Fruits*, Self-incompatibility of temperate zone fruit species, eds. Nyéki J, Soltész M, Szabó Z. Akadémiai Kiadó, Budapest, pp. 209–339. ISBN 978-963-05-8591-0.

Nyéki J, Szabó T, Szabó Z (2002) Blooming phenology and fertility of sour cherry cultivars selected in Hungary. *Int J Hortic Sci* 8:33–37. https://doi.org/10.31421/IJHS/8/2/320.

Oldén E, Nybom N (1968) On the origin of *Prunus cerasus* L. Heredity. *Hereditas* 70:3321–3323.

Olmstead JW, Iezzoni AF, Whiting MD (2007) Genotypic differences in sweet cherry fruit size are primarily a function of cell number. *J Am Soc Hortic Sci* 132:697–703. https://doi.org/10.21273/jashs.132.5.697.

Olmstead JW, Lang GA (2002) *Pmr1*, a gene for resistance to powdery mildew in sweet cherry. *HortScience* 37:1098–1099. https://doi.org/10.21273/hortsci.37.7.1098.

Ono K, Akagi T, Morimoto T, Wünsch A, Tao R (2018) Genome re-sequencing of diverse sweet cherry (*Prunus avium*) individuals reveals a modifier gene mutation conferring pollen-part self-compatibility. *Plant Cell Physiol* 59:1265–1275. https://doi.org/10.1093/pcp/pcy068.

Ordidge M, Litthauer S, Venison E, Blouin-Delmas M, Fernandez-Fernandez F et al. (2021) Towards a joint international database: Alignment of SSR marker data for european collections of cherry germplasm. *Plants* 10:1–18. https://doi.org/10.3390/plants10061243.

Öz MH, Vurgun H, Bakir M, Büyük I, Yüksel C et al. (2013) Molecular analysis of East Anatolian traditional plum and cherry accessions using SSR markers. *Genet Mol Res* 12:5310–5320. https://doi.org/10.4238/2013.November.7.6.

Ozherelieva ZE (2020) Study of frost resistance of cherry variety Turgenevka during on different rootstocks winter thaws. *Vegetable Crops of Russia (in Russian)* 5:65–70. https://doi.org/10.18619/2072-9146-2020-5-65-70.

Paltineanu C, Chitu E (2020) Climate change impact on phenological stages of sweet and sour cherry trees in a continental climate environment. *Sci Hortic* 261:109011. https://doi.org/10.1016/j.scienta.2019.109011.

Papadopoulos NT, Lux SA, Köppler K, Beliën T (2017) Invertebrate and Vertebrate Pests: Biology and Management. In: *Cherries - Botany, Production and Uses*, eds. Quero-García J, Iezzoni A, Puławska J, Lang G. CABI, Wallingford, pp. 305–364.

Patzak J, Henychová A, Paprštein F, Sedlák J (2019) Molecular S-genotyping of sweet cherry (*Prunus avium* L.) genetic resources. *Hortic Sci* 46:146–152. https://doi.org/10.17221/245/2017-HORTSCI.

Peace C, Bassil N, Main D, Ficklin S, Rosyara UR et al. (2012) Development and evaluation of a genome-wide 6K SNP array for diploid sweet cherry and tetraploid sour cherry. *PLoS One* 7. https://doi.org/10.1371/journal.pone.0048305.

Peace CP (2017) DNA-informed breeding of rosaceous crops: Promises, progress and prospects. *Hortic Res* 4:1–13. https://doi.org/10.1038/hortres.2017.6.

Pedersen BH (2006) DNA fingerprints of 51 sweet and sour *Prunus* accessions using simple sequence repeats. *J Hortic Sci Biotechnol* 81:118–124. https://doi.org/10.1080/14620316.2006.11512037.

Pereira S, Silva V, Bacelar E, Guedes F, Silva AP et al. (2020) Cracking in sweet cherry cultivars 'Early Bigi' and 'Lapins': Correlation with quality attributes. *Plants* 9:1–12. https://doi.org/10.3390/plants9111557.

Pinosio S, Marroni F, Zuccolo A, Vitulo N, Mariette S et al. (2020) A draft genome of sweet cherry (*Prunus avium* L.) reveals genome-wide and local effects of domestication. *Plant J* 103:1420–1432. https://doi.org/10.1111/tpj.14809.

Puławska J, Gétaz M, Kałużna M, Kuzmanović N, Obradović A et al. (2017) Bacterial Diseases. In: *Cherries: Botany, Production and Uses*, eds. Quero-García J, Iezzoni A, Puławska J, Lang G. CABI, Wallingford, pp. 365–385.

Qi X, Liu C, Song L, Li Y, Li M (2017) PaCYP78A9, a cytochrome P450, regulates fruit size in sweet cherry (*Prunus avium* L.). *Front Plant Sci* 8:1–14. https://doi.org/10.3389/fpls.2017.02076.

Quero-García J, Branchereau C, Barreneche T, Dirlewanger E (2022) DNA-informed breeding in sweet cherry: Current advances and perspectives. *Italus Hortus* 29:4–37. https://doi.org/10.26353/j.itahort/2022.1.C2.

Quero-García J, Iezzoni A, López-Ortega G, Peace C, Fouché M et al. (2019) Advances and Challenges in Cherry Breeding. In *Achieving Sustainable Cultivation of Temperate Zone Tree Fruits and Berries*, ed. Lang GA. Burleigh Dodds Science Publishing Limited, Cambridge, pp. 55–88. https://doi.org/10.19103/AS.2018.0040.17.

Quero-García J, Letourmy P, Campoy JA, Branchereau C, Malchev S et al. (2021) Multi-year analyses on three populations reveal the first stable QTLs for tolerance to rain-induced fruit cracking in sweet cherry (*Prunus avium* L.). *Hortic Res* 8. https://doi.org/10.1038/s41438-021-00571-6.

Quero-García J, Schuster M, López-Ortega G, Charlot G (2017) Sweet Cherry Varieties and Improvement. In: *Cherries: Botany, Production and Uses*, eds. Quero-García J, Iezzoni A, Puławska J, Lang G. CABI, Wallingford, pp. 60–94.

Radičević S, Ognjanov V, Marić S, Barac G (2021) The effect of genotype and temperature interaction on pollen performance in the pistils of autochthonous sour cherry cultivar "Feketićka." *Zemdirbyste Agric* 108. https://doi.org/10.13080/z-a.2021.108.035.

Radunic M (2010). Reproductive compatibility of sweet cherry cultivars (*Prunus avium* L.) PhD Thesis. Zagreb University Agriculture Faculty. 124 pp.

Rodríguez P, Galindo A, Collado-González J, Medina S, Corell M et al. (2018) Fruit response to water-scarcity scenarios. Water relations and biochemical changes. In: *Water Scarcity and Sustainable Agriculture in Semiarid Environment: Tools, Strategies, and Challenges for Woody Crops*, eds. Iván Francisco García Tejero and Víctor Hugo Durán Zuazo, Academic Press, London, pp. 349–375. https://doi.org/10.1016/B978-0-12-813164-0.00015-6.

Rosyara UR, Bink MCAM, van de Weg E, Zhang G, Wang D et al. (2013) Fruit size QTL identification and the prediction of parental QTL genotypes and breeding values in multiple pedigreed populations of sweet cherry. *Mol Breed* 32:875–887. https://doi.org/10.1007/s11032-013-9916-y.

Rothkegel K, Sánchez E, Montes C, Greve M, Tapia S et al. (2017) DNA methylation and small interference RNAs participate in the regulation of MADS-box genes involved in dormancy in sweet cherry (*Prunus avium* L.). *Tree Physiol* 37:1739–1751. https://doi.org/10.1093/treephys/tpx055.

Rugienius R, Frercks JB, Šikšnianienė B, Stepulaitienė I, Stanienė G et al. (2015) Characterisation of genetic diversity of the Lithuanian sour cherry (*Prunus cerasus* L.) genetic resources using microsatellite markers. *Acta Hortic* 1100:105–108. https://doi.org/10.17660/ActaHortic.2015.1100.14.

Sandefur P, Oraguzie N, Peace C (2016) A DNA test for routine prediction in breeding of sweet cherry fruit color, Pav-Rf-SSR. *Mol Breed* 36. https://doi.org/10.1007/s11032-016-0458-y.

Sansavini S, Lugli S (2008) Sweet cherry breeding programs in Europe and Asia. *Acta Hortic* 795(PART 1):41–57. https://doi.org/10.17660/actahortic.2008.795.1.

Sassa H, Kakui H, Miyamoto M, Suzuki Y, Hanada T et al. (2007) S Locus F-box brothers: Multiple and pollen-specific F-box genes with S haplotype-specific polymorphisms in apple and Japanese pear. *Genetics* 175:1869–1881. https://doi.org/10.1534/genetics.106.068858.

Schmidt H (1998) On the basis of fruit color in sweet cherry. *Acta Hortic* 468:77–81.

Schueler S, Tusch A, Schuster M, Ziegenhagen B (2003) Characterization of microsatellites in wild and sweet cherry (*Prunus avium* L.) - Markers for individual identification and reproductive processes. *Genome* 46:95–102. https://doi.org/10.1139/g02-107.

Schüller E, Fernández FF, Antanaviciute L, Anhalt-brüderl U, Spornberger A et al. (2021) Autochthonous Austrian varieties of *Prunus avium* l. Represent a regional gene pool, assessed using SSR and AFLP markers. *Genes (Basel)* 12:1–23. https://doi.org/10.3390/genes12030322.

Schuster M (2012) Incompatible (*S*-) genotypes of sweet cherry cultivars (*Prunus avium* L.). *Sci Hortic* 148:59–73.

Schuster M (2014) *S* allele studies in sour cherry. Results - Sour cherry breeding. 15–17 Semtember. COST Meeting Novi Sad, Serbia.

Schuster M (2017) Self-incompatibility (S) genotypes of cultivated sweet cherries - An overview 2017. Open Agar. 1–37 https://doi.org/10.5073/20171213-111734.

Schuster M (2019) Sour cherries for fresh consumption. *Acta Hortic* 1235:113–118. https://doi.org/10.17660/ActaHortic.2019.1235.15.

Schuster M (2020) Self-incompatibility (S) genotypes of cultivated sweet cherries - An overview update 2020. *OpenAgrar Repository* 2020:1–45. https://doi.org/10.5073/20201016-141600.

Schuster M, Schreiber H (2000) Genome investigation in sour cherry, *P. cerasus* L. *Acta Hortic* 538:375–379. https://doi.org/10.17660/ActaHortic.2000.538.66.

Sharma K, Sedlák P, Zeka D, Vejl P, Soukup J (2014) Allele-specific PCR detection of sweet cherry self-incompatibility alleles S3, S4 and S9 using consensus and allele-specific primers in the Czech Republic. *Hortic Sci (Prague)* 41:153–159. https://doi.org/10.17221/89/2014-HORTSCI.

Sharma K, Xuan H, Sedlák P (2015) Assessment of genetic diversity of Czech sweet cherry cultivars using microsatellite markers. *Biochem Syst Ecol* 63:6–12. https://doi.org/10.1016/j.bse.2015.09.013.

Shi S, Li J, Sun J, Yu J, Zhou S (2013) Phylogeny and classification of *Prunus sensu lato* (Rosaceae). *J Integr Plant Biol* 55:1069–1079. https://doi.org/10.1111/jipb.12095.

Shirasawa K, Isuzugawa K, Ikenaga M, Saito Y, Yamamoto T et al. (2017) The genome sequence of sweet cherry (*Prunus avium*) for use in genomics-assisted breeding. *DNA Res* 24:499–508. https://doi.org/10.1093/dnares/dsx020.

Silva NF, Goring DR (2001) Mechanisms of self-incompatibility in flowering plants. *Cell Mol Life Sci* 58:1988–2007.

Socias i Company RS, Kodad O, i Martí AF, Alonso JM (2015) Mutations conferring self-compatibility in *Prunus* species: From deletions and insertions to epigenetic alterations. *Sci Hortic* 192:125–131.

Sonneveld T, Robbins TP, Bošković R, Tobutt KR (2001) Cloning of six cherry self-incompatibility alleles and development of allele-specific PCR detection. *Theor Appl Genet* 102:1046–1055. https://doi.org/10.1007/s001220000525.

Sonneveld T, Robbins TP, Tobutt KR (2006) Improved discrimination of self-incompatibility S-RNase alleles in cherry and high throughput genotyping by automated sizing of first intron polymerase chain reaction products. *Plant Breed* 125:305–307. https://doi.org/10.1111/j.1439-0523.2006.01205.x.

Sonneveld T, Tobutt KR, Robbins TP (2003) Allele-specific PCR detection of sweet cherry self-incompatibility (S) alleles S1 to S16 using consensus and allele-specific primers. *Theor Appl Genet* 107:1059–1070. https://doi.org/10.1007/s00122-003-1274-4.

Sonneveld T, Tobutt KR, Vaughan SP, Robbins TP (2005) Loss of pollen-*S* function in two self-compatible selections of *Prunus avium* is associated with deletion/mutation of an *S* haplotype-specific F-box gene. *Plant Cell* 17:37–51. https://doi.org/10.1105/tpc.104.026963.

Sooriyapathirana SS, Khan A, Sebolt AM, Wang D, Bushakra JM et al. (2010) QTL analysis and candidate gene mapping for skin and flesh color in sweet cherry fruit (*Prunus avium* L.). *Tree Genet Genomes* 6:821–832. https://doi.org/10.1007/s11295-010-0294-x.

Sorkheh K (2015) Evolutionary Analysis of Basic RNase Genes from Rosaceous Species - S-RNase and Non-SRNase Genes. In: *Plants for the Future*, ed. El-Shemy H. InTech, London, p. 161–192.

Sosinski B, Gannavarapu M, Hager LD, Beck LE, KIng JJ et al. (2000) Characterization of microsatellite markers in peach [*Prunus persica* (L.) Batsch]. *Theor Appl Genet* 101:421–428. https://doi.org/10.1007/s001220051499.

Soundararajan P, Won SY, Kim JS (2019) Insight on Rosaceae family with genome sequencing and functional genomics perspective. *Biomed Res Int* 2019:7519687. https://doi.org/10.1155/2019/7519687.

Stanys V, Baniulis D, Morkunaite-Haimi S, Siksnianiene JB, Frercks B et al. (2012) Characterising the genetic diversity of Lithuanian sweet cherry (*Prunus avium* L.) cultivars using SSR markers. *Sci Hortic* 142:136–142. https://doi.org/10.1016/j.scienta.2012.05.011.

Stanys V, Stanytė R, Stanienė G, Vinskienė J (2008) S-allele identification by PCR analysis in Lithuanian sweet cherries. *Biologija* 54:22–26. https://doi.org/10.2478/v10054-008-0005-9.

Stegmeir T, Schuster M, Sebolt A, Rosyara U, Sundin GW et al. (2014) Cherry leaf spot resistance in cherry (*Prunus*) is associated with a quantitative trait locus on linkage group 4 inherited from *P. canescens*. *Mol Breed* 34:927–935. https://doi.org/10.1007/s11032-014-0086-3.

Stepulaitienė I, Žebrauskienė A, Stanys V (2013) Frost resistance is associated with development of sour cherry (*Prunus cerasus*) generative buds. *Zemdirbyste Agriculture* 100:175–178.

Stoeckel S, Grange J, Fernández-Manjarres JF, Bilger I, Frascaria-Lacoste N et al. (2006) Heterozygote excess in a self-incompatible and partially clonal forest tree species - *Prunus avium* L. *Mol Ecol* 15:2109–2118. https://doi.org/10.1111/j.1365-294X.2006.02926.x.

Struss D, Ahmad R, Southwick SM, Boritzki M (2003) Analysis of sweet cherry (*Prunus avium* L.) cultivars using SSR and AFLP markers. *J Am Soc Hortic Sci* 128:904–909. https://doi.org/10.21273/jashs.128.6.0904.

Struss D, Boritzki M, Karle R, Iezzoni AF (2002) Microsatellite markers differentiate eight Giessen cherry rootstocks. *HortScience* 37:191–193. https://doi.org/10.21273/hortsci.37.1.191.

Szikriszt B, Doğan A, Ercisli S, Akcay ME, Hegedűs A et al. (2013) Molecular typing of the self-incompatibility locus of Turkish sweet cherry genotypes reflects phylogenetic relationships among cherries and other *Prunus* species. *Tree Genet Genomes* 9:155–165. https://doi.org/10.1007/s11295-012-0543-2.

Szilágyi S, Horváth-Kupi T, Desiderio F, Bekefi Z (2022) Evaluation of sweet cherry (*Prunus avium* L.) cultivars for fruit size by FW_G2a QTL analysis and phenotypic characterization. *Sci Hortic* 292:110656. https://doi.org/10.1016/j.scienta.2021.110656.

Tao R, Iezzoni AF (2010) The S-RNase-based gametophytic self-incompatibility system in *Prunus* exhibits distinct genetic and molecular features. *Sci Hortic* 124:423–433. https://doi.org/10.1016/j.scienta.2010.01.025.

Tao R, Matsumoto D (2012) *S*-Locus mutation and self-compatibility in stone fruits. *Acta Hortic* 967:105–111. https://doi.org/10.17660/ActaHortic.2012.967.11.

Tao R, Yamane H, Sugiura A, Murayama H, Sassa H et al. (1999) Molecular typing of S-alleles through identification, characterization and cDNA cloning for S-RNases in sweet cherry. *J Amer Soc Hortic Sci* 124:224–233. https://doi.org/10.21273/JASHS.124.3.224.

Tavaud M, Zanetto A, David JL, Laigret F, Dirlewanger E (2004) Genetic relationships between diploid and allotetraploid cherry species (*Prunus avium*, *Prunus* × *gondouinii* and *Prunus cerasus*). *Heredity (Edinb)* 93:631–638. https://doi.org/10.1038/sj.hdy.6800589.

Testolin R, Marrazzo T, Cipriani G, Quarta R, Verde I et al. (2000) Microsatellite DNA in peach (*Prunus persica* L. Batsch) and its use in fingerprinting and testing the genetic origin of cultivars. *Genome* 43:512–520. https://doi.org/10.1139/gen-43-3-512.

Tobutt KR, Bosković R, Cerović R, Sonneveld T, Ružić D (2004a) Identification of incompatibility alleles in the tetraploid species sour cherry. *Theor Appl Genet* 108:775–785. https://doi.org/10.1007/s00122-003-1511-x.

Tobutt KR, Sonneveld T, Bekefi T, Bošković R (2004b) Cherry (in)compatibility genotypes - An updated cultivar table. *Acta Hortic* 663:667–672. https://doi.org/10.17660/ActaHortic.2004.663.120.

Tsukamoto T, Potter D, Tao R, Vieira CP, Vieira J et al. (2008) Genetic and molecular characterization of three novel S-haplotypes in sour cherry (*Prunus cerasus* L.). *J Expl Bot* 59:3169–3185. https://doi.org/10.1093/jxb/ern172.

Turet-Sayar M, Turkec A, Demir T (2012) Identification of sweet cherry cultivars (*Prunus avium* L.) and analysis of their genetic relationship using microsatellite DNA fingerprinting. *J Agric Sci* 4. https://doi.org/10.5539/jas.v4n8p134.

Turkoglu Z, Bilgener S, Ercisli S, Bakir M, Koc A et al. (2010) Simple sequence repeat-based assessment of genetic relationships among *Prunus* rootstocks. *Genet Mol Res* 9:2156–2165. https://doi.org/10.4238/vol9-4gmr957.

Upadysheva GY, Motyleva SM, Tumaeva TA, Mertvischeva ME, Upadyshev MT (2021) Study of frost resistance of generative buds of scion/rootstock combinations of sweet cherry by freezing in controlled conditions. *E3S Web Conf* 254:02012. https://doi.org/10.1051/e3sconf/202125402012.

Urbanovich OY, Kuzmitzkaya PV, Kazlouskaya ZA, Taranow AA (2014) Diversisty of SSR alleles the sour cherry cultivars (*Prunus cerasus*). Proccedings of the national akademy of sciences of Belarus. *Biol Sci Sr* 2:64–69.

Ushijima K, Sassa H, Dandekar AM, Gradziel TM, Tao R et al. (2003) Structural and transcriptional analysis of the self-incompatibility locus of almond. *Plant Cell* 15:771–781. https://doi.org/10.1105/tpc.009290.

Ushijima K, Yamane H, Watari A, Kakehi E, Ikeda K et al. (2004) The S haplotype-specific F-box protein gene, SFB, is defective in self-compatible haplotypes of *Prunus avium* and *P. mume*. *Plant J* 39:573–586. https://doi.org/10.1111/j.1365-313X.2004.02154.x.

Uzan Eken B, Kirdök E, Velioğlu E, Özden Çiftçi Y (2022) Assessment of genetic variation of natural populations of wild cherry (*Prunus avium* L.) via SSR markers. *Turk J Bot* 46:14–25. https://doi.org/10.3906/bot-2111-16.

Vanderzande S, Zheng P, Cai L, Barac G, Gasic K et al. (2020) The cherry 6+9K SNP array: A cost-effective improvement to the cherry 6K SNP array for genetic studies. *Sci Rep* 10:1–14. https://doi.org/10.1038/s41598-020-64438-x.

Vaughan SP, Bošković RI, Gisbert-Climent A, Russell K, Tobutt KR (2008) Characterisation of novel S-alleles from cherry (*Prunus avium* L.). *Tree Genet Genomes* 4:531–541. https://doi.org/10.1007/s11295-007-0129-6.

Vaughan SP, Russell K (2004) Characterization of novel microsatellites and development of multiplex PCR for large-scale population studies in wild cherry, *Prunus avium*. *Mol Ecol Notes* 4:429–431. https://doi.org/10.1111/j.1471-8286.2004.00673.x.

Vaughan SP, Russell K, Sargent DJ, Tobutt KR (2006) Isolation of S-locus F-box alleles in *Prunus avium* and their application in a novel method to determine self-incompatibility genotype. *Theor Appl Genet* 112:856–866. https://doi.org/10.1007/s00122-005-0187-9.

Vávra R, Suran P, Nekvindová V, Ždárská I, Skrivanová A et al. (2018) Sweet cherry resistance to spring frost damage at bloom stage. 18th International Conference on Organic Fruit-Growing: Proceedings of the Conference 19–21 February 2018, Hohenheim, Germany, pp. 183–187, ref. 6.

Vimont N, Beauvieux R, Arkoun M, Yvin JC, Dirlewanger E et al. (2019) The presence of H3K4me3 histone mark is positively correlated with expression at the DAM loci in sweet cherry during dormancy. *Acta Hortic* 1235:413–420. https://doi.org/10.17660/ActaHortic.2019.1235.57.

Wang B, Kumar V, Olson A, Ware D (2019) Reviving the transcriptome studies: An insight into the emergence of single-molecule transcriptome sequencing. *Front Genet* 10:1–11. https://doi.org/10.3389/fgene.2019.00384.

Wang C-L, Zhang Z-P, Tonosaki K, Kitashiba H, Nishio T (2013) S genotyping in Japanese plum and sweet cherry by allele-specific hybridization using streptavidin-coated magnetic beads. *Plant Cell Rep* 32:567–576. https://doi.org/10.1007/s00299-013-1388-3.

Wang D, Karle R, Iezzoni AF (2000) QTL analysis of flower and fruit traits in sour cherry. *Theor Appl Genet* 100:535–544. https://doi.org/10.1007/s001220050070.

Wang H, Zhang K, Su H (2010) Identification of the *S*-genotypes of several sweet cherry (*Prunus avium* L.) cultivars by AS-PCR and pollination. *Afr J Agric Res* 5:250–256.

Wang J, Liu W, Zhu D et al. (2020) Chromosome-scale genome assembly of sweet cherry (Prunus avium L.) cv. Tieton obtained using long-read and Hi-C sequencing. *Hortic Res* 7:122. https://doi.org/10.1038/s41438-020-00343-8.

Webster AD (1996) The Taxonomic Classification of Sweet and Sour Cherries and a Brief History of Their Cultivation. In: *Cherries: Crop Physiology, Production and Uses*. eds. Webster AD, Looney NE. CABI, Wallingford, pp. 3–24.

Wei H, Chen X, Zong X, Shu H, Gao D et al. (2015) Comparative transcriptome analysis of genes involved in anthocyanin biosynthesis in the red and yellow fruits of sweet cherry (*Prunus avium* L.). *PLoS One* 10:1–20. https://doi.org/10.1371/journal.pone.0121164.

Wenden B, Mariadassou M (2017) Sweet cherry phenology in the context of climate change: A systems biology approach. *Acta Hortic* 1162:31–37. https://doi.org/10.17660/ActaHortic.2017.1162.6.

Wheeler MJ, Franklin-Tong VE, Franklin FCH (2001) The molecular and genetic basis of touch. *New Phytol* 2001:59–84.

Wiersma PA, Erogul D, Ali S (2018) DNA fingerprinting of closely related cultivars of sweet cherry. *J Am Soc Hortic Sci* 143:282–288. https://doi.org/10.21273/JASHS04454-18.

Wiersma PA, Wu Z, Zhou L, Hampson C, Kappel F (2001) Identification of new self-incompatibility alleles in sweet cherry (*Prunus avium* L.) and clarification of incompatibility groups by PCR and sequencing analysis. *Theor Appl Genet* 102:700–708. https://doi.org/10.1007/s001220051700.

Winkler A, Peschel S, Kohrs K, Knoche M (2016) Rain cracking in sweet cherries is not due to excess water uptake but to localized skin phenomena. *J Am Soc Hortic Sci* 141:653–660. https://doi.org/10.21273/JASHS03937-16.

Wöhner TW, Emeriewen OF, Wittenberg AHJ, Schneiders H, Vrijenhoek I et al. (2021) The draft chromosome-level genome assembly of tetraploid ground cherry (*Prunus fruticosa* Pall.) from long reads. *Genomics* 113:4173–4183. https://doi.org/10.1016/j.ygeno.2021.11.002.

Wünsch A, Hormaza JI (2002) Molecular characterisation of sweet cherry (*Prunus avium* l.) genotypes using peach [*Prunus persica* (L.) Batsch] SSR sequences. *Heredity (Edinb)* 89:56–63. https://doi.org/10.1038/sj.hdy.6800101.

Wünsch A, Hormaza JI (2004) Cloning and characterization of genomic DNA sequences of four self-incompatibility alleles in sweet cherry (*Prunus avium* L.). *Theor Appl Genet* 108:299–305. https://doi.org/10.1007/s00122-003-1418-6.

Wünsch A, Tao R, Hormaza JI (2010) Self-compatibility in 'Cristobalina' sweet cherry is not associated with duplications or modified transcription levels of S-locus genes. *Plant Cell Rep* 29:715–721. https://doi.org/10.1007/s00299-010-0857-1.

Xanthopoulou A, Manioudaki M, Bazakos C, Kissoudis C, Farsakoglou AM et al. (2020) Whole genome re-sequencing of sweet cherry (*Prunus avium* L.) yields insights into genomic diversity of a fruit species. *Hortic Res* 7:1–14. https://doi.org/10.1038/s41438-020-0281-9.

Xuan H, Wang R, Büchele M, Hartmann W, Möller O (2009) Microsatellite markers (SSR) as a tool to assist in identification of sweet (*Prunus avium*) and sour cherry (*Prunus cerasus*). *Acta Hortic* 839:507–514. https://doi.org/10.17660/ActaHortic.2009.839.69.

Yamane H, Ikeda K, Hauck NR, Iezzoni AF, Tao R (2003a) Self-incompatibility (*S*) locus region of the mutated *S6*-haplotype of sour cherry (*Prunus cerasus*) contains a functional pollen *S* allele and a non-functional pistil S allele. *J Exp Bot* 54:2431–2437. https://doi.org/10.1093/jxb/erg271.

Yamane H, Ikeda K, Ushijima K, Sassa H, Tao R (2003b) A pollen-expressed gene for a novel protein with an F-box motif that is very tightly linked to a gene for S-RNase in two species of cherry, *Prunus cerasus* and *P. avium*. *Plant Cell Physiol* 44:764–769. https://doi.org/10.1093/pcp/pcg088.

Yamane H, Tao R (2009) Molecular basis of self-(in)compatibility and current status of *S*-genotyping in Rosaceous fruit trees. *J Jpn Soc Hort Sci* 78:137–157. https://doi.org/10.2503/jjshs1.78.137.

Yamane H, Tao R, Mori H, Sugiura A (2003c) Identification of a non-*S* RNase, a possible ancestral form of S-RNases, in *Prunus*. *Mol Genet Genom* 269:90–100. https://doi.org/10.1007/s00438-003-0815-5.

Yamane H, Tao R, Murayama H, Ishiguro M, Abe Y (2000a) Determining *S*-genotypes of two sweet cherry (*Prunus avium* L.) cultivars, "Takasago (Rockport Bigarreau)" and "Hinode (Early Purple)." *Engei Gakkai Zasshi* 69:29–34. https://doi.org/10.2503/jjshs.69.29.

Yamane H, Tao R, Sugiura A, Hauck NR, Iezzoni AF (2000b) Identification and characterization of *S-RNases* in tetraploid sour cherry (*Prunus cerasus*). *J Am Soc Hortic Sci* 126:661. https://doi.org/10.21273/JASHS.126.6.661.

Zambounis A, Ganopoulos I, Avramidou E, Aravanopoulos FA, Tsaftaris A et al. (2016) Evidence of extensive positive selection acting on cherry (*Prunus avium* L.) resistance gene analogs (RGAs). *Aust J Crop Sci* 10:1324–1329. https://doi.org/10.21475/ajcs.2016.10.09.p7825.

Zhang G, Sebolt AM, Sooriyapathirana SS, Wang D, Bink MC et al. (2010) Fruit size QTL analysis of an F1 population derived from a cross between a domesticated sweet cherry cultivar and a wild forest sweet cherry. *Tree Genet Genomes* 6:25–36. https://doi.org/10.1007/s11295-009-0225-x.

Zhao YY, Mgbechi-Ezeri JU, Johnson KB, Rosyara U, Iezzoni AF et al. (2014a) Identification of QTL underlying powdery mildew and bacterial canker infection in sweet cherry (*Prunus avium* L.). ASHS conference oral abstract. https://ashs.confex.com/ashs/2014/webprogram/Paper19014.html

Zhao YY, Rosyara U, Iezzoni AF, Peace C, Whiting M et al. (2014b) Identification of QTL underlying soluble solids content and titratable acidity in sweet cherry (*Prunus avium* L.). ASHS conference oral abstract.

Zheng N, Schulman BA, Song L, Miller JJ, Jeffrey PD et al. (2002) Structure of the *Cul1-Rbx1-Skp1-F box-Skp2 SCF* ubiquitin ligase complex. *Nature* 416:703–709. https://doi.org/10.1038/416703a.

Zhu M, Zhang X, Zhang K, Jiang L, Zhang L (2004) Development of a simple molecular marker specific for detecting the self-compatible *S4′* haplotype in sweet cherry (*Prunus avium* L.). *Plant Mol Biol Rep* 22:387–398. https://doi.org/10.1007/BF02772681.

8 Allele Mining in Tropical Fruit Crop *Carica papaya* L.

Current Scenario and Future Perspective

Fawad Ali and Rebecca Ford

8.1 INTRODUCTION

Papaya (*Carica papaya* L.) belongs to the family *Caricaceae*. Papaya is a dicotyledonous plant with soft-wooded herbaceous features (Bremerk et al., 1998). The papaya genome is three times Arabidopsis and shares a common lineage dating back to 72 million years ago (Wikstro et al., 2001). The *Caricaceae* family is spread over six genera. *Carica papaya* is the only specie that is used for fruit production worldwide in the tropics and subtropical climates. *Carica papaya* has a diploid genome of 372 Mb, three times that of Arabidopsis (Arumuganathan and Earle, 1991; Ming et al., 2008). Papaya genes are much involved with starch metabolism and volatile pathways; therefore, has greater importance for sensory profiling and that need allele mining for discovering key allelic variants. Due to its smaller genome size, papaya has significant potential for genomic selection by studying allele mining and learning allelic variants controlling consumer-driven fruit quality traits.

Allele mining is a pathway to genomic selection (Kumar et al., 2010) and offers a platform to increase the genetic bottleneck of a targeted crop. In papaya, the first reference genome was published in 2008 (Ming et al., 2008) of transgenic red papaya 'SunUp'. However, natural diversity in papaya stretches into the yellow-and-red papaya, of which red papaya is cultivated widely across the globe (International Plant Genetic Resources Institute, 1988). The differential characteristics of red-and-yellow papaya are fruit size and shape, flesh color, sweetness, and flesh consistency (Zhou et al., 2022). Still, the differences in tree agronomic characteristics have not been explored yet. These phenotypic variations of two distinct papaya genomes (red and yellow) need to explore. Therefore, allele mining (Kumar et al., 2010) can offer a platform for distinguishing the genetic sweeps linked to each type of papaya genome, i.e., red-and-yellow varieties.

Success in selective breeding depends on the natural diversity linked to genetic recombination in plant species deployed over time. Other than that, selection pressure exerted by humans and nature narrows down the genetic bottleneck because of the loss of several genes during human selection or domestication. Most untapped genetic variation exists in wild germplasm (DeVicente and Tanksley, 1993; Xiao et al., 1998; McCouch et al., 2007; Fujii et al., 2021), and therefore, we can always increase genetic diversity by hybridizing the cultivated plant species with wild types. Hence, broader germplasm data sets are required for creating allelic diversity. That diversity could be captured through allele mining by re-visiting papaya genomes to improve desirable agronomic and consumer-driven fruit quality traits.

To our knowledge, allele mining studies in papaya have yet to be explored. They require future findings of the larger gene pool, including the diversity panel for finding the genetic diversity of red-and-yellow papaya.

DOI: 10.1201/9781003386490-8

8.2 THE METHODS USED AND DEVELOPED FOR ALLELE MINING

The most widely used methods for discovering key allelic variants controlling fundamental gene expression have been studied across different species, including maize, rice, wheat, and brassica (Kumar et al., 2010). In tropical fruit crops, allelic composition in citrus (*Citrus* sps; Fujii et al., 2021) demonstrated the composition of carotenoids in fruit sap. In Avocado (*Persea americana* Mill; Rubinstein et al., 2019), the single nucleotide polymorphism (SNP) platform was discovered using the pools of genomic DNA using whole-genome sequencing of three race types, Guatemalan, Mexican, and West Indian. Eco-TILLING and sequencing-based allele mining has been promising approaches (Kumar et al., 2010). However, sequence-based allele mining proved to be cost-effective, less laborious, and highly precise in detecting haplotype structure (Rubinstein et al., 2019; Fujii et al., 2021). Sequence-based allele mining also offers the understanding to infer genotype-to-phenotype association based on candidate gene analysis. There is limited knowledge in papaya in discovering allelic variation linked to targeted phenotype by using similar approaches within the existing gene pool. Therefore, allele mining in papaya would help us to identify in tracing the evolutionary pattern of allelic variants of an essential gene controlling phenotypic expression coupled with identifying new haplotypes. That would allow us to develop allele-specific genetic markers (Figure 8.1) for deployment in a marker-assisted selection in the papaya breeding program.

8.3 STEPS TOWARD ALLELE MINING DISCOVERY PLATFORM IN RED-AND-YELLOW PAPAYA

8.3.1 Diversity of Papaya Germplasm

Papaya germplasm is diverse and sweeps across the red-and-yellow papaya. Both types of papaya are unique in characteristics related to tree agronomy (height to the first fruit, trunk circumference, root structure, marketable fruit, and wasted fruit) and fruit quality traits (fruit shape and

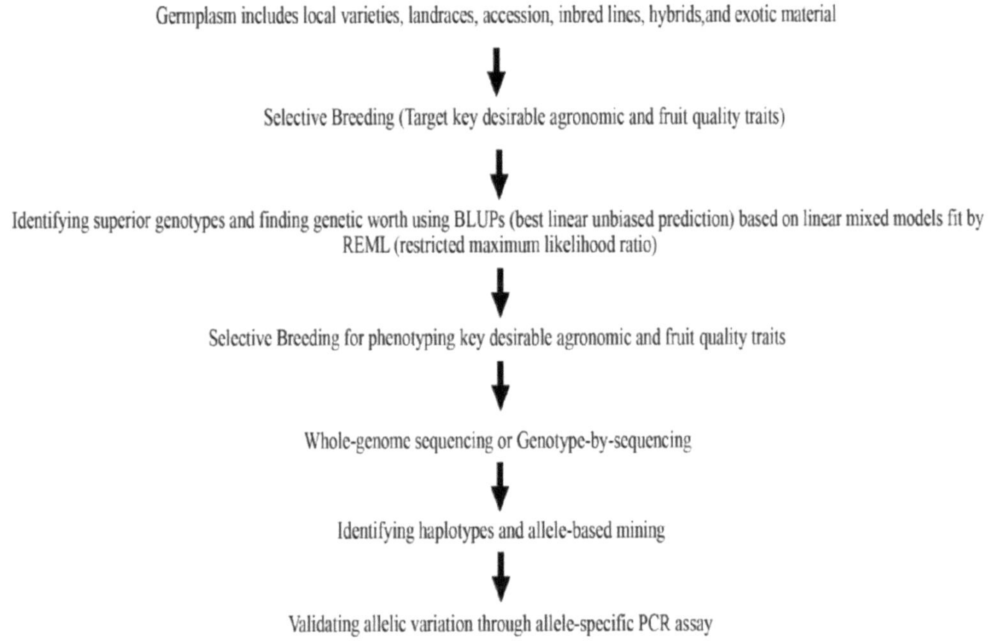

FIGURE 8.1 Selective breeding proposed methods for discovering allelic variations across the papaya genome.

size, skin color, skin glossiness, flesh color, flesh consistency, and flesh sweetness) (International Plant Genetic Resources Institute, 1988; Kanchana-udomkon and Ford, 2021). Sensory profile data also demonstrate the distinctness of red-and-yellow papaya in terms of sensory metabolites (Zhou et al., 2022). Therefore, understanding the inheritance pattern of desirable fruit quality traits in red-and-yellow papaya genotypes would help us to tease apart the phenotypic variation linked to each genetic background and would help us in entailing to link the genomic loci.

8.3.2 Plant Biology and Genetic Structure of Papaya

The chromosome number of the papaya (*Carica papaya* L.) is $2n=18$, and there are no known polyploid varieties (Chan, 2009; Nantawan et al., 2019). Considering papaya ploidy level, which is diploid (Arumuganathan and Earle, 1991; Ming et al., 2008), the natural crossing system exist with three distinct sexes: female, male, and hermaphrodites (Hofmeyr, 1938; Storey, 1941; Lange, 1961), but the sex reversal occurs under the influence of the environment (Storey, 1958). A pair of sex chromosomes distinguishes female, male and hermaphrodites (XX, XY, XY^h, respectively) (VanBuren et al., 2015; Table 8.1). A sex-determination polymerase chain reaction (PCR) assay in papaya has been successfully developed to distinguish hermaphrodites from female to male plants (Kanchana-udomkan et al., 2018). Red papaya has either a hermaphrodite tree or a male, while yellow papaya has a female or male tree. Therefore, the production pattern at the farm level in red papaya requires a hermaphrodite tree to produce fruit. However, in yellow papaya, female and male trees are needed to produce fruit (female to male ratio; 10:1; Ross et al., 2000). The success of detecting the allelic variation in papaya for differentiation of the hermaphrodite-specific region of the Y^h chromosome to the male-specific region of the Y chromosome demonstrated that the differences in the gene content are only 0.4% (VanBuren et al., 2015). Thus, establishing a platform for detecting allelic variation of key fruit quality traits in papaya could be guided by a similar approach used for sex discretion.

8.3.3 Fruit Size and Shape

Fruit size is assessed by the fruit weight (g) and by measuring the fruit length and width. Fruit shape is visually evaluated, and the most familiar shapes in red-and-yellow papaya genotypes are found as globular, round, elliptic, oval, oblong, elongate, cylindrical, pear, club, blossom end tapered, acron,

TABLE 8.1
Genetics of Sex Segregation Pattern in Papaya

		Segregation Ratios			
Numbering	Pollination Combination	Female Flower (mm)	Hermaphrodite or Bi-sexual Flower (M^Hm)	Male Flower (Mm)	Non-viable Zygotes (MM, M^H, M, M^HM^H)
1.	Mm × Mm	1		1	
2.	Mm × M^Hm	1	1		
3.	Mm (selfed)	1		2	1
4.	Mm × Mm	1		2	1
5.	M^Hm (selfed)	1	2		1
6.	M^Hm × M^Hm	1	2		1
7.	M^Hm × Mm	1	1	1	1
8.	Mm × M^Hm	1	1	1	1

Source: Hofmeyr (1938) and Chan (2009).

reniform, turbinate inferior, plum, and irregular shaped (International Plant Genetic Resources Institute, 1988; Kanchana-udomkan and Ford, 2021). Fruit size and shape that features the fruit weight are quantitative traits and bear polygenic inheritance (Ali et al., 2022). There are published examples of genetic mapping to detect quantitative trait loci (QTL) for fruit shape and size in tropical fruit crops. For instance, in citrus (*Citrus sinensis*; Curtolo et al., 2017), five QTLs were mapped on chromosomes 5 and 8; in apple (*Malus domestica* Borkh; Kostick and Luby, 2022), two QTLs were mapped on chromosomes 8 and 16; in mango (*Mangifera indica*; Bally et al., 2021), two QTLs were mapped on chromosomes (4 and 7); and two QTLs for fruit shape were mapped in peach (*Prunus persica* L.; Cirilli et al., 2021) on chromosomes 5 and 6. However, in papaya, fourteen QTLs were mapped on chromosomes 2, 3, 4, 6, 7, and 9 for fruit size and shape (Blas et al., 2010), and 21 QTLs (on chromosomes 1, 3, 4, 6, 7, 9, and 10; Nantawan et al., 2019) were underpinning the key desirable fruit size and shape, and fruit quality traits in red papaya. These studies support the evidence that natural variations in fruit size and shape in papaya (International Plant Genetic Resources Institute, 1988) are primarily controlled by significant genomic loci spread across the genome. Therefore, future studies are required in papaya using genome sequencing methods to narrow down the QTL region for candidate gene analysis and to pin out the key allelic variants of key genes linked to targeted fruit size and shape.

8.3.4 Fruit Flesh Parameters and Sweetness

Fruit quality traits include flesh color, firmness, flesh thickness, and soluble solid contents (°Brix) (Blas et al., 2010; Kanchana-udomkan et al., 2018; Ali et al., 2022). Soluble solid contents (°Brix) are a quantitatively inherited trait in papaya (Ali et al., 2022) and in other tropical fruit crops, including peach (*Prunus persica* L.; Etienne et al., 2002), apple (*Malus domestica* Borkh; Guan et al., 2015), and strawberry (*Fragaria ananassa*; Lerceteau-Köhler et al., 2006). Soluble solid contents (°Brix) affiliated with fruit sweetness and QTL were detected in red papaya in a set of bi-parental mapping populations on chromosomes 3 and 7 (Nantawan et al., 2019). These studies provide a platform to associate genetic loci with targeting phenotypes and exploring the candidate gene and allelic base variation.

8.4 SELECTING CORRECT ALLELES IN PAPAYA

Genomic selection is based on selecting the true alleles linked to target phenotypes. Determining the true alleles requires a systematic breeding approach (Figure 8.2) by understanding a targeted crop's genetic diversity, using precise phenotyping tools, and employing a robust reference genome for exploring QTL and SNP of the targeted trait. In papaya, the genetic diversity extends into the red-and-yellow papaya types; both are unique in characteristics in terms of the sex type of the plant for the production system and key desirable agronomic and consumer-driven fruit quality traits. Selecting the correct allele in papaya depends on understanding the genetic structure of red-and-yellow papaya genotypes and the type of available reference genome (Ming et al., 2008). Therefore, expanding the availability of accurate reference genomes in papaya would have the capacity to cover the polymorphic diversity in both types of papaya (red-and-yellow papaya genotypes) and establish an allele mining platform.

8.5 GENOTYPE (G) × ENVIRONMENT (E) INTERACTION FOR ALLELE STABILITY

Understanding the genotype (G)×environment (E) interaction is key to determining the robust QTL and SNP of a targeted trait that may work with changing environmental conditions under which the plants were grown. Genetic stabilization of targeted traits requires multi-environmental trials to test the effect of each ecological niche on the attributed variance component of a targeted feature. Thus, the phenotypic stability of targeted characteristics that could be measured by estimating the

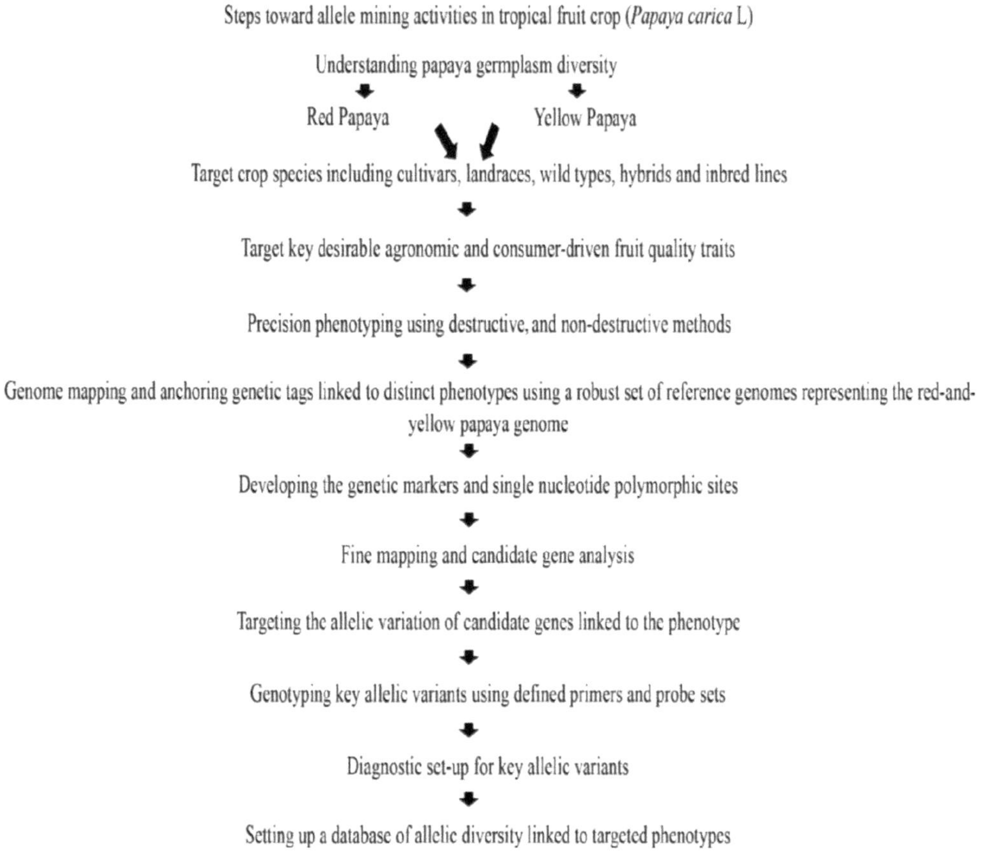

FIGURE 8.2 Steps toward allele mining discovery platform in red-and-yellow papaya.

heritability (Griffiths et al., 2000; Ali et al., 2017, 2022) is required to underpin the more significant amount of the genetic components. The traits with higher heritability (>0.60) are less likely affected by the environment; therefore, this provides an opportunity to identify the QTL and SNP to determine the genetic stabilization of key desired traits. Further, investigating the allelic variation linked to the genetic loci may help us to establish a robust allele mining platform in papaya for consumer-driven fruit quality traits.

8.6 GENETIC POLYMORPHISM DISCOVERY IN PAPAYA TOWARD MOLECULAR ASSISTED BREEDING

To shape the future papaya and improve fruit quality traits in papaya, the breeding program needs investment to explore the natural genetic variants linked to key fruit quality traits. There have been published examples underpinning the SNP using whole-genome sequencing approaches. For example, the two red papaya genomes ('Eksotika' and 'Sekaki') were aligned using 'Solo' and 'Sunup' as the reference genome for SNP discovery (Rabiatul Adawiah et al., 2016). In another study, the SNP discovery platform identified 106 ripening-related genes in red papaya cultivars ('Sekaki' and 'JS-12'; Bohry et al., 2021). No published example of SNP discovery for the yellow papaya genome exists. However, the sensory profile (Zhou et al., 2022) differentiates red-and-yellow papaya cultivars for flavor profiling and sugar-related genes. To our knowledge, there is still a gap in developing a reference genome that may truly represent the yellow papaya genome. In retrospect red papaya reference genome and SNP identified across red papaya cultivars would help us in future studies to further validate the identified SNP and develop a more robust platform of SNP discovery and allelic variation for consumer-driven fruit quality traits. The allele mining approach would facilitate efficient genetic mapping and association mapping studies to apply marker-assisted selection for selecting the superior genotypes with better fruit quality traits in the papaya breeding program.

8.7 CONCLUSION AND PROSPECTS

Due to the unique perspective and extensive use of whole-genome sequencing, allele mining paved the efficient way to link genotype to phenotype. However, phenotyping is as vital as the genotype because an efficient and accurate phenotyping platform developed across a range of papaya germplasm sets the measuring standards of a targeted trait that entails a link to genomic loci (Kumar et al., 2010). Therefore, the efficient use of advanced phenomics tools, including non-destructive methods, would facilitate the teasing apart of minor variations in targeted fruit quality traits in a gene pool of red or yellow papaya genotypes. To efficiently use ever-increasing sequencing data and crop gene banks, developing an allele mining platform in papaya for strategic selective breeding based on the targeted selection guided by either industry or consumer is imperative. To achieve this, sophisticated and targeted selection strategies must be employed in the breeding program coupled with advanced statistical genetics tools embedded with core computational skills representing the correct phenotyping and genotyping methods (both qualitative and quantitative in nature).

REFERENCES

Ali F, Ahsan M, Ali Q, Kanwal N (2017) Phenotypic stability of *zea mays* grain yield and its attributing traits under drought stress. *Front Plant Sci* 8:1397. https://doi.org/10.3389/fpls.2017.01397.

Ali F, Kanchana-udomkan C, Ford R (2022) The inheritance pattern of key desirable agronomic and fruit quality traits in elite red papaya genotypes. *Horticulturae* 8:845. https://doi.org/10.3390/.

Arumuganathan K, Earle ED (1991) Nuclear DNA content of some important plant species. *Plant Mol Biol Rep* 9:208–218.

Bally ISE, Bombarely A, Chambers AH, Cohen Y, Dillon NL, Innes DJ, Islas-Osuna MA, Kuhn DA, Muelle, LA, Ophir R, Rambani A, Sherman A, Yan H (2021) Mango genome consortium, the 'Tommy Atkins' mango genome reveals candidate genes for fruit quality. *BMC Plant Biol* 21:108. https://doi.org/10.1186/s12870-021-02858-1.

Blas AL, Ming R, Liu Z, Veatch OJ, Paull RE, Moore PH, Yu Q (2010) Cloning of the papaya chromoplast-specific lycopene β-cyclase, CpCYC-b, controlling fruit flesh colour reveals conserved microsynteny and a recombination hot spot. *Plant Physiol* 152:2013–2022.

Bohry D, Ramos HCC, Dos Santos PHD, Boechat MSB, Arêdes FAS, Pirovani AAV, Pereira MG (2021) Discovery of SNPs and InDels in papaya genotypes and its potential for marker-assisted selection of fruit quality traits. *Sci Rep* 11:292. https://doi.org/10.1038/s41598-020-79401-z.

Bremerk K, Chase MW, Stevens PF (1998) An ordinal classification for the families of flowering plants. *Ann Mo Bot Gard* 85:531–553.

Chan YK (2009) Breeding Papaya (*Carica papaya* L.). In: Jain SM, Priyadarshan PM (eds) *Breeding Plantation Tree Crops: Tropical Species*. Springer, New York. https://doi.org/10.1007/978-0-387-71201-7_4.

Cirilli M, Baccichet I, Chiozzotto R, Silvestri C, Rossini L, Bassi D (2021) Genetic and phenotypic analyses reveal major quantitative loci associated to fruit size and shape traits in a non-flat peach collection (*P. persica* L. Batsch). *Hortic Res* 8:232. https://doi.org/10.1038/s41438-021-00661-5.

Curtolo M, Cristofani-Yaly M, Gazaffi R, Takita MA, Figueira A, Machado MA (2017) QTL mapping for fruit quality in Citrus using DArTseq markers. *BMC Genomics* 18:289. https://doi.org/10.1186/s12864-017-3629-2.

DeVicente MC, Tanksley SD (1993) QTL analysis of transgressive segregation in an interspecific tomato cross. *Genetics* 134:585–596.

Etienne C, Rothan C, Moing A, Plomion C, Bodenes C, Svanella-Dumas L, Cosson P, Pronier V, Monet R, Dirlewanger E (2002) Candidate genes and QTLs for sugar and organic acid content in peach [(Prunus persica L.) Batsch. *Theor Appl Genet* 105:145–159.

Fujii H, Nonaka K, Minamikawa MF, Endo T, Sugiyama A, Hamazaki K, Iwata H, Omura M, Shimada T (2021) Allelic composition of carotenoid metabolic genes in 13 founders influences carotenoid composition in juice sac tissues of fruits among Japanese citrus breeding population. *PLoS One* 16:e0246468. https://doi.org/10.1371/journal.pone.0246468.

Griffiths JF, Griffiths AJ, Wessler SR, Lewontin RC, Gelbart WM, Suzuki DT, Miller JH (2000) Quantifying Heritability. In: *An Introduction to Genetic Analysis*, 7th ed; W.H. Freeman, New York. https://www.ncbi.nlm.nih.gov/books/ NBK21866.

Guan Y, Peace C, Rudell D, Verma S, Evans K (2015) QTLs detected for individual sugars and soluble solids content in apple. *Mol Breed* 35:135.

Hofmeyr JDJ (1938) Genetic studies of *Carica papaya* L. I. The inheritance and elation of sex and certain plant characteristics. II. Sex reversal and sex forms. *S Afr Dept and Sci Bul No 187* 1938:64.

International Plant Genetic Resources Institute (1988) *Descriptors for Papaya.* International Board for Plant Genetic Resource, Rome, Italy. ISBN. 9290431296.

Kanchana-udomkan C, Ford R (2021) *Papaya Evaluation Handbook*, 2nd Ed; (PP18000), Horticulture Innovation, Sydney, pp. 1–30.

Kanchana-udomkan C, Nantawan U, Drew R, Ford R (2018) Molecular marker-assisted papaya sex determination for improved grower efficiency. *Acta Hortic* 1205:697–704. https://doi.org/10.17660/ActaHortic.2018.1205.86.

Kostick SA, Luby JJ (2022) Apple fruit size QTLs on chromosomes 8 and 16 characterized in 'honeycrisp'-derived germplasm. *Agronomy* 12:1279. https://doi.org/10.3390/agronomy12061279.

Kumar GR, Sakthivel K, Sundaram RM, Neeraja CN, Balachandran SM, Rani NS, Viraktamath BC, Madhav MS (2010) Allele mining in crops: prospects and potentials. *Biotechnol Adv* 28:451–61. https://doi.org/10.1016/j.biotechadv.2010.02.007.

Lange AH (1961) Factors affecting sex change in the flowers of *Carica papaya* (L.). *Proc Amer Soc Hort Sci* 77:252–264.

Lerceteau-Köhler E, Moing A, Guérin G, Renaud C, Maucourt M, Rolin D (2006) QTL analysis for sugars and organic acids in strawberry fruits. *Acta Hortic* 708:573–577.

McCouch SR, Sweeney M, Li J, Jiang H, Thomson M, Septiningsih E, Edwards J, Moncada P, Xiao J, Garris A, Tai T, Martinez C, Tohme J, Sugiono M, McClung A, Yuan LP, Ahn SN (2007) Through the genetic bottleneck: *O. rufipogon* as a source of trait-enhancing alleles for *O. sativa. Euphytica* 154:317–339.

Ming R, Hou S, Feng Y, Yu Q, Dionne-Laporte A, Saw JH, Senin P, Wang W, Ly BV, Lewis KL, Salzberg SL, Feng L, Jones MR, Skelton RL, Murray JE, Chen C, Qian W, Shen J, Du P, Eustice M, Tong E, Tang H, Lyons E, Paull RE, Michael TP, Wall K, Rice DW, Albert H, Wang ML, Zhu YJ (2008) The draft genome of the transgenic tropical fruit tree papaya (*Carica papaya* Linnaeus). *Nature* 452:991–996.

Nantawan U, Kanchana-udomkan C, Bar I, Ford R (2019) Linkage mapping and quantitative trait loci analysis of sweetness and other fruit quality traits in papaya. *BMC Plant Biol* 19:449 https://doi.org/10.1186/s12870-019-2043-0.

Rabiatul Adawiah ZA, Norliza AB, Muhammad Fairuz Y, Norzihan A, Umi Kalsom AB (2016) Sequence information on single nucleotide polymorphism (SNP) through genome sequencing analysis of *Carica papaya* variety Eksotika and Sekaki. *J Trop Agric For Sci* 44: 219–222.

Ross P, Grice K, O'Hare P, Vawdrey L, Macleod N, Richards N, Kernot I, Blair A, Evans D, Astridge D, Chay-Prove P (2000) *Pawpaw Information Kit. Agrilink, your growing guide to better farming guide. Manual.* Agrilink Series QAL9910. Department of Primary Industries, Queensland Horticulture Institute. Brisbane.

Rubinstein M, Eshed R, Rozen A, Zviran T, Kuhn DN, Irihimovitch V, Sherman A, Ophir R (2019) Genetic diversity of avocado (*Persea americana* Mill.) germplasm using pooled sequencing. *BMC Genomics* 20:379. https://doi.org/10.1186/s12864-019-5672-7.

Storey WB (1941) Genetics of sex determination in papaya. *Hawaii Agr Exp Sta Ann Rep* 52–53.

Storey WB (1958) Modifications of sex expression in papaya. *Horti Adv* 2:49–60.

VanBuren R, Zeng F, Chen C, Zhang J, Wai CM, Han J, Aryal R, Gschwend AR, Wang J, Na JK, Huang L, Zhang L, Miao W, Gou J, Arro J, Guyot R, Moore RC, Wang ML, Zee F, Charlesworth D, Moore PH, Yu Q, Ming R (2015) Origin and domestication of papaya Yh chromosome. *Genome Res* 25:524–533. https://doi.org/10.1101/gr.183905.114.

Wikstro MN, Savolainen V, Chase MW (2001) Evolution of the angiosperms: calibrating the family tree. *Proc R Soc Lond B* 268:2211–2220.

Xiao J, Li J, Grandillo S, Ahn SN, Yuan L, Tanksley SD, McCouch SR (1998) Identification of trait-improving quantitative trait loci alleles from a wild rice relative, *Oryza rufipogon Genetics.* 150:899–909. https://doi.org/10.1093/genetics/150.2.899.

Zhou Z, Bar I, Ford R, Smyth H, Kanchana-udomkan C (2022) Biochemical, sensory, and molecular evaluation of flavour and consumer acceptability in Australian papaya (*Carica papaya* L.) varieties. *Int J Mol Sci.* 23:6313. https://doi.org/10.3390/ijms23116313.

9 Allele Mining for Genome Designing of Pear

Gilad Gabay

9.1 INTRODUCTION

Pear (*Pyrus* spp.) is a member of the *Rosaceae* family and subfamily *Pomoideae*. *Pyrus* is a highly diverse genus which includes 22 primary species ranging. The three predominant species are European pear, *Pyrus communis* L.; Asian sand pear, *Pyrus pyrifolia* (Burm.) Nak.; and the Chinese white pear, *Pyrus bretschneideri* Rehd (Bell, 1990). The majority of the cultivated pears are diploids ($2n=34$) consisting of 17 chromosomes (Chagné et al., 2014; Gabay et al., 2018a; Wu et al., 2013). Pear has an important economic value as a dominant deciduous fruit tree grown in temperate climates (Niu et al., 2016). It was cultivated on a vast range of climatic conditions on six continents with massive production in countries that hold the optimal climate for pear growing such as China, the United States, Italy, Argentina, and Spain (Wu et al., 2013).

Fruit tree breeding is a lengthy process with many limitations such as long juvenile periods, open pollination, self-incompatibility, and high heterozygosity levels. Some of the traits related to disease resistance and vegetative growth, such as chilling requirements, can be evaluated early in the tree life (van de Weg et al., 2018; Zurn et al., 2020; Montanari et al., 2016; Bell, 2019). Pears are perennial species with a long juvenile period (<7 years) making it difficult for pear geneticist and breeders to determine genes controlling important agricultural traits. Hence, the evolution of fruit quality traits can occur when the trees are 7 years old and in some cases even later. A traditional breeding procedure of pear normally takes at least 13–15 years. In comparison with other fruit crops, traditional pear breeding is an expensive and time-consuming process as young-seedling take longer to come into the stage of fruit formation. In addition, the juvenile trees produce many spines, and as a result, harvest and orchard management are more complicated (Brewer and Volz, 2019). Therefore, it is critical to develop genetic markers for important agronomic traits to reduce the time required to evaluate new cultivars by enabling the evaluation of genotypes at the early seedling stage (Wu et al., 2014; Zhang et al., 2021). Most of the quantitative trait loci (QTLs) that were detected in pear, are related to fruit traits, disease resistance, and vegetative traits. It is important to indicate that some of these loci are localized either within the same or in adjacent regions of a genetic linkage map, depending on years of testing, location or populations used. Therefore, those genetic markers are accurate for a specific population at a specific location. In addition, the identified QTLs are powerful tools to study the genetic mechanism which underlies the trait and can lead to discover the genes that regulate the studied trait.

In comparison with other plant species, the construction of the genetic maps in pear started relatively late (in 2001) (Wu and Qin, 2019). The completion of pear whole-genome sequencing (Wu et al., 2014; Chagné, 2015) provided a valuable genomic resource and paved the path to advances in genetic maps which also include high-resolution genetic map construction (Kumar et al., 2017; Gabay et al., 2018a; Li et al., 2017). This progress enables to carry out genome-wide association studies (GWAS) which greatly contribute to the development of molecular breeding markers and reveal the genetic mechanisms underlying important agricultural traits. The genetic maps that were

DOI: 10.1201/9781003386490-9

constructed in pear contain 17 linkage groups (LGs) which correspond to the 17 Chromosomes in pear (Kumar et al., 2017; Zurn et al., 2020; Zhang et al., 2021).

Apple and pear share large genomic similarities and therefore apple genetic studies are a valuable resource to identify genes and QTLs associated with important agricultural traits. The genomic resources of apple, including the availability of the apple sequenced genome preceded pear genomic resources (Chagné, 2015). Therefore, the identified apple genetic markers were applied in several pear studies (Celton et al., 2009). For example, the markers for genomic selection which were identified in pear for traits such as resistance to fire blight (Le Roux et al., 2012) and fruit softening (Costa et al., 2008), were first found in apple and then in the same genomic region in pear, indicating synteny between the two species (Celton et al., 2009). In some cases, traits in pears are determined by similar genes as in apple. However, other studies indicated differences in the location of the identified QTLs associated with a specific trait. Hence, there are dissimilarities in marker locations associated with important agricultural traits and therefore the genetic markers cannot be applicable between apple and pear species. These differences highlight the importance of independent pear research and the development of pear genomic resources (Yamamoto and Terakami, 2016) to develop pear-specific genetic markers (Gabay et al., 2018a).

The feasibility of next-generation sequencing (NGS) created new possibilities to explore the functions of the genes underlying identified QTLs such as the analysis of whole transcriptome (Gabay et al., 2018b; Bai et al., 2013). The related allelic variants of those genes provided the base to develop genetic markers for use in marker-assisted selection (MAS) (De Franceschi and Dondini, 2019). However, the validation of the function of the identified genes is not a common or feasible practice in pear genetic studies. Due to the long juvenile period, functional gene studies can last more than 7 years until the phenotype is evaluated. However, recent advances in transformation technologies created opportunities to carry out functional gene studies in pear such as a recent study, that showed the validation of a gene (*PbrSTONE*) function in the regulation of stone cell formation, a highly important fruit quality trait in pear (Zhang et al., 2021).

This chapter provides a review of the updated genetic knowledge and includes genes and QTLs identified in *Pyrus* species to support future breeding efforts and functional genomics studies in pear.

9.2 KEY TRAITS FOR SELECTION

One of the most important criteria to select the target trait for breeding is the heritability of the trait that indicates the breeding potential and its genotypic component relative to the environmental effect. Hence, a trait with high breeding potential is determined mostly by genetic factors (De Franceschi and Dondini, 2019). Therefore, breeders will select to target traits with high heritability. Broad-sense heritability (H^2) is accurate only for specific environmental conditions and populations (Souza et al., 1998). H^2 estimations are higher within locations, and specific to location. Hence, the genetic factor interaction with the environment (GxE) can change the performance of a specific individual between locations. However, it is likely that a certain genetic line can show similar phenotypes between years within the same location. Since QTL can be detected only for traits with significant genotype effect (Ben Sadok et al., 2013; Yamamoto et al., 2014), QTLs are valuable information for breeders. Most of QTL studies that were conducted in *Pyrus* used F1 populations. Unlike annuals where a QTL can be validated using advanced genetic populations such as heterozygous inbred families (HIFs), in deciduous trees a validation can be achieved by comparing two or more different studies using similar or different parental genetic background in different years and different locations. Therefore, in this section, QTLs which were identified in multiple studies are highlighted (Table 9.1).

TABLE 9.1
Summary of Selected Published QTLs for Pear

Fruit Quality					
LG[a]	Position (cM)[b]	LOD	Variance[c]	References	Trait
6	20,720,912 bp	15.7	13.1	Nishio et al. (2021)	Acidity
7	33,139,082 bp	6.6	9.2	Nishio et al. (2021)	Fructose
4	4.8	*0.001	**	Wu et al. (2014)	Fruit color
13	38.3	*0.001	**	Wu et al. (2014)	Fruit color
1	36.56	2.9	7.1	Zhang et al. (2013)	Fruit shape index
2	65.52	2.74	9.5	Zhang et al. (2013)	Fruit shape index
7	40.17	3.04	14.3	Zhang et al. (2013)	Fruit shape index
8	38.99	3.68	18.7	Zhang et al. (2013)	Fruit shape index
2	1	2.81	16.1	Zhang et al. (2013)	Fruit weight
7	66.56	3.89	17.2	Zhang et al. (2013)	Fruit weight
8	39.99	4.07	19.3	Zhang et al. (2013)	Fruit weight
13	99.3	2.91	14.6	Wu et al. (2014)	Fruit weight
17	16.3	3	18.5	Wu et al. (2014)	Fruit weight
11	41,197,041 bp	18.9	21.7	Nishio et al. (2021)	Glucose
5	10.5	2.94	14.2	Wu et al. (2014)	Soluble solid content
10	30.5	3.43	30	Wu et al. (2014)	Soluble solid content
14	74	5.18	23.8	Wu et al. (2014)	Soluble solid content
7	33,139,082 bp	9.9	23.5	Nishio et al. (2021)	Sucrose
5	189.89	4.45	3.5	Kumar et al. (2019)	Sweetness
Vegetative, Reproductive and Climate Adaptation Traits					
LG	Position (cM)	LOD	Variance	References	Trait
9	19.5	7.8	9.8	Gabay et al. (2018a)	Chilling requirements
8	21.4	11.5	28	Gabay et al. (2018a)	Chilling requirements
5	43.6	5.2	10.3	Gabay et al. (2018a)	G×E chilling requirements
15	16,568,005 bp	12.2	20.3	Nishio et al. (2021)	Harvest time
15	44.6	2.56	47.8	Sun et al. (2009)	Leaf length
8	45	3.57	9.9	Sun et al. (2009)	Leaf length
5	14.5	2.57	7.9	Sun et al. (2009)	Leaf length/width ratio
15	55.9	3.38	40.6	Sun et al. (2009)	Petiole length
Disease Resistance					
LG	Position	LOD	Variance	References	Trait
2	8.8	7.7	25	Gabay et al. (2021)	Fire blight
2	15	7.2	13.3	Montanari et al. (2016)	Fire blight
2	30.1	8.9	32.3	Le Roux et al. (2012)	Fire blight
2	3–19.2	3.42–11.99	15–51.2	Zurn et al. (2020)	Fire blight
3	26.7	10.54	87.1	Pierantoni et al. (2007)	Pear scab
7	60.8	10.51	86.6	Pierantoni et al. (2007)	Pear scab
8	16	5.1–9.9	22–39.1	Montanari et al. (2016)	Psylla
17	22	3.26–5.26	12.5–19.3	Dondini et al. (2015)	Psylla

(Continued)

TABLE 9.1 (*Continued*)
Summary of Selected Published QTLs for Pear

			Disease Resistance		
LG	Position	LOD	Variance	References	Trait
4	5.6–22.6	15–20	7.1–41.2	Perchepied et al. (2016)	Psylla
17	6.5–17.6	20.5–22.5	10.5–20.6	Perchepied et al. (2016)	Psylla
1	8.6–10.6	6.8–15.22	9.8–20	Perchepied et al. (2016)	Psylla

[a] LG, linkage group, correspond to 17 Pyrus chromosomes.
[b] Position of the QTL peak or QTL range is indicated in genetic position (cM) or physical position (bp) when bp is indicated.
[c] The phenotypic variance explained by the QTL.
*, Kruskal Wallis Test; **, phenotypic variance estimation is not available in this study.

9.3 FRUIT QUALITY

Traits related to fruit quality in pear are mostly regulated by multiple genes (polygenic). Hence, several loci with minor effects contribute to the final phenotype and only in few cases there is a major QTL that explained large proportion of the phenotypic variance.

Fruit quality is one of the key targets for every pear-breeding programs and is composed of a combination of characteristics such as appearance, flavor, texture and shelf life. However, there are specific demands in different regions around the world, and therefore, the breeders adjust their targets to the local market and to the climatic conditions. Environmental conditions play a major role in fruit quality (Bell and Janick, 2019), and therefore breeders considered the environmental effect to select the starting breeding material based on these effects (e.g., chilling requirements) rather than only the fruit quality traits.

Several studies have performed QTL mapping of pear fruit traits. Most fruit-related traits are polygenic and are controlled by quantitative loci. There is great variability in fruit size, shape, color, taste, and maturation date between varieties. Over the years, many studies were carried out, aiming to identify major genes and QTLs associated with fruit quality traits. In these studies, molecular markers have been developed, in order to enable MAS in breeding programs (De Franceschi and Dondini, 2019).

9.3.1 Fruit Flavor

In a QTL study performed by using pear bi-parental population for important fruit quality traits. QTLs for total soluble solids (TSS) were identified in chromosomes 5, 10 and 14 (Wu et al., 2014). TSS is a highly desirable trait in fruit quality and reflects the sugar content in the fruit. In a later study a QTL for sweetness was identified on LG 5 (Kumar et al., 2019) and in a specific sugar content level QTLs for sucrose and fructose were detected in LG 7 in a similar position (Figure 9.1), suggesting that these sugars are regulated by similar genetic factors (Nishio et al., 2021). Additional QTL associated with glucose was found in LG 11 (Table 9.1).

Another major factor to impact fruit flavor is acidity levels. The ratio between the sugar and acidy levels affect the flavor of the fruit. A QTL for acidity was detected in LG 6 (Nishio et al., 2021).

9.3.2 Fruit Shape and Weight

QTLs for fruit weight were detected in chromosomes 2, 7, 8, 13, and 17; fruit shape index in chromosomes 1, 2, 7, and 8; and fruit color in chromosomes 4, 13, and 16 (Figure 9.1) (Zhang et al., 2013; Wu et al., 2014). On a recent extensive GWAS, 37 loci associated with eight fruit quality traits

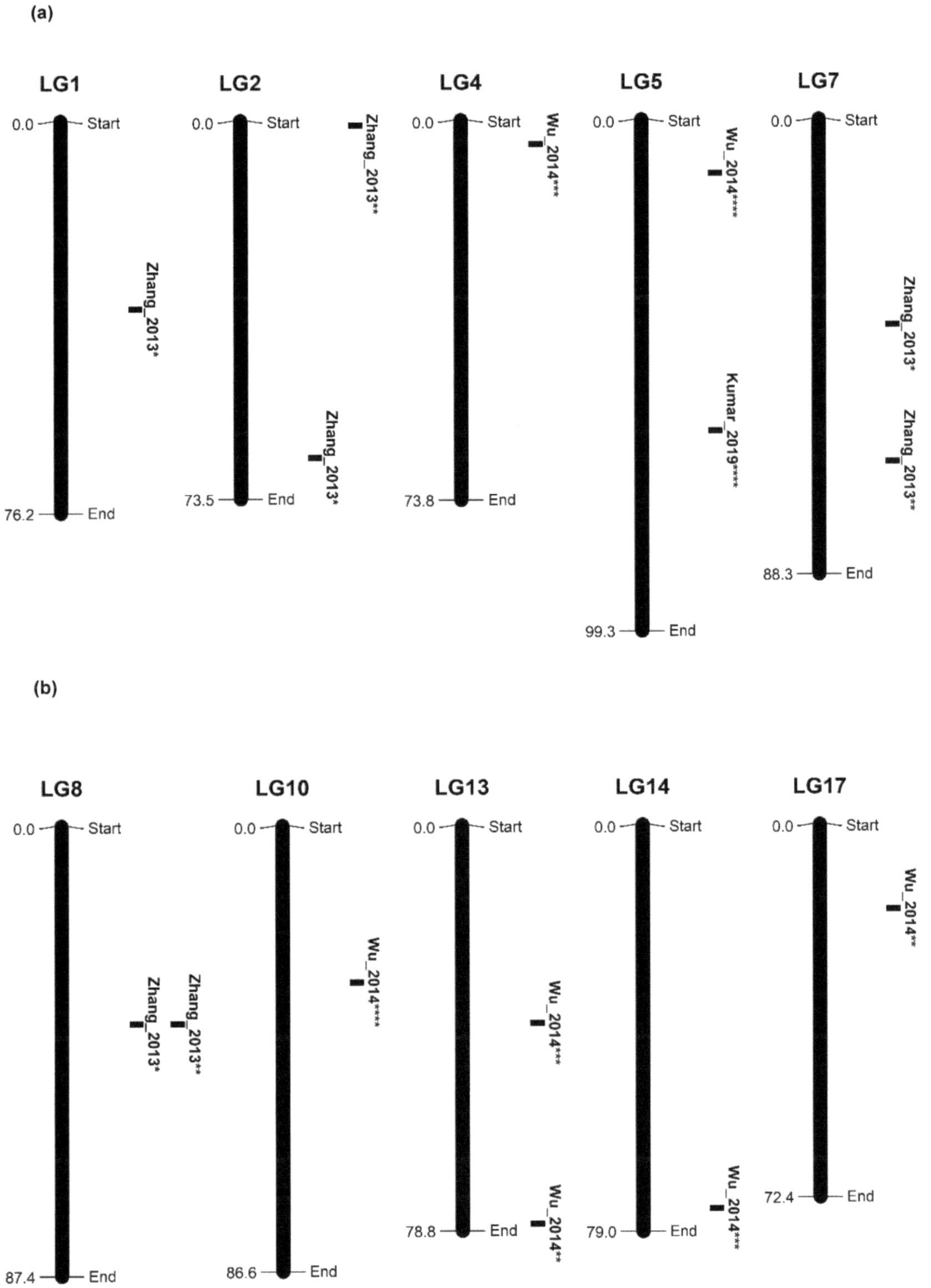

FIGURE 9.1 Positions of QTLs for fruit quality in pear. For each QTL, one-cM support intervals are shown for the average (location and years) QTL peak of the presented study. * Correspond to QTLs for fruit shape index, ** correspond to QTLs fruit weight, *** correspond to QTLs for fruit color, and **** correspond to QTLs for sugar content. Linkage group size is based on a high-resolution consensus map SPD×HS (Gabay et al., 2018a). More information about the QTLs is presented in Table 9.1.

and five loci associated with three fruit phenological traits were identified and enabled to verify the function of *PbrSTONE* gene (chromosome 11) that regulates the accumulation of stone cells, which significantly affect the quality of fruit (Zhang et al., 2021).

9.4 VEGETATIVE, REPRODUCTIVE, AND CLIMATE ADAPTATION TRAITS

9.3.1 Leaf Shape

Leaves have a critical role in light capture, gas exchange, and thermoregulation supporting optimal growth for the plant. Phenotypic variance in leaf characteristics has been correlated to adaption to a specific habitat (Vidaković et al., 2022). In a QTL study focused on leaf traits, including leaf length, leaf width (LGs 10 and 15), leaf length (LG 8, 15, and 16) and width ratio (LG 5), and petiole length (LG 4 and 15) QTLs were identified using interval mapping (Figure 9.2). The phenotypic variation explained by those QTLs ranged from 7.9% to 48.5% (Sun et al., 2009).

9.3.2 Chilling Requirements

Chilling requirements have been studied in perennial trees in the context of climate change. To ensure high fruit quality and sufficient yield, active growth following dormancy requires a certain number of chilling hours during the dormant period (Gabay and Flaishman, 2019). The recent rise in temperatures worldwide has led to a reduction in the chilling hour's accumulation. Pear is grown mainly

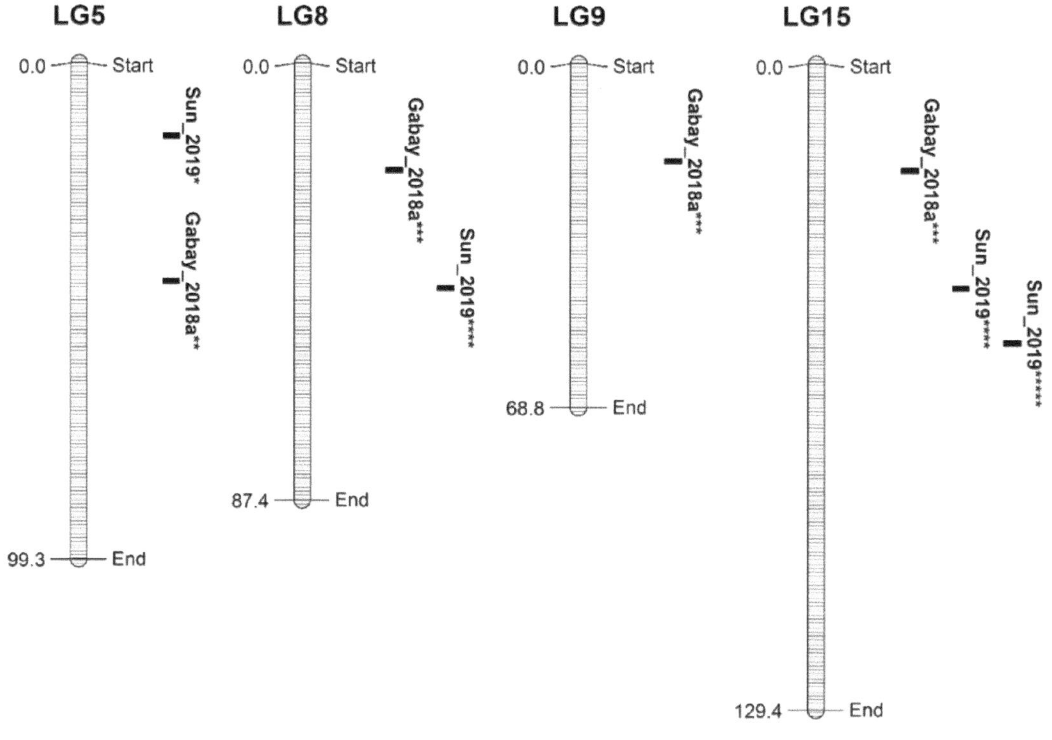

FIGURE 9.2 Positions of QTLs for vegetative, reproductive, and climate adaptation traits in pear. For each QTL, one-cM support intervals are shown for the average (location and years) QTL peak of the presented study. * Correspond to QTLs for leaf length/width ratio, ** correspond to QTLs for G×E chilling requirements, *** correspond to QTLs for chilling requirements, **** correspond to QTLs for leaf length, and ***** correspond to QTLs for petiole length. Linkage group size is based on a high-resolution consensus map SPD×HS (Gabay et al., 2018a). More information about the QTLs is presented in Table 9.1.

in temperate zones, however, due to the rise in temperatures and according to climate model predictions, the current climate change will result in a reduction in chilling units accumulation (Campoy et al., 2011). It is likely that in coming decades most pear production areas will not meet the chilling requirements of many of the commercial cultivars. In a QTL analysis of the F1 'Spadona' (low chilling requirements)×'Harrow Sweet' (high chilling requirements) pear population using genotyping by sequenced based genetic maps, major QTLs associated with chilling requirements were detected on LG 8 and on LG 9 (Figure 9.2). Additional QTLs were found on LGs 5, 13, and 15. Due to the significant genotype-by-environment (GxE) effect, novel interaction QTLs were detected on LGs 5, 8, 9, and 17. A further validation was conducted using additional 21 pear accessions to support the QTLs detected in the F1 population and significant genetic markers were detected on every LG that was detected in the F1 population (Gabay et al., 2018a). In a combined transcriptome and metabolome analysis of buds during the dormancy *12-oxophytodienoate reductase 2-like*, and the DORMANCY-ASSOCIATED MADS-BOX (DAM) genes, *PcDAM1* and *PcDAM2* were differentially expressed (DE) and selected as putative genes to orchestrated dormancy release in pear (Gabay et al., 2018b).

9.4.3 Harvest Time

There is growing interest in expending the harvest season, and the demand is both for early and late fruit marketing in the season (Saito et al., 2015). Harvest season is composed of various factors such as, flowering time and ethylene production which is correlated with shelf life and fruit ripening. The high heritability for fruit ripening date was reported before and indicates the high breeding potential for this trait (Bell et al., 1996).

The QTLs associated with harvest date were detected in LG 3 and LG 15 and were derived from the cultivar 'Taihaku' (Yamamoto et al., 2014). A recent study (Ntladi et al., 2018) has identified a QTL on LG 9, inherited from 'Flamingo' cultivar, associated with flowering time in a progeny of 'Flamingo'×'Abate Fetel' (Table 9.1). This QTL explained 30% of the phenotypic variance. Another genetic study indicated a QTL for harvest time at the same LG (15) (Nishio et al., 2021). Fruit ripening duration is the amount of days between full bloom to fruit maturation and reflects the period of time needed to ripen from the date of full bloom. This trait is important when selecting the ideal cultivar for a specific climatic region. Ripeness duration QTL was identified in LG3 (Yamamoto et al., 2014)

Several key genes controlling ethylene production in fruit were identified in pear.

Two *1-amino-cyclopropane-1-carboxylic acid (ACC)* synthase genes, *pPPACS1* and *pPPACS2*. *PPACS2* were mapped on LG 15 (Itai et al., 1999).

9.5 DISEASE RESISTANCE

Pyrus is susceptible to a vast range of diseases and pathogens (Bell et al., 1996). These diseases can be specific to a region and in some cases to a species (e.g. *P. communis*). Growers often select cultivars that are less susceptible to the diseases that are spread in their location in order to avoid loss of yield and reduce fruit quality. Therefore, developing a cultivar with high tolerance to a pest is desirable when the control practice such as applying chemical is not effective (Brewer and Volz, 2019). While many of the fruit characteristics can be evaluated only after the juvenile period, disease resistance can be tested at an early stage on a seedling <2 years old (Gabay et al., 2021). In this section, the genetic factors associated with the most important diseases that affect *Pyrus* species growth worldwide will be discussed as well as potential useful molecular markers that are already in use or can be used in a future pear-breeding programs.

9.5.1 Fire Blight

Fire blight is the most devastating disease affecting the European pear, *Pyrus communis* L. The disease is caused by the bacterium *Erwinia amylovora* (Burrill) (Bell et al., 2005; Bokszczanin et al., 2011; Calenge et al., 2005; Fahrentrapp et al., 2013; Postman, 2008), and the common practice to control the spread of the bacterium is by applying chemicals that raise public health concerns, such as antibiotics or copper-based products (McManus et al., 2002; Norelli et al., 2003; Psallidas and Tsiantos, 2000). Therefore, breeding of more tolerated cultivars is a possible solution for a growing demand for healthier alternatives. To do so, several QTL analyses were conducted to develop genetic markers for fire blight tolerance and to accelerate the breeding process. The first QTL study detected QTLs located on LGs 2a, 2b, 4, and 9 using a population derived from a 'Passe Crassane' × 'Harrow Sweet' cross (Dondini et al., 2004) and later more accurately located by using a high-resolution genetic map (Le Roux et al., 2012). QTLs were detected on LG 11 and LG 4. The QTL on LG 4 inherited from 'Doyenné du Comice', a susceptible parent, indicating that resistance alleles can be inherited from susceptible cultivars (Bokszczanin et al., 2009). A later study using the same population, detected a fire blight resistance QTL on LG 9 and additional QTLs on LG 11 (Bokszczanin et al., 2011). A major fire blight resistance QTL was identified using a high-resolution genetic maps in LG 2 (Figure 9.3) in an interspecific population ('PremP003' (*P. bretschneideri* Rehd. × *P. communis*) × 'Moonglow' (*P. communis*) (Montanari et al., 2016). On a multiple mapping population ('El Dorado' × 'Potomac', 'Old Home' × 'Bartlett' and 'NJA2R59T69' × 'Bartlett'), a stable QTL across the three different populations was identified on LG2 (Zurn et al., 2020). In a recent study, a major stable QTL (R^2 = 15.7%–38.8%) in two consecutive years was detected at 8.8 cM in LG 2 (LOD = 7.7) (Table 9.1). This population (F1) was derived from a cross between 'Spadona' (susceptible) × 'Harrow Sweet' (low susceptibility cultivar). Additional minor year-specific QTLs were detected in LGs 7, 13, and 15 and explained between 13.3% and 15.3% of the phenotypic variance (Gabay et al., 2021).

9.5.2 Resistance to Pear Scab

Two fungal species of the *Venturia* incite pear scab disease. *V. pirina* Aderh. infects *P. communis*, and *V. nashicola* infects Asian pear. QTLs for pear scab disease resistance, a polygenic trait (Chevalier et al., 2004), have been identified on LG 3 and LG 7 using a segregating F1 population derived from a cross between 'Abbé Fétel' × 'Max Red Bartlett' (Figure 9.3) (Pierantoni et al., 2007). A major gene (*Vnk*) related to scab resistance has been mapped to LG 1 (Terakami et al., 2006). Another QTL reported on LG 2, and *Rvn2 and Rvp1* were indicated as the putative casual genes (Cho et al., 2009; Bouvier et al., 2012).

9.5.3 *Pear psylla*

Three major species have major economic importance (Ouvrard, 2017; Hodkinson, 2009). *Cacopsylla pyricola* Foerster is the most common and can be found in the majority of pear growth regions such as Europe, the Middle East, and North and South America (Ouvrard, 2017). The majority of the popular commercial cultivars of *P. communis* are susceptible to pear psylla.

Therefore, the identification of resistance genes to this pest and the development of new resistant cultivars have been high priority (Sestras et al., 2009; Benedek et al., 2010).

A major QTL for control of pear psylla was detected on LG 17 of a pear accession, NY10355 (Bouvier et al., 2011), and additional QTLs located on LG 1 and LG 4. The major resistance QTL on LG 17 (Figure 9.3) has been confirmed in a different study using a different pear accession derived from different accession (NY103353) (Dondini et al., 2015) and therefore has been suggested for MAS (Table 9.1).

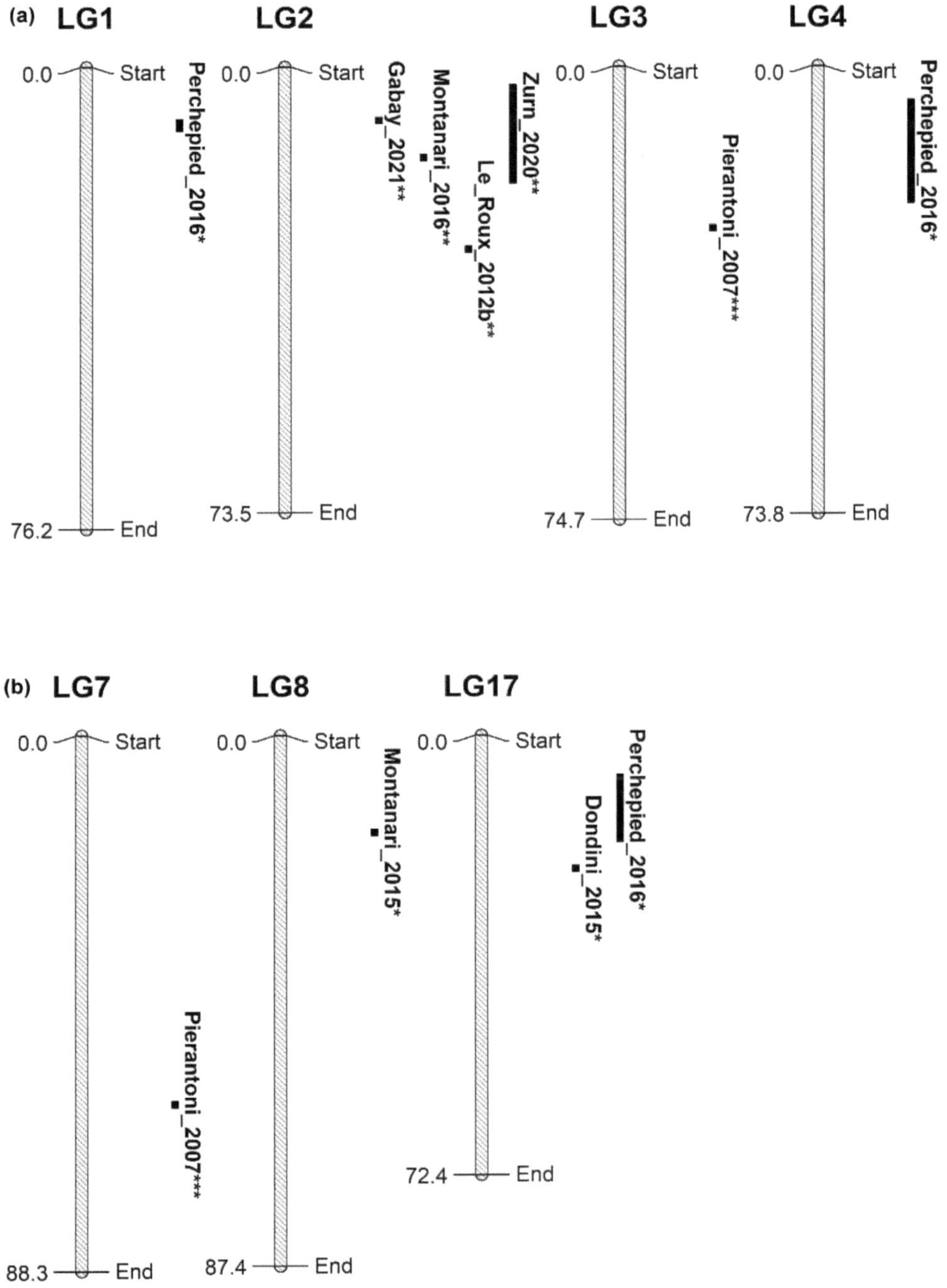

FIGURE 9.3 Positions of QTLs for disease resistance in pear. For each QTL, one-cM support intervals are shown for the average (location and years) QTL peak of the presented study. When QTL average is not indicated the range of the QTL across years and locations is shown. * Correspond to QTLs for Psylla resistance, ** correspond to QTLs for fire blight resistance, and *** correspond to QTLs for Pear Scab. Linkage group size is based on a high-resolution consensus map SPD×HS (Gabay et al., 2018a). More information about the QTLs is presented in Table 9.1.

9.6 ROOTSTOCK BREEDING

Rootstock can confer improved yield, vigorous growth, and disease resistance. In addition, the grafting procedure can also significantly reduce the juvenile period and reduce the time it takes to harvest or valuate the fruits. However, the range of clonal rootstocks is limited compared to apple rootstock variety. Rootstock for pear include several *Pyrus* species and alternatives from other closely related species, such as *Cydonia oblonga* (quince). A major trait for rootstock selection is scion dwarfing QTLs influencing scion vigor which were identified on LGs 5 and 6. The position of LG 5 is syntenic to the apple 'Malling 9' *Dw1* locus and located at a similar position of LG 5 (Knäbel et al., 2015). Adventitious roots discarding in an orchard is both time and labor consuming, and therefore, breeders aim to reduce the growth of the adventitious roots in a potential rootstock line. A QTL for adventitious root was identified on LG 7 (Knäbel et al., 2017).

9.7 CONCLUDING REMARKS AND THE POTENTIAL IMPROVEMENT

This chapter presented a large set of QTLs that are associated with trait of interest to the pear community. However, the function of the genes underlying those QTLs is still unknown.

Transcriptome analysis is already in use in pear genetic research (Tuan et al., 2017; Gabay et al., 2018b). This enables to characterize genes within the identified QTLs and select candidate genes for further validation. One possible option to validate the function of the selected genes is by testing the allelic variation of these genes in a large population of different pear accessions.

Another possible solution to overcome the long pear-breeding process is to induce or search for mutation by either naturally occurred mutation or induced mutations (e.g. gamma and ultraviolet irradiation). This practice can be used also in the breeding process. To date, there are five European and two Japanese pear cultivars registered as new mutant cultivars (Ahloowalia et al., 2004). Most of those cultivars carry natural or induced mutations that generate red fruit skin color or disease resistance (Ahloowalia et al., 2004).

The recent advances in transient transformation that have been reported in pear can dramatically affect pear functional gene studies and validate more genes of interest (Zhang et al., 2021). These discoveries can lead to more accurate development of genetic markers, and the selection can be based on the allelic variation of specific genes and not based on linked markers.

The presented studies in this chapter can assist pear breeders and researchers to develop tools for genomic selection that can contribute to reduce the pear-breeding programs and provide the knowledge for future functional genomics studies.

ACKNOWLEDGMENTS

I am grateful to Dr. Moshe Flaishman and Yardena Dhahan for the fruitful discussions and useful advice.

REFERENCES

Ahloowalia, B.S., Maluszynski, M., and Nichterlein, K. (2004). Global impact of mutation-derived varieties. *Euphytica* 135: 187–204.

Bai, S., Saito, T., Sakamoto, D., Ito, A., Fujii, H., and Moriguchi, T. (2013). Transcriptome analysis of Japanese pear (*Pyrus pyrifolia* nakai) flower buds transitioning through endodormancy. *Plant Cell Physiol.* 54: 1132–1151.

Bell, A.C., Ranney, T.G., Eaker, T.A., and Sutton, T.B. (2005). Resistance to fire blight among flowering pears and quince. *HortScience* 40: 413–415.

Bell, R. (1990). Pears (*Pyrus*). In: Moore, J.N., and Ballington, J.R. (eds) *Genetic Resources of Temperate Fruit and Nut Crops I.* International Society for Horticultural Science, Wageningen, pp. 655–697.

Bell, R.L. (2019). Genetics, genomics, and breeding for fire blight resistance in pear. In: Korban, S. (ed) *The Pear Genome*, Springer, Cham, pp. 243–264.

Bell, R.L. and Janick, J. (2019). Quantitative genetic analysis of fruit quality in pear. *J. Am. Soc. Hortic. Sci.* 115: 829–834.

Bell, R.L., Quamme, H.A., Layne, R.E.C., and Skirvin, R.M. (1996) Pear. In: Janick, J., Moore, J.N. (eds) *Fruit Breeding, Volume I: Tree and Tropical Fruits*. Wiley, New York, pp. 441–514.

Benedek, P., Szabó, T., Soltész, M., Szabó, Z., and Konrád-Németh, C. (2010). Susceptibility of European pear genotypes in a gene bank to pear psylla damage and possible exploitation of resistant varieties in organic farming. *Int. J. Hortic. Sci.* 16: 95–101.

Bokszczanin, K., Dondini, L., and Przybyla, A.A. (2009). First report on the presence of fire blight resistance in linkage group 11 of *Pyrus ussuriensis* Maxim. *J. Appl. Genet.* 50: 99–104.

Bokszczanin, K.L., Przybyla, A.A., Dondini, L., and Palucha, A. (2011). QTLs for fire blight (*Erwinia amylovora*) resistance in *Pyrus ussuriensis*. *Acta Hortic.* 896: 371–374.

Bouvier, L., Bourcy, M., Boulay, M., Tellier, M., Guérif, P., Denancé, C., Durel, C.E., and Lespinasse, Y. (2011). European pear cultivar resistance to bio-pests: Scab (*Venturia pirina*) and psylla (*Cacopsylla pyri*). *Acta Hortic.* 909: 459–470.

Bouvier, L., Bourcy, M., Boulay, M., Tellier, M., Guérif, P., Denancé, C., Durel, C.E., and Lespinasse, Y. (2012). A new pear scab resistance gene *Rvp1* from the European pear cultivar "Navara" maps in a genomic region syntenic to an apple scab resistance gene cluster on linkage group 2. *Tree Genet. Genomes* 8: 53–60.

Brewer, L. and Volz, R. (2019). Genetics and breeding of pear. In: Korban, S. (ed) *The Pear Genome*, Springer, Cham, pp. 63–101.

Calenge, F., Drouet, D., Denancé, C., Van De Weg, W.E., Brisset, M.N., Paulin, J.P., and Durel, C.E. (2005). Identification of a major QTL together with several minor additive or epistatic QTLs for resistance to fire blight in apple in two related progenies. *Theor. Appl. Genet.* 111: 128–135.

Campoy, J.A., Ruiz, D., and Egea, J. (2011). Dormancy in temperate fruit trees in a global warming context: A review. *Sci. Hortic. (Amsterdam)* 130: 357–372.

Celton, J.-M., Chagné, D., Tustin, S.D., Terakami, S., Nishitani, C., Yamamoto, T., and Gardiner, S.E. (2009). Update on comparative genome mapping between *Malus* and *Pyrus*. *BMC Res. Notes* 2: 182.

Chagné, D. (2015). *Whole Genome Sequencing of Fruit Tree Species*. Elsevier Ltd.

Chagné, D. et al. (2014). The draft genome sequence of European pear (*Pyrus communis* L. 'Bartlett'). *PLoS One* 9: 1–12.

Chevalier, M., Bernard, C., Tellier, M., Lespinasse, Y., Filmond, R., and Le Lezec, M. (2004). Variability in the reaction of several pear (*Pyrus communis*) cultivars to different inocula of *Venturia pirina*. *Acta Hortic.* 663: 177–182.

Cho, K.H., Shin, S., Kim, K.T., Suh, E.J., Hong, S.S., and Lee, H.J. (2009). Development of AFLP and CAPS markers linked to the scab resistance gene, Rvn2, in an inter-specific hybrid pear (*Pyrus* spp.). *J. Hortic. Sci. Biotechnol.* 84: 619–624.

Costa, F., Van De Weg, W.E., Stella, S., Dondini, L., Pratesi, D., Musacchi, S., and Sansavini, S. (2008). Map position and functional allelic diversity of Md-Exp7, a new putative expansin gene associated with fruit softening in apple (*Malus × domestica* Borkh.) and pear (*Pyrus communis*). *Tree Genet. Genomes* 4: 575–586.

Dondini, L., De Franceschi, P., Ancarani, V., Civolani, S., Fano, E.A., and Musacchi, S. (2015). Identification of a QTL for psylla resistance in pear via genome scanning approach. *Sci. Hortic. (Amsterdam)* 197: 568–572.

Dondini, L., Pierantoni, L., Gaiotti, F., Chiodini, R., Tartarini, S., Bazzi, C., and Sansavini, S. (2004). Identifying QTLs for fire-blight resistance via a European pear (*Pyrus communis* L.) genetic linkage map. *Mol. Breed.* 14: 407–418.

Elkins, R., Bell, R., and Einhorn, T. (2012). Needs assessment for future US pear rootstock research directions based on the current state of pear production and rootstock research. *J. Am. Pomol. Soc.* 66: 153–163.

Fahrentrapp, J., Broggini, G.A.L., Kellerhals, M., Peil, A., Richter, K., Zini, E., and Gessler, C. (2013). A candidate gene for fire blight resistance in Malus × robusta 5 is coding for a CC-NBS-LRR. *Tree Genet. Genomes* 9: 237–251.

De Franceschi, P., and Dondini, L. (2019). Molecular mapping of major genes and QTLs in pear. In: Korban, S. (ed) *The Pear Genome*, Springer, Cham, pp. 113–131.

Gabay, G., Dahan, Y., Cohen, O., Barshan, I., Itzhaki, Y., and Flaishman, M.A. (2021). Detection of QTLs in a Mediterranean climate associated with fire blight necrosis length in *Pyrus communis* using a high-resolution genetic linkage map. *Plant Breed.* 140: 703–710.

Gabay, G., Dahan, Y., Izhaki, Y., Faigenboim, A., Ben-ari, G., Elkind, Y., and Flaishman, M.A. (2018a). High-resolution genetic linkage map of European pear (*Pyrus communis*) and QTL fine-mapping of vegetative budbreak time. *BMC Plant Biol.* 18: 1–13.

Gabay, G., Faigenboim, A., Dahan, Y., Izhaki, Y., Itkin, M., Malitsky, S., Elkind, Y., and Flaishman, M.A. (2018b). Transcriptome analysis and metabolic profiling reveal the key role of α-linolenic acid in dormancy regulation of European pear. *J. Exp. Bot.* 70(3): 1017–1031.
Gabay, G., and Flaishman, M.A. (2019). Genetic and genomic analyses of vegetative budbreak in response to chilling units in European pear (Pyrus Communis L.). In: Korban, S. (ed) *The Pear Genome*, Springer, Cham, pp. 227–242.
Hodkinson, I.D. (2009). Life cycle variation and adaptation in jumping plant lice (Insecta: Hemiptera: Psylloidea): A global synthesis. *J. Nat. Hist.* 43: 65–179.
Itai, A., Kawata, T., Tanabe, K., Tamura, F., Uchiyama, M., Tomomitsu, M., and Shiraiwa, N. (1999). Identification of 1-aminocyclopropane-1-carboxylic acid synthase genes controlling the ethylene level of ripening fruit in Japanese pear (*Pyrus pyrifolia* Nakai). *Mol. Gen. Genet.* 261: 42–49.
Knäbel, M., Friend, A.P., Palmer, J.W., Diack, R., Gardiner, S.E., Tustin, S., Schaffer, R., Foster, T., and Chagné, D. (2017). Quantitative trait loci controlling vegetative propagation traits mapped in European pear (*Pyrus communis* L.). *Tree Genet. Genomes* 13: 55.
Knäbel, M., Friend, A.P., Palmer, J.W., Diack, R., Wiedow, C., Alspach, P., Deng, C., Gardiner, S.E., Tustin, D.S., Schaffer, R., Foster, T., and Chagné, D. (2015). Genetic control of pear rootstock-induced dwarfing and precocity is linked to a chromosomal region syntenic to the apple *Dw1* loci. *BMC Plant Biol.* 15: 1–16.
Kumar, S., Kirk, C., Deng, C., Wiedow, C., Knaebel, M., and Brewer, L. (2017). Genotyping-by-sequencing of pear (*Pyrus* spp.) accessions unravels novel patterns of genetic diversity and selection footprints. *Hortic. Res.* 4: 17015.
Kumar, S., Kirk, C., Deng, C.H., Shirtliff, A., Wiedow, C., Qin, M., Wu, J., and Brewer, L. (2019). Marker-trait associations and genomic predictions of interspecific pear (*Pyrus*) fruit characteristics. *Sci. Rep.* 9: 1–10.
Li, L., Deng, C.H., Knäbel, M., Chagné, D., Kumar, S., Sun, J., Zhang, S., and Wu, J. (2017). Integrated high-density consensus genetic map of *Pyrus* and anchoring of the 'Bartlett' v1.0 (*Pyrus communis*) genome. *DNA Res.* 24: 289–301.
McManus, P.S., Stockwell, V.O., Sundin, G.W., and Jones, A.L. (2002). Antibiotic use in plant agriculture. *Annu. Rev. Phytopathol.* 40: 443–465.
Montanari, S., Perchepied, L., Renault, D., Frijters, L., Velasco, R., Horner, M., Gardiner, S.E., Chagné, D., Bus, V.G.M., Durel, C.E., and Malnoy, M. (2016). A QTL detected in an interspecific pear population confers stable fire blight resistance across different environments and genetic backgrounds. *Mol. Breed.* 36: 1–16.
Nishio, S., Hayashi, T., Shirasawa, K., Saito, T., Terakami, S., Takada, N., Takeuchi, Y., Moriya, S., and Itai, A. (2021). Genome-wide association study of individual sugar content in fruit of Japanese pear (*Pyrus* spp.). *BMC Plant Biol.* 21: 1–19.
Niu, Q., Li, J., Cai, D., Qian, M., Jia, H., Bai, S., Hussain, S., Liu, G., Teng, Y., and Zheng, X. (2016). Dormancy-associated MADS-box genes and microRNAs jointly control dormancy transition in pear (*Pyrus pyrifolia* white pear group) flower bud. *J. Exp. Bot.* 67: 239–257.
Norelli, J.L., Jones, A.L., and Aldwinckle, H.S. (2003). Fire blight management in the twenty-first century: Using new technologies that enhance host resistance in apple. *Plant Dis.* 87: 756–765.
Ntladi, S.M., Human, J.P., Bester, C., Vervalle, J., Roodt-Wilding, R., and Tobutt, K.R. (2018). Quantitative trait loci (QTL) mapping of blush skin and flowering time in a European pear (*Pyrus communis*) progeny of 'Flamingo' × 'Abate Fetel.' *Tree Genet. Genomes* 14: 1–24.
Ouvrard, D. (2017). Psyl'list-The world Psylloidea database.
Pierantoni, L., Dondini, L., Cho, K.H., Shin, I.S., Gennari, F., Chiodini, R., Tartarini, S., Kang, S.J., and Sansavini, S. (2007). Pear scab resistance QTLs via a European pear (*Pyrus communis*) linkage map. *Tree Genet. Genomes* 3: 311–317.
Postman, J.D. (1994). Graft compatibility of pear with related genera. *Acta Hortic.* 367: 380.
Postman, J.D. (2008). The USDA quince and pear genebank in Oregon, a world source of fire blight resistance. *Acta Hortic.* 793: 357–362.
Psallidas, P.G. and Tsiantos, J. (2000). Chemical control of fire blight. In: Vanneste, J.L. (ed.), *Fire Blight: The Disease and Its Causative Agent, Erwinia amylovora.* CABI Publishing, Wallingford, pp. 99–234.
Le Roux, P.M.F., Christen, D., Duffy, B., Tartarini, S., Dondini, L., Yamamoto, T., Nishitani, C., Terakami, S., Lespinasse, Y., Kellerhals, M., and Patocchi, A. (2012). Redefinition of the map position and validation of a major quantitative trait locus for fire blight resistance of the pear cultivar "Harrow Sweet" (*Pyrus communis* L.). *Plant Breed.* 131: 656–664.

Ben Sadok, I., Moutier, N., Garcia, G., Dosba, F., Grati-Kamoun, N., Rebai, A., Khadari, B., and Costes, E. (2013). Genetic determinism of the vegetative and reproductive traits in an F1 olive tree progeny: Evidence of tree ontogeny effect. *Tree Genet. Genomes* 9: 205–221.

Saito, T., Tuan, P.A., Katsumi-Horigane, A., Bai, S., Ito, A., Sekiyama, Y., Ono, H., and Moriguchi, T. (2015). Development of flower buds in the Japanese pear (*Pyrus pyrifolia*) from late autumn to early spring. *Tree Physiol.* 35: 653–662.

Sestras, R., Botez, C., Ardelean, M., Oltean, I., and Sestras, A. (2009). Response of pear genotypes to *Psylla* sp. attack in central Transylvania, Romania. *Acta Hortic.* 814: 845–850.

de Souza, V.A.B., Byrne, D.H., Taylor, J.F., and De, S.V. (1998). Heritability, genetic and phenotypic correlations, and predicted selection response of quantitative traits in peach: II. An analysis of several fruit traits. *J. Am. Soc. Hortic. Sci.* 123: 598–603.

Sun, W., Zhang, Y., Le, W., and Zhang, H. (2009). Construction of a genetic linkage map and QTL analysis for some leaf traits in pear (*Pyrus* L.). *Front. Agric. China* 3: 67–74.

Terakami, S., Shoda, M., Adachi, Y., Gonai, T., Kasumi, M., Sawamura, Y., Iketani, H., Kotobuki, K., Patocchi, A., Gessler, C., Hayashi, T., and Yamamoto, T. (2006). Genetic mapping of the pear scab resistance gene Vnk of Japanese pear cultivar Kinchaku. *Theor. Appl. Genet.* 113: 743–752.

Tuan, P.A., Bai, S., Saito, T., Ito, A., and Moriguchi, T. (2017). Dormancy-associated MADS-box (DAM) and the abscisic acid pathway regulate pear endodormancy through a feedback mechanism. *Plant Cell Physiol.* 58: 1378–1390.

Vidaković, A., Šatović, Z., Tumpa, K., Idžojtić, M., Liber, Z., Pintar, V., Radunić, M., Runjić, T.N., Runjić, M., Rošin, J., Gaunt, D., and Poljak, I. (2022). Phenotypic variation in European Wild Pear (*Pyrus pyraster* (L.) Burgsd.) populations in the North-Western part of the Balkan Peninsula. *Plants* 11: 335.

van de Weg, E., Di Guardo, M., Jänsch, M., Socquet-Juglard, D., Costa, F., Baumgartner, I., Broggini, G.A.L., Kellerhals, M., Troggio, M., Laurens, F., Durel, C.E., and Patocchi, A. (2018). Epistatic fire blight resistance QTL alleles in the apple cultivar 'Enterprise' and selection X-6398 discovered and characterized through pedigree-informed analysis. *Mol. Breed.* 38: 1–18.

Wu, J. et al. (2013). The genome of the pear (*Pyrus bretschneideri* Rehd.). *Genome Res.* 23: 396–408.

Wu, J., Li, L.T., Li, M., Khan, M.A., Li, X.G., Chen, H., Yin, H., and Zhang, S.L. (2014). High-density genetic linkage map construction and identification of fruit-related QTLs in pear using SNP and SSR markers. *J. Exp. Bot.* 65: 5771–5781.

Wu, J. and Qin, M. (2019). Linkage mapping in pear. In: Korban, S. (ed) *The Pear Genome*, Springer, Cham, pp. 103–112.

Yamamoto, T. and Terakami, S. (2016). Genomics of pear and other Rosaceae fruit trees. *Breed. Sci.* 66: 148–159.

Yamamoto, T., Terakami, S., Takada, N., Nishio, S., Onoue, N., Nishitani, C., Kunihisa, M., Inoue, E., Iwata, H., and Hayashi, T. (2014). Identification of QTLs controlling harvest time and fruit skin color in Japanese pear (*Pyrus pyrifolia* Nakai). *Breed. Sci.* 64: 351.

Zhang, M.Y. et al. (2021). Genome-wide association studies provide insights into the genetic determination of fruit traits of pear. *Nat. Commun.* 12: 1–10.

Zhang, R.P., Wu, J., Li, X.G., Khan, M.A., Chen, H., Korban, S.S., and Zhang, S.L. (2013). An AFLP, SRAP, and SSR genetic linkage map and identification of QTLs for fruit traits in pear (*Pyrus* L.). *Plant Mol. Biol. Report* 31: 678–687.

Zurn, J.D., Norelli, J., Montanari, S., Bell, R., and Bassil, N. (2020). Dissecting genetic resistance to fire blight in three pear populations. *Phytopathology* 110(7): 1305–1311.

10 Molecular Markers and Allele Mining in Kiwifruit Breeding

De Mori Gloria, De Rosa Valeria, and Cipriani Guido

10.1 THE KIWIFRUIT

The genus *Actinidia* belongs to the *Actinidiaceae* family. It contains vigorous woody perennial species characterized by a climbing growth habit (Ferguson, 2016). More than 50 *Actinidia* species, originating from China, have been described in the latest taxonomic revision, divided into four sections based on morphological characteristics: *Leiocarpae*, *Maculatae*, *Stellatae*, and *Strigosae* (Table 10.1) (Li et al., 2007). Section *Leiocarpae* includes species characterized by the absence of lenticels of the fruit exocarp; species within section *Maculatae* have hairless leaves and branches; species in section *Stellatae* bear stellate trichomes, while those in section *Strigosae* simple trichomes (Li et al., 2002). Most cultivated *Actinidia* are represented by large-fruited species *A. chinensis* var. *chinensis* and *A. chinensis* var. *deliciosa* (section *Stellatae*), commonly regarded as kiwifruit. Small-fruited *A. arguta* (section *Leiocarpae*) is known as kiwiberry (or hardy kiwifruit).

Kiwifruit have been the object of domestication quite recently compared to most cultivated fruit crops. In fact, while early cultivation testimonies in China date to 1,200 years ago (Bretschneider, 1893), few domestication attempts were made until kiwifruit plants were taken overseas. The first *A. chinensis* fruits produced outside of China date back to 1910 in New Zealand (Ferguson and Allison, 2004). Among them, the green-fleshed cultivar *A. chinensis* var. *deliciosa* 'Hayward' still represents 80% of total planted area in the world. Subsequently, the domestication of yellow-fleshed cultivars of *A. chinensis* var. *chinensis* took place in China with widespread impacts on the global kiwifruit industry. Currently, cultivars 'Gold 3-ZespriGold' and 'Jintao-Jingold' constitute the main commercialized yellow-fleshed kiwifruit outside China. Recently, red-fleshed *A. chinensis* var. *chinensis* cultivars have been introduced to the market (Wang et al., 2003; Li et al., 2018).

In 2020, the top three producers worldwide of kiwi were China, Italy, and New Zealand, with more than 80% of the production gathered in Asia and Europe (Figure 10.1). Overall, more than 4.4 million tons of kiwifruit were produced in the same year (FAOSTAT, 2022, available from the Web site https://www.fao.org/faostat/en/#home).

All *Actidinia* species are functionally dioecious, meaning that female and male flowers are found on separate plants. Pistillate, female flowers present a developed ovary with many ovules, while also bearing stamens and anthers producing sterile pollen. Male flowers on the other hand are characterized by an underdeveloped ovary carrying no ovules, and many stamens producing functional pollen (McNeilage, 1991) This sexual system, which is relatively uncommon in angiosperms, greatly affects the kiwifruit market since unfruitful male plants, required for pollination, are necessarily present in orchards, thus subtracting room for fruitful plants (McNeilage and Steinhagen, 1998). Production is also highly dependent on genetic diversity among species and cultivars. *Actinidia* species are often polyploids, due to whole genome duplication events, and are found in nature as diploids, tetraploids, hexaploids, and octoploids. Polyploidy is an important resource for breeding programs because of the increased allelic diversity and variability in phenotype displays it provides. Polyploids are often more vigorous than their diploid counterparts, and the presence of several gene copies provides protection from possible deleterious effects caused by mutations. For these reasons, manipulation of ploidy in kiwifruit breeding is a valuable resource (Wu, 2012). Ploidy also appears

 DOI: 10.1201/9781003386490-10

TABLE 10.1
Taxonomic Subdivision of *Actinidia* Species within Four Sections Based on Morphological Characteristics as Described by Li et al. (2007)

Stellatae	Leiocarpae	Maculatae	Strigosae
A. chinensis[1]	*A. arguta*[5]	*A. callosa*[10]	*A. chengkouensis*
A. eriantha[2]	*A. kolomikta*	*A. chrysantha*[11]	*A. hemsleyana*[13]
A. farinosa	*A. melanandra*[6]	*A. cylindrica*	*A. henryi*
A. fulvicoma	*A. macrosperma*[7]	*A. fasciculoides*	*A. holotricha*
A. grandiflora	*A. polygama*[8]	*A. fortunatii*	*A. melliana*
A. hubeiensis	*A. tetramera*	*A. glaucocallosa*	*A. rubus*
A. lanceolata[3]	*A. valvata*[9]	*A. indochinensis*	*A. rudis*
A. latifolia[4]		*A. laevissima*	*A. vitifolia*
A. liangguangensis		*A. rubricaulis*	*A. arisanensis*[14]
A. lijiangensis		*A. rufa*[12]	
A. linguiensis		*A. sabiifolia*	
A. longicarpa		*A. trichogyna*	
A. obovata		*A. ulmifolia*	
A. persicina		*A. umbelloides*	
A. pilosula			
A. rongshuiensis			
A. rufotricha			
A. sorbifolia			
A. stellatopilosa			
A. styracifolia			
A. suberifolia			
A. venosa			
A. zhejiangensis			

(*Continued*)

TABLE 10.1 (*Continued*)
Taxonomic Subdivision of *Actinidia* Species within Four Sections Based on Morphological Characteristics as Described by Li et al. (2007)

Stellatae	Leiocarpae	Maculatae	Strigosae

The superscript numbers refer to the fruit images.

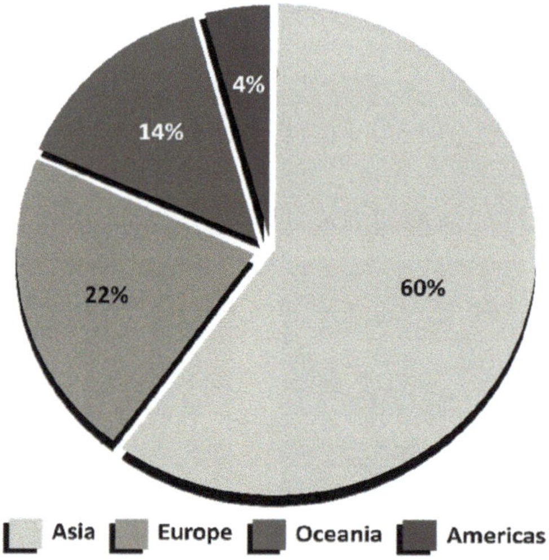

FIGURE 10.1 Production share of kiwifruit by region in 2020. (Data by FAOSTAT (2022).)

to be correlated to some fruit qualitative characteristics, such as fruit hairs and fruit flesh (Li et al., 2010). In this regard, green-fleshed, hexaploid cultivars are predominant in the industry.

10.2 ALLELE MINING FOR TRAITS OF INTEREST WITHIN KIWIFRUIT BREEDING

The 'Allele mining' can be defined as the approach used to identify naturally occurring allelic variations or suitable alleles of a candidate gene controlling key agronomic traits relevant to crop improvement (Sharma et al., 2014). The identification of the genes of interest controlling such key agronomic traits allows the association of suitable alleles at each locus. Allele-specific selection is an important tool in the hand of breeders. An ideotype includes as many positive alleles as possible, if the trait is under a polygenic control. But examples are available in kiwifruit for gender allele selection (see Section 10.4). Accordingly, genotypes selected may be used as parents for control crossing in breeding programs. Moreover, the allele mining can help in outlining the evolution of alleles, and the identification of allele-specific molecular markers for marker-assisted selection (MAS) and sequence information may potentially deploy the evolutionary relationship among the species and may be compared to derive syntenic relationships.

The following sections describe the massive breeding efforts made in the last decades, ranging from attempts to diversify the kiwifruit market and meet consumer interest, to overcoming the limits to production posed by dioicy and highly virulent pests. Moreover, this chapter provides an overview of the hard work that has gone over the years to identify suitable alleles in kiwifruit breeding.

10.2.1 Breeding of Commercial Kiwifruit Cultivars

A. chinensis var. *deliciosa* 'Hayward' represents the first cultivar to be commercialized worldwide in the 1970s, following the introduction of some seeds of the species in New Zealand in the early 20th century. The success of this cultivar is to be attributed to its larger size, general appearance, flavor, and high content in vitamin C. Its long storage life also made it suitable for export (Ferguson, 1999). The biggest change within the kiwifruit market was the commercialization of yellow-fleshed cultivars, bred in New Zealand since 1977. Crosses within accessions of *A. chinensis* var. *chinensis* allowed the selection of cultivar 'Hort16A' in 1991. Since then, a prolific breeding activity has focused on developing cultivars with varying qualities, including red-fleshed *A. chinensis* (Wang et al., 2003) (Figure 10.2).

In the context of breeding programs focused on quantitative traits such as color or size, parental genotypes need to be assessed in their ability to transfer characters of interest to the progeny (GCA, General Combining Ability). In this regard, controlled crosses performed using *A. chinensis* accessions showed that the genetic variability within kiwifruit-breeding lines, concerning fruit-related traits, is quite large (Cheng et al., 2004; Testolin, 2011). On these premises, breeding efforts will benefit from a more in-depth understanding of the molecular bases of key traits of *Actinidia*, such as flavor, fragrance, color, size, and ripening. Transcriptomic analyses have provided resources to access the genetic diversity and control of flavor and fragrance in large-fruited *Actinidia*, but also *A. arguta* and *A. eriantha* (Crowhurst et al., 2008; Wang et al., 2021). The availability of a sequenced kiwifruit genome(Wu et al., 2019b) together with the accurate definition of molecular markers associated to diverse phenotypes can guide the selection of seedlings at an early stage, facilitating and speeding up the breeding process (Testolin, 2011).

FIGURE 10.2 Diversity of fruit color between green-fleshed, yellow-fleshed, and red-fleshed *Actinidia* cultivars.

FIGURE 10.3 *Actinidia* species are apparently all functionally dioecious. Male flowers are reported on the left side, while female flower is reported on the right side.

10.2.2 Breeding of Hermaphrodite Kiwifruit

Hermaphroditism represents a compelling solution to increase orchard efficiency and productivity. The identification of 'inconstant males', male plants carrying hermaphrodite flowers and capable of producing fruits (McNeilage, 1991), paved the way to the study of sex determination and dioicy in kiwifruit (Figure 10.3). Inconstant males have been crossed in the past in the attempt to produce stably hermaphrodite plants (McNeilage and Steinhagen, 1998); however, only a small number of ovules appeared fertile (Testolin et al., 1995) and the frequency of bisexual flowers appeared seasonally variable (McNeilage, 1991). While the genetic control of sex determination in *Actinidia* plants has been recently uncovered (Akagi et al., 2018, 2019), the development of stable, absolute hermaphrodites and self-compatible kiwifruit plants has, up to now, remained elusive.

10.2.3 Breeding of Disease-Resistant Kiwifruit

Pseudomonas syringae pv. *actinidiae* (Psa), the causal agent of bacterial canker in kiwifruit, was isolated in 1984 in Japan (Takikawa et al., 1989). In 2010 this pathogen spread to New Zealand, threatening its kiwifruit industry (Everett et al., 2012). Since then, bacterial canker has become a pandemic disease for both *A. chinensis* var. *deliciosa* and *A. chinensis* var. *chinensis*, with the latter showing higher susceptibility to the disease (Scortichini et al., 2012). Symptoms of Psa infection range from necrotic leaf spots to displays of flower necrosis and blight, twig wilting and dieback, reddening of the lenticels and fruit collapse. In early spring, a reddish discoloration under the bark or rust-colored exudates can be observed (Figure 10.4) (Donati et al., 2014).

Varieties which appear more tolerant to the disease have been noted throughout the years and several breeding programs have been put in place to define the genetic bases of Psa resistance and produce Psa-resistant new cultivars. Controlled crosses between highly susceptible cultivars and others that appear tolerant in the field have indicated that resistance to Psa (see Section 10.5) is controlled by multiple QTLs (*Quantitative Trait Loci*) (Tahir et al., 2018, 2019, 2020). Moreover, a connection between Psa resistance and ploidy level has been recently hypothesized (Saei et al., 2018). *In vitro* assays showed an array of susceptibility and tolerance levels to Psa within kiwifruit genotypes, suggesting that the exploration of germplasm collections should be pursued (Wang et al., 2019).

FIGURE 10.4 PSA symptoms on 'Jintao' cultivar on leaves and bark.

10.2.4 Rootstocks Selection

While receiving generally lower attention compared to scion breeding, rootstock selection is a powerful tool that can influence the performance of many cultivated fruit crops not only by modulating plant vigor and fruit quality (Friend et al., 2014), but also by improving scion tolerance to several abiotic and biotic stresses. For example, rootstock selection has been used to promote flowering in *A. chinensis* var. *deliciosa* with 'Kaimai', or improve fruit quality in *A. chinensis* var. *chinensis* with 'Bounty 71'. More recently, the need for salt stress resistance has also led to the identification of rootstocks that might allow a better scion acclimatization to salt-rich soils (Abid et al., 2020). A solution to drought stress has also been an object of interest in recent *in vitro* screening endeavors, which have identified genotypes of species *Actinidia macrosperma* and *Actinidia valvata* as promising materials to obtain water scarcity-resistant cultivars (Zhong et al., 2018). Lastly, an important area of interest in rootstock breeding is Psa resistance. Recently, the interspecific cross *A. eriantha*×*A. deliciosa* allowed to generate rootstock 'YZ310', which appeared resistant to Psa and is considered suitable for production practice (Lei et al., 2015). *A. arguta* also appeared less susceptible to Psa, with differences depending on ploidy level (Cossio et al., 2015). Overall, these observations prove that wild *Actinidia* genotypes hold genetic resources that might reveal the key to achieve Psa resistance in cultivated varieties, which remain, at present, threatened by the disease.

10.3 TYPE OF MARKERS USED IN KIWIFRUIT BREEDING

Following some of the first recorded open-pollinated crosses starting from the late 1960s (McNeilage, 2015), the use of genetic markers has greatly evolved throughout the years to guide kiwifruit-breeding efforts more precisely and effectively. The study of isozymes represented the first resource used to detect genetic variations at individual loci. Changes at DNA sequence level can produce variations in the amino acid sequence, detectable observing banding patterns in electrophoresis analyses. Several *Actinidia* cultivars were successfully differentiated using AAT (*Aspartate Aminotransferase*), PGI (*Phosphoglucose Isomerase*), and PGM (*Phosphoglucomutase*) in a multilocus isozymes analysis (Messina et al., 1991). The isozymes repertoire was subsequently expanded and employed for phylogenetic and ploidy analyses studies (Testolin and Ferguson, 1997). Low amount of available markers and poor gel resolution limited the application of this approach. In the attempt to overcome the weaknesses of isozymes and produce better genetic maps, the potential of RFLPs (*Restriction Fragment Length Polymorphysm*) was assessed (Tanksley et al., 1989). The analysis of RFLPs depends on genomic sequence polymorphisms, due to substitutions, insertions, deletions or rearrangements, that alter the location of restriction enzyme sites. Restriction fragments are separated by means of gel electrophoresis, producing banding patterns that change depending on sequence variations in different genotypes. RFLPs analyses have been applied to *Actinidia* for phylogenetic analyses extending to chloroplastic DNA (Crowhurst et al., 1990) and mitochondrial DNA (Li et al., 2003). RFLP analysis was also coupled to PCR to produce CAPS (*Cleaved Amplified Polymorphic Sequence*) (Konieczny and Ausubel, 1993). CAPS fragments analysis was used on interspecific hybrids to test plastid inheritance in *Actinidia*, demonstrating it is strictly paternal (Cipriani et al., 1995). While restriction fragments represent a more effective tool compared to isozymes for genetic screening of traits of interest, due to their higher number and un-relatedness to environmental factors capable of altering protein profiles, they are bound to the presence of restriction sites (Chang and Meyerowitz, 1991).

A broader alternative was introduced by RAPD (*Random Amplification of Polymorphic DNA*) analysis, with the potential to detect polymorphism in unknown DNA sequences (Williams et al., 1990), extending the range of genetic markers available for fingerprinting, population genetics, and genome mapping studies. In fact, RAPD implies the use of short, arbitrary primer sequences in PCR to amplify random sequences at genome wide level. Gel separation of thus generated amplicons can produce unique profiles characteristics of different genotypes. RAPD was employed in kiwifruit fingerprinting, appearing useful in distinguishing *Actinidia* species and genotypes within the same species (Cipriani et al., 1996) and phylogenetic analyses (Li et al., 2003). Notably, RAPD analysis also provided the first examples of sex-linked DNA markers in *A. chinensis*, female sex marker SmX and male sex marker SmY (Harvey et al., 1997). To overcome the reproducibility issues of RAPD markers, SmX and SmY were subsequently converted into SCARs (*Sequence-Characterized Amplified Regions*), obtained after RAPD markers sequencing and production of target-specific primers. While being effective for sex identification screening in *A. chinensis* var. *chinensis* and *A. chinensis* var. *deliciosa*, SCARs SmX, and SmY were not detected in more distant accessions such as *A. eriantha*, *A. arguta*, *A. latifolia*, *A. setosa*, and *A. rufa* (Gill et al., 1998).

A combination of RAPD and RFLP was provided by AFLP (*Amplified Fragments Length Polymorphisms*) analysis, which implicates the digestion of genomic DNA using restriction enzymes followed by PCR amplification of the fragments using primer sequences complementary to adaptor sequences ligated to the fragments (Vos et al., 1995). Compared to RAPD and RFLP, AFLP appeared to detect polymorphisms with higher efficiency (Garcia-Mas et al., 2000). AFLP analysis was applied to kiwifruit for genome mapping and cultivar fingerprinting (Testolin et al., 1999), to search for sex-linked markers (Xiao et al., 1999), and to detect genetic variations between field-grown plants and tissue culture-derived plants (Prado et al., 2006).

Microsatellites, or SSRs (*Simple-Sequence Repeats*), are short tandemly repeated nucleotide sequences densely interspersed in eukaryote genomes. Easily amplified by PCR using primers flanking the repeated region, SSRs became valuable tools to detect polymorphisms and distinguish different plant cultivars thanks to their extensive genome coverage, and to determine the allelic contribution of each parent in hybrids because of their codominant transmission (Powell et al., 1996). SSRs high levels of polymorphism make them suited for populations genetic studies and gene mapping (Kesawat and das Kumar, 2010), and have been applied to *Actinidia* for cultivar fingerprinting (Weising et al., 1996; Huang et al., 1998). SSRs are often analyzed in conjunction with SNPs (*Single Nucleotide Polymorphisms*), which pushed forward MAS thanks to the very high abundance of these single base polymorphisms in plant genomes (Gupta et al., 2001). SSRs and SNPs are presently used for the production of genetic maps which assist and guide kiwifruit breeding through MAS, speeding up the selection process (Fraser et al., 2009; Scaglione et al., 2015b; Zhang et al., 2015).

10.4 MOLECULAR MARKERS FOR SEX DISCRIMINATION

As stated above (see Section 10.1), the genus *Actinidia* is classified as functionally dioecious. In terms of breeding efforts, male plants represent a waste of land and resources because the main focus is the selection of fruit-bearing individuals. Since kiwifruit breeders are seldom interested in selecting males in breeding programs, a great effort has gone into the search for sex markers. The ability to discard male seedlings at a very early stage through the analysis of associated markers would save labor and space in the orchard (Testolin and Cipriani, 2016).

Any cross between female and male genotypes usually produces female and male offspring in the ratio of 1:1 (Testolin et al., 1995). However, it has been observed that in populations where the male parent produces unreduced gametes the inheritance of a Y-chromosome is not 1:1 in the progeny. Moreover, rare "inconstant" male of *A. chinensis* var. *deliciosa* plants that occasionally produce fruit have been identified and fruiting males produce male and female progeny in a 3:1 ratio when selfed (Testolin et al., 1995). All these observations indicate that in *Actinidia* there is a strict genetic control of sex expression. Several studies reported that gender is apparently monofactorial and controlled by a heterogametic XY system originated approximately 20 million years ago (Testolin et al., 1995; Harvey et al., 1997; Akagi et al., 2018; Leite Montalvão et al., 2021), in which individuals carrying at least one copy of Y-chromosome are male, regardless of ploidy level, suggesting that the sex-determining region (SDR) includes an active male-determining factor.

Huang and co-workers (2013a) mapped the SDR in the subtelomeric region of a karyotypically indistinguishable chromosomes pair, linkage group 17, which corresponds to the subtelomeric region of the pseudomolecule 25 (Pilkington et al., 2018) of the male parent in the more recent genome assembly of *Actinidia chinensis* var. *chinensis* (genotype Red5), 2,166 cM from the top (Online Resources). In this region, the suppression of recombination turns out to be the key feature of sex chromosomes evolution in kiwifruit. The phenomenon explains how the two chromosomes of the pair evolved separately and the multiple genes involved to be inherited as a single genetic determinant, so that only male and female progeny are produced (Fraser et al., 2009; McNeilage et al., 2011; Pilkington et al., 2019). This aspect also explains why finding markers very close to the sex determinants is challenging and why it is hard to establish the order and the distances that separate markers and sex-genes, considering also the relatively small size of the populations studied.

As mentioned in Section 10.3, Gill and co-workers (1998) made the first attempt to find markers linked to gender working on the combination of RAPD markers with bulk segregant analysis (BSA). In this study, the screening of 500 random primers has led the identification of the two sex markers SmX and SmY which were then transformed into three SCAR markers cloning and sequencing the original RAPD and designing allele-specific primers. Recently, Chłosta and co-workers (2021) proposed that the SmY marker can be used successfully to detect Y-chromosome also in calli and regenerants. However, SmX and SmY show weakness because they are not always transferable from one cross population to another and from one species to another and if they are, they sometimes lose their polymorphism (Harvey et al., 1997; Gill et al., 1998; Testolin and Cipriani, 2016).

Other efforts based on use of SSRs markers followed, aiming at the identification of reliable markers linked with SDR (Huang et al., 1998; Fraser et al., 2009). The markers UDK096 and Ke225 together with SmX and SmY represented the closest gender set of markers reported spanning 4 cM within the SDR where the sex phenotype map. In particular, Ke225 marker is preferentially applied in breeding programs in New Zealand (http://www.plantandfood.co.nz), but the primers have not been published yet.

More recently, two different groups published two SNP-based genetic maps that allowed to improve the available genome assembly anchoring of sex-surrounding scaffolds and the isolation of new SSR markers from those scaffolds. Zhang and co-workers (2015) produced three additional SSR markers (A001, A002 and A003) within the SDR that can be used to sex type male and female plants. These markers were also successfully amplified in *A. rufa* and interspecific crosses with *A. chinensis* var. *chinensis* in comparison with those listed above. Scaglione and co-workers (2015b) applied a double digest restriction-associated DNA sequencing (ddRADseq) protocol for genotyping by sequencing that allowed to place about 120 Mbp of previously unanchored sequence of the genome assembly, and identified four new SSR markers (s_514T, s_514B, s_776T, s_776B). These markers were able to discriminate gender progeny in most analyzed crosses, whatever the ploidy level of cross parents and whatever the species essayed. However, due to the highly polymorphic nature of SSR markers, the sex-linked alleles were not always of the same length in the different crosses assayed. Accordingly, the identification of the alleles carried by the cross parents is necessary for the analysis of the allele associated with the gender in the progeny. These markers are therefore not suitable for screening germplasm collections where ahead parentage of accessions is not known. New markers associated with the SDR that are more appropriate, stable, and characterized by greater transportability among genotypes and species are therefore necessary.

QTL study associated with SDR was also performed recently in tetraploid kiwifruit by Tahir and co-workers (2020). This study provides the opportunity to confirm a strong QTL on the upper arm of the LG25 at 9.1 cM over the SNP marker Chr:2290069.

In the last years, Akagi and co-workers (2018, 2019) have accomplished significant steps forward in the characterization of the two Y-encoded sex determinants in kiwifruit. The Shy Girl (*SyGI*) gene, which is a Y-encoded cytokinin response regulator that acts as the suppressor of female development (Akagi et al., 2018), is expressed in developing flowers, specifically at the surface of the rudimentary carpels of male flowers. The Friendly Boy (*FrBy*), a fasciclin-like protein which would maintain male fertility, exhibits strong expression in tapetal cells (Akagi et al., 2019). Moreover, *FrBy* acts for the maintenance of male (M) functions, independently of *SyGI*. These two sex determinants are located at an estimated distance of 500 kb in the specific region of the Y-chromosome, while they are absent in the corresponding chromosomal female region.

A molecular protocol for Early Sex Discrimination (ESD) in *Actinidia* spp. based on these genes has been recently published (De Mori et al., 2022). ESD test involves the use of primers designed on the *SyGI* and *FrBy* genes that can discriminate between male and female plants with a simple signal of presence/absence both among the species and on genotypes within different ploidies. The test could unequivocally identify the gender of an unknown sample both in the most commercially important species of *A. chinensis* and in further 13 *Actinidia* species tested.

In the case of *A. chinensis* var. *chinensis* and *A. chinensis* var. *deliciosa*, male and inconstant male genotypes showed three peaks corresponding to the fragments of 86 base pairs (bp), 154, and 289 bp, resulting from the amplification of *SyGI*, *Ank*, and *FrBy* genes, respectively (De Mori et al., 2022), with the exception of one genotype that showed a peak of 292 bp in the case of the *FrBy* gene. *SyGI* amplicon size for the different species ranges from 86 bp of *A. rufa*, to 91 bp of *A. polygama*. The amplification of *FrBy* resulted in the amplification of a 289 bp peak in all species except *A. macrosperma*, *A. valvata*, and *A. polygama*, where the fragment size was of 292 bp and *A. eriantha* where the fragment was of 282 bp. The housekeeping gene was amplified in both male and female genotypes and pherograms showed a peak corresponding to the fragment 154 bp in size for all *Actinidia* species with the exception of *A. kolomikta* which showed a peak corresponding to a fragment of 144 bp in size in both the two female genotypes and in the male one (De Mori et al., 2022). This test represents an important tool for breeding programs and offers also the possibility of screening of a large number of plants exploiting a quick protocol of extraction. Table 10.2 reports kiwifruit sex-linked markers described in the literature. In Figure 10.5, a representation of the LG25 of the last genetic map available (Scaglione et al., 2015b), the SDR identified in the male parent and the corresponding physical region on the recent LG25 assembly of the male genotype 'Russel' (Tahir et al., 2022) and on the LG25 assembly of the female reference genome 'Red5' (Pilkington et al., 2018) with the flanking markers.

TABLE 10.2
Kiwifruit Sex-Linked Markers Reported in the Literature

Marker	Type	Forward-Primer	Reverse-Primer	At (°C)	References
SmX (b)	SCAR	AGGAGTCGGAGAGAGTAGAGAAG	AGGAGTCGGATGACCGTTGGTGA	68	Gill et al. (1998)
SmY1	SCAR	TCGCAATTCGTTAGGGATGATGCG	CATAATCAACCATCCATAAAAACCAT	58–54 td	Gill et al. (1998)
UDK-096	SSR	CCCAAAGTTCCTCACTCTTCC	TAAACAAAAACCAAATCTGGCA	55	Fraser et al. (2009)
Ke225	SSR	Not published	Not published	n.a.	Fraser et al. (2009)
s_514T	SSR	CTGGATCAGCTTCTGGACT	GGCAAAAGATGAAAAGAGTG	55	Scaglione et al. (2015b)
s_514B	SSR	TCTTTGCGATCCTATTTTGT	TGCAAACAGAAAAACAATGA	55	Scaglione et al. (2015b)
s_776T	SSR	CAATTTGACCCAAGTACCAC	ATGGCAATCAATCACTCAAT	55	Scaglione et al. (2015b)
s_776B	SSR	TGATTTGCTTGCTTTATGAAT	TCGGTTTTGTCTGTTTTAG	55	Scaglione et al. (2015b)
A001	SSR	TCAATGCATTTAGACATTCCTTTGTCCA	TGGGTAAACATAACCACATGCCAAC	n.r.	Zhang et al. (2015)
A002	SSR	TACTGACGGTCACTCCCTAATCCC	CATGGATGGAACTGGTGGAGGAAG	n.r.	Zhang et al. (2015)
A003	SSR	GCAAGCGGGGGTAAATTTGTACAG	GGATAGGAGGAGCTTTACGGACCT	n.r.	Zhang et al. (2015)
SyGI[a]	Gene	ATACGAAGTTTGAATCGGTGATTCGTC	CTAAGCCCTAGGGAAATATTTTAAAA	55 td	Akagi et al. (2018)
FrBy-UD[a]	Gene	ATTCCATTCTCATTACTAACAACTC	CCTAAAATAAACCAACCAAGAAAAC	55 td	De Mori et al. (2022)
Ank-UD[a]	Gene	TTTGCCTAAAGAAAGTAGTAGTCAT	AATTATGAATGGGAGGCTTCTT	55 td	De Mori et al. (2022)

AT, annealing temperature.
[a] Markers that are used in combination: male genotypes amplify fragments in all three markers, but female genotypes only in Ank-UD (see the text for details).

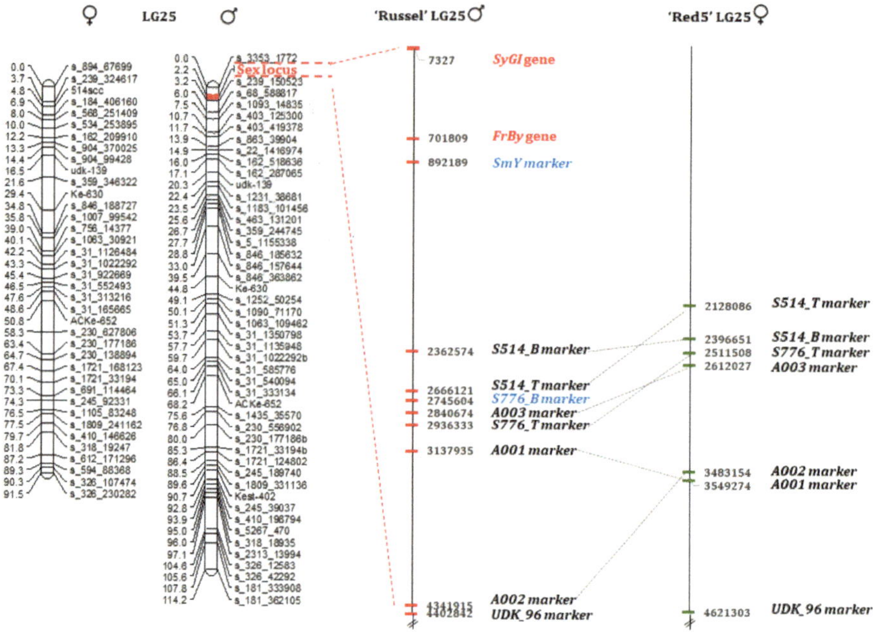

FIGURE 10.5 Representation of the sex locus in *Actinidia*. The two Y-encoded sex determinants are reported in red. The markers SmY and S776_B reported in blue were found only in male assembly. Physical positions are reported in bp. SmX marker is not reported because of multiple alignments in both male and female assembly.

10.5 MOLECULAR MARKERS LINKED TO FURTHER SIGNIFICANT QTL

Kiwifruit breeding still largely relies on phenotypic observation of cross populations growing in the field until fruiting maturity, without any selection prior to juvenility stage overcoming. Identification of desirable characters with associated loci would greatly help breeders to screen rapidly breeding populations, saving costs and labor, and favorable alleles could also be introgressed into adapted populations or accumulated in populations through crossing designs. In the literature few papers describe markers associated with traits of interest for breeders in kiwifruit. The first examples of markers associated with agronomic traits interesting for breeders were reported by McNeilage and co-workers (2011). The putative QTLs studied were fruit weight, fruit dry matter content, sucrose content, vitamin C content, and actinidin activity. A second effort was conducted by Scaglione et al. (2015a), who saturated an SSR-based framework linkage map and mapped several traits (e.g. gender, bud burst, flowering time, fruit weight, flesh color, sugar content, pH, and flesh acidity) during two consecutive seasons in a pseudo test-cross population (C8×A54.19) of diploid kiwifruit (*Actinidia chinensis* Planch.). This work led the identification of at least one major QTL for most traits analyzed with the exception of soluble solids content and fruit flesh color. Pinto and co-workers (2018) report a preliminary QTL analysis for fruit-quality traits using the same linkage map and pseudo test-cross population. The results confirm the quantitative nature of most traits analyzed, because more than one QTL was generally detected per trait in different linkage groups (LGs). However, most putative QTLs showed poor stability within the years and most of them were not identified in both parents. Regretfully, the weakness of both the works was the small progeny size that may inflate the estimated QTL effect, underestimating the number of QTLs and overestimating the effects of individual QTLs. The combination of the two could result in failure of the correct quantification of a number of QTLs.

Recently, steps forward were accomplished in the construction of a high-density linkage map for hexaploid green kiwifruit using genotyping-by-sequence (GBS) by Popowski and co-workers (2021). The limited marker-trait association has been reported so far for other polyploid kiwifruit for fruit and production traits. The linkage map obtained consists of 3,686 and 3,940 markers organized in 183 and 176 LGs for the female and male parents, respectively. Both parental linkage maps are co-linear with the reference *A. chinensis* 'Red5' genome of kiwifruit. This map was used to successfully identify QTLs for king flower number, fruit number and weight, dry matter accumulation, and storage firmness 12-weeks post-harvest. Specifically, nine QTLs were identified, two from the female parent and seven from the male parent (Popowski et al., 2021).

As described in Section 10.2.3, in the last few years, the bacterial canker, caused by Psa, has become a pathogen of economic importance since it spread wordwide (Vanneste, 2013). The sudden emergence of this disease in all of the main areas of kiwifruit production in the world suggested that it can be considered as a pandemic disease (Scortichini et al., 2012). Moreover, owing to the death and a general weakening of the infected plants, sever economic losses have occurred in the main areas of kiwifruit cultivation of the world. Bacterial diseases are notoriously difficult to control. Few products are available to control bacteria and even fewer for the control of plant pathogenic bacteria. For these reasons, the breeding programs have been focused in the last years also on the study of resistance sources and in the development of markers straightly linked with possible resistance traits showed by some kiwifruit genotypes.

Tahir and co-workers (2019) provided the first detailed view of the genetic loci modulating Psa resistance and tissue-specific response in diploid *A. chinensis*, utilizing an intensively phenotyped population of seedlings developed from a cross between Psa-susceptible "Hort16A" and a resistant breeding parent (P1) for QTL analysis. This study suggested that Psa resistance in diploid kiwifruit population is governed by several QTLs. Specifically, authors focused their attention on a QTL for control of field resistance to Psa identified in "Hort16A" on the upper arm of LG27. A SNP marker G9P1 developed from Acc30822, a gene of unknown function underlying the QTL and a multi-allelic SSR marker SSRLG27_439F4R4, contributed to 16% and 19% of the population phenotypic variance, respectively (Tahir et al., 2019). Moreover, six QTLs were identified in P1, one was located on the upper arm of LG22, three QTLs on LGs 3, 15, 24 and two QTLs on LG14 and LG28.

Breeding programs are also keen in understanding the genetic control of resilience to Psa disease in tetraploid cultivars. Different studies revealed that diploid (2×) *A. chinensis* var. *chinensis* are more susceptible to Psa infection than tetraploid (4×) (De Silva et al., 2014; Hoyte et al., 2015) which showed also low severity of disease symptoms. Recently, an overview of the genetic landscape of a tetraploid *A. chinensis*, and insights into the genomic regions regulating the quantitative form of Psa resistance in a yellow-fleshed sub-species of kiwifruit were provided (Tahir et al., 2020). QTLs analysis for resistance to Psa in the field revealed four QTLs located on LGs 1, 2, 4, and 7. Taken together, parental P2, which is the more resistant parent, has three QTLS (LGs 1, 4, and 7) which contribute to field resistance or quantitative resistance against Psa, while parental P1, which is weakly tolerant, also contributes with a minor QTL for Psa resistance in the LG2 (Tahir et al., 2020). The interesting thing is that all these genetic regions are different from those identified in the previous work on diploid *A. chinensis* (Tahir et al., 2019). As a result, the origin of Psa resistance in tetraploid species could be unrelated to any common ancestral loci in *Actinidia* species and the sources of the quantitative resistance might be different in terms of their positions in the kiwifruit genome in between the sub-species (Tahir et al., 2020) and further investigation is needed. Table 10.3 reports the putative QTLs controlling traits of interest in kiwifruit-breeding programs and associated markers described in the literature.

TABLE 10.3
Putative QTLs Controlling Traits of Interest in Kiwifruit-Breeding Programs and Associated Markers

Trait	LG Group	Parent	Nearest Marker	References
Fruit weight	2	M	Ke642	McNeilage et al. (2011)
	6	M		Scaglione et al. (2015a)
	27	M		Pinto et al. (2018)
	5F	F	LG5:3333511	Popowski et al. (2021)
	13E	M	LG13:74766	
Dry matter	19	M	Ac1192	McNeilage et al. (2011)
	24A	M	LG24:16322191	Popowski et al. (2021)
	6D	F	LG6:12592191	
Sucrose	25	F	Ke579(1)	McNeilage et al. (2011)
	23	M	Ke518	
Vitamin C	24	F	Ke529	McNeilage et al. (2011)
	26	M	Udkac038	
Actinidin	16	M	Ke801	McNeilage et al. (2011)
	16	M	Ke656(1)	
Date of bud burst	18	F		Scaglione et al. (2015a)
Date of flowering	18	F		Scaglione et al. (2015a)
pH	8	F		Scaglione et al. (2015a)
	7	F		Pinto et al. (2018)
	29	F		
Acidity	8	F		Scaglione et al. (2015a)
	11	M		
	6	M		Pinto et al. (2018)
Ripening time	1	F		Pinto et al. (2018)
	18	F		
Soluble solids content	6	F		Pinto et al. (2018)
Skin chlorophyll absorbance	21	M		Pinto et al. (2018)
	4	F		
	6	F		
	21	F		
	1	F		
	8	F		
Flesh chlorophyll absorbance	21	M		Pinto et al. (2018)
Fruit firmness	11	F		Pinto et al. (2018)
	5	M		
	7	M		
	9	M		
	21	M		
Fruit number	6D	M	LG6:2142758	Popowski et al. (2021)
	20C	M	LG20:15935274	
	5C	M	LG5:5184121	
King flowers per winter bud	20E	M	LG20:11756985	Popowski et al. (2021)
Fuit firmness-12 weeks post-harvest	15B	M	LG22:13315746	Popowski et al. (2021)

(Continued)

TABLE 10.3 (*Continued*)
Putative QTLs Controlling Traits of Interest in Kiwifruit-Breeding Programs and Associated Markers

Trait	LG Group	Parent	Nearest Marker	References
Field resistance to PSA	27	F	G9P1 SSRLG27_439F4R4	Tahir et al. (2019)
	3	M		
	15	M		
	24	M		
	14	M		
	28	M		
	1/h8	M	Chr1:14996600	Tahir et al. (2020)
	2/h2	F	Chr2:13522420	
	4/h5	M	Chr4:5917168	
	7/h5	M	Chr7:6815185	

M, male; F, female; LG, linkage group.

10.6 ROLE OF MOLECULAR MARKERS IN THE STUDY OF PHYLOGENETIC RELATIONSHIPS AMONG *ACTINIDIA* GENUS

The use of molecular analysis to determine phylogenetic relationships within the genus *Actinidia* is well documented. Molecular markers of various types have been used, from isozymes to RAPDs, microsatellites and, of course, more recently, SNPs retrieved from NGS sequencing (Crowhurst et al., 1990; Li et al., 2002). Although widely used and easy to use, most nuclear markers do not have great attractiveness among scientists because the results are not always easy to interpret. Nuclear molecular markers are, in fact, subject to genetic recombination and assigning the contribution of ancestral species is confused. Furthermore, many species of the genus *Actinidia* are polyploids, increasing the complexity of the interpretation of multi-alleles markers. The genus *Actinidia*, however, offers the rare opportunity to study the phylogeny of the species belonging to it following, independently, the maternal and paternal lines. In fact, it has been shown that the heritability of extra-nuclear DNA is different for mitochondria and chloroplasts. The former is inherited through the female parent and the latter through the male parent, respectively (Cipriani et al., 1995; Testolin and Cipriani, 1997; Chat et al., 2004). Thus, various studies have made it possible to draw a fairly precise picture of the relationships that exist between the different species of the genus *Actinidia*.

As previously described (Section 10.1), the botanical subdivision of the genus *Actinidia* has been subjected to several revisions, among which the main ones are that of Liang (1984) and the most recent of Li et al. (2007).

Fifty-one species and 21 botanical varieties are recognized, mostly originating from East Asia. The subdivision into the four sections (see Section 10.1) was made considering only morphological characters.

It has rarely been possible to carry out an extensive representative sampling of all taxa of the genus *Actinidia*. Often, a species was represented by only one individual. With this limitation, intraspecific variability could be greatly underestimated. The first phylogeny reconstructions that specifically used chloroplastic and mitochondrial DNA were made using CAPS type markers, designing primers for PCR in conserved regions of organelle genomes and appropriately selecting regions sufficiently variable between species but conserved between individuals of the same species.

The ITS regions have been also used in the phylogeny reconstruction of taxa relationship in *Actinidia* because of several favorable properties (Baldwin et al., 1995):

- It is routinely amplified thanks to its small size associated to the availability of highly conserved flanking sequences.
- It is easy to detect even from small quantities of DNA due to the high copy number of the rRNA clusters.
- It undergoes rapid concerted evolution via unequal crossing-over and gene conversion. This promotes intra-genomic homogeneity of the repeat units, although high-throughput sequencing showed the occurrence of frequent variations within plant species (Song et al., 2012).
- It has a high degree of variation even between closely related species. This can be explained by the relatively low evolutionary pressure acting on such non-coding spacer sequences.

Exon primed intron-crossing (EPIC) markers have been successfully used to reconstruct the phylogenetic relationships among five species of *Actinidia* with a most complicated sympatric distribution and reveal evidence of natural hybridization and introgression (Yao et al., 2015). The introgression due to the ease of hybridization between species that coexist and overlap in the same area leads to inconsistencies in the phylogenetic reconstruction obtained from DNA sequences of organelles and nuclear genes (Liu et al., 2013).

However, interspecific relationships have not been fully resolved or are inconsistent between studies (Li et al., 2002; Chat et al., 2004; Huang et al., 2013b), and the infrageneric classification of *Actinidia* is not well supported by the molecular data.

The application of the cheapest NGS sequencing techniques has made it possible to have numerous complete chloroplastic genomes of different species of *Actinidia* (Yao et al., 2015; Wang et al., 2016; Kim et al., 2018; Lan et al., 2018; Lin et al., 2018; Tang et al., 2019; Wu et al., 2019a; Zhang and Liu, 2019), which allow detection of thousands of nucleotide polymorphisms from intergenic spacers (IGS) regions.

Recently, aligning chloroplast genomes of eight *Actinidia* species, 24 IGS regions with high degrees of divergence included *rps16-trnQ, rps18-rpl20, trnT-trnL, trnG-trnR, atpF-atpH, petN-psbM, rpl32-trnL, ndhC-trnV, ndhF-rpl32, trnC-petN, petA-psbL, trnE-trnT, trnP-psaJ, rpoB-trnC, ndhG-ndhI, trnF-ndhJ, ycf4-cemA, psaA-ycf3, atpH-atpI, ycf3-trnS, ndhD-ccsA, psbI-trnS, rrn5-ndhF*, and *trnH-psbA* were selected. Ultimately four of the regions, i.e., *ndhF-rps32, ndhC-trnV(UAC), rps16-trnQ(UUG)*, and *trnE (UUC)-trnT(GGU)*, with high informative percentage of variable loci, were used to reconstruct the phylogenetic relationships among *Actinidia* species (Tang et al., 2019). The main findings were: (1) the phylogeny reveals two main groups within *Actinidia*; (2) the four traditionally recognized sections are not monophyletic; and (3) two sections (*Leiocarpae* and *Maculatae*) can be proposed, among which *Leiocarpae* is the earliest diverged lineage in the genus.

In conclusion, the molecular studies do not seem to confirm the traditional classification of the genus *Actinidia* and suggest a careful re-examination of the cytogenetic and morphological evidence on which the traditional taxonomy is based.

10.7 MOLECULAR MARKERS FOR FINGERPRINTING OF KIWIFRUIT

As previously described, *Actinidia* spp. are apparently all functionally dioecious. Cross pollination is therefore obligatory and, consequently, the genome is highly heterozygous.

Several different species grown commercially belong to the genus *Actinidia*, including the *A. chinensis* var. *chinensis/A. chinensis* var. *deliciosa* complex, *A. arguta, A. eriantha,* and others that are less popular. The basic chromosome series is high ($x=29$) as a consequence of several ancestral polyploidizations that occurred during evolution and, in addition, commercial genotypes can be either diploid ($2n=58$), tetraploid ($2n=116$), or hexaploid ($2n=174$). The consequences of this complex botanical and biological landscape are that it is difficult to differentiate accessions and difficult to correctly score the true alleles in polyploids.

The use of genetic information to distinguish varieties is not new in the field of plant breeding (e.g., Wünsch and Hormaza, 2002). DNA fingerprinting has been used to assess varietal purity (e.g., Nybom et al., 2014) and to enforce intellectual property rights (Wijesundara et al., 2018).

The most popular molecular markers for varietal recognition in *Actinidia* are microsatellites or SSRs. They have been widely used in both *A. chinensis* and *A. arguta*, the two most commercially important species (Weising et al., 1996; Huang et al., 1998; Kwon et al., 2013).

Most of the germplasm of the genus *Actinidia* is of Chinese origin, and most of the accessions are collected at the Wuhan Botanical Garden Wuhan Institute of Botany, The Chinese Academy of Sciences, China. Most of the cultivars in China are direct selections from wild populations of *A. chinensis* and *A. deliciosa*. It is not surprising that the most comprehensive survey of genetic variability of the genus carried out with molecular markers was accomplished by the Wuhan researchers. Zhen et al. (2004) analyzed 48 accessions of *Actinidia*, including 26 of *A. chinensis*, 19 of *A. chinensis* var. *deliciosa*, two hybrids *A. chinensis*×*A. eriantha* and an accession of *A. arguta*. Nine microsatellite markers previously developed at the University of Udine, Italy, were used (Huang et al., 1998).

Table 10.4 reports the list of 23 markers that were used for varietal identification in *Actinidia*. Included in the list shown in is a marker trio capable of differentiating gender in many species at any ploidy level via simple signals of presence/absence as detailed in Section 10.4.

10.8 NEW STEPS FORWARD IN KIWIFRUIT BREEDING

Various traditional breeding strategies were applied over the years aiming to improve several crop traits, including kiwifruit (e.g. conventional crossing breeding, marker-assisted breeding as stated in paragraphs above, chemical – and radiation mediated mutation breeding and genetic engineering breeding). Today, many breeders are focusing on available modern molecular technologies. The clustered regularly interspaced short palindromic repeats CRISPR/Cas9 technology has recently become the most popular tool for crop improvement due to its versatility. Indeed, compared with other techniques, the CRISPR/Cas9 system is more efficient and straightforward for genome editing since it is dictated by nucleotide complementarity of the guide RNA to a specific sequence without complex protein engineering. Moreover, genome editing using this technique can provide also the possibility to create predictable and inheritable mutations in specific sites of genome, with the lowest probability of off-target and no integration of exogenous gene sequences.

This interesting technique has found application also in kiwifruit breeding. Varkonyi-Gasic and co-workers (2019) at Plant & Food of New Zealand reported the first gene editing in *Actinidia*. In the study, the authors used CRISPR/cas9 approach to manipulate and perform functional analysis of kiwifruit CEN-like genes *AcCEN4* and *AcCEN*. Mutation of these genes transformed a climbing woody perennial, which develops auxiliary inflorescences after many years of juvenility, into a compact plant with rapid terminal flower and fruit development (Varkonyi-Gasic et al., 2019). This result offers the opportunity to cultivate kiwifruit as an annual crop and the fast life cycle acquired by these edited plants holds the potential to dramatically increase targeted breeding, whereas optimization of growing conditions for faster growth and indoor farming could increase productivity and enable cropping in any geographical and climatic conditions (Varkonyi-Gasic et al., 2019). Moreover, the plants generated in this study offered the opportunity to be used as model plants in following studies conducted by Akagi and co-workers (2018, 2019). Indeed, these authors accomplished the discovery of genes controlling dioecy and its application to produce hermaphroditic selections using CRISPR/Cas9 editing technology on early flowering plants of diploid kiwifruit.

A step forward has been done targeting *SyGI* gene in order to induce a stable gene knock-out also in two tetraploid male accessions of *Actinidia chinensis* var. *chinensis* (De Mori et al., 2020).

TABLE 10.4
Kiwifruit Markers Reported in the Literature for Varietal Identification

Marker	Forward-Primer	Reverse-Primer	At (°C)	Core Repeat	LG[a]	Primers Position from Kiwifruit Genome[a]	RG (bp)
Ank-UD[1b]	TTTGCCTAAAGAAAGTAGTAGTCAT	AATTATGAATGGGAGGCTTCTT	55 td	—	1	18,540,922–18,541,055	144/154
UDK-407[2]	CGTTCGAGGGTAAGGAGTTG	CCTAACCCGACCGTAGTGTG	55	(AG)21	3	18,345,676–18,345,768	n.a.
UDK-414[2]	GCCATTTCAAGAACATTTTTTG	TGGCTATATTTTGCAAGCCC	55	(AG)20	4	6,948,734–6,948,845	n.a.
UDK-409[2]	ATGACCTATTGCCAAGTGGC	TTGTGTGTACCCACCACCC	55	(AG)19	6	10,953,400–10,953,501	n.a.
NZK-769[3]	ACAGCTGAGTTGGCAATATG	AGGAATCACCTGTGTTAGTG	55	(CTT)8	7	12,607,609–12,607,709	n.a.
UDK-030[2]	TCATGTTTGTGGTTGAGTTGTG	AGCAATAAACTCAAGCGCGT	55	(AG)9(AC)15	8	25,249,446–25,249,530	104/113
UDK-026[2]	CGCTGACCAGATTCTGATGA	TTGAAAATCACTGAGCACAACC	55	(AC)14	11	11,429,289–11,429,413	n.a.
UDK-037[2]	CATTTCAGTGTTAACTCTTTCG	CCCTCTTTCAAAGGTTCTCT	55	Complex	13	16,259,889–16,259,942	73/75
UDK-040[2]	TCGAGTTACCTAGCTACTCCGC	CAAGGGAAGAAAATGTTGAACC	55	(AC)15(AG)21	13	15,362,425–15,362,546	143/183
UDK-448[2]	CTTCGCCATCTACGGTAACA	GCCTATGTTCCACCGGTC	55	GA	13	7,888,439–7,888,544	124/132
UDK-034[2]	TTATATGGTGCGGCATGCTA	TGAATGCAGAAGGCAATCAG	55	(AC)12(AG)17	19	13,948,118–13,948,288	188/194
UDK-413[2]	GCACTTTACACCTCTCAATCAA	AAAGATGTTTTTGAACCATGCA	55	(AG)21	19	12,450,772–12,450,875	n.a.
NZK-766[3]	CATCGAAGCATATGAAACAG	CTCCTGAGGTTAAGACGGAG	55	Complex	21	17,116,088–17,116,218	n.a.
UDK-001[2]	GAATCGCGTAATGATTGATGG	GTTTCCCACTCTGCAAAAGC	55	(AC)10(TC)16	21	13,141,818–13,142,069	270/-
UDK-035[2]	AAGAGCCATAGCTTATTCACCG	AAGTAAAGCCATTGTCATTGCA	55	(AC)17	23	16,931,135–16,931,253	n.a.
UDK-015[2]	CCGAGTCATGATCGAGTTGA	GGCTCAACTTGGAGAAGTGG	55	(AC)21	24	12,646,215–12,646,312	n.a.
S-514T[4]	CTGGATCAGCTTCTGGACT	GGCAAAAGATGAAAAGAGTG	55	(GAA)13	25	2,128,086–2,128,298	212/229
SyGI[5b]	ATACGAAGTTTGAATCGGTGATTCGTC	CTAAGCCCTAGGGAAATATTTTAAAA	55 td	—	25	n.a.	86/91
FrBy-UD[1b]	ATTCCATTCTCATTACTAACAACTC	CCTAAAATAAACCAACCAAGAAAAC	55 td	—	25	n.a.	282/292
UDK-092[2]	ATCACATGAATACGTGGATTGC	TGTGTGTTCATTTGTCATCCTT	55	AG	26	10,291,960–10,292,119	n.a.
UDK-039[2]	GGTTTGATCGGTCTTCGAAA	ATAAATGTGTGCCAGTGCGA	55	(AC)10	27	5,071,117–5,071,251	n.a.
NZK-768[3]	GATCCGGGTTTTAAGAAG	ATGGAGTTCCTATGCTTCTG	55	(CTT)5	28	28: 847,046–847,239	n.a.
NZK-767[3]	GATCAGTTGAGGCAAGAAG	GGAGTCATCTATGGAACTTG	55	(CTT)5	n.a.	n.a.	n.a.

[1] De Mori et al. (2022), [2] Huang et al. (1998), [3] Weising et al. (1996), [4] Scaglione et al. (2015), [5] Akagi et al. (2018).
RG, reference genotype (alleles in base pairs). The reference genotype is 'Hongyang' (diploid) (unpublished data); AT, annealing temperature; n.a., not available.
[a] Kiwifruit genome assembly: GCA_003024255.1.
[b] Marker used in combination applying td protocol = touch down: –0.5°C per step for 10 cycles.

Voogd and co-workers (2022) studied the regulation of expression of a MADS-box gene with similarity to flowering locus C (FLC) and functionally characterized this gene using overexpression and mutagenesis though CRISPR/Cas9. This study determines the role of this FLC-like gene in kiwifruit phenology, demonstrating functional divergence despite the conserved aspects of FLC regulation, and represents another goal because these results have implications for the future management and breeding of perennials for resilience to changing climate (Voogd et al., 2022).

Recently, an up-date of CRISPR/Cas9 technique has been developed. CRISPR-drive is a recent and robust tool that allows durable genetic manipulation of plant diseases. CRISPR-drive contains a similar process of transferring new genes to plants by CRISPR/Cas9. The target of the technique is susceptible gene associated with disease susceptibility (Tek and Budak, 2022). CRISPR-drives are theoretically valid to develop homozygous S-mutant plants against pathogens by only routine pollination after CRISPR-drive transformation to just one parental line. This way, breeders could use this parental line in different crossing combinations without reapplying the genome-editing technique or backcrossing (Tek and Budak, 2022). Because of the potential of this technique, CRISPR-drive could represent a key future also in kiwifruit breeding thinking of future discovery about gene/s associated with PSA disease or other emergent disease like kiwifruit vine decline syndrome (KVDS).

REFERENCES

Abid M, Zhang YJ, Li Z, Bai DF, Zhong YP, and Fang JB (2020). Effect of salt stress on growth, physiological and biochemical characters of four kiwifruit genotypes. *Sci Hortic* 271, 109473. doi:10.1016/J.SCIENTA.2020.109473.

Akagi T, Henry IM, Ohtani H, Morimoto T, Beppu K, Kataoka I, et al. (2018). A Y-encoded suppressor of feminization arose via lineage-specific duplication of a cytokinin response regulator in kiwifruit. *Plant Cell* 30, 780–795. doi:10.1105/tpc.17.00787.

Akagi T, Pilkington SM, Varkonyi-Gasic E, Henry IM, Sugano SS, Sonoda M, et al. (2019). Two Y-chromosome-encoded genes determine sex in kiwifruit. *Nat Plants* 5, 801–809. doi:10.1038/s41477-019-0489-6.

Baldwin BG, Sanderson MJ, Porter JM, Wojciechowski MF, Campbell CS, and Donoghue MJ (1995). The its region of nuclear ribosomal DNA: A valuable source of evidence on angiosperm phylogeny. *Ann Mo Bot Gard* 82, 247–277. doi:10.2307/2399880.

Bretschneider EV (1893). *Botanicum Sinicum. Notes on Chinese Botany from Native and Western Sources.* Shanghai: Kelly & Walsh, Ltd. doi.org/10.5962/bhl.title.4313.

Chang C, and Meyerowitz EM (1991). Plant genome studies: Restriction fragment length polymorphism and chromosome mapping information. *Curr Opin Biotechnol* 2, 178–183. doi:10.1016/0958-1669(91)90007-R.

Chat J, Jáuregui B, Petit RJ, and Nadot S (2004). Reticulate evolution in kiwifruit (*Actinidia*, *Actinidiaceae*) identified by comparing their maternal and paternal phylogenies. *Am J Bot* 91, 736–747. doi:10.3732/AJB.91.5.736.

Cheng CH, Seal AG, Boldingh HL, Marsh KB, Macrae EA, Murphy SJ, et al. (2004). Inheritance of taste characters and fruit size and number in a diploid *Actinidiachinensis* (kiwifruit) population. *Euphytica* 138, 185–195. doi:10.1023/B:EUPH.0000046802.28347.41.

Chłosta I, Kwolek D, Sliwinska E, Góralski G, and Popielarska-Konieczna M (2021). Sex-linked molecular markers identify female lines in endosperm-derived kiwifruit callus and in regenerants. *Plants* 10(3), 526. doi:10.3390/PLANTS10030526.

Cipriani G, Di Bella R, and Testolin R (1996). Screening RAPD primers for molecular taxonomy and cultivar fingerprinting in the genus *Actinidia*. *Euphytica* 90, 169–174. doi:10.1007/BF00023855.

Cipriani G, Testolin R, and Morgante M (1995). Paternal inheritance of plastids in interspecific hybrids of the genus *Actinidia* revealed by PCR-amplification of chloroplast DNA fragments. *Mol Gen Genet MGG* 247, 693–697. doi:10.1007/BF00290400.

Cossio F, Debersaques F, and Latocha P (2015). Kiwiberry (*Actinidia arguta*): New perspectives for a great future. *Acta Hortic* 1096, 423–434. doi:10.17660/ACTAHORTIC.2015.1096.51.

Crowhurst RN, Gleave AP, MacRae EA, Ampomah-Dwamena C, Atkinson RG, Beuning LL, et al. (2008). Analysis of expressed sequence tags from *Actinidia*: Applications of a cross species EST database for gene discovery in the areas of flavor, health, color and ripening. *BMC Genomics* 9, 1–26. doi:10.1186/1471-2164-9-351/FIGURES/12.

Crowhurst RN, Lints R, Atkinson RG, and Gardner RC (1990) Restriction fragment length polymorphisms in the genus *Actinidia* (*Actinidiaceae*). *Plant Syst E* 172, 193–203.

De Mori G, Testolin R, and Cipriani G (2022). A molecular protocol for early sex discrimination (ESD) in *Actinidia* spp. *J Berry Res* 12, 249–266. doi:10.3233/JBR-211530.

De Mori G, Zaina G, Franco-Orozco B, Testolin R, De Paoli E, and Cipriani G (2020). Targeted mutagenesis of the female-suppressor *SyGI* gene in tetraploid kiwifruit by CRISPR/CAS9. *Plants* 10, 62. doi:10.3390/plants10010062.

De Silva NH, Gea L, and Lowe R (2014). Genetic analysis of resistance to *Pseudomonas syringae* pv. *actinidiae* (Psa) in a kiwifruit progeny test: An application of generalised linear mixed models (GLMMs). *Springerplus* 3, 547. doi:10.1186/2193-1801-3-547.

Donati I, Buriani G, Cellini A, Mauri S, Costa G, and Spinelli F (2014). New insights on the bacterial canker of kiwifruit (*Pseudomonas syringae* pv. *actinidiae*). *J Berry Res* 4, 53–67. doi:10.3233/JBR-140073.

Everett KR, Pushparajah IPS, and Vergara MJ (2012). *Pseudomonas syringae* pv *actinidiae* on surfaces in the orchard. *NZ Plant Protect* 65, 19–24. doi:10.30843/NZPP.2012.65.5419.

Ferguson AR (1999). Kiwifruit cultivars: Breeding and selection. *Acta Hortic* 498, 43–51. doi:10.17660/ACTAHORTIC.1999.498.4.

Ferguson AR (2016). Botanical Description. In: Testolin R, Huang H-W, Ferguson AR (eds) *The Kiwifruit Genome*. Springer International Publishing, Beijing, pp. 1–13.

Ferguson AR, and Allison A (2004). 1904-the year that kiwifruit (*Actinidia deliciosa*) came to New Zealand. *NZ J Crop Hortic Sci* 32, 3–27. doi: 10.1080/01140671.2004.9514276.

Fraser LG, Tsang GK, Datson PM, De Silva HN, Harvey CF, Gill GP, et al. (2009). A gene-rich linkage map in the dioecious species *Actinidia chinensis* (kiwifruit) reveals putative X/Y sex-determining chromosomes. *BMC Genomics* 10, 102. doi:10.1186/1471-2164-10-102.

Friend AP, Palmer JW, Seymour SM, and Diack RN. (2014). Potential of clonal rootstocks for devigoration and enhanced fruit characteristics in kiwifruit orchards. *Acta Hortic* 1058, 429–434. doi:10.17660/ACTAHORTIC.2014.1058.52.

Garcia-Mas J, Oliver M, Gómez-Paniagua H, and de Vicente MC (2000). Comparing AFLP, RAPD and RFLP markers for measuring genetic diversity in melon. *Theor Appl Genet* 101, 860–864. doi:10.1007/S001220051553.

Gill GP, Harvey CF, Gardner RC, and Fraser LG (1998). Development of sex-linked PCR markers for gender identification in *Actinidia*. *Theor Appl Genet* 97, 439–445. doi:10.1007/s001220050914.

Gupta PK, Roy JK, and Prasad M (2001). Single nucleotide polymorphisms: A new paradigm for molecular marker technology and DNA polymorphism detection with emphasis on their use in plants. *Curr Sci* 80, 524–535. https://about.jstor.org/terms.

Harvey CF, Gill GP, Fraser LG, and McNeilage MA (1997). Sex determination in *Actinidia*. 1. Sex-linked markers and progeny sex ratio in diploid *A. chinensis*. *Sex Plant Reprod* 10, 149–154. doi:10.1007/s004970050082.

Hoyte S, Reglinski T, Elmer P, Mauchline N, Stannard K, Casonato S, et al. (2015). Developing and using bioassays to screen for psa resistance in New Zealand Kiwifruit. *Acta Hortic* 1095, 171–180. doi:10.17660/ACTAHORTIC.2015.1095.21.

Huang S, Ding J, Deng D, Tang W, Sun H, Liu D, et al. (2013a). Draft genome of the kiwifruit *Actinidiachinensis*. *Nat Commun* 4, 2640. doi:10.1038/ncomms3640.

Huang HW, Zhong CH, Jiang ZW, Li XW, Yao XH, Li DW, et al. (2013b). *Actinidia: Taxonomy, Resources, Domestication, Cultivation*. Beijing: Science Press.

Huang WG, Cipriani G, Morgante M, and Testolin R (1998). Microsatellite DNA in *Actinidia chinensis*: isolation, characterisation, and homology in related species. *Theor Appl Genet* 97, 1269–1278. doi.org/10.1007/s001220051019.

Kesawat MS, and das Kumar B (2010). Molecular markers: It's application in crop improvement. *J Crop Sci Biotechnol* 12, 169–181. doi:10.1007/S12892-009-0124-6.

Kim SC, Lee JW, Baek SH, Lee MW, and Kang YJ. (2018). The complete chloroplast genome sequence of *Actinidia rufa* (*Actinidiaceae*). *Mitochondrial DNA B* 3, 564–565. doi:10.1080/23802359.2018.1450676.

Konieczny A, and Ausubel FM (1993). A procedure for mapping *Arabidopsis* mutations using co-dominant ecotype-specific PCR-based markers. *Plant J* 4, 403–410. doi:10.1046/J.1365-313X.1993.04020403.X.

Kwon SJ, Lee GA, Kwack YB, Lee HS, Cho GT, Ko HC, et al. (2013). Development of 34 new microsatellite markers from *Actinidia arguta*: Intra-and interspecies genetic analysis. *Plant Breed Biotechnol* 1, 137–147. doi:10.9787/PBB.2013.1.2.137.

Lan Y, Cheng L, Huang W, Cao Q, Zhou Z, Luo A, et al. (2018). The complete chloroplast genome sequence of *Actinidia kolomikta* from north China. *Conserv Genet Resour* 10, 475–477. doi:10.1007/S12686-017-0852-8/FIGURES/2.

Lei YH, Jing ZB, and Li L (2015). Selection and evaluation of a new kiwifruit rootstock hybrid for bacterial canker resistance. *Acta Hortic* 1096, 413–420. doi:10.17660/ACTAHORTIC.2015.1096.50.

Leite Montalvão AP, Kersten B, Fladung M, and Müller NA (2021). The diversity and dynamics of sex determination in dioecious plants. *Front Plant Sci* 11, 2280. doi:10.3389/FPLS.2020.580488.

Li D, Zhong C, Liu Y, and Huang H (2010). Correlation between ploidy level and fruit characters of the main kiwifruit cultivars in China: Implication for selection and improvement. *NZ J Crop Hortic Sci* 38(2), 137–145. doi:10.1080/01140671.2010.482966.

Li J, Huang H, and Sang T (2002). Molecular phylogeny and infrageneric classification of *Actinidia* (*Actinidiaceae*). *Syst Bot* 27, 408–415.

Li JQ, Li XW, and Soejarto DD (2007). *Actinidiaceae*. In: Wu ZY, Raven PH, Hong DY (eds) *Flora of China*, 12, pp. 334–362. Science Press, Beijing.

Li MZ, Wang L H, Zhuang Q G, and Xie, Y (2018). New red- and yellow-fleshed kiwifruit cultivars. *Acta Hortic* 1218, 135–138. doi:10.17660/ACTAHORTIC.2018.1218.17.

Li Z, Huang H, Jiang Z, Li J, and Kubisiak TL (2003). Phylogenetic relationships in *actinidia* as revealed by RAPDs and PCR-RFLPs of mtDNA. *Acta Hortic* 610, 387–396. doi:10.17660/ACTAHORTIC.2003.610.51.

Liang CF (1984) *Actinidia*. In: Feng KM (ed) *Flora Reipublicae Popularis Sinicae*. Science Press, Beijing, 49, pp. 196–324.

Lin M, Qi X, Chen J, Sun L, Zhong Y, Fang J, et al. (2018). The complete chloroplast genome sequence of *Actinidia arguta* using the PacBio RS II platform. *PLoS One* 13, e0197393. doi:10.1371/JOURNAL.PONE.0197393.

Liu L, Yao XH, and Huang HW (2013). Development of EPIC markers in *Actinidia* and their application for phylogenetic analysis of genus *Actinidia*. *Acta Hortic Sin* 40, 1162–1168.

McNeilage MA (1991). Gender variation in *Actinidia deliciosa*, the kiwifruit. *Sex Plant Reprod* 4, 267–273. doi:10.1007/BF00200546.

McNeilage MA (2015). Genetics and breeding of kiwifruit. *Acta Hortic* 1096, 185–190. doi:10.17660/ACTAHORTIC.2015.1096.19.

McNeilage MA, Fraser LG, Tsang GK, Datson PM, De Silva HN, Crowhurst RN, et al. (2011). Molecular genetics and genomics and kiwifruit breeding. *Acta Hortic* 913, 63–70. doi:10.17660/ActaHortic.2011.913.5.

McNeilage MA, and Steinhagen S (1998). Flower and fruit characters in a kiwifruit hermaphrodite. *Euphytica* 101, 69–72. doi:10.1023/A:1018398924819.

Messina R, Testolin R, and Morgante M (1991). Isozymes for cultivar identification in kiwifruit. *HortScience* 26, 899–902. doi:10.21273/HORTSCI.26.7.899.

Nybom H, Weising K, and Rotter B (2014). DNA fingerprinting in botany: Past, present, future. *Investig Genet* 5, 1–35. doi:10.1186/2041-2223-5-1.

Pilkington SM, Crowhurst R, Hilario E, Nardozza S, Fraser L, Peng Y, et al. (2018). A manually annotated *Actinidia chinensis* var. *chinensis* (kiwifruit) genome highlights the challenges associated with draft genomes and gene prediction in plants. *BMC Genomics* 19, 1–19. doi:10.1186/s12864-018-4656-3.

Pilkington SM, Tahir J, Hilario E, Gardiner SE, Chagné D, Catanach A, et al. (2019). Genetic and cytological analyses reveal the recombination landscape of a partially differentiated plant sex chromosome in kiwifruit. *BMC Plant Biol* 19, 172. doi:10.1186/s12870-019-1766-2.

Pinto C, Testolin R, and Cipriani G (2018). Analysis of components of genetic variance and breeding value of kiwifruit (*Actinidia chinensis* Planch.) using best linear unbiased prediction (BLUP) and pedigree information. *Acta Hortic* 1218, 73–80. doi:10.17660/ActaHortic.2018.1218.9.

Popowski E, Thomson SJ, Knäbel M, Tahir J, Crowhurst RN, Davy M, et al. (2021). Construction of a high-density genetic map for hexaploid kiwifruit (*Actinidia chinensis* var. *deliciosa*) using genotyping by sequencing. *G3 Genes Genomes Genet* 11, 7. doi:10.1093/G3JOURNAL/JKAB142.

Powell W, Machray GC, and Proven J (1996). Polymorphism revealed by simple sequence repeats. *Trends Plant Sci* 1, 215–222. doi:10.1016/1360-1385(96)86898-1.

Prado MJ, Gonzalez MV, Romo S, and Herrera MT (2006). Adventitious plant regeneration on leaf explants from adult male kiwifruit and AFLP analysis of genetic variation. *Plant Cell Tiss Org Cult* 88, 1–10. doi:10.1007/S11240-006-9116-0.

Saei A, Hoeata K, Krebs A, Sutton P, Herrick J, Wood M, et al. (2018). The status of *Pseudomonas syringae* pv. *actinidiae* (Psa). In the New Zealand kiwifruit breeding programme in relation to ploidy level. *Acta Hortic* 1218, 293–298. doi:10.17660/ACTAHORTIC.2018.1218.40.

Scaglione D, Fornasiero A, Cattonaro F, Mesa K, Pinto C, Shiny P, et al. (2015a). A RAD-based linkage map of kiwifruit (*Actinidia chinensis* Pl.) and map position of genetic determinants of traits of interest to breeders. *Acta Hortic* 1096, 191–198. doi:10.17660/ACTAHORTIC.2015.1096.20.

Scaglione D, Fornasiero A, Pinto C, Cattonaro F, Spadotto A, Infante R, et al. (2015b). A RAD-based linkage map of kiwifruit (*Actinidia chinensis* Pl.) as a tool to improve the genome assembly and to scan the genomic region of the gender determinant for the marker-assisted breeding. *Tree Genet Genomes* 11, 1–10. doi:10.1007/s11295-015-0941-3.

Scortichini M, Marcelletti S, Ferrante P, Petriccione M, and Firrao G (2012). *Pseudomonas syringae* pv. *actinidiae*: A re-emerging, multi-faceted, pandemic pathogen. *Mol Plant Pathol* 13, 631–640. doi:10.1111/J.1364-3703.2012.00788.X.

Sharma MK, Singh S, and Sengar RS (2014). Allele mining and its applications. *Biotech Today* 4, 34. doi:10.5958/2322-0996.2014.00016.7.

Song J, Shi L, Li D, Sun Y, Niu Y, Chen Z, et al. (2012). Extensive pyrosequencing reveals frequent intra-genomic variations of internal transcribed spacer regions of nuclear ribosomal DNA. *PLoS One* 7, e43971. doi:10.1371/JOURNAL.PONE.0043971.

Tahir J, Brendolise C, Hoyte S, Lucas M, Thomson S, Hoeata K, et al. (2020). QTL Mapping for resistance to cankers induced by *Pseudomonas syringae* pv. *actinidiae* (Psa) in a tetraploid *Actinidia chinensis* kiwifruit population. *Pathogens* 9(11), 967. doi:10.3390/PATHOGENS9110967.

Tahir J, Crowhurst R, Deroles S, Hilario E, Deng C, Schaffer R, et al. (2022). First chromosome-scale assembly and deep floral-bud transcriptome of a male kiwifruit. *Front Genet* 13, 852161. doi:10.3389/FGENE.2022.852161.

Tahir J, Gardiner SE, Bassett H, Chagné D, Deng CH, and Gea L (2018). Tolerance to *Pseudomonas syringae* pv. *actinidiae* in a kiwifruit breeding parent is conferred by multiple loci. *Acta Hortic* 1203, 67–70. doi:10.17660/ACTAHORTIC.2018.1203.10.

Tahir J, Hoyte S, Bassett H, Brendolise C, Chatterjee A, Templeton K, et al. (2019). Multiple quantitative trait loci contribute to resistance to bacterial canker incited by *Pseudomonas syringae* pv. *actinidiae* in kiwifruit (*Actinidia chinensis*). *Hortic Res* 6, 1–18. doi:10.1038/s41438-019-0184-9.

Takikawa Y, Serizawa S, Ichikawa T, Tsuyumu S, and Goto M (1989). *Pseudomonas syringae* pv. *actinidiae* pv. nov.: The causal bacterium of canker of kiwifruit in Japan. *Ann Phytopathol Soc Jpn* 55, 437–444.

Tang P, Xu Q, Shen R, and Xiaohong Y (2019). Phylogenetic relationship in *Actinidia* (*Actinidiaceae*) based on four noncoding chloroplast DNA sequences. *Plant Syst E* 305, 787–796. doi:10.1007/s00606-019-01607-0.

Tanksley SD, Young ND, Paterson AH, and Bonierbale MW (1989). RFLP mapping in plant breeding: New tools for an old science. *Bio/Technology* 7, 257–264. doi:10.1038/nbt0389-257.

Tek MI, and Budak K (2022). A new approach to develop resistant cultivars against the plant pathogens: CRISPR drives. *Front Plant Sci* 13, 1099. doi:10.3389/FPLS.2022.889497.

Testolin R (2011). Kiwifruit breeding: From the phenotypic analysis of parents to the genomic estimation of their breeding value (GEBV). *Acta Hortic* 913, 123–130. doi:10.17660/ActaHortic.2011.913.14.

Testolin R, and Cipriani G (1997). Paternal inheritance of chloroplast DNA and maternal inheritance of mitochondrial DNA in the genus *Actinidia*. *Theor Appl Genet* 94(6), 897–903. doi:10.1007/S001220050493.

Testolin R, and Cipriani G (2016). Markers, Maps, and Marker-Assisted Selection. In: Testolin R, Huang HW, Ferguson AR (eds) *The Kiwifruit Genome*. Springer, Beijing, pp. 85–99. doi:10.1007/978-3-319-32274-2_7.

Testolin R, Cipriani G, and Costa G (1995). Sex segregation ratio and gender expression in the genus *Actinidia*. *Sex Plant Reprod* 8, 129–132. doi:10.1007/BF00242255.

Testolin R, and Ferguson AR (1997). Isozyme polymorphism in the genus *Actinidia* and the origin of the kiwifruit genome. *Syst Bot* 22, 685–700.

Testolin R, Huang WG, and Cipriani G (1999). Towards a linkage map in kiwifruit (*Actinidia chinensis* Planch.) based on microsatellites and saturated with aflp markers. *Acta Hortic* 498, 79–84. doi:10.17660/ACTAHORTIC.1999.498.8.

Vanneste JL (2013). Recent progress on detecting, understanding and controlling *Pseudomonas syringae* pv. *actinidiae*: A short review. *NZ Plant Protect* 66, 170–177.

Varkonyi-Gasic E, Wang T, Voogd C, Jeon S, Drummond RSM, Gleave AP, et al. (2019). Mutagenesis of kiwifruit *CENTRORADIALIS*-like genes transforms a climbing woody perennial with long juvenility and axillary flowering into a compact plant with rapid terminal flowering. *Plant Biotechnol J* 17, 869–880. doi:10.1111/pbi.13021.

Voogd C, Brian LA, Wu R, Wang T, Allan AC, and Varkonyi-Gasic E (2022). A MADS-box gene with similarity to FLC is induced by cold and correlated with epigenetic changes to control budbreak in kiwifruit. *New Phytol* 233, 2111–2126. doi:10.1111/nph.17916.

Vos P, Hogers R, Bleeker M, Reijans M, van de Lee T, Hornes M, et al. (1995). AFLP: A new technique for DNA fingerprinting. *Nucleic Acids Res* 23, 4407–4414. doi:10.1093/NAR/23.21.4407.

Wang F, Li J, Ye K, Liu P, Gong H, Jiang Q, et al. (2019). An in vitro *Actinidia* bioassay to evaluate the resistance to *Pseudomonas syringae* pv. *actinidiae*. *Plant Pathol J* 35, 372–380. doi:10.5423/PPJ.OA.02.2019.0035.

Wang M, Li M, and Meng A (2003). Selection of a new red-fleshed kiwifruit cultivar "Hongyang." *Acta Hortic* 610, 115–117. doi:10.17660/ACTAHORTIC.2003.610.13.

Wang WC, Chen SY, and Zhang XZ (2016). Chloroplast genome evolution in *Actinidiaceae*: clpP loss, heterogenous divergence and phylogenomic practice. *PLoS One* 11, e0162324. doi:10.1371/JOURNAL.PONE.0162324.

Wang X, Zeng Y, Nieuwenhuizen NJ, and Atkinson RG (2021). TPS-b family genes involved in signature aroma terpenes emission in ripe kiwifruit. *Plant Signal Behav* 16, 11. doi:10.1080/15592324.2021.1962657.

Weising K, Fung RW M, Keeling DJ, Atkinson RG, and Gardner RC (1996). Characterisation of microsatellites from *Actinidia chinensis. Mol Breed* 2, 117–131. doi:10.1007/BF00441427.

Williams JGK, Kubelik AR, Livak KJ, Rafalski JA, and Tingey SV (1990). DNA polymorphisms amplified by arbitrary primers are useful as genetic markers. *Nucleic Acids Res* 18, 6531–6535. doi:10.1093/NAR/18.22.6531.

Wijesundara WWMUK, Ananda GKS, Ranaweera LT, Udawela UAKS, Weebadde CK Sooriyapathirana SDSS (2018). Intellectual property rights (IPRs) for plant breeders: A review on theoretical framework and employment of DNA-based varietal authentication for claiming IPR. *J. Agric. Sci. - Sri Lanka* 4(2), 53–70, 2018. doi: http://doi.org/10.4038/sljfa.v4i2.64

Wu H, Li M, Wang D, Liu H, and Xu X (2019a). The complete chloroplast genome sequence of *Actinidia Callosa* Var. Henryi. *Mitochondrial DNA Part B* 4(1), 652–653. doi:10.1080/23802359.2018.1561223.

Wu H, Ma T, Kang M, Ai F, Zhang J, Dong G, et al. (2019b). A high-quality *Actinidia chinensis* (kiwifruit) genome. *Hortic Res* 6, 117. doi:10.1038/S41438-019-0202-Y/41964298/41438_2019_ARTICLE_202.PDF.

Wu JH (2012). Manipulation of ploidy for kiwifruit breeding and the study of *Actinidia* genomics. *Acta Hortic* 961, 539–546. doi:10.17660/ACTAHORTIC.2012.961.71.

Wünsch A, and Hormaza JI (2002). Cultivar identification and genetic fingerprinting of temperate fruit tree species using DNA markers. *Euphytica* 125, 59–67. doi.org/10.1023/A:1015723805293.

Xiao XG, Zhang LS, Li SH, Wang B, Testolin R, and Cipriani G (1999). First step in the search for aflp markers linked to sex in *Actinidia. Acta Hortic* 498, 99–104. doi:10.17660/ACTAHORTIC.1999.498.11.

Yao X, Tang P, Li Z, Li D, Liu Y, and Huang H (2015). The first complete chloroplast genome sequences in *Actinidiaceae*: Genome structure and comparative analysis. *PLoS One* 10, e0129347. doi:10.1371/JOURNAL.PONE.0129347.

Zhang J, and Liu H (2019). The complete choloroplast genome sequence of *Actinidia lanceolate. Mitochondrial DNA Part B* 4, 1187–1188. doi:10.1080/23802359.2019.1591200.

Zhang Q, Liu C, Liu Y, Vanburen R, Yao X, Zhong C, et al. (2015). High-density interspecific genetic maps of kiwifruit and the identification of sex-specific markers. *DNA Res* 22(5), 367–375. doi:10.1093/dnares/dsv019.

Zhen Y, Li Z, Huang H, and Wang Y (2004). Molecular characterization of kiwifruit *(Actinidia)* cultivars and selections using SSR markers. *J Am Soc Hortic Sci* 129, 374–382. doi:10.21273/JASHS.129.3.0374.

Zhong YP, Li Z, Bai DF, Qi XJ, Chen JY, Wei CG, et al. (2018). In vitro variation of drought tolerance in five *Actinidia* species. *J Am Soc Hortic Sci* 143, 226–234. doi:10.21273/JASHS04399-18.

11 Allele Mining in Apricot (*Prunus armeniaca* L.) Breeding
Current State and Future Prospect

Filippo Geuna, Juan Alfonso Salazar, and Pedro Martinez-Gomez

11.1 INTRODUCTION

11.1.1 Worldwide Apricot Cultivation

Apricot (*Prunus armeniaca* L.), $2n=16$, is an important fruit crop originating in China where it was originally domesticated and nowadays grown all over the word. This species is at the basis of the food tradition in the Mediterranean Basin, and an element of the Mediterranean diet. Apricot is one the most important fruit species in the world for direct human consumption. In the year 2020, apricot had an average annual harvest around 3.8 million metric tonnes, produced on over 580,000 hectares in 69 countries, thus accounting for some 0.43% of the total annual fresh fruit production (FAOSTAT 2020, http://faostat3.fao.org; accessed October 2022).

More than 80% of apricot production comes from the Mediterranean and Middle East areas and is concentrated in a period of 30–50 days, mainly centered around June. Turkey is the world's largest producer of both fresh and dried apricot, accounting for a total 833,398 metric tonnes (in the year 2020). Apricots are grown in nearly all parts of Turkey, with the exception of the very humid regions around the Black Sea and in the high plateaus of the East Anatolian Region. In the same year, Uzbekistan ranked second with 529,109 metric tonnes, while Iran was the third-world producer with 334,408 tonnes. Among the European countries, Italy ranked first with 173,380 tonnes followed by Spain with 128,700 tonnes and France with 85,830 tonnes.

Apricot culture is greatly restricted by climatic conditions, especially those related to chill accumulation in several growing areas, with a significant influence on productivity. In addition, blossoming times, fruit quality traits, and yield parameters are key variables that restrict the suitability of apricot varieties for cultivation in diverse edaphic conditions (Ledbetter, 2008).

11.1.2 Geographical Distribution of Genetic Resources

Apricot diversification occurred in two geographic centers, one is central China and the other is in Western Central Asia, namely in the Fergana valley, at the borders of Kyrgyzstan, Tajikistan, and Uzbekistan (Vavilov, 1951; Faust et al., 1998). In these regions, which represent diversification areas of this species, apricot production is focused on the development of cultivars for fresh market, kernel production, and ornamental use.

In the literature up to now, different studies have been published that address the differentiation and geographic distribution of apricot genetic resources. In an attempt to investigate the routes of apricot diffusion from the main differentiation centers, Bourguiba et al. (2012) analyzed 207 accessions from the eastern to the southwestern Mediterranean Basin representative of France (49), Tunisia (42), Morocco (34), Turkey (32), Algeria (23), Italy (19), and Spain (8), using 25 simple sequence repeat (SSR) marker loci. The accessions were selected for local variability in each country and excluding those derived from breeding programs. Moreover, while apricots are mainly

 DOI: 10.1201/9781003386490-11

vegetatively propagated, in oasis agroecosystems of North African countries it is possible to find traditional local germplasm propagated by seeds. The structure analysis placed the genotypes in three main gene pools: 'Irano-Caucasian', 'North Mediterranean Basin', and 'South Mediterranean Basin'. Moreover, the authors concluded that, based on the proportion of specific alleles between the northern Mediterranean and the southern route, apricot diffusion occurred along two main routes: the first one through countries north of the Mediterranean Sea and the second one probably through North African countries brought by the Arabs. Thus, their data seem to contradict the results of Kostina (1969) who hypothesized only one route of diffusion in the Mediterranean countries, while supporting the conclusions by Faust et al. (1998) and Mehlenbacher et al. (1990). In addition, the study depicted a scenario of domestication and diffusion revealing a significant reduction of genetic diversity, ranging from about 37% to 49% from the secondary apricot diversification zone ('Irano-Caucasian') to the southwestern Mediterranean Basin.

In another large study (Bourguiba et al., 2020), 890 apricot accessions representative of the global apricot germplasm were analyzed through 25 SSR loci and revealed a total of 609 alleles with a mean of 24.36 alleles per locus. Compared with results obtained in previous studies focusing on apricot diversity at a large scale, these data show the highest level of diversity within a worldwide apricot collection. The structure analysis identified seven groups: "Eastern Asia", "Central Asia", "Irano-Caucasian", "Continental Europe", "Mediterranean Europe", "North Africa", and "America." A marked differentiation was indicated between "Eastern Asia" and "North Africa" groups (F_{ST}=0.187). Nevertheless, a low differentiation was observed between the "America" group and both the "Irano-Caucasian" and the "Mediterranean Europe" groups (F_{ST}=0.032 and 0.036, respectively) as well as between the "Mediterranean Europe" and "North Africa" groups (F_{ST}=0.038).

The structure analysis clearly assigned 590 accessions of the 890 analyzed (66.29%) in one of the seven clusters, while the remaining 300 accessions showed an admixed ancestry, with the "America" geographical group enclosing the highest number of accessions (27/45; 60%).

In the attempt to produce an overview of the whole genetic diversity of the germplasm analyzed, a core collection was also constructed based on 306 accessions (34.38% of total) and representative of the 100% of the total alleles found.

Apricot is propagated through grafting, although it is also seed-propagated, mainly in some oasis agroecosystems. In Europe, clonal propagation through cuttings and grafting was used for a long time. Nevertheless, most of its residual variability occurred from a mix between seed- and grafting-propagated apricots, especially in North Africa.

A survey of genetic diversity in a wild relative of apricot, wild *Prunus sibirica*, was performed comprising 22 populations representative of geographic Mongolia and other Chinese provinces (Wang et al., 2014). The accessions included in the study were collected at elevations spanning a large interval (87–1,334 m) and distributed from latitudes of N40°02′ to N46°34′ and longitudes of E112°22′ to E130°17′. The greatest geographic distance between populations reached 1,526.5 km. The genetic analysis was based on a set of 31 SSR loci totaling 599 different alleles from 672 genotyped individuals. A genetic subdivision using F-statistics estimated a moderate inbreeding coefficient (0.173) and an F_{ST} value of 0.071 across all loci, indicating moderate genetic differentiation among the sites. In addition, the Mantel test showed that genetic distance did not significantly correlate with geographic distance.

The extensive analysis of a large set of apricot samples (271 cultivated accessions and 306 wild trees) collected between central China and the Caucasus regions and analyzed through 48 SSR markers made it possible to reconstruct the differentiation process between cultivated apricot and related species (Liu et al., 2019). The study, by also employing an approximate Bayesian computation (ABC), revealed that at least three independent domestication events occurred, with extensive gene flow among the cultivated groups, arguably due to both ancient (through commercial exchanges along the Silk Roads) and modern trade-dependent diffusion. Moreover, high genetic differentiation was found between the different Armeniaca species, *P. armeniaca*, *Prunus sibirica*, and *Prunus mume*. Surprisingly, the authors found two highly differentiated genetic clusters within the samples identified in the field as "*Prunus sibirica*", with one cluster inferred possibly resulting from an ancient hybridization between *P. armeniaca* and the "pure" *P. sibirica*. They speculated that systematic delimitation of *P. armeniaca* and *P. sibirica* should probably be revised, and predicted a better understanding to be achievable by adding morphological data and denser genetic markers. Another study exploring the genetic structure of Chinese apricots by molecular markers supported the existence of five major ecological groups, including those from North China, Northwest China, Northeast China, Southeast China, and Xinjiang, with frequent germplasm exchanges occurring among these groups (Zhang et al., 2021).

The recognized role of the Himalayan Mountains uplift in the creation of geographical barriers and in climate changes, which in turn modeled plant diversification (Beer et al., 2008; Yu et al., 2018), might also be at the base of differentiation within apricot and apricot relatives. Interestingly, compared to previous studies, an additional genetic cluster of cultivated apricots was recognized, represented by cultivated apricots from Central Asia and the Xinjiang near-Western Chinese province. Due to the limited number of individuals considered in the ABC analysis, future studies are deemed necessary to test whether it was due to an additional domestication event followed by hybridization with apricots from Central Asia. *Prunus mume*, a species used for ornamental purposes, was finally placed in a distinct genetic cluster. To infer its origin, further cultivated and wild samples need to be included for further comparison.

Central Asia and China, from Kashmir to Tien-shan are major regions for the cultivation of apricot accessions; here cultivars are often characterized by high chilling requirements (CRs) (Decroocq et al., 2016; Bourguiba et al., 2020). Central Asia shows the richest variability. While in the Mediterranean countries apricot is largely cultivated in Turkey, Italy, Spain, and France, in Asia it is mostly produced in North-Western Himalayan regions. In particular, apricot is often reported to grow wildly in desert areas of Tibet under cold climate conditions, in regions of Southern China and in some parts of Northern India, which include the temperate areas of Jammu and Kashmir, Uttarakhand, and Himachal Pradesh, at elevations between 2,000 and 2,500 m above sea level. In the province of Jammu and Kashmir, apricot presents a rich genetic diversity, with both local seed-propagated accessions and some exotic cultivars propagated by grafting onto traditional local cultivars.

With the aim of characterizing the genetic richness and structure of the North-Western Himalayan Provinces of India, Sheikh et al. (2021) analyzed 50 apricot genotypes using 46 SSR markers. In total, 179 alleles were detected with a number of alleles (Na) ranging from 1 to 6, with an average value of 3.89. The accessions were grouped in four subpopulations. Out of these, populations 1, 2, and 4 contained 36 genotypes, all of indigenous origin, while subpopulation 3 included only one indigenous genotype and the remaining 13 were exotic.

A relatively high amount of genetic diversity and distance in the germplasm examined was revealed by the large range of allele sizes found, consistently with Vavilov's hypothesis who considered this zone a rich area of diversity (Vavilov, 1951).

The Central Asian group being the oldest and more diverse group includes local apricots from Xinjiang, Afghanistan, Baluchistan, Pakistan, and Northern India (Kumar et al., 2009). Wani et al. (2021) have addressed the investigation of genetic diversity in the regions of Kashmir and Ladakh in India by characterizing 120 apricot genotypes through 21 SSR markers. Samples analyzed were 78 genotypes from the Kashmir region and 42 genotypes from the Ladakh region and spanning the following geographic coordinates: latitudes 33°47′01.33″ N to 34°39′47.93″ N and longitudes 74°48′32.85″ E to 76°31′34.98″ E. A total number of 52 alleles with an average of 2.476 alleles per locus were detected. Based on the cluster analysis, the genotypes under study were placed into three major clusters, revealing significant genetic diversity. Many of the genotypes were grouped in cluster 1 (58 cultivars), whereas the second cluster, the smallest among the three, contained only 29 genotypes in which most of the genotypes were from the Kashmir region. Similarly, 12 genotypes of Ladakh and 21 genotypes of Kashmir were grouped together in cluster 3.

Eastern and central Europe host part of the migration routes from Asia to the Mediterranean Basin, and its genetic resources are expected to represent the diversity and genetic drift process associated with domestication. The investigation of phenotype of 443 apricot cultivars and selections in collections allowed the identification of the most appropriate parents to produce bred lines addressing late blossoming, late fruit ripening, higher frost resistance of flower buds, resistance against *Cytosporacincta*, *Stigmina carpophila*, *Sphaerotheca pannosa* var. *persicae*, content of dry matter, overall sugar, and vitamin C (Balan and Ivascu, 1994). Large and long-lasting breeding programs allowed to collect, characterize, and breed new cultivars featuring local alleles for resistance to various pathogens and fruit quality traits (Krška and Vachůn, 2016; Krška, 2018; Golubev, 2019).

11.2 MOLECULAR TOOLS OF ALLELE MINING

A decade ago, any study at the genome or transcriptome level in *Prunus* species was based on the use of the peach (*Prunus persica* L.) reference genome in its version 1 (IPGI, 2013) and later version 2 (Verde et al., 2017). However, in just a few years, the novel massive sequencing technologies have made significant strides with the Whole-Genome Sequencing (WGS) of *Prunus* species of agronomic relevance such as almond (Sánchez-Pérez et al., 2019), apricot (Jiang et al., 2019), European plum (Callahan et al., 2021), and Japanese plum (Huang et al., 2021).

By browsing the sequence databases, there is evidence that the apricot community can now rely on a growing valuable sequencing information data set which could be useful to any porpoise. In particular, in the National Center for Biotechnology Information (NCBI) database, it is now possible to find out seven assembled apricot genomes, being five of them available and 170,716 nucleotide sequences at the genome level; 29,824 genes identified at the transcriptome level and, as for proteome, 182,483 proteins and 100,255 identical protein groups (https://www.ncbi.nlm.nih.gov).

Moreover, the NCBI database contains valuable descriptive information included in the figure of bioprojects reaching in apricot a total of 94, which correspond to 94 independent apricot assays. In addition, these bioprojects provide 2,836 Sequence Read Archive (SRA), which correspond to sequences of each specific bioproject from WGS, RNA-seq (RNA sequencing), or BSA-seq (Bulked Segregant Analysis), among others. As for genomes data, apricot species is the most relevant after peach (Figure 11.1).

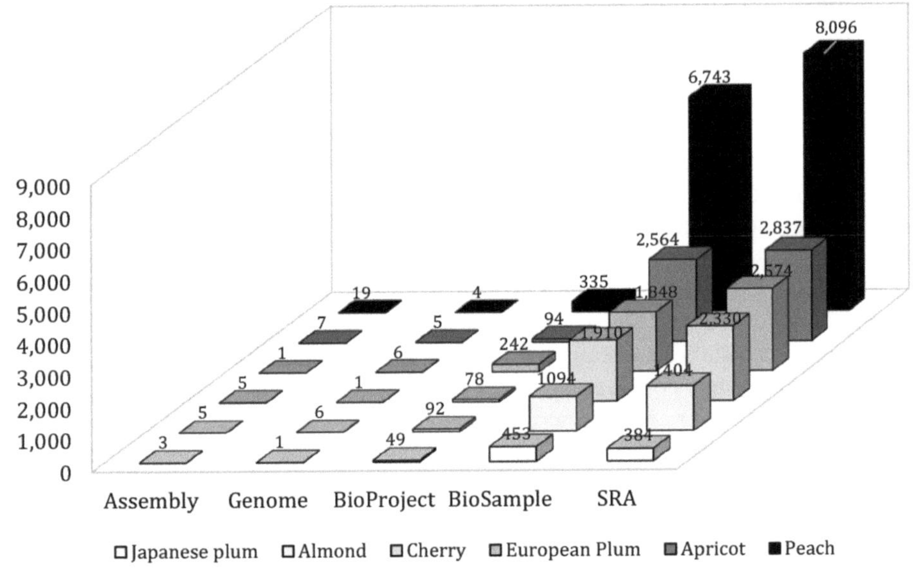

FIGURE 11.1 Summary of the apricot (dark gray) genomics resources (data counts) available at NCBI in comparison to other relevant *Prunus* species (black to white).

TABLE 11.1
Summary of Some of the Most Relevant Marker-Trait Associations (QTLs) in Different Apricot Progenies

Trait	Population	References
Sharka resistance	'Polonais' × 'Stark Early Orange'	Lambert et al. (2007)
Sharka resistance	'Lito' × 'BO81604311'	Dondini et al. (2011)
Xanthomonas arboricola resistance	'Harostar' × 'Rouge de Mauves'	Socquet-Juglard et al. (2013a)
Powdery mildew resistance	'Bergeron' × 'Currot'	Salazar et al. (2016a)
Powdery mildew resistance	'Goldrich' × 'Currot'	Salazar et al. (2016a)
Apple chlorotic leaf spot virus	'Bergeron' × 'Currot'	Salazar et al. (2017)
Chilling requirement	'Perfection' × 'A.1740'	Olukolu et al. (2009)
Flowering time	'Z506–07' × 'Currot'	Campoy et al. (2011)
Ripening time and fruit development	'Bergeron' × 'Currot'	Salazar et al. (2016b)
Ripening time and fruit development	'Goldrich' × 'Currot'	Salazar et al. (2016b)
Chilling and heat requirements	'Nanko' × 'SC'	Kitamura et al. (2018)
Chilling and heat requirements	'Ellching' × 'Nanko'	Kitamura et al. (2018)
Ripening time	'Fiamma' × 'BO93623033'	Baccichet et al. (2022)
Ripening time	'Lito' × 'BO81604311'	Baccichet et al. (2022)
Architecture and fruit quality traits	Harostar' × 'Rouge de Mauves'	Socquet-Juglard et al. (2013b)
Fruit quality traits	'Z701-1' × 'Palsteyn'	Salazar et al. (2013)
Skin color and soluble solids content	'Bergeron' × 'Currot'	García-Gómez et al. (2019)
Skin color and soluble solids content	'Goldrich' × 'Currot'	García-Gómez et al. (2019)
Fruit quality traits	'Chuanzhihong' × 'Saimaiti'	Zhang et al. (2022)
Acidity	'Lito' × 'BO81604311'	Dondini et al. (2022)

11.2.1 Genomic Resources

Genomics is the main discipline in charge of finding new solutions and novel molecular tools for plant breeding. For this reason, in the last decades, genomics has been especially addressed in order to move forward toward molecular marker-assisted selection (MAS), which will serve to increase the plant breeding efficiency in the current *Prunus* breeding programs.

In this context, in *Prunus* species such as apricot, molecular markers for genetic mapping and quantitative trait loci (QTL) identification have been extensively applied in the last few years. However, the identification of marker-trait association through genetic mapping has proven to be useful only in some few cases. We will go through some of them, including traits such as disease resistance, phenology, or fruit quality (Table 11.1).

In terms of disease resistance, Sharka virus is one of the most devastating pathogens affecting apricot fruit quality. Genetic maps were constructed to focus Sharka disease in a segregant apricot progeny including Sharka resistant and susceptible genotypes. To address this problem, QTL analysis was carried out in a population derived from the cross of the susceptible cultivar 'Polonais' and resistant 'Stark Early Orange' apricot population (Lambert et al., 2007). The results showed a major QTL mapping in the linkage group (LG1) of 'Stark Early Orange' reaching close to 60% of phenotypic explanation variance (PEV), which means that most of the PEV is given in the LG1. Subsequently, Dondini et al. (2011) studied the genetic determinants of Plum pox virus (PPV) resistance in a F1 apricot population derived from 'Lito' (Sharka resistant). In this study they phenotyped 118 seedlings for three years by inoculation with both the 'Dideron' and 'Marcus' strains, identifying the same QTL previously recognized by Lambert et al. (2007).

Another example of disease addressed by QTL identification is the major QTL of *Xanthomonas arboricola* pv. *pruni* identified by Socquet-Juglard et al. (2013a). This bacterium causes spots on stone fruit and determines severe yield losses in apricot production. The major QTL identified reached a LOD score over 15 explaining a phenotypic variation over 50% in the LG5. On the other side, we can also find QTLs linked to fungal diseases. For instance, powdery mildew, caused by *Sphaerotheca pannosa* var. *persicae* is another relevant disease in different apricot production regions. A study of two apricot progenies 'Bergeron' × 'Currot' and 'Goldrich' × 'Currot' for two years served for mapping several QTLs linked to different levels of infection especially in the LG2 and LG3 (Salazar et al., 2016a). Few years later, the transmission of resistance to *Apple chlorotic leaf spot virus* (ACLSV) was evaluated in a 'Bergeron' × 'Currot' apricot population for 3 years. This disease is also known as apricot pseudo pox, "viruela" (in Spain) and "butteratura" (in Italy) and causes an enormous impact on fruit production. The results yielded new QTLs linked to this disease especially in LG5 and LG7 in alternative years and in LG2 and LG4 all years (Salazar et al., 2017; Rubio et al., 2019).

As for phenology traits, there are several works focused on genetic mapping in different apricot progenies, particularly based on CR and budbreak, flowering or ripening time in apricot. Two apricot cultivars selected for high and low CR were used to generate a F1 population for genetic mapping using SSR and amplified fragment length polymorphism (AFLP) markers. Finally, a total of 12 putative QTLs were identified by composite interval mapping and multiple regression, explaining close to 60% of the total phenotypic variance and suggesting candidate genes involved in CR (Olukolu et al., 2009). Later, Campoy et al. (2011) explored flowering time on a BC1 apricot progeny derived from the cross 'Z506–07' × 'Currot'. In this work, alleles from two microsatellites (UDAp-423 and AMPA-105) were found to be tightly linked to flowering time (Campoy et al., 2011). On the other side, Salazar et al. (2016b) were able to identify significant QTLs linked to flowering time, fruit development period, as well as ripening time, which are partly due to the apricot CRs to start the budbreak and therefore flowering. The importance of these QTLs is capital in any breeding program because the early or late character of the seedlings is crucial in order to extend the fruit calendar window. In this work, two apricot populations 'Bergeron' × 'Currot' and 'Goldrich' × 'Currot' were evaluated for 3 years using 187 and 200 seedlings, respectively. Subsequently, genetic maps were constructed combining SSRs and single nucleotide polymorphisms (SNPs) markers in both populations. The results yielded consistent QTLs linked to fruit development period and ripening time especially in the LG4 being confirmed for 3 years of phenotyping and explaining a percentage of the variation (PEV) value over 40%. These major QTLs opened the possibility of early selection by molecular markers for harvest season in apricot.

The first application of genotyping-by-sequencing (GBS) in apricot was by Gürcan et al. (2016) and resulted in the application of 1162 SNP markers to investigate the genetic relation and population structure of 90 accessions, many of which are of Turkish origin. To elucidate the reliability of the SNPs that were obtained by GBS, the SNP analysis in the apricot germplasm was compared with *S* locus and SSR genotyping on the same set of accessions.

More recently, Kitamura et al. (2018) have studied a F1 population of *Prunus mume*, especially focusing on four traits: CR and heat requirement (HR) for dormancy release, blooming date, and leafing date (LD) over several years. In this work, the GBS technique was implemented to generate saturated genetic maps. All evaluated traits showed segregation and significant QTLs controlling CR, HR, and LD linked to a region in linkage group 4 were identified. These results, in spite of *Prunus mume* having more an ornamental interest conversely to *Prunus armeniaca*, prove that these traits are deeply controlled in a locus of the LG4 and related to early and late character of *Prunus* species. Recently, this finding has been confirmed in apricot by Baccichet et al. (2022). In this work, 32,492 polymorphic sites were genotyped in 128 apricot genotypes using a novel technique called Single primer enrichment technology (SPET). Genome-wide association studies (GWAS) revealed maturity date (MD) QTLs on several chromosomes, especially in the linkage groups 1 and 4, confirming major QTLs obtained in previous works and providing a high performance and reliability of the SPET technique in the dissection of the MD trait in stone fruits.

Moreover, in order to cover some of the most important traits from the tree or pomological point of view, it is important to highlight architecture and fruit quality traits. In this sense, genetic mapping, QTL analyses and marker-trait association have been focused to elucidate these important traits in the current breeding programs in *Prunus* species. Some examples in apricot include tree architecture traits such as tree diameter, total branch number or tree shape, as well as fruit quality traits such as fruit color, shape, or soluble solid content (Socquet-Juglard et al., 2013b). In this work, several QTLs linked to these traits were mapped in a F1 population. However, QTLs from architecture-related traits were intermittent between years, conversely to QTLs identified on LG1 and LG4 for soluble solid content or fruit color on LG3 being consistent for two consecutive years. Later, Salazar et al. (2013) evaluated a F1 population 'Z701-1' × 'Palsteyn' in terms of fruit quality traits. In this work, 12 pomological traits were analyzed including, among the most significant, fruit weight, stone weight, flesh color, malic acid, and soluble solid content. These results, after the construction of a low-coverage genetic map using SSRs, allowed the identification of many QTLs distributed along the apricot genome, but controlling a scarce PEV. This could be explained for the polygenic character of these traits and therefore for their quantitative inheritance. Currently, García-Gómez et al. (2019) have also studied important fruit quality-related traits in two apricot populations and combining SSRs and SNPs being able to dissect significant QTLs for fruit color in the LG3 and soluble solid content in the LG4 in the same loci linked to MD. These findings confirm the great importance of the LG4 in apricot where many biological processes linked to fruit maturation take place. Nowadays, other attempts to achieve new findings to elucidate fruit quality traits at genomic level have been addressed. Zhang et al. (2022) evaluated 169 seedlings of a F1 apricot population derived from crossing 'Chuanzhihong' × 'Saimaiti'. In this work, they mapped over 6,000 high-quality SNPs, obtaining a density map lower than 1 cM between markers. However, despite such a great saturation, they were only able to map minor QTLs along the genome linked to fruit weight, fruit height, fruit lateral width, fruit ventral width, soluble solids content, and fruit firmness in intermittent years. In the same line, Dondini et al. (2022) have identified new QTLs linked to acidity level in the population 'Lito' × 'BO81604311' distributed in five linkage groups such as LG4, LG5, LG6, LG7, and LG8. This demonstrates partly, on the one hand, the difficulty of obtaining significant and persistent QTLs between years for quantitative traits, on the other hand, that it is not necessary so many markers to obtain a reliable result, since half of the work also depends on the phenotyping and genetic background. Having summarized, some of the most important findings linked to agronomic traits related to alleles at genomic level, show the difficulty to find molecular tools useful for the assisted selection of interesting genotypes at nursery level due to the polygenic nature of the majority quality traits. Therefore, there is still a long and tedious way to go in order to address the use of efficient molecular markers in the current breeding programs. It is therefore necessary to address and complement important areas such as Omics, Phenomics, Genomics, and Transcriptomics to achieve it. Thus, a good start would be to delve into those alleles related to major genes in order to look for a useful application, exploring in the new apricot genome.

In the last few years, the new sequencing technologies have allowed to address new challenges. One way has been posed by assembling heterozygous genomes as apricot, through gamete binning as a proof of principle (Campoy et al., 2020). Gamete binning is a method for chromosome-level, haplotype-resolved genome assembly independent of parental genomes or recombinant progenies. The segregation of sequence variation in the gamete genomes enables a straightforward phasing of all variants into two haplotypes, which subsequently allows for genetic mapping and sorting of WGS reads into distinct read sets, each representing a different haplotype. Haplotype-resolved genome assemblies result from assembling these independent read sets, which can be scaffolded to a chromosome level using a genome-derived genetic map.

The publication of the first apricot genome (Jiang et al., 2019) has paved the way for a new level of study of this species. In addition to studies on its evolution and domestication, an accurate analysis of the variability at the genomic and allelic level is now possible.

For example, a comparative analysis of the assembled apricot genome against those of other *Prunus* species showed that, during its evolution, 1,324 gene families expanded, producing 2,300 apricot-specific genes while 981 families contracted. The genes from the expanded families were mostly enriched in phenylpropanoid and flavonoid biosynthesis. In addition, the citrate synthase family appears to have undergone expansion.

Overall, a transcriptome-based prediction accounts for a total of 30,436 protein-coding genes, with an average transcript length of 1,641 bp. The average gene density of apricot resulted in 137 genes per Mb, a value higher than any other one of the Prunus species under comparison.

A comparative study reported genomic information for 168 wild and cultivated apricot accessions (Li et al., 2020). The authors provided insights into the dissemination of *P. armeniaca* as a cultivated species. During domestication from wild to cultivated, apricot clearly shows signatures of selection, combined with differentially expressed genes (DEGs), underlying distinct fruit traits, including sugars, organic acids, aromas, and carotenoids. Moreover, a population structure analysis confirmed the hypothesized mixed genetic background of apricot in different ecological groups.

The largest genomics-based apricot germplasm characterization to date was based on 600 *Armeniaca* apricot genomes plus four map-anchored high-quality assemblies (Groppi et al., 2021). The study disclosed a high genetic diversity between Chinese and European apricots, resulting from independent domestication events followed by gene flow. Among the large set of supporting data, it also addresses a comparative analysis of chloroplast genomes and a proposed reconstruction of the *Armeniaca* phylogeny.

11.2.2 Transcriptomic Resources

Transcriptomics have also been an important source for searching candidate genes linked to any trait of interest. In apricot trees, the shelf-compatibility is another important trait to consider since this ensures fruit set and higher yields production in apricot cultivation. In this sense, the gametophytic self-incompatibility (GSI) system was proposed as controlled by two genes located in the *S* locus, S-RNase, and F-box gene (SFB) (Romero et al., 2004; Herrera et al., 2018). This study also revealed that S1- and S2-RNases were expressed in style tissues, while SFB1 and SFB2 were only expressed in pollen. Later, Vilanova et al. (2006) evidenced that self-incompatible cultivars show a 358-bp insertion in the SFB_C gene, resulting in the expression of a truncated protein.

On the other side, Schurdi-Levraud et al. (2006) investigated the changes in the expression profiles in the apricot cultivar 'Goldrich' by inoculation with PPV using cDNA-AFLP. Some of the expressed patterns were analyzed by RT-PCR or hybridization-based methods. Five of these genes were evaluated in the partially resistant 'Goldrich' cultivar and a susceptible genotype. Finally, these candidate genes seemed to play an important role in the virus invasion or quantitative resistance of apricot to PPV.

According to endodormancy in woody plants, *MADS box* gene was identified as an expressed gene in the endodormant lateral buds of *Prunus mume.* In this work, *PmDAMs* showed to be repressed during prolonged cold exposure maintaining a low level of expression. Conversely, the response of *PmDAM4* to *PmDAM6* to a short cold exposure period was related to differences between low- and high-chill genotypes being its expression negatively correlated with endodormancy release (Sasaki et al., 2011). Other studies for a suitable gene selection linked to flower development in *Prunus mume* evaluated miRNA expression using qRT-PCR (Wang et al., 2014). In this work, they considered three samples including bud development and flowering, fully open flowers, and all *P. mume* samples. Finally, they found *PP2A-2* and *UBC* genes involved in bud development and flowering.

As for fruit quality-related traits, Zhang et al. (2019a) focused studies on carotenoid content linked to apricot nutritional values. The main goal of this work was to identify key genes associated with carotenoid metabolism in apricot fruit by transcriptome sequencing. The RNA-seq study resulted in the identification of over 40,000 unigenes being close to 7,000 DEGs during apricot ripening. This analysis revealed a wide range of genes linked to carotenoid metabolism and therefore with fruit color including *NCED1*, *CCD1/4*, *PIF3/4*, *HY5*, *ERF003/5/12*, *RAP2–12*, *AP2*, *AP2-like*, *BZR1*, *MADS14*, *NAC2/25*, *MYB1R1/44*, *GLK1/2*, and *WRKY6/31/69*. This approach served to propose a network model of apricot carotenoid metabolism, identifying key genes in this process, which will facilitate molecular MAS.

On the other side, important organoleptic traits such as taste and aroma were considered during apricot fruit ripening by Zhang et al. (2019b). In this study, authors were focused on searching candidate genes for flavor to construct a regulatory network using related transcription factors. In this work, they characterized the transcriptome of 'Jianali' apricot in three maturation stages, being able to identify 1,237 DEGs. *SUS1*, *SPS2*, and *INV1* genes were correlated in the sugar biosynthesis pathway, while *NADP-ME4* had a considerable effect on organic acid metabolism, as well as *CCD1* and *FAD2* were involved in aroma volatiles and lactone biosynthesis, respectively. In addition, several sugar transporters were involved in the sugar transport, along with organic acids and volatiles. Finally, all this information served to construct a regulatory network to elucidate molecular mechanisms involved in taste and aroma evolution in apricot.

Through a similar approach, García-Gómez et al. (2019) were also able to monitor the gene expression of important ripening-related genes in two contrasting genotypes of the apricot population 'Goldrich' × 'Currot' for three different maturation stages. In this work, these genes were extracted from major QTLs previously identified in this population validating important candidate genes involved in fruit color (*MYB10*), carotenoid biosynthesis (*LOX 2*, *CCD1*, and *CCD4*), anthocyanin biosynthesis (*ANS*, *UFGT*, and *F3'5'H*), organic acid biosynthesis (*NAD ME*), or ripening-related genes (*NAC*).

Some of these genes including carotenoid cleavage dioxygenase 4 (*CCD4*) or sucrose synthase (*SuSy*) play an important role in the ripening process related to the changes experienced in skin and flesh color, as well as soluble sugar content (García-Gómez et al., 2020). These genes open out the opportunity of using them for MAS in apricot breeding programs.

Another study proposes the use of multiple linear regression models to predict anthocyanins or carotenoids content, considering Chalcone synthase (*CHS*) as the best predictor for anthocyanins, Carotenoid cleavage dioxygenase (*CCD4*) and Zeta-carotene desaturase (*ZDS*) for carotenoids, and Lipoxygenase 2 (*LOX2*) and *MADS box* linked to the ripening process (García-Gómez et al., 2022).

More recently, a whole-genome transcriptome analysis was carried out in apricot (cv. 'Goldrich') following application of ethylene-modulating compounds (Salazar et al., 2022). In this assay, fruits from different apricot varieties were monitored during its shelf-life period after 1-MCP (ethylene-inhibitor) and Ethrel (ethylene-precursor) treatments. 'Goldrich' showed the greatest difference in terms of fruit softening and ethylene emission between treatments. In this study, after RNA-seq analysis, the interaction within DEGs was studied by the Kyoto Encyclopedia of Genes and Genomes (KEGG) database and as protein–protein interaction network, which revealed photosynthesis and plant hormone signal transduction as the most significant metabolic pathways involved in the ripening process. Some of the most important candidate genes were pectinesterase and 1-aminocyclopropane-1-carboxylate (*ACC*), both upregulated by Ethrel, and auxin-responsive protein (*IAA*), aquaporin *PIP*, photosystem I subunit V *psaG*, GDP-L-galactose and ferredoxin, being all of them upregulated by 1-MCP. On the other hand, GDP-L-galactose and ferredoxin could be playing an important role in the fruit antioxidant activity, slowing down 1-MCP treatment during the fruit senescence period.

Finally, during the last 20 years we have seen the evolution of the sequencing technologies and the availability of annotated genomes useful for further studies at molecular level. In this sense different studies at transcriptomic level have provided a bunch of information linked to any biological process linked to fruit trees paying special attention to resistance or susceptibility to diseases, phenology, fruit quality, or even traits linked to fruit senescence.

11.3 ALLELES-RELATED TRAITS AND THEIR APPLICATION IN CURRENT BREEDING PROGRAMS

As we have stated, in the last years we have experienced an exponential increase of genomic and transcriptomic-related works to broaden the knowledge about the main biological processes involved in the most significant agronomics traits in the current breeding programs. However, in spite of there being plenty of related studies at genomic and transcriptomic level, in apricot there are only few reliable molecular tools to apply for assisted selection at nursery level, which would be the main application to increase the efficiency of an apricot breeding program.

One of the main traits addressed has been the self-compatibility (SC) alleles through the APRFBC8 marker (Halász et al., 2010), which is able to identify SC genotypes through the SFBC-allele (fragment size ≈ 500 bp). Although self-incompatibility mechanisms are still being studied in different genetic backgrounds, this molecular marker is useful to increase the efficiency of apricot breeding programs.

On the other hand, the study of resistance to Sharka has been and is being a long and tedious work, due to its devastating consequences in apricot fruits. After previous studies by other authors, Soriano et al. (2012) were able to associate three SSR markers, *PGS1.21*, *PGS1.23*, and *PGS1.24* linked to allele variants of this disease. Another study was focused on studying the feasibility of these molecular markers in MAS using 80 apricot genotypes including 25 commercial cultivars and 55 new releases (Rubio et al., 2014). This study showed that *PGS1.21* and *PGS1.24* were the most related to Sharka virus resistance. However, the use of these molecular markers does not ensure a whole effectiveness in terms of Sharka resistant genotype selection, which could be due to the presence of a second locus involved in PPV resistance. Despite that, the use of these markers provides a greater efficiency in the current apricot breeding programs. Subsequently, Decroocq et al. (2014) used 11 apricot mapping populations in three different locations to confirm the effectiveness of PGS markers. This study provided evidence that Sharka resistant genotype selection only using these PGS markers is not completely reliable, since genotype-phenotype discrepancies occurred. In this sense, new studies have been addressed using Genome-Wide Association (GWA) analysis, which has served to validate a major QTL of PPV resistance proposing new candidate genes (Mariette et al., 2016). After that, new and accurate molecular markers have been designed including resistant and susceptible alleles to improve the genomic selection (GS) for potentially Sharka resistant or susceptible apricot genotypes (Zuriaga et al., 2018; Gürcan et al., 2019).

Finally, most recently, other strategies were planned to discern PPV resistance (De Mori et al., 2019). In this study 'Lito' was sequenced using the PacBio technology (at a 98X genome equivalent coverage), which allowed it to identify structural variants and chromosome-specific allele expression. According to these results, *MATHd* genes could have a possible role in PPV resistance. Moreover, a list of other candidate genes was provided spanning the peak of QTL for resistance.

On the other hand, other possibilities for MAS in apricot species arise from the important major QTLs described in chromosome 4 related to the fruit ripening and therefore with the early or late character of apricot seedlings. In this sense, there are several species of *Prunus* in which the same QTL has been identified, among them the peach tree (Pirona et al., 2013), Japanese plum (Salazar et al., 2017), or cherry tree (Calle and Wünsch, 2020). Another confirmation was provided by Groppi et al. (2021) who identified significantly enriched sweeps associated with QTLs in the central region of chromosome 4 (from 7 to 18 Mbp coordinates), indicating a potential hotspot of human selection targets. The traits being affected comprise phenology (fruit maturity and bloom date) and fruit quality (ripening, aroma, and firmness). The extensive genome-wide analysis revealed subsets of 232 and 44 genes under selection in the European and Chinese germplasm, respectively.

Another interesting QTL mentioned in apricot, is linked to fruit color on chromosome 3, which has also been identified in Japanese plum (Salazar et al., 2017) and cherry (Sooriyapathirana et al., 2010) in MYB family genes. In the Japanese plum species, there are already works in which markers have been developed in the MYB10 gene of chromosome 3 for skin and flesh color (Fiol et al., 2021, 2022).

The development of markers related to the fruit color, and therefore with antioxidant compounds such as carotenoids in apricot are also key targets in the genetic breeding programs of this species.

Besides, the postharvest performance is also considered another main goal, since to incorporate new molecular markers to select fruit with a longer shelf-life period would be interesting to improve fruit marketing and quality (Moustafa and Cross, 2019). A study in Japanese plum has confirmed minor QTLs for fruit firmness evolution using destructive and non-destructive methods in the LG4 and LG5, which opens the possibility to a possible selection of genotypes of these species according to their postharvest performance (Salazar et al., 2020). Thereby, although there is much work to be done, new molecular markers useful for assisted selection of traits such as the early or late ripening character or linked to the apricot fruit color and quality are expected to appear.

11.4 APRICOT ONLINE RESOURCES

The European *Prunus* network (ECP/GR: European Cooperative Programme for Crop Genetic Resources, https://www.ecpgr.cgiar.org/) under the authority of IPGRI (International Plant Genetic Resource Institute) had initiated a European *Prunus* Database (EPDB) for the management of the *Prunus* genetic resource collections. It covers a great number of *Prunus* genetic resource collections in most of the European countries.

EPDB is managed by INRAE (Institut National de la Recherche Agronomique) in Bordeaux. It contains passport data and primary characterization descriptors. The passport data were harmonized with the other European plant genetic resource databases. The list of agronomic and morphological descriptors is periodically updated. The database contains data of European collections of all *Prunus* species, cultivated stone fruits and their related species as well as the wild ones. The EPDB portal provides access to five distinct databases providing passport, characterization, and evaluation data. The current apricot database (http://www.bordeaux.inra.fr/euapricotdb/) contains 1982 accessions, physically represented by 1947 individuals maintained by 18 institutes from 13 countries. One of the final objectives of the database was to sort out the valuable material that originated from Europe in the *Prunus* genetic resources, comprising a list of *Prunus* European "core collections". In addition, EPDB was realized to permit the exchange of information and a rational and systematic conservation of genetic resources (Zanetto et al., 2002). Nevertheless, the resource appears outdated and a better integration with other existing repositories should be sought.

The Community plant variety rights (CPVR) system provides European intellectual property protection for new plant cultivars of all botanical genera and species, including hybrids between genera and species; the system is operated by the Community Plant Variety Office (CPVO; https://cpvo.europa.eu/en, accessed in October 2022).

As regards apricot genetic resources, the EURISCO database (Maggioni and Engels, 2014; Weise et al., 2017, http://eurisco.ecpgr.org, accessed October 2022) lists 2,571 accessions in national inventories distributed among 15 European countries and including Switzerland, Ukraine, and United Kingdom.

In the United States, the USDA-operated National Plant Germplasm System (NPGS) provides access to data and resources through the Germplasm Resources Information Network (GRIN, https://npgsweb.ars-grin.gov/gringlobal/search.aspx) which currently lists 1,813 apricot entries.

The Plant Variety Database (PLUTO; http://www.upov.int/pluto/en), which belongs to the International Union for the Protection of New Varieties of Plants (UPOV, http://www.upov.int, accessed October 2022) lists 2,005 entries with a search for the apricot ("PRUNU_ARM") UPOV code, including data on the variety name, breeder's reference, and trade name.

The Genome Database for Rosaceae (GDR, https://www.rosaceae.org) is the central repository and main data-mining resource for genomics, genetics, and breeding data of the *Rosaceae* family, including apricot between stone fruits. Since 2003, GDR provides web interfaces and analysis tools for genetic and genomic data such as genus-specific EST unigene sets, linkage maps and genetic markers. The data sets hosted in the GDR have markedly expanded with the availability of whole-genome sequence, phenotypic and genotypic data (Jung et al., 2019).

11.5 CONCLUDING REMARKS AND FUTURE PERSPECTIVES

Central Asia is a center of diversity for many globally valued fruit and nut tree species, including apple, apricot, walnut, pear, plum, almond, pomegranate, and grape (Vinceti et al., 2022). Forests throughout these regions are a source of wild relatives of these and other species still growing spontaneously and often used as rootstock for trees planted in home gardens and orchards. They are better adapted to local conditions and typically feature better resistance to biotic stresses, such as pests and diseases, than improved, modern varieties. A large portion of apricot genetic variability, especially in these areas, still remains unexplored and represents an important repository of interesting alleles for breeding.

At present, as we have been highlighting, new sequencing technologies have allowed the sequencing of genomes and transcriptomes of almost all interesting fruit species, which together with the previous work on phenotyping varieties and progenies has served for the development of new molecular tools improving the efficiency of plant breeding programs. However, the results obtained seem scarce, especially after so much effort. Therefore, by knowing the difficulties inherent in dealing with quantitative traits, other strategies should be considered to assist in the genetic improvement of fruit species in the short or medium term.

On the other hand, beyond genotyping, in the last ten years, massive phenotyping strategies are being increasingly adopted, which provide us with a lot of data on fruit yield, disease resistance, tolerance to biotic and abiotic stresses, and, ultimately, a bunch of data on crop management (Jangra et al., 2021; Okura, 2022). However, although the use of these technologies is being particularly useful in crops such as cereals (Aharon et al., 2020), in fruit species such as apricot, it is difficult to attain and manage homogeneity as in herbaceous crops. In fact, large extensions of orchards are required to obtain relevant and accurate data. In spite of this, the new imaging technologies are improving at a constant pace, and further implementation is expected in fruit trees (Huang et al., 2020; Danilevicz et al., 2021).

Therefore, with the amount of information gathered by the several phenotyping methods increasing continuously, huge data sets need to be evaluated. Moreover, the portfolio of variables potentially being phenotyped largely exceeds the requirements of genotyping techniques. For instance, the combined use of weather stations and cameras collecting images is a good way for testing phenology models through 'machine learning' in *Prunus* species (Zhu et al., 2022), providing very valuable information from the point of view of crop management.

As an additional method in the breeder's toolbox, GWAS and GS could speed up the efficiency of breeding programs, by combining the knowledge of genotype and phenotype of populations with unrelated genetic backgrounds. The large-scale study of relationships among germplasm and phenotypic traits permits to obtain the breeding value of traits of interest such as fruit weight, sugar content, or titratable acidity in a straightforward way by exploiting the overall genomic breeding value (Desta and Ortiz, 2014; Newell and Jannink, 2014; Aranzana et al., 2019; Zahid et al., 2022).

Nevertheless, other breeding alternatives exist independent from molecular markers, and based on conventional genetic engineering or genome editing, which would permit a precise introduction of deletions or insertions in a gene with known function, obtaining new ideotypes featuring the desired traits. However, their adoption still appears problematic in many *Prunus* spp., especially due to the poor efficiency of genetic transformation and regeneration (Ricci et al., 2020; Vergara et al., 2021).

While allele mining is traditionally associated with the conventional exploration of medium- to large-scale mRNA and protein profile changes and variants, important investigation fields like non-coding RNAs and epigenomics deserve greater consideration and investments (Miryeganeh, 2021).

The increasing number of genetic and genomic resources and data accumulated in the last decade is changing the scenario of apricot breeding. While several DNA and protein data are accumulating at an increasing pace, there is a growing perception of the importance of collecting, organizing, and maintaining biological repositories and core collections as invaluable stores of precious alleles. Beside genome and protein data, there is a plethora of phenotyping tools that are emerging as companion instruments to correlate phenotypes with the underlying genotypes (Isobe and Ninomiya, 2022). A rational, shared and agreed upon use of data banks and field resources is a prerequisite to address this scenario right now and for the years to come (Lai et al., 2015; Morales et al., 2020). For example, molecular markers-based data are still largely not provided as supplementary data at publication time. This implies a limited tuition of data gathered and, over time, a loss of precious information, not only for the global community but, ultimately, for the investigators responsible for the generation of such data themselves. This paradigm shift is only possible from a combined effort of both the global research community and editors to stimulate a mutual, virtuous change.

REFERENCES

Aharon S, Peleg Z, Argaman E, Ben-David R, Lati R (2020) Image-based high-throughput phenotyping of cereals early vigor and weed-competitiveness traits. *Remote Sens* 12:3877. https://doi.org/10.3390/rs12233877.

Aranzana MJ, Decroocq V, Dirlewanger E, Eduardo I, Gao ZS, Gasic K, Iezzoni A, Jung S, Peace C, Prieto H, Tao R, Verde I, Abbott AG, Arús P (2019) Prunus genetics and applications after *de novo* genome sequencing: achievements and prospects. *Hortic Res* 6:58. https://doi.org/10.1038/s41438-019-0140-8.

Baccichet I, Chiozzotto R, Scaglione D, Bassi D, Rossini L, Cirilli M (2022) Genetic dissection of fruit maturity date in apricot (*P. armeniaca* L.) through a single primer enrichment technology (SPET) approach. *BMC Genomics* 23(1):712. https://doi.org/ 10.1186/s12864-022-08901-1.

Balan V, Ivascu A (1994) Genetic resources in apricot and peach trees in Romania. In: Schmidt H, Kellerhals M (eds) *Progress in Temperate Fruit Breeding. Developments in Plant Breeding*, vol 1. Springer, Dordrecht, pp. 439–445. https://doi.org/10.1007/978-94-011-0467-8_88.

Beer R, Kaiser F, Schmidt K, Arnmann B, Carraro G, Grisa E, Tinner W (2008) Vegetation history of the walnut forests in Kyrgyzstan (Central Asia): natural or anthropogenic origin? *Quater Sci Rev* 27(5–6):621–632. https://doi.org/10.1016/j.quasc irev.2007.11.012.

Bourguiba H, Audergon JM, Krichen L, Mamouni A, Trabelsi S, D'Onofrio C, Asma BM, Santoni S, Khadari B (2012) Loss of genetic diversity as a signature of apricot domestication and diffusion into the Mediterranean Basin. *BMC Plant Biol* 12:49. https://doi.org/10.1186/1471-2229-12-49.

Bourguiba H, Scotti I, Sauvage C, Zhebentyayeva T, Ledbetter C, Krška B, Remay A, D'Onofrio C, Iketani H, Christen D, Krichen L, Trifi-Farah N, Liu W, Roch G, Audergon JM (2020) Genetic structure of a worldwide germplasm collection of *Prunus armeniaca* L. reveals three major diffusion routes for varieties coming from the species' center of origin. *Front Plant Sci* 11:638. https://doi.org/10.3389/fpls.2020.00638.

Callahan AM, Zhebentyayeva TN, Humann JL, Saski CA, Galimba KD, Georgi LL, Scorza R, Main D, Dardick CD (2021) Defining the 'HoneySweet' insertion event utilizing NextGen sequencing and a *de novo* genome assembly of plum (*Prunus domestica*). *Hortic Res* 8:8. https://doi.org/10.1038/s41438-020-00438-2.

Calle A, Wünsch A (2020) Multiple-population QTL mapping of maturity and fruit-quality traits reveals LG4 region as a breeding target in sweet cherry (*Prunus avium* L.). *Hortic Res* 7:127. https://doi.org/10.1038/s41438-020-00349-2.

Campoy JA, Ruiz D, Egea J, Rees DJG, Celton JM, Martínez-Gómez P (2011) Inheritance of flowering time in apricot (*Prunus armeniaca* L.) and analysis of linked quantitative trait loci (QTLs) using simple sequence repeat (SSR) markers. *Plant Mol Biol Rep* 29:404–410. https://doi.org/10.1007/s11105-010-0242-9.

Campoy JA, Sun H, Goel M, Jiao WB, Folz-Donahue K, Wang N, Rubio M, Liu C, Kukat C, Ruiz D, Huettel B, Schneeberger K (2020) Gamete binning: chromosome-level and haplotype-resolved genome assembly enabled by high-throughput single-cell sequencing of gamete genomes. *Genome Biol* 21:306. https://doi.org/10.1186/s13059-020-02235-5.

Community Plant Variety Office. Available online: https://cpvo.europa.eu/en.

Danilevicz MF, Bayer PE, Nestor BJ, Bennamoun M, Edwards D (2021) Resources for image-based high-throughput phenotyping in crops and data sharing challenges. *Plant Physiol* 187:699–715. https://doi.org/10.1093/plphys/kiab301.

De Mori G, Falchi R, Testolin R, Bassi D, Savazzini F, Dondini L, Tartarini S, Palmisano F, Minafra A, Spadotto A, Scalabrin S, Geuna F (2019) Resistance to Sharka in apricot: comparison of phase-reconstructed resistant and susceptible haplotypes of 'Lito' chromosome 1 and analysis of candidate genes. *Front Plant Sci* 10:1576. https://doi.org/10.3389/fpls.2019.01576.

Decroocq S, Chague A, Lambert P, Roch G, Audergon J-M, Geuna F, Chiozzotto R, Bassi D, Dondini L, Tartarini S, Salava J, Krška B, Palmisano F, Karayiannis I, Decroocq V (2014) Selecting with markers linked to the PPVres major QTL is not sufficient to predict resistance to Plum pox virus (PPV) in apricot. *Tree Genet Genomes* 10:1161–1170. https://doi.org/10.1007/s11295-014-0750-0.

Decroocq S, Cornille A, Tricon D, Babayeva S, Chague A, Eyquard JP, Karychev R, Dolgikh S, Kostritsyna T, Liu S, Liu W, Geng W, Liao K, Asma BM, Akparov Z, Giraud T, Decroocq V (2016) New insights into the history of domesticated and wild apricots and its contribution to Plum pox virus resistance. *Mol Ecol* 25(19):4712–29. https://doi.org/10.1111/mec.13772.

Desta ZA, Ortiz R (2014) Genomic selection: genome-wide prediction in plant improvement. *Trends Plant Sci* 19(9):592–601. https://doi.org/10.1016/j.tplants.2014.05.006.

Dondini L, Domenichini C, Dong, Y, Gennari F, Bassi D, Foschi S, Lama M, Adami M, De Franceschi P, Cervellati C, Bergonzoni L, Alessandri S, Tartarini S (2022) Quantitative trait loci mapping and identification of candidate genes linked to fruit acidity in apricot (*Prunus armeniaca* L.). *Front Plant Sci* 13:838370. https://doi.org/10.3389/fpls.2022.838370.

Dondini L, Lain O, Vendramin V, Rizzo M, Vivoli D, Adami M, Guidarelli M, Gaiotti F, Palmisano F, Bazzoni A, Boscia D, Geuna F, Tartarini S, Negri P, Castellano M, Savino V, Bassi D, Testolin R (2011) Identification of QTL for resistance to plum pox virus strains M and D in Lito and Harcot apricot cultivars. *Mol Breed* 27:289–299. https://doi.org/10.1007/s11032-010-9431-3.

FAOSTAT (2020). Available online: http://faostat3.fao.org; accessed October 2022.

Faust M, Surányi, D, Nyujtó F (1998) Origin and dissemination of apricot. *Hort Rev* 22:225–266.

Fiol A, Garcia S, Dujak C, Pacheco I, Infante R, Aranzana MJ (2022) An LTR retrotransposon in the promoter of a PsMYB10.2 gene associated with the regulation of fruit flesh color in Japanese plum. *Hortic Res* 9:uhac206. https://doi.org/10.1101/2022.01.26.477575.

Fiol A, García-Gómez BE, Jurado-Ruiz F, Alexiou K, Howad W, Aranzana MJ (2021) Characterization of Japanese plum (*Prunus salicina*) PsMYB10 alleles reveals structural variation and polymorphisms correlating with fruit skin color. *Front Plant Sci* 12:655267. https://doi.org/10.3389/fpls.2021.655267.

García-Gómez B, Salazar J, Dondini L, Martinez-Gomez P, Ruiz D (2019) Identification of QTLs linked to fruit quality traits in apricot (*Prunus armeniaca* L.) and biological validation through gene expression analysis using qPCR. *Mol Breed* 39(2):28. https://doi.org/10.1007/s11032-018-0926-7.

García-Gómez BE, Ruiz D, Salazar JA, Rubio M, Martínez-García PJ, Martínez-Gómez P (2020) Analysis of metabolites and gene expression changes relative to apricot (*Prunus armeniaca* L.) fruit quality during development and ripening. *Front Plant Sci* 11:1269. https://doi.org/10.3389/fpls.2020.01269.

García-Gómez BE, Salazar JA, Egea JA, Rubio M, Martínez-Gómez P, Ruiz D (2022) Monitoring apricot (*Prunus armeniaca* L.) ripening progression through candidate gene expression analysis. *Int J Mol Sci* 23(9):4575. https://doi.org/10.3390/ijms23094575.

Golubev AM (2019) The creation of disease-resistant apricot varieties in the Lower Volga. *Pomicult Small Fruits Culture Russia* 59:266–271. (In Russ.) https://doi.org/10.31676/2073-4948-2019-59-266-271.

Groppi A, Liu S, Cornille A, Decroocq S, Bui QT, Tricon D, Cruaud C, Arribat S, Belser C, Marande W, Salse J, Huneau C, Rodde N, Rhalloussi W, Cauet S, Istace B, Denis E, Carrère S, Audergon J-M, Roch G, Lambert P, Zhebentyayeva T, Liu W-S, Bouchez O, Lopez-Roques C, Serre R-F, Debuchy R, Tran J, Wincker P, Chen X, Pétriacq P, Barre A, Nikolski M, Aury J-M, Abbott AG, Giraud T, Decroocq V (2021) Population genomics of apricots unravels domestication history and adaptive events. *Nat Commun* 12(1):3956. https://doi.org/10.1038/s41467-021-24283-6.

Gürcan K, Çetinsağ N, Pınar H, Macit T (2019) Molecular and biological assessment reveals sources of resistance to Plum pox virus-Turkey strain in Turkish apricot (*Prunus armeniaca*) germplasm. *Sci Hort* 252:348–353. https://doi.org/10.1016/j.scienta.2019.04.003.

Gürcan K, Teber S, Ercisli S, Yilmaz KU (2016) Genotyping by sequencing (GBS) in apricots and genetic diversity assessment with GBS-derived single-nucleotide polymorphisms (SNPs). *Biochem Genet* 54(6):854–885. https://doi.org/10.1007/s10528-016-9762-9.

Halász J, Pedryc A, Ercisli S, Yilmaz KU, Hegedus A (2010) S-genotyping supports the genetic relationships between Turkish and Hungarian apricot germplasm. *J Am Soc Hortic Sci* 135:410–417. https://doi.org/10.21273/JASHS.135.5.410.

Herrera S, Lora J, Hormaza JI, Herrero M, Rodrigo J (2018) Optimizing production in the new generation of apricot cultivars: self-incompatibility, S-RNase allele identification, and incompatibility group assignment. *Front Plant Sci* 9:527. https://doi.org/10.3389/fpls.2018.00527.

Huang Y, Ren Z, L, D, Liu X (2020) Phenotypic techniques and applications in fruit trees: a review. *Plant Methods* 16:107. https://doi.org/10.1186/s13007-020-00649-7.

Huang Z, Shen F, Chen Y, Cao K, Wang L (2021) Chromosome-scale genome assembly and population genomics provide insights into the adaptation, domestication, and flavonoid metabolism of Chinese plum. *Plant J* 108(4):1174–1192. https://doi.org/110.1111/tpj.15482.

Isobe S, Ninomiya S (2022) Digital phenotyping and data analysis for plant breeding. *Breed Sci* 72(1):1. https://doi.org/10.1270/jsbbs.72.1.

Jangra S, Chaudhary V, Yadav RC, Yadav NR (2021) High-throughput phenotyping: a platform to accelerate crop improvement. *Phenomics* 1(2):31–53. https://doi.org/10.1007/s43657-020-00007-6.

Jiang F, Zhang J, Wang S, Yang L, Luo Y, Gao S, Zhang M, Wu S, Hu S, Sun H, Wang Y (2019) The apricot (*Prunus armeniaca* L.) genome elucidates Rosaceae evolution and beta-carotenoid synthesis. *Hortic Res* 6:128. https://doi.org/10.1038/s41438-019-0215-6.

Jung S, Lee T, Cheng CH, Buble K, Zheng P, Yu J, Humann J, Ficklin SP, Gasic K, Scott K, Frank M, Ru S, Hough H, Evans K, Peace C, Olmstead M, DeVetter LW, McFerson J, Coe M, Wegrzyn JL, Staton ME, Abbott AG, Main D (2019) 15 years of GDR: new data and functionality in the Genome Database for Rosaceae. *Nucleic Acids Res* 47(D1):D1137–D1145. https://doi.org/10.1093/nar/gky1000.

Kitamura Y, Habu T, Yamane H, Nishiyama S, Kajita K, Sobue T, Kawai T, Numaguchi K, Nakazaki T, Kitajima A, Tao R (2018) Identification of QTLs controlling chilling and heat requirements for dormancy release and bud break in Japanese apricot (*Prunus mume*). *Tree Genet Genomes* 14:33. https://doi.org/10.1007/s11295-018-1243-3.

Kostina KF (1969) The use of varietal resources of apricots for breeding. *Trud Nikit Bot Sad* 40:45–63.

Krška B (2018) Genetic apricot resources and their utilisation in breeding. In: Soneji JR, Nageswara-Rao M (eds) *Breeding and Health Benefits of Fruit and Nut Crops*. IntechOpen, London, pp. 158. ISBN 978-1-78923-272-1 https://doi.org/10.5772/intechopen.77125.

Krška B, Vachůn Z (2016) Apricot breeding at the faculty of horticulture in Lednice. *Agronomy* 6(2):27. https://doi.org/10.3390/agronomy6020027.

Kumar M, Mishra GP, Singh R, Kumar J, Naik PK, Singh SB (2009) Correspondence of ISSR and RAPD markers for comparative analysis of genetic diversity among different apricot genotypes from cold arid deserts of trans-Himalayas. *Physiol Mol Biol Plants* 15(3):225–236. https://doi.org/10.1007/s12298-009-0026-6.

Lai K, Lorenc MT, Edwards D (2015) Molecular marker databases. *Methods Mol Biol* 1245:49–62. https://doi.org/10.1007/978-1-4939-1966-6_4.

Lambert P, Dicenta F, Rubio M, Audergon JM (2007) QTL analysis of resistance to sharka disease in the apricot (*Prunus armeniaca* L.) 'Polonais' × 'Stark Early Orange' F1 progeny. *Tree Genet Genomes* 3:299–309. https://doi.org/10.1007/s11295-006-0069-6.

Ledbetter CA (2008) Apricots. In: Hancock JF (ed) *Temperate Fruit Crop Breeding*. Springer, Dordrecht, pp. 39–82. https://doi.org/10.1007/978-1-4020-6907-9_2.

Li W, Liu L, Wang Y, Zhang Q, Fan G, Zhang S, Wang Y, Liao K (2020) Genetic diversity, population structure, and relationships of apricot (Prunus) based on restriction site-associated DNA sequencing. *Hortic Res* 7:69. https://doi.org/10.1038/s41438-020-0284-6

Liu S, Cornille A, Decroocq S, Tricon D, Chague A, Eyquard JP, Liu WS, Giraud T, Decroocq V (2019) The complex evolutionary history of apricots: species divergence, gene flow and multiple domestication events. *Mol Ecol* 28(24):5299–5314. https://doi.org/10.1111/mec.15296.

Maggioni L, Engels J (2014) Networking for plant genetic resources. *Pan Eur Networks Sci Technol* 10:285–297.

Mariette S, Wong Jun Tai F, Roch G, Barre A, Chague A, Decroocq S, Groppi A, Laizet Y, Lambert P, Tricon D, Nikolski M, Audergon J-M, Abbott AG, Decroocq V (2016) Genome-wide association links candidate genes to resistance to Plum Pox Virus in apricot (*Prunus armeniaca*). *New Phytol* 209(2):773–784. https://doi.org/10.1111/nph.13627.

Mehlenbacher S, Cociu V, Hough F (1990) Apricots. In: Moore JN, Ballington JR (eds) *Genetic Resources of Temperate Fruits and Nuts Crops*. International Society of Horticultural Science, Wageningen, pp. 65–107.

Miryeganeh M (2021) Plants' epigenetic mechanisms and abiotic stress. *Genes* 12(8):1106. https://doi.org/10.3390/genes12081106.

Morales N, Bauchet GJ, Tantikanjana T, Powell AF, Ellerbrock BJ, Tecle IY, Mueller LA (2020) High density genotype storage for plant breeding in the Chado schema of Breedbase. *PLoS One* 15(11):e0240059. https://doi.org/10.1371/journal.pone.0240059.

Moustafa K, Cross J (2019) Production, pomological and nutraceutical properties of apricot. *J Food Sci Technol* 56(1):12–23. https://doi.org/10.1007/s13197-018-3481-7.

Newell MA, Jannink JL (2014) Genomic selection in plant breeding. *Methods Mol Biol* 1145:117–130. https://doi.org/10.1007/978-1-4939-0446-4_10.

Okura F (2022) 3D modeling and reconstruction of plants and trees: a cross-cutting review across computer graphics, vision, and plant phenotyping. *Breed Sci* 72(1):31–47. https://doi.org/10.1270/jsbbs.21074.

Olukolu BA, Trainin T, Fan S, Kole C, Bielenberg DG, Reighard GL, Abbott AG, Holland D (2009) Genetic linkage mapping for molecular dissection of chilling requirement and budbreak in apricot (*Prunus armeniaca* L.). *Genome* 52(10):819–828. https://doi.org/10.1139/g09-050.

Pirona R, Eduardo I, Pacheco I, Da Silva Linge C, Miculan M, Verde I, Tartarini S, Dondini L, Pea G, Bassi D, Rossini L (2013) Fine mapping and identification of a candidate gene for a major locus controlling maturity date in peach. *BMC Plant Biol* 13:166. https://doi.org/10.1186/1471-2229-13-166.

Ricci A, Sabbadini S, Prieto H, Padilla IM, Dardick C, Li Z, Scorza R, Limera C, Mezzetti B, Perez-Jimenez M, Burgos L, Petri C (2020) Genetic transformation in peach (*Prunus persica* L.): challenges and ways forward. *Plants* 9(8):971. https://doi.org/10.3390/plants9080971.

Romero C, Vilanova S, Burgos L, Martínez-Calvo J, Vicente M, Llácer G, Badenes ML (2004) Analysis of the S-locus structure in *Prunus armeniaca* L. identification of S-haplotype specific S-RNase and F-box genes. *Plant Mol Biol* 56(1):145–157. https://doi.org/10.1007/s11103-004-2651-3.

Rubio M, Ruiz D, Egea J, Martínez-Gómez P, Dicenta F (2014) Opportunities of marker assisted selection for Plum pox virus resistance in apricot breeding programs. *Tree Genet Genomes* 10:513–525. https://doi.org/10.1007/s11295-014-0700-x.

Rubio M, Salazar J, Dicenta, F, Ruiz D, Martinez-Gomez P, Martínez-García P (2019) Identification of quantitative trait loci (QTLs) linked to Apple chlorotic leaf spot virus (ACLSV) resistance in apricot. *Euphytica* 215:170. https://doi.org/10.1007/s10681-019-2481-7.

Salazar JA, Ruiz D, Egea J, Martínez-Gómez P (2013) Transmission of fruit quality traits in apricot (Prunus armeniaca L.) and analysis of linked quantitative trait loci (QTLs) using simple sequence repeat (SSR) markers. *Plant Mol Biol Report* 31(6):1506–1517. https://doi.org/10.1007/s11105-013-0625-9.

Salazar JA, Batnini M-A, Trifi-Farah N, Ruiz D, Martínez-Gómez P, Rubio M (2016a) Quantitative trait loci (QTLs) identification and the transmission of resistance to powdery mildew in apricot. *Euphytica* 211:245–254. https://doi.org/10.1007/s10681-016-1734-y.

Salazar JA, Pacheco I, Shinya P, Zapata P, Silva C, Aradhya M, Velasco D, Ruiz D, Martínez-Gómez P, Infante R (2017) Genotyping by sequencing for SNP-based linkage analysis and identification of QTLs linked to fruit quality traits in Japanese plum (*Prunus salicina* Lindl.). *Front Plant Sci* 8:1–14. https://doi.org/10.3389/fpls.2017.00476.

Salazar JA, Pacheco I, Zapata P, Shinya P, Ruiz D, Martínez-Gómez P, Infante R (2020) Identification of loci controlling phenology, fruit quality and post-harvest quantitative parameters in Japanese plum (*Prunus salicina* Lindl.). *Postharv Biol Technol* 169:111292. https://doi.org/10.1016/j.postharvbio.2020.111292.

Salazar JA, Ruiz D, Campoy JA, Tartarini S, Dondini L, Martínez-Gómez P (2016b) Inheritance of reproductive phenology traits and related QTL identification in apricot. *Tree Genet Genomes* 12:71. https://doi.org/10.1007/s11295-016-1027-6.

Salazar JA, Ruiz D, Zapata P, Martínez-García PJ, Martínez-Gómez P (2022) Whole transcriptome analyses of apricots and Japanese plum fruits after 1-MCP (ethylene-inhibitor) and Ethrel (ethylene-precursor) treatments reveal new insights into the physiology of the ripening process. *Int J Mol Sci* 23(19):11045. https://doi.org/10.3390/ijms231911045.

Sánchez-Pérez R, Pavan S, Mazzeo R, Moldovan C, AieseCigliano R, Del Cueto J, Ricciardi F, Lotti C, Ricciardi L, Dicenta F, López-Marqués RL, Møller BL (2019) Mutation of a bHLH transcription factor allowed almond domestication. *Science* 364(6445):1095–1098. https://doi.org/10.1126/science.aav8197.

Sasaki R, Yamane H, Ooka T, Jotatsu H, Kitamura Y, Akagi T, Tao R (2011) Functional and expressional analyses of PmDAM genes associated with endodormancy in Japanese apricot. *Plant Physiol* 157:485–497. https://doi.org/10.1104/pp.111.181982.

Schurdi-Levraud Escalettes V, Hullot C, Wawrzy'nczak D, Mathieu E, Eyquard JP, Le Gall O, Decroocq V (2006) Plum pox virus induces differential gene expression in the partially resistant stone fruit tree *Prunus armeniaca* cv. Goldrich. *Gene* 374:96–103. https://doi.org/10.1016/j.gene.2006.01.021.

Sheikh ZN, Sharma V, Shah RA, Raina S, Aljabri M, Mir JI, AlKenani N, Hakeem KR (2021) Elucidating genetic diversity in apricot (*Prunus armeniaca* L.) cultivated in the North-Western Himalayan provinces of India using SSR Markers. *Plants* 10(12):2668. https://doi.org/10.3390/plants10122668.

Socquet-Juglard D, Christen D, Devènes G, Gessler C, Duffy B, Patocchi A (2013a) Mapping architectural, phenological, and fruit quality QTLs in apricot. *Plant Mol Biol Rep* 31:387–397. https://doi.org/10.1007/s11105-012-0511-x.

Socquet-Juglard D, Duffy B, Pothier JF, Christen D, Gessler C, Patocchi A (2013b) Identification of a major QTL for *Xanthomonas arboricola* pv. *pruni* resistance in apricot. *Tree Genet Genomes* 9:409–421. https://doi.org/10.1007/s11295-012-0562-z.

Sooriyapathirana SS, Khan A, Sebolt AM, Wang D, Bushakra JM, Lin-Wang K, Allan AC, Gardiner SE, Chagné D, Iezzoni AF (2010) QTL analysis and candidate gene mapping for skin and flesh color in sweet cherry fruit (*Prunus avium* L.). *Tree Genet Genomes* 6:821–832. https://doi.org/10.1007/s11295-010-0294-x.

Soriano JM, Domingo ML, Zuriaga E, Romero C, Zhebentyayeva T, Abbott AG, Badenes ML (2012) Identification of simple sequence repeat markers tightly linked to plum pox virus resistance in apricot. *Mol Breed* 30:1017–1026. https://doi.org/10.1007/s11032-011-9685-4.

The International Peach Genome Initiative (2013) The high-quality draft genome of peach (*Prunus persica*) identifies unique patterns of genetic diversity, domestication and genome evolution. *Nat Genet* 45:487–494. https://doi.org/10.1038/ng.2586.

Vavilov NI (1951) Phytogeographic basis of plant breeding - the origin, variation, immunity and breeding of cultivated plants. *Chron Bot* 13:1–366. https://doi.org/10.1097/00010694-195112000-00018.

Verde I, Jenkins J, Dondini L, Micali S, Pagliarani G, Vendramin E, Paris R, Aramini V, Gazza L, Rossini L, Bassi D, Troggio M, Shu S, Grimwood J, Tartarini S, Dettori MT, Schmutz J (2017) The Peach v2.0 release: high-resolution linkage mapping and deep resequencing improve chromosome-scale assembly and contiguity. *BMC Genomics* 18:225. https://doi.org/10.1186/s12864-017-3606-9.

Vergara R, Olivares F, Olmedo B, Toro C, Muñoz M, Zúñiga C, Mora R, Plantat P, Miccono M, Loyola R, Aguirre C, Prieto H (2021) Gene editing in *Prunus* spp.: the challenge of adapting regular gene transfer procedures for precision breeding. In: Küden AB, Küden A (eds) *Prunus - Recent Advances*. IntechOpen, Rijeka, Ch. 5. ISBN 978-1-83969-582-7.

Vilanova S, Badenes ML, Burgos L, Martínez-Calvo J, Llácer G, Romero C (2006) Self-compatibility of two apricot selections is associated with two pollen-part mutations of different nature. *Plant Physiol* 142(2):629–41. https://doi.org/10.1104/pp.106.083865.

Vinceti B, Elias M, Azimov R, Turdieva M, Aaliev S, Bobokalonov F, Butkov E, Kaparova E, Mukhsimov N, Shamuradova S, Tugunbaev K, Azizova N, Loo J (2022) Home gardens of Central Asia: Reservoirs of diversity of fruit and nut tree species. *PLoS One* 17(7):e0271398. https://doi.org/10.1371/journal.pone.0271398.

Wang Z, Kang M, Liu H, Gao J, Zhang Z, Li Y, Wu R, Pang X (2014) High-level genetic diversity and complex population structure of Siberian apricot (*Prunus sibirica* L.) in China as revealed by nuclear SSR markers. *PLoS One* 9(2):e87381. https://doi.org/10.1371/journal.pone.0087381.

Wani AA, Hussain K, Zargar SA, Ahmad F, Mahajan R, Zargar SM, Khuroo AA (2021) Assessment of the genetic diversity and population structure of apricot (*Prunus armeniaca* L.) germplasm of the Northwestern Himalaya using SSR markers. *Plant Genet Resour Characteris Utilis* 19(5):384–393. https://doi.org/10.1017/S1479262121000459.

Weise S, Oppermann M, Maggioni L, vanHintum T, Knüpffer H (2017) EURISCO: the European search catalogue for plant genetic resources. *Nucleic Acids Res* 45 (D1):D1003–D1008. https://doi.org/10.1093/nar/gkw755.

Yu Y, Fu J, Xu Y, Zhang J, Ren F, Zhao H, Tian S, Guo W, Tu X, Zhao J, Jiang D, Zhao J, Wu W, Wang G, Ma R, Jiang Q, Wei J, Xie H (2018) Genome re-sequencing reveals the evolutionary history of peach fruit edibility. *Nat Commun* 9(1):5404. https://doi.org/10.1038/s41467-018-07744-3.

Zahid G, Aka Kaçar Y, Dönmez D, Küden A, Giordani T (2022) Perspectives and recent progress of genome-wide association studies (GWAS) in fruits. *Mol Biol Rep* 49:5341–5352. https://doi.org/10.1007/s11033-021-07055-9.

Zanetto A, Maggioni L, Tobutt KR, Dosba F (2002) Prunus genetic resources in Europe: achievement and perspectives of a networking activity. *Genet Resour Crop E* 49:331–337. https://doi.org/10.1023/A:1015524630456.

Zhang L, Zhang Q, Li W, Zhang S, Xi W (2019a) Identification of key genes and regulators associated with carotenoid metabolism in apricot (*Prunus armeniaca*) fruit using weighted gene coexpression network analysis. *BMC Genomics* 20:876. https://doi.org/10.1186/s12864-019-6261-5.

Zhang Q, Feng C, Li W, Qu Z, Zeng M, Xi W (2019b) Transcriptional regulatory networks controlling taste and aroma quality of apricot (*Prunus armeniaca* L.) fruit during ripening. *BMC Genomics* 20:45. https://doi.org/10.1186/s12864-019-5424-8.

Zhang Q, Liu J, Liu W, Liu N, Zhang Y, Xu M, Liu S, Ma X, Zhang Y (2022) Construction of a high-density genetic map and identification of quantitative trait loci linked to fruit quality traits in apricots using specific-locus amplified fragment sequencing. *Front Plant Sci* 13:798700. https://doi.org/10.3389/fpls.2022.798700.

Zhang Q, Zhang D, Yu K, Ji J, Liu N, Zhang Y, Xu M, Zhang Y-J, Ma X, Liu S, Sun W-H, Yu X, Hu W, Lan S-R, Liu Z-J, Liu W (2021) Frequent germplasm exchanges drive the high genetic diversity of Chinese-cultivated common apricot germplasm. *Hortic Res* 8:215. https://doi.org/10.1038/s41438-021-00650-8.

Zhu Y, Chen M, Gu Q, Zhao Y, Zhang X, Sun Q, Gu X, Zheng K (2022) Machine learning methods for efficient and automated *in situ* monitoring of peach flowering phenology. *Comput Elect Agri* 202:107370. https://doi.org/https://doi.org/10.1016/j.compag.2022.107370.

Zuriaga E, Romero C, Blanca JM, Badenes ML (2018) Resistance to Plum Pox Virus (PPV) in apricot (*Prunus armeniaca* L.) is associated with down-regulation of two MATHd genes. *BMC Plant Biol* 18:25. https://doi.org/10.1186/s12870-018-1237-1.

12 Allele Mining in Melon Crop Improvement

Rakesh Kumar

12.1 INTRODUCTION

Melon, a member of the cucurbit family, includes cucumbers, watermelons, squash, pumpkins, and gourds. This important vegetable crop is grown worldwide for its sweet taste, flavor, nutritive value, and medicinal value. Melon is widely grown in temperate, subtropical to tropical regions globally. It is a very diverse crop in terms of typology: Cantaloupe, Honeydew, Yellow Canary, Piel de Sapo, Charentais, Italo-American, Hami, and Ananas are some of the major types (Figure 12.1). The diversity in melon opens up avenues for genetic improvement through the leveraging of traits across typologies.

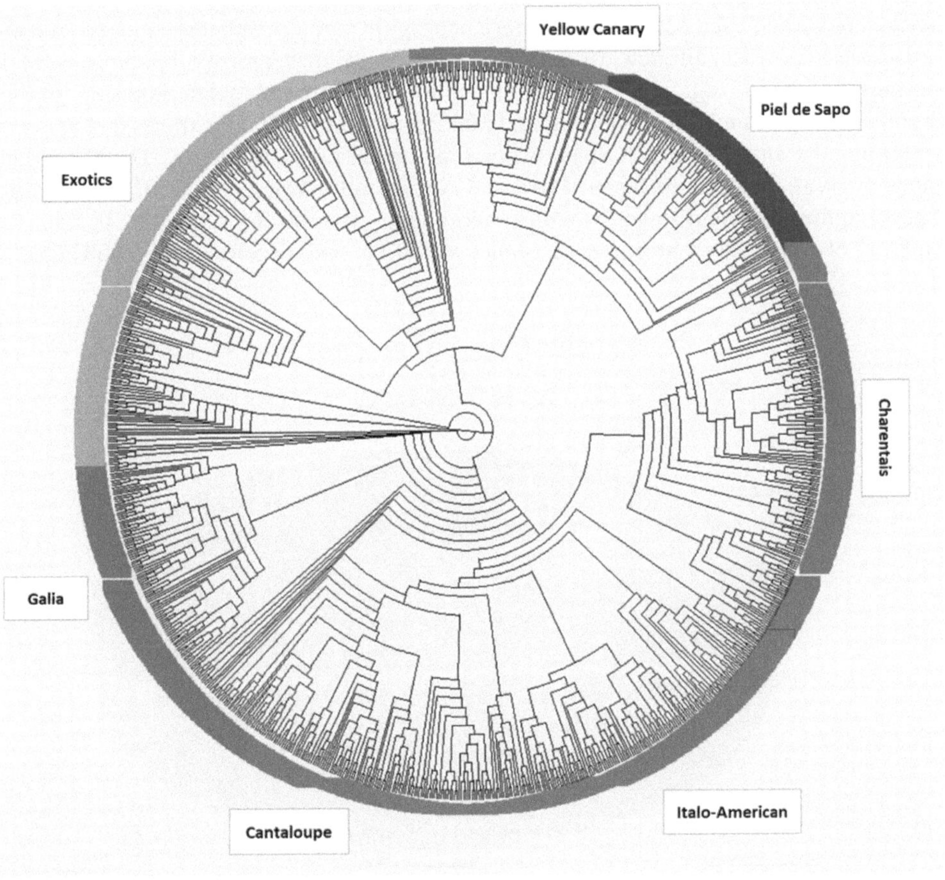

FIGURE 12.1 Genetic diversity in the melon germplasm.

DOI: 10.1201/9781003386490-12

Melons were domesticated in Asia and Africa for their seeds, which are rich sources of protein and fats (Zeven and De Wet 1982). However, over the centuries, melons were improved for other horticultural traits, like flavor, aroma, and total soluble solids. Manchali et al. (2021) characterized diverse melon cultivars for their nutrition and healthfulness. Carotenoids are major compounds in melon flesh, which is a rich source of vitamin A. In addition, melons contain sugars, protein, and antioxidants like phenolic compounds and ascorbic acid (Chang et al. 2015; Gómez-García et al. 2020). Sugar content is the major criterion for judging maturity and quality in melon. Sucrose is the major source of melon's sweetness, and glucose and fructose are other contributing sugars (Schaffer et al. 1987; Burger and Schaffer 2007).

Melons, which are grown worldwide, are valuable sources of biodiversity for sustainable agriculture. In 2021, 28 million tons of melon were produced on 1.07 million hectares (Food and Agriculture Organization 2023). Asian countries account for 77% of the total acreage, followed by the Americas (11.5%), Europe (6.9%), and Africa (3.9%; Figure 12.2). China is major producer of melons (14 million tons), followed by Turkey (1.6 million tons) and India (1.5 million tons). Kazakhstan, Afghanistan, Guatemala, Iran, Spain, Brazil, and Italy are other major producers of melons (Food and Agriculture Organization 2023). Each country has a different product profile (typology) for consumers. Cantaloupes are mainly grown in the Americas and Europe, Ananas in the Middle East, Piel de Sapo in Spain, Yellow Canary in Brazil, and Charentais in France.

This means that melons have been selected for different traits based on the region in which they are grown. Because melons are very sensitive to temperature, day length, humidity, light intensity, and abiotic and biotic stresses, they must adapt to the agro-climatic region in which they are grown. Most breeding efforts are focused on traits desirable to the grower, including high yield; high sugars; field shelf life; vine strength; dense netting (in cantaloupe); resistance to fruit cracking; and resistance to powdery mildew, fusarium wilt, downy mildew, and viral pathogens. Melons are often exported from one region to another. For example, most winter melon produced for the United States comes from Guatemala, Honduras, and Mexico. Similarly, Brazil exports melon to Europe in the winter months. In the summer, Spain exports melon to France, Germany, The Netherlands, and Germany; Australia exports to Japan (Food and Agriculture Organization 2023). Breeding melon for export requires different traits for withstanding long-distance shipping (e.g., post-harvest shelf life, stability of sugars, flesh firmness, resistance to rotting, and discoloration). The United States is

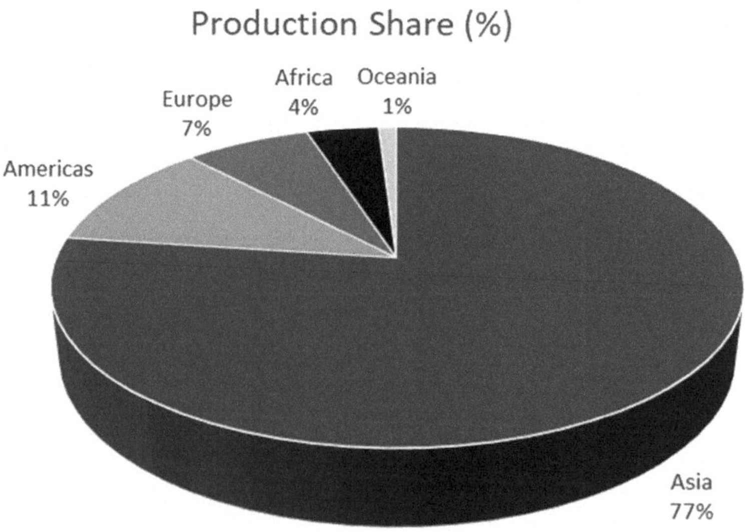

FIGURE 12.2 Production share of melon by region (Food and Agriculture Organization 2023).

the top importer of melon, followed by The Netherlands, France, the United Kingdom, and Canada. Spain is the top exporter, followed by Guatemala, Brazil, and Honduras (Food and Agriculture Organization 2023). Hence, any breeding design must satisfy trait requirements throughout the value chain (e.g., of growers, shippers, retail, and finally end consumers).

12.2 ALLELE MINING

Breeding designs rely on genetic diversity in the crop. Just like diamonds and fossil fuels can be mined from vast landmasses, superior alleles can be mined from vast reservoirs of germplasm diversity in a process called *allele mining.* Significant numbers of novel alleles were left behind during domestication and crop improvement and could be tapped to produce new cultivars for sustainable crop production. Mutations are evolutionary forces that result in allelic diversity. Initially, allele mining mostly focused on single-nucleotide polymorphisms (SNPs) and indels because they are present in coding regions (Gokidi et al. 2017). However, mutations in noncoding regions like introns, the 5′ UTR, and the 3′ UTR also play a role in trait expression. Some examples of this are the *rubi3* gene in rice (Samadder et al. 2008) and *VRN-1* in barley and wheat (Fu et al. 2005).

As many horticulturally important traits are complex, the availability and utilization of multi-parental resources is required for comprehensive genetic dissection of natural variation and detection of a repertoire of favorable alleles for crop improvement. Current melon cultivars are derived from wildtype and landraces through recombination and selection. They have been altered significantly for adaptation, size, shape, flavor, aroma, shelf life, netting density, flesh color, and other traits, which has resulted in a loss of allelic variation (Sun et al. 2022). The key is to select a set of accessions representing total diversity in the crop and characterize for traits of interest as well as sequence them.

Melon breeding has made significant progress since the release of the first melon genome 10 years ago (Garcia-Mas et al. 2012), followed by extensive resequencing of hundreds of melon accessions (Zhao et al. 2019; Liu et al. 2020; Dafna et al. 2021). Five additional high-quality melon genome assemblies that can be used as references for this species have been published over the past 4 years (Zhang et al. 2019; Yang et al. 2020; Yano et al. 2020; Ling et al. 2021; Pichot et al. 2022). However, the recently introduced pan-genome approach, which is based on multiple genomes that capture the diversity within a species, is required for allele mining in cases of vast genetic diversity (Liu et al. 2020; Sun et al. 2022; Oren et al. 2022).

Different approaches have been proposed for allele mining in crop plants (Shinde et al. 2022), including the following:

1. Targeting induced local lesions in genomes (TILLING)
2. Sequence-based allele mining

12.2.1 TILLING

TILLING is a reverse genetics technique that can identify allelic variation (point mutation) due to induced mutations in a gene of known sequence by heteroduplex analysis. It is also used to study gene functions. Dahmani-Mardas et al. (2010) constructed a mutant collection of 4,023 melon M2 TILLING families to screen for 11 genes, four of which were involved in ethylene production. Using TILLING, they found the *G194D* mutant on ACC oxidase 1, which improved shelf life.

EcoTILLING is a variation on TILLING that relies on natural variation in the gene of interest. The rest of the procedure is the same as TILLING. In melon, a new allele of the *eIF4E* gene was identified for resistance to *Melon Necrotic Spot Virus* (MNSV) using this approach (Nieto et al. 2006).

12.2.2 Sequence-Based Allele Mining

Sequence-based allele mining involves amplifying alleles of genes in different accessions and then sequencing them to identify SNPs. In melon, the *MLO* gene was mined for resistance to powdery mildew using sequencing (Hong et al. 2015).

12.3 ORIGIN AND TAXONOMY

12.3.1 Origin

Although melon was domesticated 4,000 years ago (Janick et al. 2007; Paris 2016), controversy surrounds its center of origin (Pitrat 2016). Both wild and cultivated accessions can be found growing in both Africa and Asia. The current consensus among researchers is that Africa is the center of origin (Kirkbride 1993), whereas Asia is the area of diversification. Two domestication events have been proposed, one in Africa and the other in Asia (Endl et al. 2018; Pichot et al. 2022). However, another study proposed three independent domestication events, two in India and one in Africa, likely leading to all melon botanical groups (Zhao et al. 2019). This is based on a resequencing of 1,175 accessions, representing the global diversity of the species.

Melon is a diploid and dicotyledonous species. It has $2n=2x=24$ chromosomes with a supposedly small genome (454 Mb). Although it is cross-pollinated, little heterosis has been reported (Dogimont and Sari 2022). Phylogenetically speaking, *Cucumis melo* is closely associated with *Cucumis trigonus* and *Cucumis picrocarpus* (Endl et al. 2018). These two species grow in India and Australia, respectively. Most domesticated melons can be attributed to two wild backgrounds: *Cucumis melo* ssp. *melo* and *Cucumis melo* ssp. *meloides*. The subspecies *meloides* is widely found in Africa, whereas the subspecies *melo* is found in Asia. Most modern cultivars grown worldwide originated from this *melo* subspecies (Chomicki et al. 2020). Cultivars in Asia are also derived from *Cucumis melo* ssp. *agrestis* (Lian et al. 2021).

Pichot et al. (2022) reported that based on structural relationships, melon has retained its ancestral genome structure. The difference in genome size between the *melo* and *agrestis* subspecies is significant (Oren et al. 2022) and implies the possible evolutionary significance of within-species variation in genome size (Biémont 2008). The variation in genome size and the consistently differential, large chromosome inversions imply that these two subspecies evolved separately under cultivation, with little or no interaction. A parallel evolution under domestication for fruit quality traits was described for the two cultivated subspecies of another cucurbit, *Cucurbita pepo* L. (Paris et al. 2012).

12.3.2 Classification

Initially, based on ovary pubescence, melon was classified into two subspecies: *Cucumis melo* subsp. *melo* and *Cucumis melo* subsp. *agrestis* (Pitrat et al. 2000). The two subspecies were further divided into 16 groups, five in *agrestis* and 11 in *melo*. Pitrat (2016) again described 19 groups of wild, feral, and domesticated melons, many of which have subgroups (Table 12.1).

12.4 MELON GENOMICS

12.4.1 Genetic Mapping

Linking genotype with phenotype is a fundamental goal in biology and requires robust data for both. A high-quality accurate genome is critical for allele mining and analyses of loci. The very first linkage map was reported by Pitrat (1991). Quite a few maps have been released using few amplified fragment length polymorphic markers (Wang et al. 1997), random amplified polymorphic DNA (Baudracco-Arnas and Pitrat 1996), and simple sequence repeats (Gonzalo et al. 2005).

TABLE 12.1
Classification of Melon in Major Groups and Subgroups

Group	Subgroup	Accessions	Distribution
Agrestis		HS	Africa, Asia, and Australia
Kachri		AHK 119, AH120	India
Chito			Central America and Caribbean Islands
Tibish		Seinat	Sudan
Acidulus		PI313970, PI164323; PI164723	India and Sri Lanka
Momordica		MR-1 (PI124111), PI414723	India, Southeast Asia
Conomon		Tokyo Wase Shiro Uri, Ko Shiro Uri, PI266935	China, Japan
Makuwa			Far East
	Ogon	Ogon 9, Mi Tang Thing, Showa Kogane Nahi Makuwa	
	Nash-uri	PI420150, Nyumelon	
	Yuki	Ginsen Makuwa, LJ904436	
	Kanro	Yantai, Wasada Uri, Nagata Kim Makuwa	
	Ginmakuwa	Chenggam, Chang Bougi	
	Seikan	Lanzhou	
Chinensis		PI161375, PI255479, PI266934	Far East
Flexuosus			India, Northern Mediterranean
	Adjour	Fakouss CM4, Fegouss 1	
	Tara	PI222187	
	Arya	AMES 20662	
Chate		Carosello, Dolmalik	
Dudaim		PI117362, Queen Anne's Pocket	
Chandalak			Central Asia, India
	Zami	Zaami 610, Mag Assil, Varanasi local	
	Tachmi	Persia 202, Samsouri green, Semosouri Varamin, Shahd Shiraz	
	Garma	PI124552	
	Bucharici		
Indicus		Local landraces	Central India
Ameri			Asia
	Ananas	Ananas Yokneam, Ananas Dokki, PI534538, Gorgab 1	
	Maculati	Shakar Palaki, Ak Uruk, Chanjo, Cotton Rose Melon, Baba Kharman	
	Bargi	Evankey 2, Koktcha 588, Twaghermez Isfahan	
	Mashhadi	Ghasri, Khagkhani, Jharbezeh Mashadi	
Cassaba			Western to Central Asia
	Kirkagac	Hidir, Kirkagac 1, Mustapha Kemal Pacha 2	
	Hassanbey	Hasan Bey, Yuva	

(Continued)

TABLE 12.1 (*Continued*)
Classification of Melon in Major Groups and Subgroups

Group	Subgroup	Accessions	Distribution
	Kuscular	Golden Beauty Casaba, Gare Ghez, Cumra	
Ibericus			Spain
	Piel de Sapo	Piel de Sapo T111	
	Amarillo	Amarillo Oro	
	Tendral	Negro, Tendral Verde	
	Rochet	Sigura, Cavaillon Espagnol, Escrito	
	Branco	Branco de Ribateja	
Inodorus			Asia, Americas
	Honeydew	TamDew, Angel Dew	
	Earl's	Chung Hsing, Earl's	
Cantalupensis			Europe, Americas
	Prescott	Noir des Carmes, Cantaloup de Bellegarde, Iopor Menyhert, Prescott a Chassis	
	Saccharinus	Gris de Rennes, Aarhus Torg, Madelenen, Leknicki Cukrovi	
	Charentais	Nantais Oblong, Vedrantais	
	Ogen	Hemed, Muskotaly, Noy Yizre'el	
	American Western	Top Mar, Hale's Best, PMR45, Jacumba, Perlita	
	American Eastern	Harvest Queen, New Yorker, Delicious 51, Iroquois	

Recent advances in genome sequencing have expanded comparisons among multiple related individuals. Since 2012, mapping studies in melon have benefited from the availability of a reference genome (Garcia-Mas et al. 2012) that enables the physical mapping and annotation of quantitative trait loci (QTL) intervals. This assembly contained up to 375 Mb of sequences assembled into 1,594 scaffolds, with an N50 of 4.68 Mb. This was followed by a high-resolution map (Argyris et al. 2017) to anchor up to 98.2% of the scaffold assembly to melon chromosomes and improve the orientation of the scaffolds of the previous assembly and precisely measure the gap content (Ruggieri et al. 2018). However, there were still 41 Mb of unassigned sequences. Using PacBio single-molecule real-time sequencing technology, 90% of gaps were filled and unassigned sequences, in particular repetitive and centromeric areas, were drastically reduced (Castanera et al. 2020; Figure 12.3). Such regions may affect the expression of genes linked to traits of interest.

12.4.2 A Pan-Genome Approach

The amount of within-species genomic structural variation that has been discovered indicates that a single reference genome cannot represent the complete sequence diversity of a species; this realization has led to an increased emphasis on a pan-genome approach. Pan-genome approaches encompass the core genes shared by all accessions of a crop and variable genes found in some individuals. Essentially, these approaches represent the genomic variation in the germplasm or gene pool of the species. Variable genes help to generate diverse traits among different varieties and may provide resources for allele mining. The resequencing of hundreds of melon accessions (Zhao

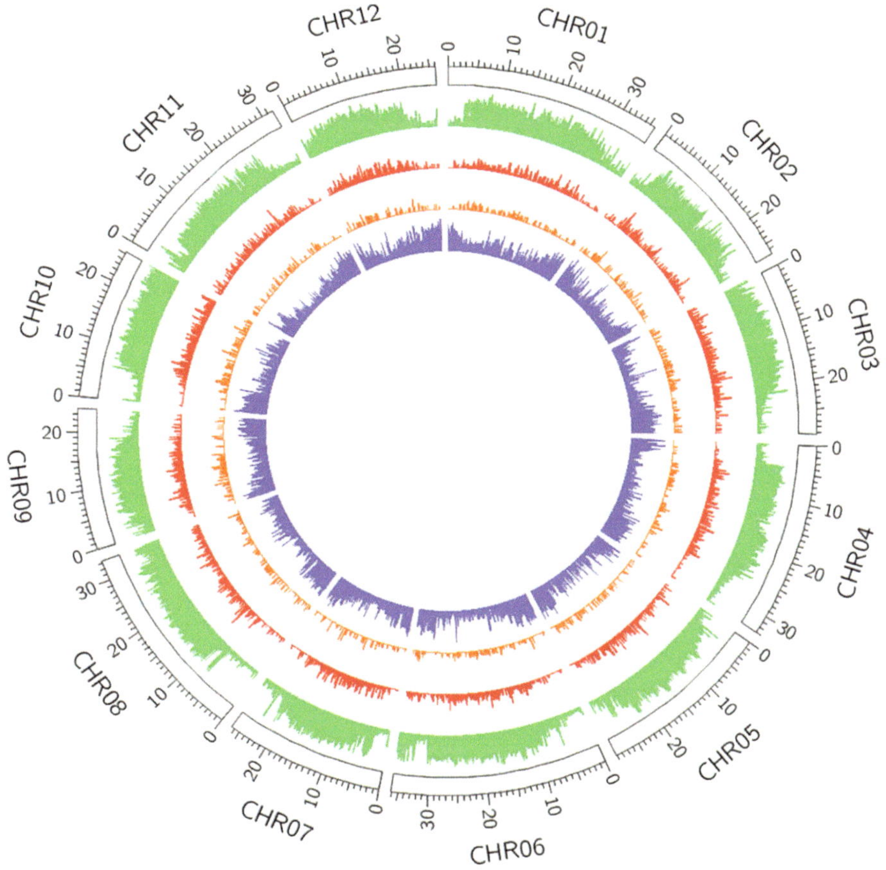

FIGURE 12.3 Melon reference genome. Distribution of transposable elements and genes across v4.0 pseudomolecules. Green shows the density of REPET features per window (6,000 windows in total). Red shows the density of full-length long terminal repeat (LTR) retrotransposons annotated in v4.0 that were absent in the v3.6.1 assembly. Orange shows the density of polymorphic LTR retrotransposons with insertion time less than 2 Mya (Castanera et al. 2020). Purple shows gene density.

et al. 2019; Liu et al. 2020) in recent years has further advanced the genomic analysis of candidate genes, causative genes, and allele mining. Assemblies of diverse melon accessions reveal variation in genome size ranging from 358 to 374 Mb, with an average of 365 Mb (Oren et al. 2022). There is a difference of 16 Mb (4.5%) between the largest and smallest genomes across melon accessions. Liu et al. (2020) sequenced 297 wild, landrace, and improved cultivars and obtained 2,045,412 high-quality SNPs, which helped with the genomic history of melon trait improvement. Using data from the same 297 melon accessions, Sun et al. (2022) used a pan-genomic approach to identify resistance gene analogs, analyze their distribution and gene presence–absence variation selection among populations.

12.5 GENE BANKS

Gene banks, which hold large diversity of germplasm, are rich sources of novel genes. They are great resources for mining superior alleles for traits of interest. As many as 7.4 million accessions are maintained in 1,700 gene banks in different international institutions globally (Godiki et al. 2017). These resources, which are untapped and underutilized (at less than 1%; Upadhyaya et al. 2006), could be systematically incorporated into breeding programs. Systematic and consistent

TABLE 12.2
Number of Germplasm Resources in Major Gene Banks (Accessed January 11, 2023)

Gene Bank	Accessions Listed	Available
USDA-GRIN	5,777	1981
INRA	102 (+2,232 private)	–
AVDRC	178	–

utilization of germplasm from a gene bank is crucial for breeders to identify, extract, and exploit this extensive diversity. Genome-wide characterization of untapped genetic resources using modern genome sequencing techniques provides new opportunities for sustainable breeding and efficient usage of genetic material.

Wild relatives of melon contain large allelic diversity. Some of this diversity has been lost in modern cultivars as a result of improvement. Wild melons have been found in Asia and Africa (Pitrat 2016). Liu et al. (2020) reported a clear separation of melon landraces from improved cultivars based on phylogenetic mapping, LD decay, and population structure analysis. They reported a significant decrease in genetic diversity during melon crop improvement.

A lot of accessions are available in different gene banks across the world. These accessions could be requested for allele mining for crop improvement (Table 12.2).

12.6 ADVANCES IN MELON BREEDING

Melon has been under cultivation for thousands of years (Janick et al. 2007; Paris 2016). During domestication, it went through a severe process of diversification. This high genetic diversity is of particular interest to breeders. Most cultivated melons belong to the *melo* and *agrestis* groups. However, many of their feral and wild forms still exist in centers of diversification in Asia and Africa, and are resources for allele mining. Current melon-breeding programs focus on improving agronomic traits (yield, size, and shape), fruit quality traits (total soluble solids, flesh firmness, and shelf life), and resistance to biotic and abiotic stresses (powdery mildew, fusarium wilt, viral pathogens, and cracking).

Although melon is cross-pollinated, it does not display significant heterosis. In the past few decades, melon has been improved using traditional breeding techniques like pedigree breeding, single-seed descent, and backcross breeding. These breeding efforts focus on crossing genotypes, followed by phenotypic selection for desirable traits in parental lines. This is followed by the development of molecular markers for trait selection to assist traditional breeding efforts.

In the past decade, low-cost sequencing and digital platforms have revolutionized melon breeding. The sequencing of a large number of melon accessions has opened the door for better mapping of traits (Garcia-Mas et al. 2012; Liu et al. 2020; Sun et al. 2022). It has also helped untangle the evolutionary history of melon to help in allele mining efforts (Diaz et al. 2017; Pichot et al. 2022). There has also been a revolution in high-resolution digital phenotyping tools available to breeders. These tools help breeders collect large amounts of data using machine-learning approaches. This shift toward selection based on the genome (genomic selection) rather than traditional marker-assisted trait selection has increased genetic gain (Figure 12.4). Early breeding efforts in melon focused on traits controlled by one or a few major genes, like disease resistance, shell color, sutures, and so on. However, with new tools, it is now possible to dissect complex traits like fruit yield.

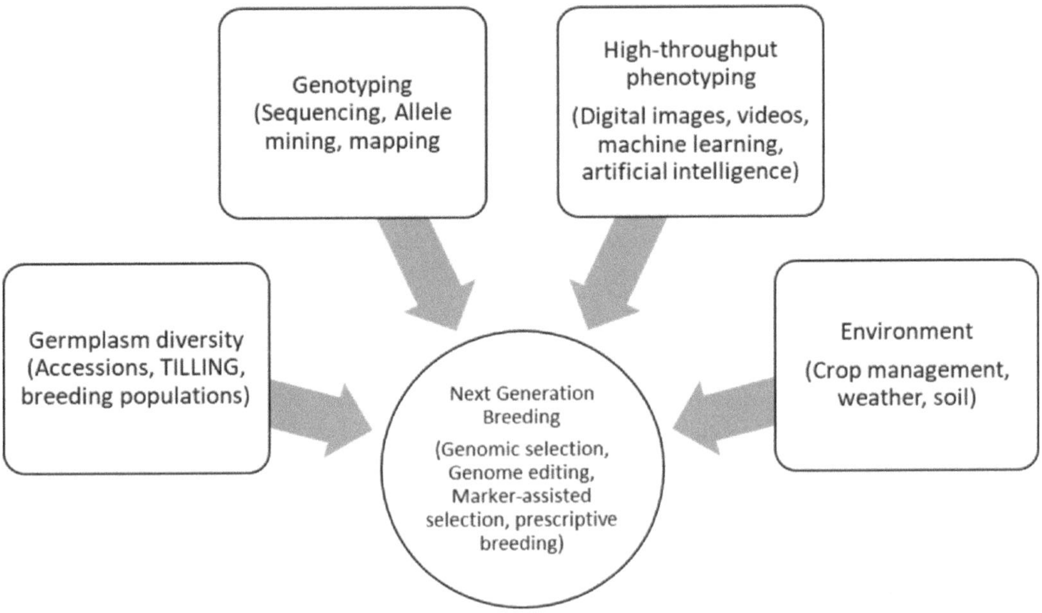

FIGURE 12.4 Components of next-generation breeding.

12.7 TRAITS

Melon is very diverse in terms of plant architecture, fruit size, fruit shape, netting density, flesh color, and resistance to biotic and abiotic stresses (Stepansky et al. 1999; Mliki et al. 2001). Current cultivars are bred from wildtype melons after domestication and artificial selection have led to changes in their yield, color, flavor, and other traits, which differ from those of the wildtypes (Pitrat 2013).

The primary focus of breeding is to improve agronomic traits like sex expression, fruit yield, strength, and resistance to diseases and pests to make sure farmers can grow the crop successfully. With climate change, there has been an even greater focus on these traits with shifts in growing areas as well as challenges from emerging diseases and pests. Because of urbanization, melon is being exported long distances, and hence traits like shelf life and consistency are becoming important. Fruit quality traits like sugar, aroma, and flavor are more complex, but next-generation breeding using high-throughput sequencing and phenotyping is enabling improvement in these traits.

12.7.1 Sex Expression

Melons are monoecious, andromonoecious, gynoecious, and hermaphroditic (Aamir et al. 2021). However, cultivated ones are mostly monoecious and andromonoecious. Sex is determined by two major genes: the andromonoecious (*a*) and gynoecious (*g*) genes. The andromonoecious (*a*) gene controls expression of monoecious (*A-G-*) and andromonoecious (*aaG-*) sex expression (Poole and Grimball 1939; Wall 1967). This gene, which codes for ACC synthase gene, *CmACS7*, has been finely mapped and cloned. The transition between monoecy and andromonoecy is conferred by a single substitution in *CmACS7*, which leads to an inactive form of this key enzyme (1-aminocyclopropane-1-carboxylic acid synthase) in ethylene biosynthesis (Boualem et al. 2008). Monoecious (*A-G-*) and andromonoecious (*aaG-*) individuals have male flowers on the main stem and female or hermaphrodite flowers on axillary branches, respectively, whereas gynoecious (*AAgg*) and hermaphrodite (*aagg*) individuals only have female and hermaphrodite flowers, respectively (Boualem et al. 2008). Molecular markers linked to the gene (Noguera et al. 2005; Sinclair et al. 2006; Kim et al. 2010) and within the gene have been published (Boualem et al. 2008).

The gynoecious (*g*) gene controls the transition from monoecious sex expression to gynoecious expression (i.e., carrying only female flowers). Epigenetic mechanisms play an important role in sex expression in melon. The transition from male to female flowers in the gynoecious genotype is a result of epigenetic changes in the *CmWIP1* promoter caused by insertion of the transposon Gyno-Hat (Martin et al. 2009; Piferrer 2013). The female-promoting gene *CmACS11* inhibits expression of the male-promoting gene *CmWIP1* through deposition of H3K27me3 (Latrasse et al. 2017). This demonstrates that three genes (*CmACS7*, *CmACS11*, and *CmWIP1*), along with H3K9ac and H3K27me3, control sex expression epigenetically (Aamir et al. 2021). An additional gene, the gynomonoecious (*gy*) gene, interacts with the andromonoecious and gynoecious genes to produce stable gynoecious sex expression in WI998 (Kenigsbuch and Cohen 1987, 1990). However, an additional locus with a recessive *m* allele (*CmACS11*) also produces a stable gynoecious phenotype (Boualem et al. 2022).

12.7.2 Male Sterility

Five recessive genes for male sterility have been reported in melon: *ms-1* to *ms-5* (Pitrat 1991, 2002). All are located on different chromosomes. Park et al. (2009) reported the *ms-3* gene on linkage group 9 from Deltex×TGR-1551. The *ms-5* gene was mapped to a 30 kb candidate region on chromosome 9 comprising six genes, and ABORTED MICROSPORE (*AMS*) was reported to be candidate gene (Sheng et al. 2017; Wang et al. 2021).

12.7.3 Fruit Shape

Fruit shape plays an important role in breeding design. Most commercial cantaloupe are round, although Piel de Sapo, Yellow Canary, and Hami are elongated. An elongated fruit shape is dominant over a round fruit shape and is controlled by the single dominant gene *O*. A similar pattern has also been reported for fruit shape where the *sp* gene controls round fruit shape and it is recessive to obtuse fruit shape (Bains and Kang 1963). Normally, monecious (*A-G-*) cultivars are elongated and andromonoecious (*aaG-*) ones are round (Rosa 1928). This is further supported by a report in which a fruit shape QTL colocalized with the andromonoecious locus on chromosome 2. It was also reported that the same gene (*CmACS7*) involved in the ethylene biosynthesis pathway controls both monecious sex expression and elongated fruit shape in chromosome 2 (Boualem et al. 2022).

Fruit shape is also influenced by the number of carpels. Pentamerous fruit tend be rounder than trimerous fruit. QTL colocalization with the *p* gene (pentamerous) on chromosome 12 produces rounder fruit than 3-carpel fruit. The *CmCLV3* gene, which determines the number of carpels, affects fruit shape in andromonoecious fruit (Boualem et al. 2008; Liu et al. 2020). A gene encoding OVATE family protein (OFP) was identified as a candidate gene for fruit shape in QTL mapped on chromosome 8. Gene expression was elevated due to retrotransposon insertion in the promoter. This was associated with fruit development in a flat-round HP22 line (Martínez-Martínez et al. 2021; Ma et al. 2022).

12.7.4 *Rind*

The color of a melon's shell is an important factor for consumer acceptability in different regions. Melons have light, dark, green, yellow, orange, and white rinds. These colors can be uniform but can also be mixed with secondary colors to form mottled, speckled, dotted, or striped varieties (Pitrat 2016). Melon rind is influenced by chlorophyll, carotenoids, and flavonoids like naringenin chalcone.

Light shell is dominant over dark shell and is controlled by a single dominant gene. In one study, the accumulation of naringenin chalcone (yellow shell) was inherited as a monogenic dominant trait in the cross Noy Amid (yellow rind)×Tendral Verde Tardio (dark green rind); the accumulation of chlorophyll and carotenoids segregates jointly as a single dominant gene independent of the accumulation of naringenin chalcone (Tadmor et al. 2010). A Kelch domain-containing F-box protein-coding gene (*CmKFB*) on chromosome 10 was identified as causative for the accumulation of naringenin chalcone in melon fruit rind (Feder et al. 2015). However, Zhao et al. (2019) reported that green rind was dominant epistatic to white and yellow rind (non-green rind) in a segregating population. The *CmAPRR2* gene was identified as regulating fruit pigmentation in melon and watermelon (Oren et al. 2019). It could be an attractive target for TILLING in melon.

Melon rind has additional variations, like speckles, mottling, and stripes, the inheritance of which is complex. Mottled rind (*Mt*) was reported dominant to uniform color in a cross between Annamalai and Makuwa (Ganesan 1988). This is further supported by different studies that mapped a single dominant gene on chromosome 2 (Pereira et al. 2018; Lv et al. 2018) and chromosome 4 (Liu et al. 2019). In another study, two dominant genes (*CmMt1* and *CmMt2*) epistatic to each other were reported in a cross between Songwan Charmi and Mi Gua (Shen et al. 2021). Another gene (*st3*) for striped rind has also been reported (Liu et al. 2019). The *CmAPP2* gene reported previously is a candidate gene for *CmMt1*. The position of *st3* is closer to *CmAPP2*. MELO3C026282 encoding an AtCPSFL1 homolog is the most likely candidate gene for *CmMt2* (Shen et al. 2021).

The inheritance of melon rind color is complex, and many genes have been reported. Rind color plays an important role in consumer perceptions. Many candidate genes have been reported that could be suitable for allele mining.

Charentais and Italian melons display sutures on their rinds. A single recessive gene, sutures (*s*), controls this trait (Bains and Kang 1963).

12.7.5 Fruit Netting

Fruit netting is a highly sought-after trait in American cantaloupes. Dense netting is desired for protection against bruising and nicks during packing and transport. It is particularly useful when melon need to be shipped long distance. A single dominant gene (*Rn*) for rind netting, mapped to linkage group II, has been proposed (Herman et al. 2008; Dogimont and Sari 2022). A major QTL was mapped on chromosome 2 in RIL populations derived from crosses between a smooth rind honeydew line (Tam Dew, Noy Amid or PI 414723) and a densely netted, medium-size reticulatus line (Dulce), together with additional interacting QTLs on chromosomes 2 and 9 (Oren et al. 2020). Herman et al. (2008) also reported additional minor loci for net density.

12.7.6 Fruit Yield and Size

A high yield and uniform shape are prerequisites for a melon to be introduced commercially. Fruit yield and fruit size (weight) are positively correlated (Zalapa et al. 2007). Ramamurthy and Waters (2015) reported a small segment on linkage group 8 responsible for yield and yield-related traits. This region could be very important for further studies. This is further supported by the fact that several QTLs were reported in linkage group 8 for fruit weight and fruit number in a RIL population from USDA 846-1×Top Mark (Zalapa et al. 2007). Fruit yield and associated traits are very complex, and advances in sequencing and phenotyping could help breed high-yielding varieties.

12.7.7 Total Soluble Solids

Total soluble solids (sugar content) is one of the most important traits for improvement in melon. Sweet melon is highly desired by the consumer. Many customers have set a threshold for minimum sugar to accept melon in their stores. Sucrose, fructose, and glucose are the three major sugars in melon, of which sucrose is dominant. Rapid accumulation of sucrose during the final stages of maturity leads to sweet melon. Understanding the genetics of sugar accumulation is a top priority for breeders.

Several QTLs associated with sugar content have been reported over many years in diverse genetic types (Monforte et al. 2004; Sinclair et al. 2006; Park et al. 2009; Harel-Beja et al. 2010). Burger et al. (2002) reported that a single major gene, *suc*, controls sucrose accumulation. The metabolic pathway of sucrose accumulation involves several enzymes, like invertases, sucrose synthase, and sucrose phosphate synthase (Xu et al. 2022). Final sugar accumulation is a function of sucrose synthesis and breakdown. A total of 42 genes control the sugar metabolism pathway in melon (Dai et al. 2011). Very few candidate genes have been identified so far.

Recently, three important clusters related to sugar content were identified on chromosomes 4, 5, and 7 in a cross between Piel de Sapo and Sogwhan Charmi. Among them, a major QTL, *SUCQSC5.1*, reduced soluble solids and sucrose content, and *MELO3C014519* was considered the candidate gene (Argyris et al. 2017). In addition, some new differentially expressed genes responsible for sucrose content were detected in melon, such as *CmINH3*, *CmTPP1*, *CmTPS5*, *CmTPS7*, and *CmTPS9* (Schemberger et al. 2020). Sugars are made in mesophyll cells and transported to other parts of the plant. Therefore, sugar accumulation in melon fruit depends on sugar synthesis as well as sugar transport (Xu et al. 2022). The tonoplast sugar transporter *CmTST2* plays an important role in sugar accumulation in melon fruit (Cheng et al. 2018).

Although many QTLs have been detected in melon, the fact that populations have such diverse genetic backgrounds makes it difficult to select candidates for validation. Moreover, sugar accumulation is a very complex trait and is influenced by genetic background and the environment.

12.7.8 Firmness

Flesh firmness is directly responsible for field shelf life and post-harvest shelf life. Melon can be categorized as having a medium, extended, or long shelf life. Melons with firm flesh are suitable for shipping long distance anywhere from a few weeks to a month. However, for melons that are consumed locally, medium flesh firmness is acceptable.

Two QTLs for flesh firmness were identified in a RIL population from PI414723 × Dulce (Harel-Beja et al. 2010). Dahmani-Mardas et al. (2010) constructed a mutant collection of 4,023 melon M_2 TILLING families and cloned genes for ethylene production. Two missense mutations, *L124F* and *G194D*, of the ethylene biosynthetic enzyme ACC oxidase 1 were identified. *G194D* mutant fruit showed delayed ripening with an enhanced shelf life. Eleven SNPs on chromosomes 6, 8, 11, and 12 were associated with fruit firmness in a genome-wide association study (GWAS) of 125 accessions as well as an F_2-segregating population between MR-1 and Hale's Best Jumbo (Nimmakayala et al. 2016). Gene annotation revealed that some of the SNPs were located in β-D-xylosidase, glyoxysomal malate synthase, chloroplastic anthranilate phosphoribosyl transferase, and histidine kinase. These genes were previously characterized for fruit ripening and softening in other crops.

Moreno et al. (2008) used near-isogenic lines derived from the non-climacteric melon parental lines PI 161375 and Piel de Sapo to map QTLs colocalizing with candidate genes involved in ethylene regulation, biosynthesis, and cell wall degradation. These included *CmACS5* on linkage group III, *CmEXP3* and *CmXTH5* on linkage group 10, *CmEIL1*, *CmEIL3*, and *CmEIL4* on linkage groups II and III.

12.7.9 Flesh Color

Melon is very diverse when it comes to flesh color. The primary flesh colors in commercial varieties are orange, white, and green. Some magenta color cultivars also exist. Two major genes, white flesh (*wf*) and green flesh (*gf*), govern flesh color (Hughes 1948; Iman et al. 1972). Clayberg (1992) confirmed that green and white flesh colors are recessive to orange and indicated that *gf* and *wf* interact epistatically. This was further supported by Perin et al. (2002) and Fukino et al. (2008) when they reported single recessive gene (*gf*) segregation in crosses Vedraintais (orange flesh)×P1161375 (green flesh) and AR5 (orange flesh)×Harukai N°3. This gene was mapped to linkage group IX. Three QTLs associated with color variation (white, green, orange) with putative epistatic interaction were identified in a cross between the Chinese line Q3-2-2 (white flesh) and Top Mark (orange flesh) (Cuevas et al. 2009, 2010).

Recently, the two major genes for flesh color in melon were identified. The candidate gene *CmOr* associated with carotenoid accumulation in melon fruit flesh was discovered, which is identical to the previously described *gf* locus (Tzuri et al. 2015). Combining the QTL with results from GWAS, researchers detected a 96 kb overlapping interval containing 11 protein-coding genes (Zhao et al. 2019). However, a previously reported candidate gene *MELO3C003069* for *Wf* is 202 kb away from the 96 kb interval, and thus *MELO3C003097* is considered a strong candidate for the *Wf* locus (Galpaz et al. 2018).

12.7.10 Powdery Mildew

Powdery mildew is an important fungal pathogen responsible for the largest crop losses in melon worldwide. It is mainly caused by *Podosphaera xanthii* and *Golovinomyces cichoracearum* (Xu et al. 2022; Dogimont and Sari 2022). There are seven races reported in melon *Podosphaera xanthii* and two races in *Golovinomyces cichoracearum* based on climatic condition. Over many years, several genes have been reported to promote resistance to powdery mildew. However, their relationship is unknown. *Pm-1*, *Pm-2*, *Pm-3*, *Pm-4*, *Pm-5*, *Pm-6*, *Pm-7*, *Pm-8*, *Pm-E*, *Pm-F*, *Pm-G*, *Pm-T*, *Pm-S*, *Pm-w*, *Pm-x*, *Pm-y*, *Pm-z*, *PmV.1*, *PmXII.1*, *PMRI*, and *pmrs* have been reported in different accessions (Dogimont and Sari 2022). However, additional wild accessions and landraces may also be resistant. Discovering the relationships among these genes could be interesting. Hong et al. (2015) found that mutant *CmMLO2* resulted in broad-spectrum resistance to powdery mildew in melon. Using BSASeq, researchers identified a single QTL for resistance to *P. xanthii* in chromosome 12 in the resistant breeding line wm-6. A gene encoding an ankyrin repeat-containing protein is the most likely candidate gene to confer resistance to powdery mildew (Cao et al. 2021).

12.7.11 Fusarium Wilt

Fusarium wilt is a fungal disease caused by *Fusarium oxysporum* (*Fom*). This soil-borne fungus affects almost all cucurbit crops worldwide. There are primarily four races: race 0, race 1, race 2, and race 1.2. Resistance to these races is controlled by different genes (Dogimont and Sari 2022). The single dominant gene *Fom-1* from cultivar Doublon controls resistance to race 0 and race 2 (Risser et al. 1976). *Fom-1* was mapped to linkage group IX (Perin et al. 2002) and encodes TIR-NLR protein, which may share a promoter region with the *Prv* gene (resistance to *Papaya Ring Spot Virus*; Brotman et al. 2013). *Fom-2* confers resistance to races 0 and 1 (Joobeur et al. 2004). *Fom-2* was mapped to linkage group XI (Perin et al. 2002). This gene encodes NLR protein of the CC-NLR subfamily (Joobeur et al. 2004). *Fom-3* confers the same resistance as *Fom-1* (races 0 and 2), but the resistance comes from the cultivar Perlita (Zink and Gubler 1985). Another gene, *fom-4*, which confers resistance to races 0 and 2, is a recessive gene closely linked to *Fom-1* in the Spanish accession Tortuga (Oumouloud et al. 2010). Recently, another single dominant gene for resistance to fusarium wilt was reported and defined as *Fom-5* (Deol et al. 2022).

There has been a report of a new melon pathogen, *Fusarium oxysporum* f. sp. *radicis-cucumerinum* (Forc), which kills plants prematurely and leads to crown rot. Resistance was controlled by two genes, one dominant (*Forc-1*) and the other recessive (*forc-2*), in a cross of the highly resistant Hemed and susceptible US cultivar Dulce genotype (Elkabetz et al. 2016).

12.7.12 Viral Disease

A diverse array of viral pathogens affects melon crop production. Among them, Cucurbit Yellow Stunting Disorder Virus (CYSDV), Tomato Leaf Curl New Delhi Virus (CMV), Squash Mosaic Virus, MNSV, Cucumber Vein Yellowing Virus (CVYV), and Cucumber Green Mottle Mosaic Virus are important ones.

Translation factors from the 4E and 4F protein families play important parts in several RNA plant viruses in crops. Eukaryotic translation factors (eIF) play an important role in virus resistance in melon (Xu et al. 2022). In a study using EcoTILLING, a new allelic variant of eIF4E was resistant to MNSV (Nieto et al. 2006). In another study, eIF4E knockdown resulted in resistance to MNSV, CVYV, Moroccan Watermelon Mosaic Virus, and Zucchini Yellow Mosaic Virus (ZYMV) (Rodriguez-Hernandez et al. 2012). Additional forms of resistance have also been reported, including CmVPS41 for CMV and the single dominant gene Zym in PI414723 for ZYMV (Pascual et al. 2019; Adler-Berke et al. 2021; Xu et al. 2022).

12.7.13 Insect Resistance

Aphids, cucumber beetles, and white flies are major insects that affect the melon crop. They are also vectors of many viruses, like CYSDV, CMV, and ZYMV. A single dominant gene, *Ag*, confers tolerance to aphids and freedom from leaf curling in LJ 90634 and PI414723 (Kishaba et al. 1971, 1976).

Resistance to viruses transmitted by *Aphis gossypii* is controlled by a single gene, called *Vat* (virus aphid transmission). The *Vat* locus was mapped to linkage group V (Brotman et al. 2002; Perin et al. 2002). A single dominant gene (*Lt*) confers resistance to leafminer (*Liriomyza trifolii*) in the cultivar Nantais Oblong (Dogimont et al. 1999). Two recessive genes, *dc-1* and *dc-2*, confer resistance to melon fruit fly by complementary gene action (Sambandam and Chelliah 1972).

In general, most nucleotide-binding leucine-rich repeat (NB-LRR) genes are specialized in resistance. NB-LRR genes are responsible for resisting many diseases and insects in melon (Marone et al. 2013). *Vat* is one such NB-LRR gene that confers resistance to aphids (Tian et al. 2003).

REFERENCES

Aamir M, Karmakar P, Singh VK, Kashyap SP, Pandey S, et al. (2021) A novel insight into transcriptional and epigenetic regulation underlying sex expression and flower development in melon (*Cucumis melo* L.). *Physiol Plant* 173:1729–1764.

Adler-Berke N, Goldenberg Y, Brotman Y, Kovalski I, Gal-On A, et al. (2021) The melon *zym* locus conferring resistance to ZYMV: High resolution mapping and candidate gene identification. *Agronomy* 11(12):2427.

Argyris JM, Diaz A, Ruggieri V, Fernandez M, Jahrmann T, et al. (2017) QTL analyses in multiple populations employed for the fine mapping and identification of candidate genes at a locus affecting sugar accumulation in melon (*Cucumis melo* L.). *Front Plant Sci* 8:1679.

Bains MS, Kang US (1963) Inheritance of some flower and fruit characters in muskmelon. *Indian J Genet Plant Breed* 23:101–106.

Baudracco-Arnas S, Pitrat M (1996) A genetic map of melon (*Cucumis melo* L.) with RFLP, RAPD, isozyme, disease resistance and morphological markers. *Theor Appl Genet* 93:57–64.

Biémont C (2008) Genome size evolution: Within-species variation in genome size. *Heredity* 101:297–298.

Boualem A, Berthet S, Sureshbhai Devani R, Camps C, Fleurier S, et al. (2022) Ethylene plays a dual role in sex determination and fruit shape in cucurbits. *Curr Biol* 32(11):2390–2401.

Boualem A, Fergany M, Fernandez R, Troadec C, Martin A, et al. (2008) A conserved mutation in an ethylene biosynthesis enzyme leads to andromonoecy in melons. *Science* 321:836–838.

Brotman Y, Normantovich M, Goldenberg Z, Zvirin Z, Kovalski I, et al. (2013) Dual resistance of melon to *Fusarium oxysporum* races 0 and 2 and to papaya ring-spot virus is controlled by a pair of head-to-head-oriented NB-LRR genes of unusual architecture. *Mol Plant* 6(1):235–238.

Brotman Y, Silberstein L, Kovalski I, Perin C, Dogimont C, et al. (2002) Resistance gene homologues in melon are linked to genetic loci conferring disease and pest resistance. *Theor Appl Genet* 104:1055–1063.

Burger Y, Saar U, Katzir N, Paris HS, Yeselson Y, et al. (2002) A single recessive gene for sucrose accumulation in *Cucumis melo* fruit. *J Amer Soc Hort Sci* 127:938–943.

Burger Y, Schaffer AA (2007) The contribution of sucrose metabolism enzymes to sucrose accumulation in *Cucumis melo*. *J Amer Soc Hort Sci* 132:704–712.

Cao Y, Diao Q, Chen Y, Jin H, Zhang Y, et al. (2021) Development of KASP markers and identification of a QTL underlying powdery mildew resistance in melon (*Cucumis melo* L.) by bulked segregant analysis and RNA-Seq. *Front Plant Sci* 11:1–11.

Castanera R, Ruggieri V, Pujol M, Garcia-Mas J (2020) An improved melon reference genome with single-molecule sequencing uncovers a recent burst of transposable elements with potential impact on genes. *Front Plant Sci* 10:1–10.

Chang CI, Chou CH, Liao MH, Chen TM, Cheng CH, et al. (2015) Bitter melon triterpenes work as insulin sensitizers and insulin substitutes in insulin-resistant cells. *J Funct Foods* 13:214–224.

Cheng J, Wen S, Xiao S, Lu B, Ma M, et al. (2018) Overexpression of the tonoplast sugar transporter CmTST2 in melon fruit increases sugar accumulation. *J Exp Bot* 69:511–523.

Chomicki G, Schaefer H, Renner SS (2020) Origin and domestication of Cucurbitaceae crops: Insights from phylogenies, genomics and archaeology. *New Phytol* 226:1240–1255.

Clayberg CD (1992) Interaction and linkage test of flesh colour genes in *Cucumis melo* L. *Cucurbit Genet Coop Rpt* 15:53.

Cuevas H, Staub JE, Simon PW (2010) Inheritance of beta-carotene-associated mesocarp color and fruit maturity of melon (*Cucumis melo* L.). *Euphytica* 173:129–140.

Cuevas HE, Staub JE, Simon PW, Zalapa JE (2009) A consensus linkage map identifies genomic regions controlling fruit maturity and beta-carotene-associated flesh color in melon (*Cucumis melo* L.). *Theor Appl Genet* 119:741–756.

Dafna A, Halperin I, Oren E, Isaacson T, Tzuri G, et al. (2021) Underground heterosis for yield improvement in melon. *J Exp Bot* 72:6205–6218.

Dahmani-Mardas F, Troadec C, Boualem A, Lévêque S, Alsadon AA, et al. (2010) Engineering melon plants with improved fruit shelf life using the TILLING approach. *PLoS ONE* 5:e15776.

Dai N, Cohen S, Portnoy V, Tzuri G, Harel-Beja R, et al. (2011) Metabolism of soluble sugars in developing melon fruit-a global transcriptional view of the metabolic transition to sucrose accumulation. *Plant Mol Biol* 76:1–18.

Deol JK, Sharma SP, Rani R, Kalia A, Chhuneja P, *et al.* (2022) Inheritance analysis and identification of SSR markers associated with fusarium wilt resistance in melon. *J Hortic Sci Biotechnol* 97:66–74.

Diaz A, Martin-Hernandez AM, Dolcet Sanjuan R, Garces-Claver A, Alvarez JM, et al. (2017) Quantitative trait loci analysis of melon (*Cucumis melo* L.) domestication-related traits. *Theor Appl Genet* 130:1837–1856.

Dogimont C, Bordat D, Pages C, Boissot N, Pitrat M. (1999) One dominant gene conferring the resistance to the leafminer, *Liriomyza trifolii* (Burgess) diptera: Agromyzidae in melon (*Cucumis melo* L.). *Euphytica* 105:63–67.

Dogimont C, Sari N (2022) 2022 gene list for melon. *Cucurbit Genet Coop* 45:1–47.

Elkabetz M, Paris HS, Burger Y, Hanan A, Cohen R (2016) Two genes for resistance to *Fusarium oxysporum* f. sp. *radicis-cucumerinum* in melon (*Cucumis melo*, Cucurbitaceae). *Sci Hortic* 201:57–60.

Endl J, Achigan-Dako EG, Pandey AK, Monforte AJ, Pico B, et al. (2018) Repeated domestication of melon (*Cucumis melo*) in Africa and Asia and a new close relative from India. *Am J Bot* 10:1662–1671.

Feder A, Burger J, Gao S, Lewinsohn E, Katzir N, et al. (2015) A Kelch domain-containing F-box coding gene negatively regulates flavonoid accumulation in *Cucumis melo* L. *Plant Physiol* 169:1714–1726.

Food and Agriculture Organization (2023) Database: production. Available online: https://www.fao.org/faostat/en/#data/QCL/visualize (accessed on January 9, 2023).

Fu D, Szucs P, Yan L, Helguera M, Skinner JS, et al. (2005) Large deletions within the first intron in *VRN-1* are associated with spring growth habit in barley and wheat. *Mol Genet Genom* 273:54–65.

Fukino N, Ohara T, Monforte AJ, Sugiyama M, Sakata Y, et al. (2008) Identification of QTL for resistance to powdery mildew and SSR markers diagnostic for powdery mildew resistance genes in melon (*Cucumis melo* L.). *Theor Appl Genet* 118:165–175.

Galpaz N, Gonda I, Shem-Tov D, Barad O, Tzuri G, et al. (2018) Deciphering genetic factors that determine melon fruit-quality traits using RNA-Seq-based high-resolution QTL and eQTL mapping. *Plant J* 94:169–191.

Ganesan J (1988) Genetic studies on certain characters of economic importance in muskmelon (*Cucumis melo* L.). Annamalai University, India.

Garcia-Mas J, Benjak A, Sanseverino W, Bourgeois M, Mir G, et al. (2012) The genome of melon (*Cucumis melo* L.). *Proc Natl Acad Sci USA* 109:11872–11877.

Gokidi Y, Bhanu N, Chandra K, Singh MN, Hemantaranjan A (2017) Allele mining-an approach to discover allelic variation in crops. *J Plant Sci Res* 33(2):167–180.

Gómez-García R, Campos DA, Aguilar CN, Madureira AR, Pintado M (2020) Valorization of melon fruit (*Cucumis melo* L.) by-products: Phytochemical and biofunctional properties with emphasis on recent trends and advances. *Trends Food Sci Technol* 99:507–519.

Gonzalo MJ, Oliver M, Garcia-Mas J, Monfort A, Dolcet-Sanjuan R, et al. (2005) Simple-sequence repeat markers used in merging linkage maps of melon (*Cucumis melo* L.). *Theor Appl Genet* 110:802–811.

Harel-Beja R, Tzuri G, Portnoy V, Lotan-Pompan M, Lev S, et al. (2010) A genetic map of melon highly enriched with fruit quality QTL and EST markers, including sugar and carotenoid metabolism genes. *Theor Appl Genet* 121:511–533.

Herman R, Zvirin Z, Kovalski I, Freeman S, Denisov Y, et al. (2008) Characterization of *Fusarium* race 1.2 resistance in melon and mapping of a major QTL for this trait near a fruit netting locus. In: Pitrat M (ed) *Cucurbitaceae 2008: Proceedings of the IXth Eucarpia Meeting on Genetics and Breeding of Cucurbitaceae*, INRA, Avignon, France. pp. 149–156.

Hong C, Weiping K, Junfen L, Jiping L (2015) Analysis of powdery mildew resistance in wild melon *MLO* mutants. *Hortic Plant J* 1(3):165–171.

Hughes M (1948) The inheritance of two characters of *Cucumis melo* and their interrelationship. *Proc Amer Soc Hortic Sci* 52: 399–402.

Iman MK, Abo-Bakr MA, Hanna HY (1972) Inheritance of some economic characters in crosses between sweet melon and snake cucumber. I. Inheritance of qualitative characters. *Assiut J Agric Sci* 3:363–380.

Janick J, Paris HS, Parrish DC (2007) The cucurbits of Mediterranean antiquity: Identification of taxa from ancient images and descriptions. *Ann Bot London* 100:1441–1457.

Joobeur T, King JJ, Nolin SJ, Thomas CE, Dean RA (2004) The Fusarium wilt resistance locus Fom-2 of melon contains a single resistance gene with complex features. *Plant J* 39:283–297.

Kenigsbuch D, Cohen Y (1987) Inheritance of gynoecious sex type in muskmelon. *Cucurbit Genet Coop Rpt* 10:47–48.

Kenigsbuch D, Cohen Y (1990) The inheritance of gynoecy in muskmelon. *Genome* 33:317–320.

Kim H, Baek J, Choi YO, Lee JH, Sung SK, et al. (2010) Identification of a cluster of oligonucleotides repeat sequences and its practical implication in melon (*Cucumis melo* L.) breeding. *Euphytica* 171:241–249.

Kirkbride JH Jr (1993) *Biosystematic Monograph of the Genus Cucumis (Cucurbitaceae)*. Parkway, Boone.

Kishaba AN, Bohn GW, Toba HH (1971) Resistance to *Aphis gossypii* in muskmelon. *J Econ Entomol* 64:935–937.

Kishaba AN, Bohn GW, Toba HH (1976) Genetic aspects of antibiosis to *Aphis gossypii* in *Cucumis melo* line from India. *J Amer Soc Hort Sci* 101:557–561.

Latrasse D, Rodriguez-Granados NY, Veluchamy A, Mariappan KG, Bevilacqua C, et al. (2017) The quest for epigenetic regulation underlying unisexual flower development in *Cucumis melo*. *Epigenetics Chromatin* 10:22.

Lian Q, Fu Q, Xu Y, Hu Z, Zheng J, et al. (2021) QTLs and candidate genes analyses for fruit size under domestication and differentiation in melon (*Cucumis melo* L.) based on high resolution maps. *BMC Plant Biol* 21:126.

Ling J, Xie X, Gu X, Zhao J, Ping X, et al. (2021) High-quality chromosome level genomes of *Cucumis metuliferus* and *Cucumis melo* provide insight into *Cucumis* genome evolution. *Plant J* 107(1):136–148.

Liu L, Sun T, Liu X, Guo Y, Huang X, et al. (2019) Genetic analysis and mapping of a striped rind gene (st3) in melon (*Cucumis melo* L.). *Euphytica* 215:1–12.

Liu S, Gao P, Zhu Q, Zhu Z, Liu H, et al. (2020) Resequencing of 297 melon accessions reveals the genomic history of improvement and loci related to fruit traits in melon. *Plant Biotechnol J* 18:2545–2558.

Lv J, Fu Q, Lai Y, Zhou M, Wang H (2018) Inheritance and gene mapping of spotted to non-spotted trait gene CmSp-1 in melon (*Cucumis melo* L. var. *chinensis* Pangalo). *Mol Breed* 38:105.

Ma J, Li C, Zong M, Qiu Y, Liu Y, et al. (2022) CmFSI8/CmOFP13 encoding an OVATE family protein controls fruit shape in melon. *J Exp Bot* 73(5):1370–1384.

Manchali S, Kotamballi N, Murthy C, Vishnuvardana, Patil BS (2021) Nutritional composition and health benefits of various botanical types of melon (*Cucumis melo* L.). *Plants (Basel)* 10(9):1755.

Marone D, Russo MA, Laidò G, De Leonardis AM, Mastrangelo AM (2013) Plant nucleotide binding site-leucine-rich repeat (NBS-LRR) genes: Active guardians in host defense responses. *Int J Mol Sci* 14:7302–7326.

Martin A, Troadec C, Boualem A, Rajab M, Fernandez R, et al. (2009) A. transposon-induced epigenetic change leads to sex determination in melon. *Nature* 461:1135–1138.

Martínez-Martínez C, Gonzalo MJ, Sipowicz P, Campos M, Martínez-Fernández I, et al. (2021) A cryptic variation in a member of the ovate family proteins is underlying the melon fruit shape QTL fsqs8.1. *Theor Appl Genet* 135(3):785–801.

Mliki A, Staub JE, Zhangyong S, Ghorbel A (2001) Genetic diversity in melon (*Cucumis melo* L.): An evaluation of African germplasm. *Genet Resour Crop E* 48:587–597.

Monforte AJ, Oliver M, Gonzalo MJ, Alvarez JM, Dolcet-Sanjuan R, et al. (2004) Identification of quantitative trait loci involved in fruit quality traits in melon (*Cucumis melo* L.). *Theor Appl Genet* 108:750–758.

Moreno E, Obando JM, Dos-Santos N, Fernández-Trujillo JP, Monforte AJ, et al. (2008) Candidate genes and QTLs for fruit ripening and softening in melon. *Theor Appl Genet* 116:589–602.

Nieto C, Morales M, Orjeda G, Clepet C, Monfort A, et al. (2006) An eIF4E allele confers resistance to an uncapped and non-polyadenylated RNA virus in melon. *Plant J* 48(3):452–62.

Nimmakayala P, Tomason YR, Abburi VL, Alvarado A, Saminathan T, et al. (2016) Genome-wide differentiation of various melon horticultural groups for use in GWAS for fruit firmness and construction of a high-resolution genetic map. *Front Plant Sci* 7:1437.

Noguera FJ, Capel J, Alvarez JI, Lozano R (2005) Development and mapping of a codominant SCAR marker linked to the *andromonoecious* gene of melon. *Theor Appl Genet* 110:714–720.

Oren E, Dafna A, Tzuri G, Halperin I, Isaacson T (2022) Pan-genome and multi-parental framework for high-resolution trait dissection in melon (*Cucumis melo*). *Plant J* 112:1525–1542.

Oren E, Tzuri G, Dafna A, Meir A, Kumar R, et al. (2020) High-density NGS-based map construction and genetic dissection of fruit shape and rind netting in *Cucumis melo*. *Theor Appl Genet* 133:1927–1945.

Oren E, Tzuri G, Vexler L, Dafna A, Meir A, et al. (2019) The multi-allelic APRR2 gene is associated with fruit pigment accumulation in melon and watermelon. *J Exp Bot* 70:3781–3794.

Oumouloud A, Arnedo-Andrés M, González-Torres R, Alvarez JM (2010) Inheritance of resistance to *Fusarium oxysporum* f. sp. *melonis* races 0 and 2 in melon accession Tortuga. *Euphytica* 176:183–189.

Paris HS (2016) Overview of the origins and history of the five major cucurbit crops: Issues for ancient DNA analysis of archaeological specimens. *Veg Hist Archaeobot* 25:405–414.

Paris HS, Lebeda A, Kristkova E, Andres TC, Nee MH (2012) Parallel evolution under domestication and phenotypic differentiation of the cultivated subspecies of *Cucurbita pepo* (Cucurbitaceae). *Econ Bot* 66(1):71–90.

Park SO, Hwang HY, Crosby KM (2009) A genetic linkage map including loci for male sterility, sugars, and ascorbic acid in melon. *J Amer Soc Hortic Sci* 134:67–76.

Pascual L, Yan J, Pujol M, Montforte AJ, Pico B, et al. (2019) *CmVPS41* is a general gatekeeper for resistance to *Cucumber mosaic virus* phloem entry in melon. *Front Plant Sci* 10:1219.

Pereira L, Ruggieri V, Pérez S, Alexiou KG, Fernández M, et al. (2018) QTL mapping of melon fruit quality traits using a high-density GBS-based genetic map. *BMC Plant Biol* 18:324.

Perin C, Hagen LS, De Conto V, Katzir N, Danin-Poleg Y, et al. (2002) A reference map of *Cucumis melo* based on two recombinant inbred line populations. *Theor Appl Genet* 104:1017–1034.

Pichot C, Djari A, Tran J, Verdenaud M, Marande M, et al. (2022) Cantaloupe melon genome reveals 3D chromatin features and structural relationship with the ancestral Cucurbitaceae karyotype. *iScience* 25(1):103696.

Piferrer F (2013) Epigenetics of sex determination and gonadogenesis. *Dev Dyn* 242:360–370.

Pitrat M (1991) Linkage groups in *Cucumis melo* L. *J Hered* 82:406–411.

Pitrat M (2002) 2002 gene list for melon. *Cucurbit Genet Coop Rpt* 25:76–93.

Pitrat M (2013) Phenotypic diversity in wild and cultivated melons (*Cucumis melo*). *Plant Biotechnol* 30:273–278.

Pitrat M (2016) Melon genetic resources: Phenotypic diversity and horticultural taxonomy. In: Grumet R, Katzir N, Garcia-Mas J (eds) *Genetics and Genomics of Cucurbitaceae*. Springer Science+Media, New York. pp. 25–60.

Pitrat M, Hanelt P, Hammer K (2000) Some comments on infraspecific classification of cultivars of melon. *Acta Hortic* 510:29–36.

Poole CF, Grimball PC (1939) Inheritance of new sex forms in *Cucumis melo* L. *J Hered* 30:21–25.

Ramamurthy RK, Waters BM (2015) Identification of fruit quality and morphology QTL in melon (*Cucumis melo*) using a population derived from *flexuosus* and *cantalupensis* botanical groups. *Euphytica* 204(1):163–177.

Risser G, Banihashemi Z, Davis DW (1976) Proposed nomenclature of *Fusarium oxysporum* f. sp. *melonis* races and resistance genes in *Cucumis melo*. *Phytopathology* 66:1105–1106.

Rodriguez-Hernandez AM, Gosalvez B, Sempere RN, Burgos L, Aranda MA, et al. (2012) Melon RNA interference (RNAi) lines silenced for Cm-eIF4E show broad virus resistance. *Mol Plant Pathol* 13:755–763.

Rosa JT (1928) The inheritance of flower types in *Cucumis* and *Citrullus*. *Hilgardia* 3:233–250.

Ruggieri V, Alexiou KG, Morata J, Argyris J, Pujol M, et al. (2018) An improved assembly and annotation of the melon (*Cucumis melo* L.) reference genome. *Sci Rep* 8:8088.

Samadder P, Sivamani E, Lu J, Li X, Qu R (2008) Transcriptional and post-transcriptional enhancement of gene expression by the 5′ UTR intron of rice rubi3 gene in transgenic rice cells. *Mol Gen Genom* 279:429–439.

Sambandam CN, Chelliah S (1972) *Cucumis callosus* (Rottl.) Logn., a valuable material for resistance breeding in muskmelons. In: 3rd International Symposium on Subtropical Tropical Horticulture, Bangalore, India pp. 63–68.

Schaffer AA, Aloni B, Fogelman E (1987) Sucrose metabolism and accumulation in developing fruit of *Cucumis. Phytochemistry* 26:1883–1887.

Schemberger MO, Stroka MA, Reis L, Los KKS, de Araujo GAT, et al. (2020) Transcriptome profiling of non-climacteric "yellow" melon during ripening: Insights on sugar metabolism. *BMC Genom* 21:262.

Shen J, Xu X, Zhang Y, Niu X, Shou W (2021) Genetic mapping and identification of the candidate genes for mottled rind in *Cucumis melo* L. *Front Plant Sci* 12:2563.

Sheng Y, Wang Y, Jiao S, Jin Y, Peng J, et al. (2017) Mapping and preliminary analysis of ABORTED MICROSPORES (AMS) as the candidate gene underlying the male sterility (MS-5) mutant in melon (*Cucumis melo* L.). *Front Plant Sci* 8:902.

Shinde AV, Thakur NR, Deshmukh, AS (2022) Allele mining: An approach for crop improvement. In: Shinde YA (ed) *Advances in Agricultural and Horticultural Sciences.* Shri Vaishnav Vidyapeeth Vishwavidyalaya, Indore. pp. 235–245.

Sinclair JW, Park SO, Lester G, Yoo KS, Crosby K (2006) Identification and confirmation of RAPD markers and andromonoecious associated with quantitative trait loci for sugars in melon. *J Amer Soc Hortic Sci* 131:360–371.

Stepansky A, Kovalski I, Perl-Treves R (1999) Intraspecific classification of melons (*Cucumis melo* L.) in view of their phenotypic and molecular variation. *Plant Syst E* 217:313–332.

Sun Y, Wang J, Li Y, Jiang B, Wang X, et al. (2022) Pan-genome analysis reveals the abundant gene presence/absence variations among different varieties of melon and their influence on traits. *Front Plant Sci* 13:835496.

Tadmor Y, Burger J, Yaakov I, Feder A, Libhaber SE, et al. (2010) Genetics of flavonoid, carotenoid, and chlorophyll pigments in melon fruit rinds. *J Agric Food Chem* 58:10722–10728.

Tian D, Traw M, Chen J, Kreitman M, Bergelson J (2003) Fitness costs of R-gene-mediated resistance in *Arabidopsis thaliana. Nature* 423:74–77.

Tzuri G, Zhou X, Chayut N, Yuan H, Portnoy V, et al. (2015) A "golden" SNP in CmOr governs the fruit flesh color of melon (*Cucumis melo*). *Plant J* 82(2):267–279.

Upadhyaya HD, Gowda CLL, Pundir RPS, Reddy VG, Sube S (2006) Development of core subset of ginger millet germplasm using geographical origin and data on 14 quantitative traits. *Genet Resour Crop E* 53:679–685.

Wall JR (1967) Correlated inheritance of sex expression and fruit shape in *Cucumis. Euphytica* 16:199–208.

Wang L, Dai DY, Wu X, Sheng YY, Ji P, et al. (2021) Genes regulating the aborted microspores (Ams)-mediated male sterility networks in melon (*Cucumis melo* L.). *Hortic Sci Technol* 39(5):645–659.

Wang YH, Thomas CE, Dean RA (1997) A genetic map of melon (*Cucumis melo* L.) based on amplified fragment length polymorphism (AFLP) markers. *Theor Appl Genet* 95:791–798.

Xu L, He Y, Tang L, Xu Y, Zhao G (2022) Genetics, genomics, and breeding in melon. *Agronomy* 12:2891.

Yang J, Deng G, Lian J, Garraway J, Niu Y, et al. (2020) The chromosome-scale genome of melon dissects genetic architecture of important agronomic traits. *iScience* 23:101422.

Yano R, Ariizumi T, Nonaka S, Kawazu Y, Zhong S, et al. (2020) Comparative genomics of muskmelon reveals a potential role for retrotransposons in the modification of gene expression. *Commun Biol* 3:1–13.

Zalapa JE, Staub JE, McCreight JD, Chung SM, Cuevas H (2007) Detection of QTL for yield-related traits using recombinant inbred lines derived from exotic and elite US western shipping melon germplasm. *Theor Appl Genet* 114:1185–1201.

Zeven AC, De Wet JM (1982) *Dictionary of Cultivated Plants and Their Regions of Diversity: Excluding Most Ornamentals, Forest Trees and Lower Plants*, PUDOC, Wageningen. pp. 21–196.

Zhang H, Li X, Yu H, Zhang Y, Li M, et al. (2019) A high quality melon genome assembly provides insights into genetic basis of fruit trait improvement. *iScience* 22:16–27.

Zhao G, Lian Q, Zhang Z, Fu Q, He Y, et al. (2019) A comprehensive genome variation map of melon identifies multiple domestication events and loci influencing agronomic traits. *Nat Genet* 51:1607–1615.

Zink FW, Gubler WD (1985) Inheritance of resistance in muskmelon to fusarium wilt. *J Amer Soc Hortic Sci* 110:600–604.